Laboratory Manual
for Anatomy
and Physiology

Laboratory Manual for Anatomy and Physiology

WITH FETAL PIG DISSECTIONS

Patricia J. Donnelly
George A. Wistreich

East Los Angeles College

HarperCollinsCollegePublishers

Acquisitions Editor: Bonnie Roesch
Developmental Editor: Meryl R. G. Muskin
Project Coordination and Text Design: York Production Services
Cover Design: Kay Petronio
Cover Photo: © 1989 Howard Sochurek, The Stock Market
Compositor: Modern Graphics Inc.
Printer and Binder: Von Hoffmann Press
Cover Printer: The Lehigh Press, Inc.

An attempt has been made to obtain permission from all suppliers of photographs used in this edition. Some sources have not been located, but permission will be requested from them upon notification to us of their ownership of the material. All material without a credit line is from *Laboratory Manual for Anatomy and Physiology: With Cat Dissections,* Third Edition.
LABORATORY MANUAL FOR ANATOMY AND PHYSIOLOGY: With Fetal Pig Dissections

Library of Congress Cataloging-in-Publication Data

Donnelly, Patricia J.
    Laboratory manual for anatomy and physiology : with fetal pig dissections / Patricia J. Donnelly,
George A. Wistreich.
        p.   cm.
    Includes bibliographical references and index.
    ISBN 0-06-500905-3
    1. Physiology—Laboratory manuals.   2. Anatomy—Laboratory manuals.   3. Swine—Fetuses—
Dissection—Laboratory manuals.
    A.   II.   Title.
    QP44.D665   1993
    612'.0078—dc20   93-6653
                                                                                      CIP

93 94 95 96 9 8 7 6 5 4 3 2 1

CONTENTS

SECTION I

Microscopy, Cellular Structure, Organization, and Function   1

EXERCISE

SECTION II

An Introduction to the Organ and Systems Levels of Organization   87

EXERCISE

SECTION III

The Integumentary System   97

EXERCISE

SECTION IV

The Support System   109

EXERCISE

SECTION V

The Muscle System   171

EXERCISE

v

SECTION VI

Nervous System Coordination and Control   319

EXERCISE

SECTION VII

The Endocrine System   365

EXERCISE

SECTION VIII

The Senses   387

EXERCISE

SECTION IX

Fetal Pig Dissection: An Introduction to the Visceral Organs   447

EXERCISE

SECTION X

The Cardiovascular System   459

EXERCISE

SECTION XI

Blood and the Immune System   527

EXERCISE

SECTION XII

The Respiratory System   581

EXERCISE

SECTION XIII

The Digestive System   607

EXERCISE

SECTION XIV

The Urinary and Reproductive Systems 653

EXERCISE

APPENDICES

*T*he purpose of *Laboratory Manual for Anatomy and Physiology: With Fetal Pig Dissections,* is to provide students with functional and effective experiences that will enable them not only to observe but also to understand fundamental principles and concepts of human anatomy and physiology. The exercises in this manual are organized into complete units of instruction so that students study each organ system and observe and recognize fundamental similarities and differences among the systems.

As the manual's subtitle indicates, we have used the fetal pig as the primary dissection animal. Tables in the exercises on the muscular system and certain other systems relate information learned in the dissection procedure on the fetal pig to human anatomy. The choice of this animal as a dissection specimen was made for two reasons: first, human cadavers are difficult to obtain and maintain; second, we feel that a student's own dissection has more instructional value than mere observation of demonstration material. In designing the physiology experiments we kept in mind the need to provide modern and functional approaches to demonstrate physiological concepts and techniques. The experiments and studies use relatively new, inexpensive, and easily obtainable specimens and laboratory setups. It is our feeling that basic principles of physiology can be demonstrated with simple techniques and a minimum of complex equipment.

ORGANIZATION OF THE MANUAL

The 51 laboratory exercises in this manual include a variety of special features that are designed to stimulate interest in the subject matter, to involve students in the learning process, and to guide them through specifically planned activities.

The manual offers a broad range of exercises from which instructors may choose. It may be used for a one- or two-semester course in anatomy and physiology at the undergraduate level. Some of the exercises presume a knowledge of basic high school biology and chemistry. Although the manual was designed for courses related to allied health curricula, it can be used equally well in courses for general mammalian anatomy and physiology.

The 51 exercises are arranged in 14 sections as follows:

Section I introduces the cellular and tissue levels of organization. Exercises are designed to show the features of cellular structure and function and the chemical and physical aspects of cellular permeability. This section begins with a functional exercise emphasizing the proper use and care of the microscope.

Section II surveys anatomical orientation and terminology and serves as an introduction to many terms used in subsequent exercises.

Section III demonstrates the general features of the skin, nails, and hair. Particular attention is given to fingerprints and the general appearance of blood vessels in response to different temperatures.

Section IV familiarizes students with the structure and functions of the skeletal system. Exercises in this section provide many color drawings of the support system in the human body.

Section V contains numerous exercises designed to demonstrate fetal pig muscle anatomy and the structure and properties of mammalian muscle. The fetal pig is the laboratory animal, and there are black-and-white photographs with accompanying line drawings and color photographs to assist the students in their dissections. Other exercises cover the structure of muscle and the chemistry of muscular contraction.

Section VI provides students with a comparative and functional treatment of nervous system coordination and control. The anatomy of the human system is stressed and is used to demonstrate cranial and spinal reflex responses. The sheep brain is dissected.

Section VII provides an overall view of the human endocrine system. The interrelationships, locations, and functions of the glands are covered in sufficient detail to emphasize the importance and special properties of the endocrine system.

Section VIII contains numerous simple experiments that emphasize the importance and special properties of the senses.

Section IX familiarizes students with basic dissection procedures and points out the organ systems that can easily be seen without more extensive dissection.

Section X presents the major anatomy and physiology of the cardiovascular system. There are exercises on the heart, lymphatic system, and blood structure. An exercise demonstrating the clinical features of coronary circulation is also included.

Section XI presents coverage of the components of blood, the immune system, and various aspects of diagnostic hematology, including factors involved with coagulation, hemoglobin determinations, blood typing, cross-matching, and the immune response.

Section XII familiarizes students with the basic anatomical and physiological features of the respiratory system. Procedures for conducting pulmonary function tests are incorporated in this section.

Section XIII treats the structure of the digestive system. Exercises dealing with enzymatic action, digestion chemistry, teeth, and cholesterol testing are important parts of this section.

Section XIV deals with the anatomy and physiological features of both the urinary and reproductive systems. It includes an examination of the urogenital organs and demonstrates the principles associated with dialysis, fluid and electrolyte balance, and acid–base balance. An exercise dealing with routine diagnostic urinalysis is contained within this portion of the manual. In addition, attention is given to gamete formation, ovulation prediction, and pregnancy testing.

LEARNING AIDS

A number of learning aids have been incorporated to make this manual an effective tool for students. These include the following:

Section overviews. Each major section of the laboratory manual is preceded by a description of the range of topics, exercises, and experiments to be covered. A list of general references is included to provide background information and a basis for further study.

Illustrations. Photographs and diagrams have been included to provide a greater understanding of anatomical relationships and physiological principles, techniques, and test results. Photographs showing important structures, reactions, and results are used throughout the manual. Special sections containing color photographs of fetal pig muscles and histology are presented to simplify identification.

Procedure diagrams. Exercises that introduce basic techniques contain step-by-step illustrations of the procedure to show the proper handling of equipment, animals for dissection, and related items.

Laboratory review. At the end of each exercise there are a number of selected completion, multiple-choice, matching, and other forms of questions to test students' comprehension of the major principles and results of the exercises performed. In addition, this section encourages students to think critically.

Index. An index has been included in the manual. Students should consult this index to facilitate location of anatomical parts, tests, and procedures.

INSTRUCTOR'S MANUAL

An instructor's manual to accompany the laboratory manual is available. It contains specific directions for the preparation of laboratory materials and descriptions of alternative procedures and the use of supplementary aids. There are suggested sources for audiovisual aids, laboratory equipment and supplies, and questions for laboratory examinations and quizzes.

ACKNOWLEDGMENTS

We thank the following colleagues who read and reviewed the manuscript versions of our work and whose comments we found constructive:

Margaret Folsom, Methodist College

Kristi Sather-Smith, Hinds Community College

Carol Haspel, LaGuardia Community College

The authors also gratefully acknowledge the Schleicher & Schnell Corporation; IsoLab, Inc.; ChemTrak Corporation, and the OvuGen Corporation for providing samples of their products for use in developing this manual.

Special thanks to College Biological Supply Company, Escondido, California, for supplying the dissection specimens: Dr. Joel Schechter of the Department of Anatomy, University of Southern California Medical School, and Leonard D. Dank, President of Medical Illustrators, Inc., for art work; and Dr. Daniel Fertig of East Los Angeles College and Elbogen Medical Photography, Los Angeles for photographic assistance.

We are also deeply grateful to the members of the editorial and production staffs of HarperCollins Publishers.

P. J. Donnelly
G. A. Wistreich

TO THE STUDENT

LABORATORY OPERATING PROCEDURE

In the laboratory you have an opportunity to study and apply anatomical and physiological concepts and principles at close range and to see for yourself the details of structures, processes, and relationships characteristic of living systems. Your observations should provide the answers to most questions that arise concerning the material under study. Your laboratory instructor will offer helpful suggestions, but will not be a source of easy answers for unprepared students. The following rules should assist you in getting the most out of laboratory sessions.

LABORATORY PROCEDURES AND REGULATIONS

1. Laboratory work begins promptly at the time designated in the schedule.
2. Laboratory demonstrations, which include general directions designed to make your work easier and more efficient, usually will be given at the beginning of laboratory sessions.
3. You will be assigned a microscope, a laboratory drawer with other necessary equipment, and a laboratory work space for the duration of the class. It is your responsibility to keep all equipment issued to you in good working condition. Report damaged or missing items to your instructor immediately. You are expected to read each laboratory exercise and carry out suggested reading assignments before you come to class. The instructor will designate in advance which parts of each exercise will be undertaken during each class period.
4. Try to acquire self-reliance in performing procedures. Read all directions carefully and follow them closely. You may find it helpful to underline procedural steps. Learn to observe your own work critically so that you can correct and improve your performance as you do each procedure. However, if you have any doubts about any portion of the exercise, ask your instructor for help.
5. Place items such as extra textbooks, coats, and other personal belongings in the area specified by your instructor. Never let such items clutter your operating space.
6. In exercises requiring the use of chemicals, take only the quantity of materials you need.
7. Leave all laboratory facilities and equipment in good order at the close of each period.
8. Conserve electricity, water, and gas. Turn off water faucets and gas burners when they are not needed.

LABORATORY RECORDS

1. Your lab notebook is a valuable record. Keep it up to date, using procedures outlined by your instructor.
2. As you perform your experiments, record your data. Always perform exercises in the sequence assigned.

3. The illustrations in this manual suggest the way drawings should be made, spaced, and labeled. Try to draw plain, clear-cut outlines of structures and other items studied. Use pencil. Coloring or shading will be specified as needed.

4. The labeled drawings supplied in this manual are important objects of study and should be used as guides in locating anatomical parts and as an aid in recalling concepts and principles.

GENERAL LABORATORY SAFETY PRECAUTIONS

1. Your instructor will show you the location and demonstrate the use of emergency equipment such as fire extinguishers, fire blankets, and first-aid kits. Know their locations and how to use them.

2. Never eat, store food, or smoke in the laboratory.

3. To prevent damage to street clothes, always wear a laboratory coat or apron during the class period. Remove this protective apparel before leaving the laboratory.

4. Tie back long hair to prevent it from being a laboratory fire hazard.

5. Double-check the label of any chemical you use.

6. When diluting acids, always add acid (AAA) to water.

7. Never heat an inflammable liquid over an open flame.

8. When heating materials in a test tube over an open flame, never point the open end toward anyone. Hold the test tube at an angle, not upright, over the flame, heating the sides and bottom. Keep moving the test tube so it does not overheat. Use only glassware marked Pyrex or Kimex. Other glassware will shatter when heated.

9. Do not inhale the fumes of any substance directly. When required to note the odor of a chemical, first take a deep breath of fresh air and exhale normally. Then with your hand waft the vapors toward your nose and sniff slowly.

10. Do not pipette chemicals by mouth.

11. Do not use your fingers as a stopper for a tube or a bottle when shaking the tube to mix its contents.

12. Never eat, smoke, drink, taste chemicals or lick labels in the laboratory. DO NOT BRING FOOD TO THE LABORATORY.

13. If you get a chemical on your skin or in your eyes, immediately flood with water and call for help.

14. Report any fire or personal injury such as a cut or burn to your instructor immediately.

15. Never remove equipment, solutions, or other materials associated with the exercise from the laboratory.

16. Dispose of chemicals, biological materials, and paper as indicated by your instructor. Do not put paper or dissected materials in the sink.

17. In cleaning dissection instruments, always hold sharp edges away from you.

18. Wash your hands thoroughly with detergent and water before leaving the laboratory. Carry out this procedure even in cases of coffee and health breaks.

19. If you are in doubt about a personal health problem, always ask your instructor first before you attempt to do anything.

PRECAUTIONS FOR THE HANDLING OF BLOOD OR OTHER BODY FLUIDS

The *Centers for Disease Control* has developed a series of guidelines for health care workers in order to prevent the transmission of human immunodeficiency viruses (HIVs) as well as other blood-related infectious diseases such as hepatitis

B virus infection. The guidelines in the form of laboratory precautions and procedures are clearly specified in the appropriate exercises of this manual. Following the specific steps will enable you to work with body fluids safely and effectively and without any danger to your safety and well being.

MATERIALS TO BE PROVIDED BY THE STUDENT

1. One clean laboratory coat or other type of protective apparel
2. Textbook and laboratory manual
3. One 4H drawing pencil
4. One eraser (not art gum)
5. Colored pencils—red, blue, yellow, and green
6. One wax marking pencil or a washable-ink felt pen
7. One package of glass slides and cover slips
8. One package of lens paper
9. Two packages of matches
10. One clean, lint-free, soft cloth
11. Slide box (optional)
12. Surgical or rubber gloves for protection against stains and contact with chemicals
13. Safety goggles or glasses (optional)
14. One dissection kit or individual instruments to include:
    a. One pair of coarse scissors
    b. One pair of fine forceps
    c. One scalpel with at least three disposable blades
    d. Two dissecting needles

PRONUNCIATION GUIDE

In the exercises relating to bones and muscles, you will find *phonetic pronunciations* for some anatomical terms. Taking time to sound out new terms and to say them aloud once or twice will help you master one of the tasks of the anatomy and physiology course—learning its specialized vocabulary. The following key explains the system used for the pronunciations.

1. The strongest accented syllable appears in capital letters (e.g., bī-LAT-er-al, dī-ag-NŌ-sis). A syllable that has a secondary accent is marked by a single quote mark (e.g., kon'-sti-TOO-shun, fiz'-ē-OL-ō-jē).
2. Vowels pronounced with long sounds are indicated by a line above the vowel and are pronounced as in the following common words.
   ā as in *māke*
   ē as in *bē*
   ī as in *īvy*
   ō as in *pōle*
3. Vowels not marked for long sounds are pronounced with the short sound, as in the following words.
   e as in *bet*
   i as in *sip*
   o as in *not*
   u as in *bud*
4. Other phonetic symbols are used to indicate the following sounds.
   a as in *above*
   soo as in *sue*
   kyoo as in *cute*
   oy as in *oil*

Through the use of this laboratory manual you should gain a deeper understanding of human anatomy and physiology. Such knowledge will demand not only a genuine effort but also an organized coordination of text reading and laboratory experimentation. A solid foundation in the basic principles of anatomy and physiology will make the effort well worthwhile.

Microscopy, Cellular Structure, Organization, and Function

*P*hysiologic systems usually exhibit two characteristic properties: *transduction* and *translocation*. The first of these properties involves the changing of energy from one form to another, while the other property, translocation, is associated with the movement of either material or information from one location to another. Because the cell is considered to be the most basic component of a physiologic system, the properties of such units and the way in which they are organized into higher systems, such as tissues and organs, determine their overall function. Furthermore, the performance of cellular structures, as well as the coordination of their activities, helps to maintain the state of dynamic equilibrium in the body's internal environment; this is known as *homeostasis*.

Among the different tools essential for studying cells and their activities are microscopes. The primary function of a microscope is to enable the observer to distinguish structures and points that are separated by short distances. This function is *resolution* or *resolving power*.

Resolution is more important than magnification: it is not always desirable to obtain the largest image possible, but it is necessary to obtain sharp detail. Resolution is dependent on many factors, including illumination and the nature of the specimens. The resolving power of a microscope is inversely related to the wavelength of light used—that is, the shorter the wavelength, the greater the resolving power.

Two categories of microscopes in general use are the *bright-field instrument*, which contains optical lenses and uses visible light as the source of illumination, and the *electron microscope*, which contains electromagnetic lenses and uses a beam of electrons as the source of illumination. Because the electron beam has a considerably shorter wavelength than ordinary light, the resolving power and magnification obtainable with electron microscopes far exceeds those obtained with other optical systems (Table I-1). Two general types of electron microscope are currently in use: the transmission and the scanning instruments. Another

Table I-1 Comparison of Microscope Types

Type	Magnification Range	Resolution (in nm)	Applications Include
Compound bright field	10–2,000	300	Used extensively for the examination of various cell types, either stained or unstained
Fluorescence	10–3,000	200	Employs fluorescent dyes for staining; useful in many medical diagnostic procedures for identifying cells, including cancer and microorganisms and/or their parts
Phase contrast	10–1,500	200	Used to examine and study cellular structures; staining of specimens is not required
Transmission electron (TEM)	200–1,000,000	1	Used to examine and study the ultrastructure of cells and viruses and certain biochemicals
Scanning electron (SEM)	200–500,000 and greater	5.0–10.0	Used to examine and study surface features of cells and nonbiological materials; images are presented in three-dimensional form
Scanning transmission electron (STEM)	Combination of TEM and SEM	Combination of TEM and SEM	Used for applications similar to those listed for TEM and SEM; also used for analysis of chemical elements in specimens; equipped with computer for measurements and calculations

instrument, the scanning transmission electron microscope, combines many of the properties of both instruments.

Several elaborate techniques are used in electron microscopy in preparing specimens and providing specimen contrast. Such techniques include ultrathin sectioning (slicing) and staining (Figure I-1a), the use of electron stains, and depositing the atoms of heavy metals in a vacuum onto the surfaces of specimens. The microscope used is a deciding factor in the selection of the preparation technique. The scanning electron microscope is a particularly useful instrument because it provides three-dimensional views of specimens without the need for extensive preparation techniques (Figure I-1b). Such views provide a basis for understanding complex relationships within cells, tissues, and organs and enable observers to gain a perspective as to the true locations and organization of cellular structures.

Specimens examined by electron microscopy must be viewed in a vacuum.

(a)

(b)

Figure I-1(a) Electron micrographs: (a) a transmission micrograph showing Golgi apparatus **(G)**, mitochondria **(Ms)**, endoplasmic reticulum **(ER)**, nucleus **(N)**, and nucleolus **(Nu)**; (b) a scanning micrograph providing an internal three-dimensional view of an eukaryotic cell—cellular organelles that can be seen include the endoplasmic reticulum **(ER)**, with ribosomes attached; Golgi apparatus **(G)**; nucleus **(N)**; and individual mitochondria **(M)**. The original magnification is 26,000×. (From K. Tanaka and A. Mitsushima, *Journal of Microscopy,* 133: 213–222, 1984.)

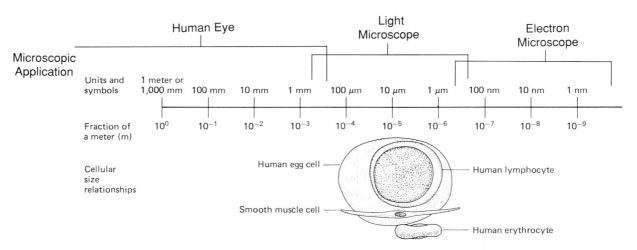

Figure I-2 Metric units of length commonly used in the biological sciences. 1 meter = 1000 millimeters **(mm)**; 1 mm = 1000 micrometers **(μm)**; 1 μm = 1000 nanometers **(nm).** The relative sizes of various types of cells and the respective magnification ranges of different microscopes are shown.

Furthermore, specimens viewed or photomicrographs taken of specimens lack color, which is a major means of obtaining contrast in light microscopes. Exercise 1 presents the features of the light microscope, emphasizing that it incorporates optical lenses and visible light as the source of illumination.

Sizes and shapes are important identifying properties of cells. In the biological sciences, several common metric units are used to express the dimensions of cells and their parts. The common metric units of measurement include the millimeter **(mm)**, the micrometer **(μm)**, and the nanometer **(nm)**. The relationships of the units and their application to cells are shown in Figure I-2.

Cells from different forms of life and from different regions of a single multicellular organism differ greatly in various features, including shape, size, rate of metabolic activity, and functions. While all cells share certain basic structural features, some have specialized structures called *organelles.* The exercises in this section not only emphasize numerous properties of cells and their organization but also demonstrate that they are efficient basic units of life.

GENERAL REFERENCES

Burrells, W. *Microscope Techniques: A Comprehensive Handbook for General and Applied Microscopy.* New York: Halsted Press, 1976.

Cargille, J. J. *Immersion Oil and the Microscope.* Technical Reprint 10-51. Cedar Grove, NJ: R. R. Cargil Laboratories, 1975.

de Duve, C. *Blueprint for a Cell: The Nature and Origin of Life.* Burlington, NC: Neil Paterson, 1991.

DeRobertis, E. D. P., and E. M. F. DeRobertis, Jr. *Cell and Molecular Biology,* 8th ed. Philadelphia: Lea & Febiger, 1987.

Gartner, L. P., and J. L. Hiatt. *Atlas of Histology.* Baltimore: Williams & Wilkins, 1987.

Mollring, F. K. *Microscopy from the Very Beginning.* Oberkochem, Federal Republic of Germany: Carl Zeiss.

Tortora, G. J., and S. R. Grabowski. *Principles of Anatomy and Physiology,* 7th ed. New York: HarperCollins, 1993.

SPECIFIC REFERENCES

Dautry-Varsat, A., and H. F. Lodish. "How Receptors Bring Proteins into Cells." *Scientific American,* 250: 52–58, 1984.

Liotta, L. A. "Cancer Cell Invasion and Metastasis." *Scientific American,* 266: 54–63, 1992.

Sandberg, A. A. *The Chromosomes in Human Cancer and Leukemia,* 2nd ed. New York: Elsevier, 1990.

The Microscope and Specimen Preparation

After completing this exercise, you should be able to

1. Identify the parts of the microscope
2. Explain the functions of the parts of the microscope
3. Demonstrate the proper method of focusing, changing objectives, carrying the microscope, and cleaning the microscope
4. Calculate the total magnification of any ocular and objective combination
5. Prepare a temporary wet mount for the examination of a specimen
6. Define and understand the following terms and concepts: *resolving power, parfocal, working distance, depth of field,* and *magnification*
7. Identify several types of human blood cells and their parts

The compound microscope is a precision instrument. It is essentially a system of accurately ground lenses arranged to give sharp, clear, magnified images of minute objects. Magnification that may be obtained with different lens systems ranges from approximately 10 to 2000 times ($\times$) the diameter of the specimen being observed. It should be noted that for each lens system, the maximum magnification usually is obtained by highly trained microscopists using auxiliary equipment under ideal conditions.

The optical lenses of the microscope include the *ocular* or eyepiece, usually three objectives with different magnifying powers, and the substage condenser. All other parts of the microscope are accessory to the main purpose of magnification by these lenses. They are designed to hold the object to be examined steadily in place, to regulate the amount of light necessary for clear viewing, and to aid in the focusing by adjusting the space relationship between the ocular, the objective, and the specimen on the slide.

Microscopes differ greatly from laboratory to laboratory. Your instructor will demonstrate and explain the general construction and manipulation of the particular instrument you will use. There is, however, a set of basic rules that should always be observed for all instruments, to ensure maximum efficiency and proper maintenance.

USE AND CARE OF THE MICROSCOPE

1. Always carefully remove the microscope from its place of storage by holding it firmly by the arm with one hand and supporting it at the base with the other. Keep the instrument in an upright position.
2. Place the microscope at least 6 inches from the edge of your laboratory table.
3. Do not tamper with or remove any parts. If the microscope does not seem to be functioning properly, call your instructor immediately. Other laboratory sections may use the same instrument, and any part damaged but not reported by someone else might be charged to you.
4. Do not handle the lenses with your fingers. Perspiration contains fatty acids and other substances that can mar the lens glass. Always use the specified type of lens paper for cleaning the optical system. *Never use paper towels or cloths.* These materials will scratch the delicate lens surfaces.
5. Wipe the lenses of the instrument before and after laboratory use.
6. Use a clean, soft, linen cloth to remove dirt, dust, or other materials from parts other than the lenses.

7. Do not allow liquids, particularly acids and alcohol, to come in contact with any part of the microscope.
8. Always use a cover slip when examining objects or organisms mounted in water or other fluids.
9. Remove immersion oil from the microscope with lens paper.
10. Always raise the body tube sufficiently before either placing a slide on or removing a slide from the microscope stage.
11. Before putting the microscope away, always put the low-power objective in working position (i.e., in line with the body tube), open the diaphragm, and raise the condenser to its highest fixed position.
12. When disconnecting the light source, do not pull on the wire; instead, grasp the plug firmly to remove it from the socket.
13. If a cover is provided for the microscope, be certain that it covers the entire instrument.

PRINCIPLES OF MICROSCOPY

To serve its purpose, a microscope must provide adequate magnifying power so that the finest details of a specimen are separated enough to be visible to the eye. The magnifying power of microscope lenses usually is indicated by markings on objectives and eyepieces (e.g., $45\times$, $100\times$). The total magnification obtainable is the product of the magnifications produced by the lenses individually.

While magnification is an important microscope property, the limiting factor that determines the usefulness of the light microscope is its resolving power. The resolving power of an optical system is its ability to produce separate images of small parts of a specimen that are only a short distance apart—that is, the ability to distinguish fine detail. In practice, the limit of resolution obtainable with a compound light microscope is about 0.2 μm, or roughly 1000 times the resolving power of the unaided human eye.

The quality of the image obtained or its definition depends on the degree of contrast between the specimen and its surroundings and between parts of a specimen. Such contrast is due in part to differences in light absorption, and various techniques that affect absorption are used to improve contrast. For example, the use of chemical stains results in color differences. Also, reducing the intensity of the light illuminating a specimen makes it easier to distinguish cells that have a refractive index similar to the fluid in which they are suspended.

Direct examination of living specimens can be extremely useful in determining size and shape relationships, motility, and reactions to various chemicals. One of the methods in general use for studies of this type is the wet-mount technique. This method maintains the natural shape of cells and reduces the distorted effects that can occur when specimens are dried and preserved.

In this exercise, the wet-mount technique is used to emphasize particular properties of the light microscope and to introduce features of living cells.

Identification of Microscope Components

Although compound microscopes from different manufacturers vary in appearance or design, they consist of the same basic components. The parts of a typical microscope and their functions are described in Table 1-1. Use this information as you examine and locate specific parts of the microscope.

Materials

1. Compound microscope with an appropriate light source (one per student).
2. Large cut-away diagram of a representative microscope for class use.

Procedure

This procedure is to be performed by students individually.

1. Carefully remove the microscope assigned to you from the cabinet by firmly grasping the microscope's arm with one hand and placing the other hand under the instrument's base for support.
2. Place the microscope approximately 6 inches from the edge of the laboratory table, with the microscope arm facing you.
3. Examine your microscope, and with the aid of Figure 1-1, locate and learn the functions of each component listed in Table 1-1.
4. Answer items pertaining to this portion of this exercise in the laboratory review section.

Temporary Wet-Mount Specimen Preparation

Specimens for microscopic examination can be prepared in several ways. The temporary wet-mount method described here is sufficient for materials to be examined for a short time period. (*Note:* The specimens prepared in the following two procedures will be used for viewing in a later section of this exercise.)

Materials

1. The following materials should be provided per four students:

Table 1-1 Microscope Components and Their Functions

Microscope Component	Location or Description	Functions
Ocular (eyepiece)	Uppermost series of lenses	Magnification
Body tube (barrel)	Main cylindrical part	Holds oculars, conducts light rays from specimen
Nosepiece	Movable, usually circular plate at bottom of body tube	Holds objectives
Low-power objective	Shortest objective, usually magnifies $10\times$	Magnification
High-power (dry) objective	Intermediate-sized objective, usually magnifies $43\times$ to $45\times$	Magnification
Oil immersion objective	Longest objective; marked with either etched, red, or black circles; usually magnifies $100\times$	Magnification
Condenser	Lens system located below central stage opening	Concentrates and directs light beam through specimen
Stage	Platform on which specimens for examination are placed	Specimen support
Mechanical stage	Control devices that permit movement of slides from left to right and forward and backward on stage	Used to move specimen
Iris diaphragm	Located beneath stage in association with condenser unit, controlled with a lever	Regulates brightness or intensity of light passing through lenses
Coarse adjustment knob[a]	Generally located below stage	Used for preliminary and coarse focusing by raising or lowering body tube
Fine adjustment knob[a]	Generally located below stage	Used for final or fine focusing by raising or lowering body tube
Condenser adjustment knob	Control knob located below stage	Used to obtain full illumination by raising or lowering condenser
Base	Heavy, bottom portion on which instrument rests	Microscope support
Arm	Somewhat curved portion of microscope, used in carrying instrument	Microscope support
Mirror[b]	Double-faced mirror for reflecting light through specimen	Reflects light from source through lens systems

[a]On several instruments, the coarse and fine adjustments are combined in one knob device.
[b]Found on older instruments.

a. One eyedropper
b. One small beaker of water
c. One pair of scissors
d. One dissecting needle
e. One bottle of methylene blue stain
f. Four wooden applicator sticks
g. Container with disinfectant for applicator sticks
2. The following materials should be provided for class use:

a. Microscope slides and cover slips
b. Newspaper
c. Lens paper

Procedure 1: Letter "e"

This procedure is to be performed by students individually.

Figure 1-1 A cut-away diagram of a compound monocular microscope showing major components. Many instrument components are combined. Note that the fine-adjustment and coarse-adjustment knobs are combined and that the condenser adjustment is permanently fixed in position. (Courtesy of Reichert Scientific Instruments.)

1. Carefully clean a glass slide and a cover slip, as indicated by the instructor. Place a drop of water on the center of the microscope slide (see Figure 1-2).
2. Cut a row of a few letters from newspaper. Be certain that at least one of the letters is a small "e."
3. Place the string of letters in the drop of water on the slide, and make certain that the liquid covers them.
4. Holding the cover slip at a 45° angle, place one edge on the slide in contact with the drop of water and then slowly lower it into place over the letters (see Figure 1-2).
5. Save this preparation for later viewing.

1. Place drop of liquid on slide   2. Add specimen   3. Put cover slip in place

Figure 1-2 Steps in the preparation of a temporary wet mount.

Procedure 2: Cheek Cells

This procedure is to be performed by students individually.

The squamous epithelial lining of the human cheek provides a readily available source of cells for microscopic examination. These cells will serve as an introduction to general properties of animal cells.

1. Place 1 drop of water and 1 drop of methylene blue on the center of a clean glass slide.
2. Gently scrape the inside of your cheek with the end of a clean wooden applicator stick. (*Note:* Force is not necessary to dislodge cells.)
3. Gently mix the scraped material with the drops of fluid on the slide.
4. Apply the cover slip as described earlier and shown in Figure 1-2.
5. Save this preparation for viewing in the next part of this exercise.
6. Dispose of the applicator stick, as indicated by your instructor.

Use of the Microscope

This portion of the exercise is designed to familiarize you with basic features of microscope's parts and operation. The procedures described are for instruments with movable body tubes.

Materials

1. The following materials should be provided for class use:
   a. Temporary wet mounts of the letter "e" and of squamous epithelial cells prepared in Procedures 1 and 2 of the preceding section of this exercise.
   b. Lens paper

2. Prepared slides of crossed colored silk threads (one per four students).

Procedure 1: Microscope Operation

This procedure is to be performed by students individually.

1. Wipe all lenses with lens paper.
2. Move the low-power objective into position under the body tube. You will feel a click when it is correctly in place.
3. Connect your light source, and turn it on.
4. Place the letter "e" slide on the mechanical stage, and secure it with the slide holder or other means provided. The finger-control lever for the slide holder is generally located at the rear left-hand corner of the mechanical stage. Forward pressure on the lever will cause the slide holder to open, and the slide may be inserted into the holding area. The bottom of the slide should rest on the microscope stage and not on the mechanical stage apparatus. Release the finger-control lever to allow the slide holder to close and to hold the slide within the mechanical stage.
5. Using the control knobs on the mechanical stage, move the slide until the specimen on the slide is over the center of the condenser opening in the stage. One control knob moves the slide left to right, and the other moves it front to back.
6. Use the coarse-adjustment knob to lower the low-power objective to approximately ¼ inch above the cover slip.
7. Use the condenser adjustment knob to raise the condenser as far as it will go.
8. While looking into the ocular, move the iris diaphragm lever back and forth. Note the change in illumination. The intensity or brightness of illu-

mination is controlled by adjusting the iris diaphragm.

9. Locate your specimen by using the low-power objective and coarse adjustment first. Focus *upward*—that is, move the objectives upward—with the fine-adjustment knob until the specimen comes into clear view. In a microscope with a fixed body tube, the objectives remain in a fixed position. Focus is controlled by moving the stage *downward* until the specimen comes into view. If you have difficulty in locating your specimen, look for a region with a color or intensity different from other portions of the slide.

10. In order to change objectives (switch from low power to high power), do not raise the body tube; simply turn the nosepiece to bring the desired objective into place. The specimen will generally stay in focus with any objective, but a slight turn of the fine-adjustment knob may be necessary. Most microscopes are *parfocal*—that is, once the specimen is in focus, it will remain so when the objective lens is changed.

11. When looking through the microscope, keep both eyes open at all times. This may be difficult at first with a monocular microscope, but it will become easier with practice. Keeping both eyes open prevents eye strain and enables you to observe and draw microscopic views of specimens without moving your head.

12. When the microscope is no longer needed, the following steps should be performed:
    a. Remove any slide from the stage.
    b. Clean the instrument, including the lenses, to remove any fluid from specimens.
    c. Open the diaphragm.
    d. Raise the condenser to its highest position.
    e. Move the low-power objective into position over the stage.
    f. Wrap the electrical cord around the base.
    g. Return the microscope to its proper storage place.

Procedure 2: Examination of Temporary Wet Mounts

This procedure is to be performed by students individually.

Letter "e"

1. Examine the letter "e" on the slide, as it appears to the unaided eye in a normal reading position.

Make a sketch of the letter in the results and observations section, following Procedure 3. All that is necessary is an outline of the "e" because we are interested only in the orientation or position of the letter rather than detail. (Refer to Color Plate 1.)

2. Focus and examine the preparation under the low-power objective following the steps listed in Procedure 1.

3. Slowly move the slide, first side to side and then forward (up) and backward (down), with the mechanical-stage-control knobs. Compare this view of the orientation of the "e" with the view with the unaided eye.

4. Sketch a representative microscopic field in the results and observations section.

5. Change to the high-power objective. Adjust the light, and carefully focus with the fine-adjustment knob. A portion of the letter "e" will be off center.

6. Center the portion of the letter while you have it under high power, and then turn to low power, to determine how far off center the "e" is in your particular microscope. This is important and should aid you in all your future microscope work.

7. Practice shifting from low to high power and back again. (Note that only a small amount of manipulation with the fine-adjustment knob is necessary to bring the letter into proper focus whenever the other objective is swung into position.)

8. Sketch the representative high-power microscopic field in the results and observations section.

9. On your sketch of the "e," as observed under low power, draw a circle around the portion of the letter seen under high power.

10. Remove, clean, and return the slide to a tray or other location indicated by your instructor.

11. Answer the laboratory-review questions pertaining to this portion of the exercise.

Squamous Epithelial Cells

Note that these cells are thin and flat and somewhat irregular in shape. The edges of the cells often become folded over in the process of preparation, and the cells tend to clump together in groups. Some cells may be covered by small, almost dotlike structures. These are usually bacteria from the oral region.

1. Examine your cheek-cell preparation under the low-power objective. Make the necessary adjustments for light intensity.

2. Select a microscopic field for study, in which the cells are separate from each other and exhibit some-

what centrally located and darker-staining nuclei. Refer to Color Plate 2.

3. Using first low power and then high power, draw several squamous cells in the space provided in the results and observations section. Label the following: cytoplasmic (cell) membrane, cytoplasm, and nucleus. If bacteria (small spherical or sticklike forms) are present, indicate them also.

Procedure 3: Depth of Field

This procedure is to be performed by students individually.

To focus any microscope, the specimen being examined must be brought to the proper distance from the objective lens. This is done as indicated in Procedure 1, by the vertical movement of the body tube. In this portion of the exercise, you will consider the depth of field (focus) or the thickness of a specimen that is in focus at any one time. By use of this technique, determinations can be made as to what structures and location details are in a specimen. In addition, knowledge of the width, length, and depth of a specimen can be used to understand the three-dimensional feature of objects.

1. Place a prepared slide containing three different colored threads on the microscope stage, so that the point where the threads cross is over the center of the light opening in the stage.
2. Using the low-power objective, find the location on the slide where the three threads cross.
3. Use the coarse adjustment to raise the body tube slowly until the place where the threads cross is barely in focus. The colors should be visible, but the image will be fuzzy and blurred.
4. Slowly lower the body tube until the edge of the first thread appears clear. What is the color of the first thread to appear?
5. Use the fine-adjustment knob to focus on the second and third threads. Determine the color of the threads in the middle and on the bottom. Enter your findings in Table 1-2 of the results and observations section.
6. Obtain an unknown thread slide from your instructor. Examine it in the same way you examined the known slide. Determine the color on top, in the middle, and on the bottom, and enter your findings in the results and observations section. Check with the instructor to see whether your observations are correct.

RESULTS AND OBSERVATIONS

Letter "e"

1. Sketches

Unaided eye view

Low power

High power

2. List three ways in which the microscope changes the appearance of the letter "e."

a. _____

b. _____

c. _____

Squamous Epithelial Cells

Low power

High power

Depth of Focus

1. Unknown thread slide number: _____
2. Complete Table 1-2.

Working Distance and the Oil Immersion Lens

The distance between a specimen and an objective lens is known as the *working distance*. Magnification increases as this distance decreases, and more light is needed for better and more accurate viewing of a spec-

imen. Figure 1-3 shows the relationship between the working distances of objectives and the iris-diaphragm adjustments needed to provide adequate illumination. In addition, the more a specimen is magnified, the greater is the chance for the specimen and the lens of the objective to come into contact. Great care should be exercised when the high-power (oil immersion) objective is in position. Improper use of objectives can result in broken slides and damaged objectives.

The objective of highest power on compound light microscopes is usually the oil-immersion objective. The objective must always be immersed in cedar oil or some other substance with about the same index of refraction as glass. The refraction index of a substance determines the angle at which light is bent when it passes through that substance. If a glass slide has air above and below it, as when the high- or low-power objective is used, light is bent upon entering the glass and again when it emerges into air. Because of this refraction, image details become blurred, and some light is lost as it passes by rather than enters the objective (Figure 1-4). With the oil-immersion lens, light passing through the slide proceeds in a straight path through the immersion oil. Immersion oil thus increases the resolving power of the objective so that finer details can be seen and more of the incident light is used.

In this portion of the exercise, a prepared blood smear is used in order to demonstrate the properties of the oil-immersion objective. The general microscopic features of human blood cells are given in Table 1-3, Figure 1-5, and Color Plate 3.

Materials

The following materials should be provided for class use:

1. Prepared human-blood smears with coverslips
2. Charts and other reference materials showing blood-cell types
3. Immersion oil
4. Lens paper and xylene

Table 1-2

Thread Layer	Color of Thread Layer in Known Slide	Color of Thread Layer in Unknown Slide
Top		
Middle		
Bottom		

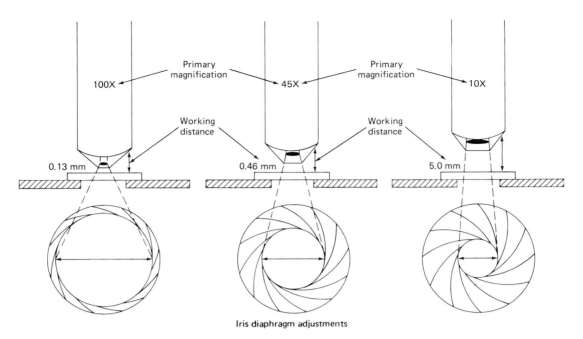

Iris diaphragm adjustments

Figure 1-3 The relationship between the working distances of the low (10×), high (dry) (45×), and oil-immersion (100×) objectives, and the iris-diaphragm adjustments necessary to obtain satisfactory specimen illumination. The primary magnification and working distance (in millimeters) are given for each objective.

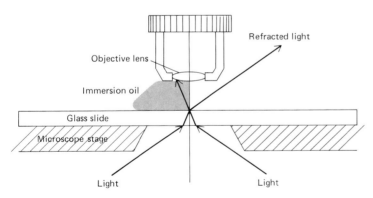

Figure 1-4 A diagrammatic comparison of the effects of immersion in oil and in air on the path of light leaving a microscope glass slide.

Procedure

This procedure is to be performed by students individually.

1. Place a blood-smear slide into position on the microscope stage.
2. Rotate the low-power objective into position.
3. Focus and locate a field showing both red and white blood cells.
4. Swing the low-power objective aside.
5. Put one drop of immersion oil on the slide region over the light opening in the stage.

6. Move the oil-immersion objective into position over the slide. The objective should touch the oil and be just short of touching the slide.
7. Look through the ocular, and move the iris-diaphragm lever to make any necessary light adjustments. If the light is too bright, it will be difficult to distinguish the cells from the background.
8. Focus upward with the fine-adjustment knob until the blood cells come clearly into view. Refer to Color Plates 3, 36, and 37.
9. Examine the slide, and identify all the blood cell types described in Table 1-3.

Table 1-3 General Microscopic Features of Human Blood Cells

Blood Cell	General Description
Erythrocyte (red blood cell)	Circular, biconcave disc appearance No nucleus in mature state Most numerous of all types of cells found in the peripheral blood (bloodstream) Usually measures approximately 7 µm in length and 2 µm in thickness
Leukocytes (white blood cells) Granulocyte (granular leukocyte)	Contains distinct cytoplasmic granules, which, upon staining, react with dyes Granulocytes also have irregular and multilobed (lobose) nuclei
Eosinophil	Granules react with dyes to yield characteristic red color
Basophil	Granules stain blue
Neutrophil	Granules stain neutral or orange
Agranulocyte	Does not contain granules Has rounded, rather than lobose, nucleus
Lymphocyte	Has individual nucleus that occupies major portion of cell
Monocyte	Has kidney-shaped nucleus; generally larger than lymphocyte
Platelet (thrombocyte)	A circulating, nonnucleated fragment of a large bone-marrow cell (megakaryocyte), which participates in blood clotting

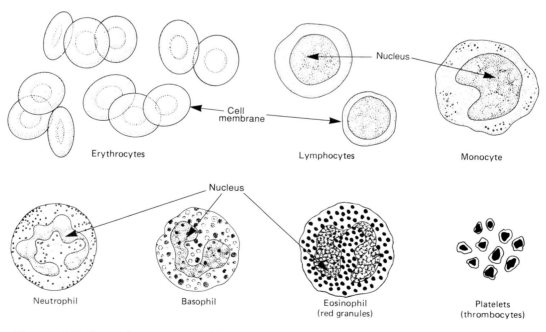

Figure 1-5 General appearance of blood cell types.

10. Identify the cells indicated in the blood smear in Figure 1-6 in the following results and observations section.
11. Answer the laboratory-review questions pertaining to this portion of this exercise.
12. After completing this procedure, remove the slide, and clean it with the xylene and lens paper. (*Note:* xylene should be used with slides having fixed cover slips.)
13. Wipe the oil from the oil-immersion objective, and swing the low-power objective into position.

RESULTS AND OBSERVATIONS

1. Identify all labeled cells in Figure 1-6.

   a. _____

   b. _____

   c. _____

   d. _____

   e. _____

   f. _____

Figure 1-6 Representative human blood smear preparation.

LABORATORY REVIEW *1*

THE MICROSCOPE AND SPECIMEN PREPARATION

COMPLETION AND FILL-IN

1. Label the microscope parts in Figure Q1-1.

Figure Q1-1 (Courtesy of Carl Zeiss, Inc.)

2. What is the magnification of the following objectives without an ocular?

   a. Low power _____   b. High power _____   c. Oil immersion _____
3. Which of the objectives comes closest to a specimen and is most likely to break a

   slide if proper precautions are not taken? _____

ANSWERS

a. _____

b. _____

c. _____

d. _____

e. _____

f. _____

g. _____

h. _____

i. _____

j. _____

k. _____

4. Complete Table Q1-1.

Table Q1-1

Microscope Part	Functions
Ocular	a.
Low power, high power, and oil-immersion objectives	b.
c.	Lens, located beneath the stage opening, which controls the light-beam size
d.	Controls the intensity of light
e.	Platform on which slides or other materials to be examined are studied
Coarse-adjustment knob	f.
g.	Adjustment device used for final focusing
Condenser adjustment knob	h.
i.	Used for providing support during transport

5. What materials should be used for the cleaning of microscope lenses? _____

6. a. Distinguish between resolving power and magnification. _____

   _____

   b. Which of these is the more important property of a microscope? _____

   c. List the factors that contribute to resolving power. _____

   _____

7. Table Q1-2 lists some common problems encountered in microscope usage. Indicate the proper procedure for correcting each of them.

Table Q1-2

Problem	Correction Procedure
Oil-immersion lens does not enter immersion oil on slide	a.
Microscope field is too bright	b.
Specimen appears blurred after proper light and focusing adjustments are made	c.
Microscope view does not change as slide is moved	d.
Only half of the field appears to be illuminated	e.

8. Indicate which microscope part in Figure Q1-1 is used in the situations described in Table Q1-3.

Table Q1-3

Situation	Microscope Part
Fine focusing of a specimen	a.
Securing a slide for viewing	b.
Adjusting the brightness of the light	c.
Providing full illumination for viewing	d.
Preliminary focusing of a specimen	e.
Changing of objectives	f.

ANSWERS

1. —————
2. —————
3. —————
4. —————
5. —————
6. —————

MULTIPLE CHOICE

Insert the correct answer in the spaces provided.

1. What property of an ordinary laboratory microscope enables a person to change objectives without major focusing adjustments?
   a. resolving power
   b. magnification
   c. parfocal
   d. focusing
   e. resolution

2. How does closing the diaphragm affect image brightness?
   a. decrease
   b. increase
   c. no effect

3. What is the total magnification obtainable using the oil-immersion objective and a $10 \times$ ocular?
   a. 100 ×
   b. 450 ×
   c. 45 ×
   d. 100 ×
   e. 1000 ×

4. Which of the following objectives has the shortest working distance?
   a. 2 ×
   b. 10 ×
   c. 45 ×
   d. 100 ×

5. When changing objectives from low power to high power, it is generally necessary to
   a. lower the condenser
   b. raise the condenser to its maximum
   c. open the diaphragm
   d. close the diaphragm
   e. a and d (only)

6. Changing from the low power to the high power will enable a person to see _____ of a specimen.
   a. less
   b. more
   c. about the same portion

The Cellular Level
of Organization

After completing this exercise,
you should be able to

1. Distinguish between prokaryotic and eukaryotic forms of cellular organization
2. List the major organelles found in eukaryotic animal cells
3. Give the functions of major eukaryotic organelles
4. Locate and identify the basic parts of animal cells

Living organisms are composed of basic units of structure and function termed *cells*. Furthermore, cells contain subcellular parts, referred to as *organelles* (or little organs), that exhibit distinctive structural and functional properties (see Figure 2-1).

The use of electron microscopes and associated techniques for the preparation of biological specimens has demonstrated clearly the existence of two fundamentally different types of cellular organization—namely, the *prokaryotic* (primitive nucleus) and *eukaryotic* (true nucleus).

The prokaryotic cell, which is the unit of cellular organization for bacteria, is distinguished by its small size and the absence of many of the organelles normally found in eukaryotic cells (Figures 2-1 and 2-2). All living organisms other than bacteria have a eukaryotic type of cellular organization (Table 2-1).

Dyes (coloring agents) termed *biologic stains* frequently are used to distinguish organelles and/or chemical substances found in and between cells. Several categories of biologic stains are recognized; however, tissue stains are the most commonly used in microscopic work. In procedures involving these stains, one, two, and occasionally three dyes are used to differentiate (distinguish) the nucleus from the cytoplasm of cells and to permit the differentiation among different tissue types. Stains may be used to color both living and preserved tissue. In addition, special techniques employing enzymes or acids can be used to identify specific chemicals in cellular structures. For example, the nucleic acids deoxyribonucleic acid (DNA) and ribonucleic acid (RNA) are identified with the use of the enzymes deoxyribonuclease (DNase) and ribonuclease (RNase) and acids such as trichloroacetic acid and perchloric acid.

In this exercise, major attention is given to the basic parts of eukaryotic cells that are readily visible with the compound bright-field microscope. Specimens for the exercise include stained representatives of animal eukaryotic cells, such as blood and cheek cells, and several of their organelles. Examples of enzyme-treated cells also are considered.

An Introduction to Cellular Organelles

Materials

1. Six demonstration microscopes, one for each of the following prepared, stained slides:
   a. Cross-section of frog *(Rana pipiens)* liver
   b. Frog or fish blood smear
   c. Smooth muscle cells (teased preparation or a cross-section of intestinal tissue)
   d. Human blood smear
   e. Squamous epithelial cheek cells
   f. Ciliated epithelial cells
2. Unknown slide preparation, one per student
3. Lens paper

Procedure

This procedure is to be performed by students individually.

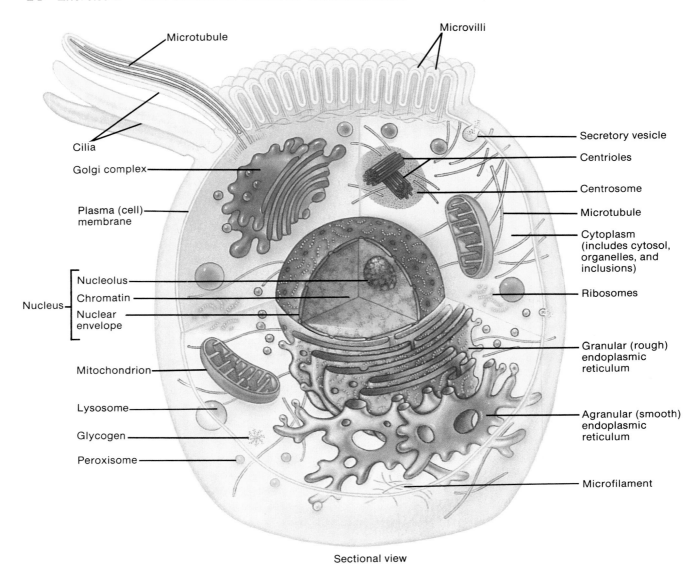

Figure 2-1 The cellular organization of a eukaryotic cell. (From G. J. Tortora, *Principles of Human Anatomy,* 6th ed., New York: HarperCollins, 1992.)

1. Examine the preparations provided under low and high powers. Refer to Figure 2-3 and Color Plates 2, 3, 4, 5, 6, 7, and 8 as guides.
2. Look for the various organelles associated with eukaryotic cells.
3. Pay particular attention to the size and shape of the various cell types.
4. Complete the table and questions in the following results and observations section.
5. After you have examined the six demonstration slides, obtain an unknown slide from your instructor. Examine the slide, and identify the specific type of cell it contains. The preparation will be similar to one of those studied in the exercise. Enter your findings on the report form in the results and ob-

servations section. Check with your instructor to see whether your identification is correct.

RESULTS AND OBSERVATIONS

Introduction to Cellular Organelles

1. List four organelles of animal cells.

   a. _____

   b. _____

   c. _____

   d. _____

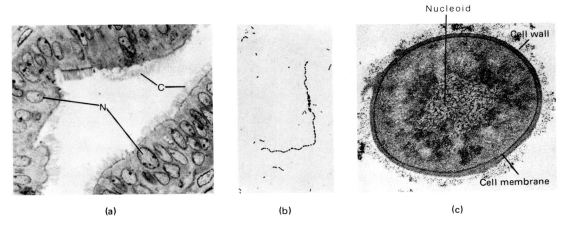

(a) (b) (c)

Figure 2-2 Prokaryotic and eukaryotic cells. (a) Eukaryotic cells taken from a human oviduct; note the well-defined nuclei **(N)** and cilia **(C);** such details are obtainable with properly prepared specimens (from I. A. Brosens and G. Vasquez, *Journal of Reproductive Medicine,* 16: 171, 1976). (b) The appearance of the much smaller prokaryotic cells; these bacterial forms are *cocci* (spherical shape). (c) An electron micrograph of a prokaryotic cell clearly showing the absence of membrane-bound internal structures (organelles) and a well-defined nucleus (nucleoid). (Figures (b) and (c) are from C. M. Ward, Jr., and G. W. Claus, *Journal of Bacteriology,* 114: 378–389, 1973.)

Table 2-1 Eukaryotic Organelles and Their Functions and Activities

Organelle	Associated Functions and Activities
Cell plasma membrane	Transport of substances into and out of cells Phagocytosis (engulfment of foreign material) in some cells Pinocytosis (uptake of fluid)
Cell wall (found only in plants and certain microorganisms)	Imparts shape and strength to the cell Protection against some osmotic imbalances
Chloroplast	Photosynthesis
Cilium	Motion or movement of substances past the ciliated cell
Endoplasmic reticulum	Protein synthesis
Flagellum	Propulsion
Golgi complex	Storage and sorting structure for cell products Transport of proteins, lipids, and other cell products to the cell surface
Microbody or peroxisome	Enzymatic activities
Microtubule	Transport of materials within cell Development and maintenance of cell shape Cell division Part of internal structure of cilia and flagella
Mitochondrion	Synthesis of the energy-rich compound adenosine triphosphate (ATP)

Table 2-1 *(Continued)*

Organelle	Associated Functions and Activities
Nucleolus	Major site for the formation of ribosomal components
Nucleus	Control of cellular physiological process Transfer of hereditary factors to subsequent generations
Ribosome	Protein synthesis
Vacuole	Storage site for water; for some amino acids, carbo-hydrates, and proteins; for cell products; or for cell wastes

Figure 2-3 Squamous epithelial animal cells: Note that animal cells show variation in size and shape. (The original magnification is 450×.)

2. Compare the various cell types provided in this exercise, as to the presence of organelles listed in Table 2-2. Indicate the presence of an organelle by a " + " and its absence by a " − ", and briefly describe the shape of the cell under study—for example, round, oval, or spindle.
3. State the type of cell that the unknown slide preparation contains: _____

Demonstration of Organelles by Special Stains and Treatment

Materials

1. *Ascaris* egg in mitosis, showing centriole (iron hematoxylin and orange G stain)

Table 2-2

Organelle and Cell Shape	Frog Blood Red	Frog Blood White	Human Blood Red	Human Blood White	Frog Liver	Smooth Muscle	Squamous Epithelium
Cell membrane							
Nucleolus							
Nucleus							
Vacuole							
Shape of cell							

2. *Amphiuma* liver cells, showing mitochondria (iron hematoxylin stain)
3. Golgi apparatus preparation (silver stain)
4. *Necturus* liver cells showing distinctive nucleus and nucleoli (azure B stain)
5. Liver tissue stained with iron hematoxylin and eosin
6. Liver tissue stained for DNA with Feulgen reaction
7. DNase-treated liver tissue and stained with Feulgen reaction
8. RNase-treated liver tissue

Procedure

This procedure is to be performed by students individually.

1. Examine the preparations provided under the high-power and oil-immersion objectives. Refer to Figure 2-3 and Color Plates 9 and 10 as guides.
2. Look for the various organelles or reactions indicated on each slide. Use the following descriptions to aid you in either finding specific organelles or interpreting reactions.
   a. *Asters*—strands extending from and around a central body
   b. *Mitochondria*—small, individual, threadlike strands, generally blue within the cytoplasm; some are short and sticklike (rod) in appearance
   c. *Golgi complex*—brown or black strands forming a network around a cell's nucleus
   d. *Nucleolus*—small, darkly stained bodies within a cell's nucleus
3. Sketch a representative view of each specially stained organelle, as seen under oil immersion, in the spaces provided in the following results and observations section. Include all organelles observed and the general outline of a cell.
4. Next, examine the liver cells stained by the Feulgen reaction under low and high power. This staining procedure shows the presence of DNA. Refer to

Color Plates 9 and 10. Use these color plates for comparison purposes.
5. Now examine the enzyme-treated preparations. Based on your observations of the stained cells, determine what portion(s) of these cells was (were) affected by the respective enzyme treatments.
6. Complete Table 2-3, and answer the questions in the following results and observations section.

RESULTS AND OBSERVATIONS

1. Sketch the organelles and their locations.

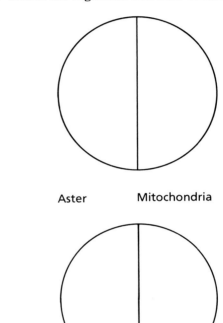

Aster   Mitochondria

Golgi Complex   Centriole

Table 2-3

Treatment	Cellular Components			
	Cell Membrane	Nucleus	Nucleolus	Cytoplasm
No treatment				
RNase treatment				
DNase treatment				

2. Compare the different enzyme-treated cells, as to the presence of the cellular components listed in Table 2-3. Indicate the presence or absence of a component by a " + " or " − " sign, respectively.

LABORATORY REVIEW *2*

THE CELLULAR LEVEL OF ORGANIZATION

COMPLETION AND FILL-IN

1. Identify the cell parts shown in Figure Q2-1.

(a)

(b)

Figure Q2-1 (a) Frog blood cells; (b) ciliated epithelial cells. (Courtesy of Carolina Biological Supply Company.)

(a) FROG BLOOD CELLS

a. _____

b. _____

c. _____

(b) CILIATED EPITHELIAL CELLS

a. _____

b. _____

c. _____

2. Using your textbook or other references, complete Table Q2-1 by giving the functions of the organelles listed.

Table Q2-1

Organelle	Functions
Nucleus	a.
Nucleolus	b.
Endoplasmic reticulum	c.
Golgi complex	d.
Mitochondrion	e.
Centriole	f.
Cilia	g.
Ribosome	h.
Plasma membrane	i.

3. Is the cell in electron micrograph Figure Q2-2 *prokaryotic* or *eukaryotic?*

Figure Q2-2 Prokaryotic or eukaryotic?

4. Identify the organelles in Figure Q2-3.

    a. _____

    b. _____

    c. _____

    d. _____

    e. _____

    f. _____

Figure Q2-3 Eukaryotic cellular organization.

5. State the appropriate metric equivalent units for the following:

    a. 11 millimeters = _____ μm

    b. 100 nanometers = _____ mm

    c. 1 μm = _____ mm

    d. 17 μm = _____ mm

    e. 10 nm = _____ mm

    f. 3 μm = _____ nm

6. State the function or purpose of the following in the examination of cells:

    a. Biologic stains _____

    _____

b. DNase _____

_____

ANSWERS

c. RNase _____

_____

1. _____
2. _____
3. _____
4. _____
5. _____
6. _____
7. _____
8. _____
9. _____
10. _____
11. _____
12. _____
13. _____
14. _____
15. _____

MULTIPLE CHOICE

Insert the correct answers in the spaces provided.

1. Subcellular parts that have distinctive and functional properties are called
   a. prokaryotic cells
   b. organelles
   c. small organs
   d. cells
   e. eukaryotic cells
2. Which of the following serves as the external boundary of a cell, through which substances enter and leave?
   a. flagellum
   b. nucleus
   c. endoplasmic reticulum
   d. microfilament
   e. plasma membrane
3. The specific part of the cell that holds or contains hereditary information is known as the
   a. nucleolus
   b. chloroplast
   c. mitochondrion
   d. nucleus
   e. endoplasmic reticulum
4. The cellular structure responsible for or associated with movement is known as a
   a. Golgi complex
   b. cilium
   c. flagellum
   d. nucleus
   e. b and c (only)
5. The cellular sites known for their production of ATP are which of the following?
   a. Golgi complex
   b. endoplasmic reticulum
   c. mitochondria
   d. chromosomes
   e. lysosomes
6. Protein synthesis in a typical cell occurs specifically in or on sites known as the
   a. Golgi complex
   b. cell membrane
   c. centrosome
   d. nucleolus
   e. ribosomes
7. Which of the following organelles direct most if not all activities of the cell?
   a. mitochondria
   b. nucleolus
   c. endoplasmic reticulum
   d. nucleus
   e. cell membrane

8. Short, hairlike organelles that extend from certain cells and may be used to move substances across surfaces are
   a. flagella
   b. microtubules
   c. cell membranes
   d. cilia
   e. cytoskeletons

9. Specific organelles described as being flattened sacks of membranes that collect, package, and distribute molecules within the cell are
   a. chromosomes
   b. lysosomes
   c. centrioles
   d. flagella
   e. the Golgi complex

10. The network of protein microfilaments in a eukaryotic cell's cytoplasm that maintains the shape of the cell, anchors its organelles, and is associated with movements is known as which of the following?
    a. flagella
    b. cytoskeleton
    c. Golgi complex
    d. lysosomes
    e. centrioles

11. _____ are the organelles responsible for protein synthesis.
    a. mitochondria
    b. lysosomes
    c. chromosomes
    d. peroxisomes
    e. ribosomes

12. The term used for membrane-bound structures found in the cytoplasm of eukaryotes and known to perform various functions is
    a. organ
    b. organelle
    c. nucleus
    d. cell membrane
    e. cytoplasm

13. The phospholipid bilayer with embedded proteins that regulates cell permeability is the
    a. nucleus
    b. nuclear envelope
    c. plasma membrane
    d. nucleolus
    e. mitochondrion

14. The powerhouses of a eukaryotic cell, which are prokaryoticlike inclusions, are known as which of the following?
    a. nucleus
    b. ribosomes
    c. Golgi complex
    d. mitochondrion
    e. endoplasmic reticulum

15. The organelle associated with protein synthesis that lacks ribosomes is known as which of the following?
    a. cell membrane
    b. microtubules
    c. smooth endoplasmic reticulum
    d. rough endoplasmic reticulum
    e. mitochondrion

Cellular Reproduction (Mitosis and Cytokinesis)

1. Describe the sequence of events in the process of mitosis
2. Distinguish between mitosis and cytokinesis
3. Identify specific mitotic stages in prepared slides
4. Describe the general features of a cell cycle
5. Indicate the position of mitosis in the cell cycle

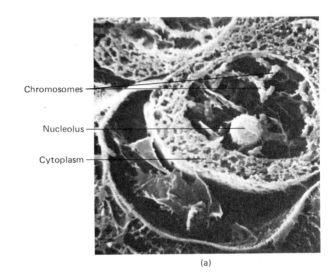

(a)

As indicated in Exercise 2, a eukaryotic cell's internal organization is a complicated assembly of chemical systems designed to perform a number of basic activities and functions to maintain its life on a day-to-day basis. However, cells under various conditions become injured, diseased, or worn and then die. Thus, new cells are needed for purposes of replacement and growth. The nucleus of each cell generally contains a precise set of genetic information built into more or less elongated structures known as *chromosomes* (Figure 3-1). A cell's genetic information is vital to its proper functioning and to determining the properties of the next generation of cells. Nuclear division resulting in the precise distribution of duplicated chromosomes into two newly formed cells is *mitosis. Meiosis,* a specialized type

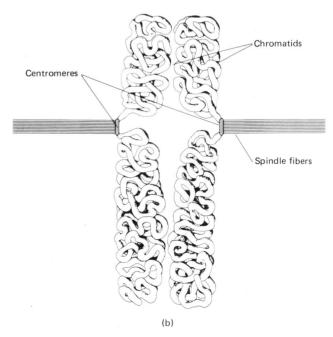

(b)

Figure 3-1 Cells during division: (a) scanning micrograph showing chromosomes and a nucleolus within a nucleus (reproduced by permission of the National Research Council of Canada from E. D. P. Whelan and others, *Canadian Journal of Botany,* 52:1438–1440, 1974); (b) chromatids attached to spindle fibers at centromeres.

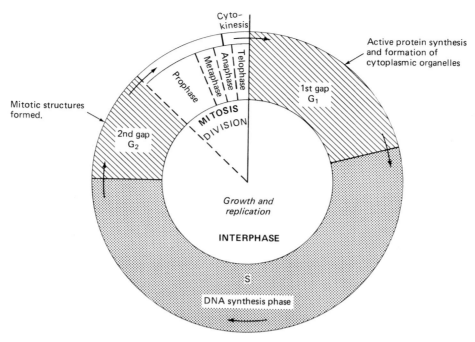

Figure 3-2 The phases of a cell cycle: The relationship of the four phases (G_1, S, G_2, and mitosis) of the cycle are shown; events such as cytoplasmic organelle production, DNA synthesis, and cellular and nuclear division are related to cell phases.

of cell division that occurs during the formation of *gametes* (mature sex cells), is described in Section XIV.

During a cell's existence, it passes through a regular, organized sequence of physiological events known as the *cell cycle* (Figure 3-2). The cycle is complex and is divided into four phases: G_1 (gap), S (synthesis), G_2 (gap), and *mitosis*. As Figure 3-2 shows, G_1, S, and G_2 compose the interphase. During the G_1 phase, active protein synthesis occurs, as well as the formation of new mitochondria, Golgi apparatus, endoplasmic reticulum, and other cytoplasmic materials. Cell size also increases in the G_1 phase. In the S phase, DNA synthesis takes place, and chromosomes are duplicated. The DNA in a eukaryotic nucleus, together with a large amount of protein and a small portion of RNA, forms chromatic strands. In the G_2 phase, structures directly related to mitosis are produced. Mitosis follows the G_2 phase. During mitosis, chromatin strands in the nucleus condense by coiling and folding to form compact chromosomes (Figure 3-1b). Microscopically, each chromosome has a characteristic shape and appears to consist of two chromatin strands termed *chromatids* (see Figure 3-1b). Those chromatids are attached firmly to each other at a specific region termed a *centromere* (Figure 3-1b). Mitosis usually occurs in close association with the division of the cytoplasm *(cytokinesis)*. During this event, the various duplicated cytoplasmic organelles in a parent cell are divided between two newly formed cells.

Mitosis is an extremely efficient process that provides for the even distribution of hereditary material during cell division. In describing the events taking place in mitosis, it is convenient to divide the process into four stages: *prophase, metaphase, anaphase,* and *telophase*. The period between successive stages is *interphase*. It should be noted that mitosis is actually a continuous process and does not occur in discrete steps. It is not always readily apparent when one stage ends and another begins. Table 3-1 summarizes the main events of mitosis.

Normal mitosis during the development of a human from a fertilized egg cell (ovum) produces only cells with the number of chromosomes characteristic of a human. Thus, all cells of the body should contain the same number. There are, however, situations in which abnormalities occur. These include development and mutation.

The stages of mitosis are not limited to normal cells. Abnormal types of cell growth, spoken of as cancers, occur when growth is no longer regulated by the processes that control normal cell growth (Figure 3-3). Any cell in the human body has the potential of becoming a cancer cell. Cancer cells grow and divide more rapidly than normal cells and collectively form a mass termed a *malignant tumor*. Such growths tend to invade surrounding tissues and disrupt the normal structure and function of body organs.

Table 3-1   Cell Division and Mitosis

Stage	General Events	Diagrammatic Representation
Interphase	1. Chromosomes are uncoiled. 2. Intense cellular metabolic activity includes DNA synthesis (replication).	Centrioles
Prophase	1. Individual chromosome structure becomes visible and recognizable as identical chromatids held together by centromeres. 2. Centrioles of cells separate and migrate toward opposite poles of cells. 3. Nucleolus and nuclear membrane disappear. 4. Spindle appears. 5. Asters are present in animal cells only.	Centromere, Centrioles, Chromosome (2 chromatids)
Metaphase	Chromosomes line up by their centromeres at the equator or spindle.	Centrioles, Chromosomes
Anaphase	1. Chromatids separate and move to opposite poles of the cell. 2. Each chromatid becomes an independent chromosome. 3. Cytokinesis begins.	Chromatid, Centriole
Telophase	1. Chromosome movement is completed. 2. Chromosomes assume the appearance they had in interphase. 3. New nucleolus and nucleus appear. 4. Cell plate appears only in plants.	Chromosome, Centrioles, Nuclear membrane

Materials

The following materials should be provided for class use:

1. Prepared slides of whitefish (*Coregonus*) blastulas showing mitosis
2. Models or charts showing stages of mitosis

Procedure

This procedure is to be performed by students individually.

1. Use both low- and high-power objectives to examine the prepared slide showing mitotic stages in the whitefish cells. The slide should contain several

Figure 3-3 Micrographs of malignant cells dividing: The stages of division are different visually and physiologically from those of normal cells. Note the general appearance of chromosomes. (Reproduced with permission from Freidman and others, "Needle Aspiration of Metastatic Melanoma," Acta *Cytologica,* 24: 715, 1980.)

embryonic cells exhibiting different stages of mitosis. (Refer to Color Plates 7 and 11.)

2. With the aid of Table 3-1, identify the stages of mitosis, and sketch each one in the following results and observations section. Label all significant parts of your sketches.

3. Find a clear (good) cell in metaphase, and count the number of chromosomes present. Do the same for a cell in anaphase.

4. Answer the questions pertaining to this portion of the exercise in the following results and observations section.

RESULTS AND OBSERVATIONS

Mitosis

1. Sketch and label the stages of mitosis.

2. In which stage of mitosis is an equatorial plate evident? _____

3. How many chromosomes were present during metaphase? _____

4. Were the number of chromosomes present during anaphase the same as the number during metaphase? Explain your answer. _____

___ _____

_____

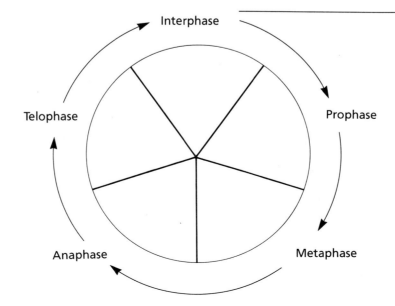

LABORATORY REVIEW *3*

NAME _____

LAB SECTION _____ DATE _____

CELLULAR REPRODUCTION (MITOSIS AND CYTOKINESIS)

COMPLETION, FILL-IN, AND DEFINITION

1. List and explain the four phases of the cell cycle.

   a. _____

   b. _____

   c. _____

   d. _____

2. Identify all parts of the animal cell undergoing metaphase in mitosis, as shown in Figure Q3-1.

   a. _____

   b. _____

   c. _____

   d. _____

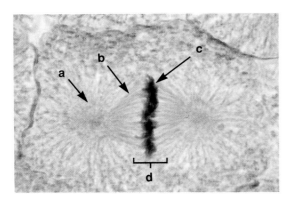

Figure Q3-1

3. List and briefly describe the stages of mitosis in their correct sequence.

   a. _____

   b. _____

   c. _____

   d. _____

   e. _____

4. Identify the stages of mitosis shown in Figure Q3-2.

   a. _____

   b. _____

   c. _____

   d. _____

(a)

(b)

(c)

(d)

Figure Q3-2 Stages of mitosis.

5. Define or explain the following:

   a. Nucleolus _____

   b. DNA _____

   _____

   c. Meiosis _____

   d. Centromere _____

   _____

   e. Chromatid _____

MULTIPLE CHOICE

ANSWERS

Insert the correct answers in the spaces provided.

1. Which of the following organelles divides and organizes spindle fibers during mitosis?
   a. nucleus
   b. chromosomes
   c. microtubules
   d. nucleolus
   e. centrioles
2. Which of the following move chromosomes during cell division?
   a. vesicles
   b. nucleus
   c. microtubules
   d. Golgi complex
   e. ribosomes
3. The structures that are condensed forms of DNA, found in the nuclear area and associated with proteins and RNA, are called which of the following?
   a. microtubules
   b. cilia
   c. centrioles
   d. ribosomes
   e. chromosomes
4. During which of the following phases of mitosis does the nuclear membrane of a cell disappear?
   a. interphase
   b. prophase
   c. metaphase
   d. anaphase
   e. telophase
5. During which phase of mitosis do new nuclear membranes form?
   a. interphase
   b. prophase
   c. metaphase
   d. anaphase
   e. telophase
6. The replication of DNA takes place during which of the following phases of mitosis?
   a. interphase
   b. prophase
   c. metaphase
   d. anaphase
   e. telophase
7. Sister chromosomes separate and move to opposite poles of a cell during which phase?
   a. interphase
   b. prophase
   c. metaphase
   d. anaphase
   e. telophase
8. Spindle fibers form in which of the following phase(s)?
   a. interphase
   b. prophase
   c. metaphase
   d. anaphase
   e. telophase

1. _____
2. _____
3. _____
4. _____
5. _____
6. _____
7. _____
8. _____
9. _____
10. _____

9. Chromosomes become arranged on an equatorial plate in the phase known as what?
   a. interphase
   b. prophase
   c. metaphase
   d. anaphase
   e. telophase
10. During which phase of mitosis do asters form from centrioles?
    a. interphase
    b. prophase
    c. metaphase
    d. anaphase
    e. telophase

Fluid Transport and Molecular Activity: Brownian Movement, Filtration, Diffusion, and Osmosis

After completing this exercise, you should be able to

1. List three physical processes by which materials move into and out of cells
2. Perform simple experiments to demonstrate filtration, diffusion, and osmosis
3. Distinguish between the general process of diffusion and the more specific process of osmosis
4. Define or explain *Brownian movement, solution, solute,* and *solvent*

All the chemical reactions in a cellular environment take place in solution. Before a substance can enter a cell, it must dissolve. The term *solution* can refer to the situation in which a substance is dissolved more or less evenly in a liquid. If, for example, the carbohydrate glucose is mixed thoroughly with water, the solid crystals of the substance dissolve and are dispersed throughout the liquid. The resulting mixture—a solution—is composed of the liquid *solvent* and the dissolved *solute.* Substances that dissolve are *soluble,* while those that do not dissolve are *insoluble.* The digestive process in humans and other animals generally involves turning a variety of large compounds into small compounds that are soluble in water and can be carried into and out of cells.

Substances move into and out of cells by several physical processes, including (1) *diffusion,* resulting from a concentration gradient (movement of molecules from an area of higher concentration to a lower one); (2) *osmosis,* the diffusion of water molecules across a semipermeable membrane; and (3) *filtration,* produced by hydrostatic pressure differences.

Within cells, numerous minute particles appear to move in an erratic, zigzag motion known as *Brownian movement.* The movement occurs as small particles suspended within the cytoplasm are bumped about by the molecules of the cell's internal environment.

A cell membrane consists mainly of protein and phospholipid molecules arranged into a delicate, thin layer (see Figure 4-1). Most of its functions are associated with regulating the passage of materials into and out of cells. A membrane that allows molecules to pass through it is called *permeable.* Because cell membranes allow certain ions and certain molecules to pass through them, but do not allow the passage of others, they are said to be *selectively permeable, differentially permeable,* or *semipermeable.*

Movement of dissolved inorganic and organic substances within solutions or through the cell membrane involves a natural physical process termed *diffusion.* Diffusion can be demonstrated easily by placing a small dye crystal (the solute) in a volume of water (the solvent). The solute moves from an area of higher solute concentration to an area of lower concentration.

In *osmosis,* a type of diffusion, a solvent (the water in which substances are dissolved) moves from a solution of lower solute concentration to a solution of higher solute concentration when the solutions are separated by a semipermeable membrane.

Filtration is described as the movement of solvents and dissolved substances across a semipermeable membrane, under the influence of mechanical pressure. Such movement is always from an area of higher pressure to an area of lower pressure and continues as long as a difference in pressure exists.

This exercise will demonstrate these various phenomena and processes. (*Dialysis,* a process involving the separation of small molecules from larger ones by diffusion of the smaller molecules through a semipermeable membrane, is discussed in Section XIV.)

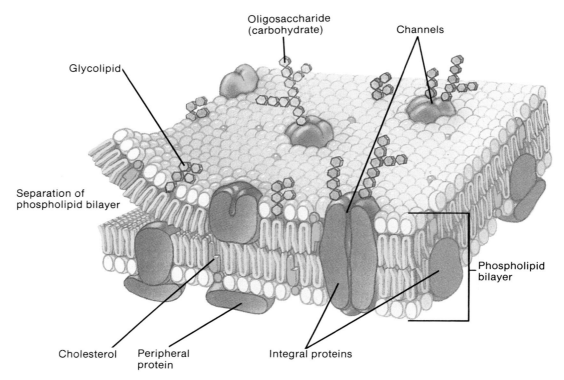

Figure 4-1 A diagrammatic enlargement of a cell (plasma) membrane showing the relationship of the phospholipid bilayer and its protein molecules. The bilayer is shown in a separated form, to illustrate the relationships among the various types of molecules found with a membrane. (From G. J. Tortora, *Principles of Human Anatomy,* 6th ed., New York: HarperCollins, 1992.)

Brownian Movement

Materials

The following materials should be provided for class use:

1. Dilute solution of India ink
2. Eyedropper and rubber bulb
3. Glass slides and cover slips
4. Microscopes

Procedure

This procedure is to be performed by students individually.

1. Place a small drop of the dilute India-ink solution on a glass slide, and add a cover slip.
2. Observe the preparation under the high-power objective of your microscope.
3. Reduce the light intensity, and note the direction of movement of the India-ink particles. Indicate the type of movement seen in the results and observations section.

Filtration

Chemical compounds in solution may behave in one of two ways. The molecules may remain intact—that is, undissociated—or they may dissociate—that is, separate into components termed *ions.* This process is *ionization.* Compounds that behave in this manner are known as *electrolytes;* compounds made up of molecules that do not dissociate are called *nonelectrolytes.*

Ions carry an electrical charge. For example, when the electrolyte sodium chloride is dissolved in water, it provides positively charged sodium ions (Na^+) and negatively charged chloride ions (Cl^-).

Both nonelectrolytes and electrolytes are used in this exercise, to show how dissolved substances can be forced through physical barriers.

Materials

1. The following materials should be provided per four students:
   a. One funnel (65 mm diameter)
   b. One ring stand and clamp

c. One 50-ml beaker
d. One sheet of filter paper of appropriate diameter
e. Two test tubes
2. The following materials should be provided for class use.
   a. 5 percent starch solution
   b. 10 percent sodium chloride (NaCl) solution
   c. Iodine reagent
   d. Silver nitrate (AgNO$_3$) solution (0.1 molar)
   e. Powdered wood charcoal
   f. 10-ml graduated cylinders
   g. Wooden tongue depressors

Procedure

This procedure is to be performed by students in groups of four.
1. Use a graduated cylinder to measure 2 ml of the 10 percent sodium chloride (NaCl) solution. Pour the solution into a test tube. Add 2 drops of the silver nitrate (AgNO$_3$). Observe and record your findings in the results and observations section. This is a test for the presence of chloride ions (Cl$^-$).
2. Put 3 ml of the 5 percent starch solution into the other test tube. Add 3 drops of the iodine reagent. Observe and record your findings in the results and observations section. This is a test for the presence of starch.
3. Rinse out both test tubes well, and save for a later portion of this exercise.
4. Fold one piece of filter paper in quarters, and open the folded paper to form a cone. Place the cone in a funnel.
5. Set up a filtration system similar to the one shown in Figure 4-2.
6. Prepare a mixture for filtering by mixing the following items in a small beaker:
   a. 4.5 ml starch solution
   b. 2 ml sodium chloride solution
   c. A small amount of the powdered charcoal
   d. 10 ml water
7. Place a test tube under the funnel, and slowly pour 10 ml of water through the system. Collect the water, and discard it. The filter paper should now be firmly in place.
8. Place another test tube temporarily under the funnel, and pour the mixture through the system.
9. Collect the *filtrate* (the solution that passes through) in a beaker.
10. Divide the filtrate between two tubes. Perform the test for starch with one tube and the test for chloride ions with the other.
11. Record your findings, and answer the questions in the results and observations section.

Figure 4-2 A filtration system. Be certain that all parts of the system are firmly in place.

Diffusion

Materials

1. The following materials should be provided per four students:
   a. One plastic Petri plate containing a layer of agar
   b. One pair of forceps
   c. A millimeter ruler
2. The following materials should be provided for class use:
   a. Filter paper discs, approximately 5 mm in diameter
   b. 1 percent crystal violet solution
   c. 1 percent methylene blue solution
   d. 1 percent safranin O solution
   e. Wax pencil or other marking device
   f. Paper towels

Procedure

This procedure should be performed by students in groups of four.

1. Measure the diameter of the filter paper discs. Enter your measurement in the results and observations section.
2. Obtain a Petri plate containing a thin layer of gelled agar. Mark the bottom of the plate with a wax pencil, to divide the agar preparation into three sectors.
3. Holding a filter paper disc with the forceps, touch it to the surface of the crystal violet solution until the dye penetrates throughout the disc.
4. Remove the Petri plate top, and place the disc in the center of one of the marked sectors. Press the disc gently with the forceps, so that a good contact is made with the agar surface. Close the plate.
5. Wipe the forceps free of the dye solution. Repeat Steps 3 and 4 with the methylene blue and safranin O solutions, placing the discs in the centers of their respective sectors of the Petri plate.
6. Examine the discs on the plate at the beginning of the experiment (0 minutes, time) and at 15, 30, 45, and 60 minutes. Measure the diameters of the developing diffusion zones. (See Color Plate 12.)
7. Enter your findings, and answer the questions in the results and observations section.

Osmosis

The membrane used in this portion of the exercise is a piece of synthetic material often used as a substitute for a true biological membrane. Such synthetic material contains pores or holes that are approximately the same size as those in biological membranes but are more uniform.

Materials

1. The following materials should be provided per four students:
   a. One thistle tube (bell-shaped funnel tube)
   b. One dialysis membrane square (about 10 mm square)
   c. One rubber band
   d. One 250-ml beaker and one 500-ml beaker
   e. One ring stand and clamp or similar support system
   f. A millimeter ruler
   g. One marking pen
2. The following materials should be provided for class use:
   a. Four sucrose (cane sugar) solutions having concentrations of 1, 10, 20, and 40 percent sucrose, by weight, respectively
   b. 1 percent methylene blue solution in a dropper bottle
   c. Marking pen or similar device
   d. Transparent tape

Procedure

This procedure is to be performed by students in groups of four.

An osmometer system will be constructed, with which to determine the rates at which water diffuses through a synthetic membrane separating solutions with differing sucrose concentrations.

1. The instructor will assign a specific sucrose concentration to each student group, such that all four concentrations will be used. Be certain to examine all other systems and the results obtained by other members of the laboratory, so that the significance and influence of different sucrose concentrations can be noted.
2. Using the assigned sucrose concentration, prepare an osmometer system, as follows (see Figure 4-3):
   a. Moisten one dialysis membrane square thoroughly.
   b. Put about 50 ml of the assigned sucrose solution and enough methylene blue to color the preparation into the 250-ml beaker.
   c. While one laboratory partner places a finger firmly over the bottom stem opening of the thistle tube, pour the sucrose solution slowly into the tube until the head (funnel portion) is full.
   d. Allow a small amount of the solution to escape from the stem opening by releasing the finger at the stem opening.
   e. Place the wet membrane square over the funnel end and wind the rubber band around it several

Figure 4-3 An osmometer system.

times to be sure that the system is secure—that is, that it will not leak.

3. Invert the tube so that the funnel portion is facing down, then attach the thistle tube to a ring stand and clamp or similar support.

4. Fill the 500-ml beaker with water, and place it under the osmometer.

5. Lower the osmometer into the water to the depth as shown in Figure 4-3.

6. Allow 5 minutes for the system to equilibrate. Then locate the top level (meniscus) of the sucrose solution in the tube, and mark it.

7. Using the ruler, measure the level of the meniscus in terms of millimeters moved from the starting point. Measure it again at 15, 30, 45, 60, and 90 minutes.

8. Record your findings, and answer the questions in the results and observations section.

9. Determine the rate of diffusion for each of the sucrose solutions used in the class. (This will require getting the results of others.) To calculate the rate, select a specific time interval (for example, 0–15 minutes), and note the distance covered by the fluid. The rate of diffusion, expressed in millimeters/minute, is equal to the distance divided by the time.

RESULTS AND OBSERVATIONS

Brownian Movement

1. Use arrows to indicate the type of movement of India-ink particles.

Filtration

1. Indicate the appearance of the solutions after the reagents were added.

Iodine added to starch: _____

Silver nitrate added to sodium chloride: _____

2. Which substances were present in the filtrate?

_____

_____

3. What happened to the charcoal? _____

_____

Diffusion

1. Enter your measurements in Table 4-1.

2. Were the diffusion zones clear and sharp? _____

_____

3. Did the dyes diffuse down through the agar? _____

_____

Osmosis

1. Enter your measurements in Table 4-2.
2. Enter the diffusion rates of the different solutions in Table 4-3.
3. Which of the solutions had a faster rate of diffusion?

_____

Explain your answer. _____

_____

Table 4-1

Time (in minutes)	Zone Diameters (in mm)		
	Crystal Violet	Methylene Blue	Safranin O
0			
15			
30			
45			
60			

Table 4-2

Time	Meniscus Level
0	
15	
30	
45	
60	
90	

Table 4-3

Sucrose Concentration	Rate of Diffusion (in mm/minute)
1	
10	
20	
40	

LABORATORY REVIEW *4*

FLUID TRANSPORT AND MOLECULAR ACTIVITY

COMPLETION AND DEFINITION

1. Distinguish between the items in the following pairs:

   a. Osmosis and diffusion _____
   _____
   _____

   b. Filtration and diffusion _____
   _____

2. What is active transport? _____
   _____

3. Which cellular structures are associated with the following molecular activities?

   a. Osmosis _____

   b. Diffusion _____

   c. Active transport _____

   d. Pinocytosis _____

4. Explain Brownian movement. _____
   _____

5. What is dialysis? _____
   _____

6. Distinguish between electrolytes and nonelectrolytes.
   _____
   _____

ANSWERS

1. _____

2. _____

3. _____

4. _____

5. _____

6. _____

7. _____

MULTIPLE CHOICE

Insert the correct answers in the spaces provided.

1. The transport of a solute across a membrane from an area of low concentration to a region of high concentration, using a carrier protein and the expenditure of chemical energy is known as what?
   a. active transport
   b. proton pump
   c. facilitated diffusion
   d. chemiosmosis
   e. Brownian movement

2. The transport of molecules across a cell membrane from a region of high concentration to a region of low concentration, using a carrier protein is known as what?
   a. active transport
   b. diffusion
   c. facilitated diffusion
   d. chemiosmosis
   e. filtration.

3. A term for cell drinking is
   a. active transport
   b. proton pump
   c. facilitated diffusion
   d. chemiosmosis
   e. pinoyctosis.

4. The bulk transport of materials from a cell by extruding from vesicles at the cell surface is called which of the following?
   a. active transport
   b. exocytosis
   c. facilitated diffusion
   d. chemiosmosis
   e. endocytosis

5. The uptake of material into cells by inclusion within an invagination of the plasma membrane forming a vesicle is known as what?
   a. active transport
   b. osmosis
   c. phagocytosis
   d. chemiosmosis
   e. pinocytosis

6. The ingestion of particles, fragments of cells, or even entire cells is called
   a. active transport
   b. proton pump
   c. diffusion
   d. phagocytosis
   e. exocytosis.

7. The channels in the membrane that the cell uses to concentrate metabolites and ions is which of the following?
   a. Na–K pump
   b. proton pump
   c. ATP
   d. coupled channels
   e. chemiosmosis.

Selective Permeability and the Structural Integrity of Cell Membranes

After completing this exercise, you should be able to

1. Describe the effects of osmotic imbalances on cell membranes
2. Explain the differences among isotonic, hypotonic, and hypertonic solutions
3. Define *crenation* and *hemolysis*

The cell membrane is an important and dynamic component of cells. In addition to serving as the container for the various cellular structures, compounds, and related materials, it regulates the flow of materials into and out of the cell. All substances, functional or nonfunctional, must pass across the cell's membrane. Several important physical processes associated with such movement were described in Exercise 4.

In Exercise 4, it was shown that when a concentrated sugar solution and water are separated by a membrane permeable to water, the water molecules pass through the membrane to the sugar solution, thereby raising its level. A permeable membrane offers no resistance to the passage of water molecules. Therefore, water molecule movement occurs in both directions. However, water molecules will travel in greater numbers per unit of time from the region where their concentration is highest to where it is lowest.

The force exerted by the water molecules as they flow across a semipermeable membrane is referred to as *osmotic pressure*. Every solution has a potential osmotic pressure. The osmotic pressure of a solution is proportional to the number of solute particles (substance dissolved) per unit volume of a solution. Thus, the more particles in a solution the greater is the osmotic pressure or *osmolality* of the solution. Moreover,

the greater the osmotic pressure of a solution, the greater is its pulling power for water. Water moves toward the area of greater osmolality. It is important to note that the number of particles in solution constitutes the critical factor in determining osmotic pressure. A solution containing an ionizing substance, such as sodium chloride, will exert a greater osmolality than one containing a comparable concentration of a nonionizing substance (refer to the filtration experiment in Exercise 4). The diffusing particles may be ions, molecules, or aggregates of molecules.

From a physiological standpoint, the osmotic properties of solutions can be determined by comparing their effects on red blood cells to the effect of blood plasma (the liquid portion of blood) on the cells. If blood cells are placed in any fluid other than normal plasma, they may remain unaltered, may shrink, or may swell. If they remain unchanged, the solution surrounding them is said to have the same osmotic properties (osmolality) as blood plasma and is called *isotonic* or *isosmotic*. The osmotic pressure of the cell's external environment is equal to that inside the cell (Figure 5-1a, b).

If the cells swell, the solution has lower osmolality than blood plasma and is called *hypotonic* or *hypoosmotic*. The osmotic pressure of the cell's external environment is less than that inside the cell (Figure 5-1c, d). If the pressure does not reach a balance, the cell will continue to swell until it bursts. Such a rupture of red blood cells is known as *hemolysis*.

If the cells shrink, the surrounding solution has higher osmotic properties than plasma and is called *hypertonic* or *hyperosmotic*. The osmotic pressure exerted by the external fluid is greater than that inside the cell. The shrinking and wrinkling of red blood cells in a hypertonic solution is known as *crenation* (Figure 5-1e, f).

In this exercise, yeast and red blood cells are used to demonstrate selected properties of cell membranes.

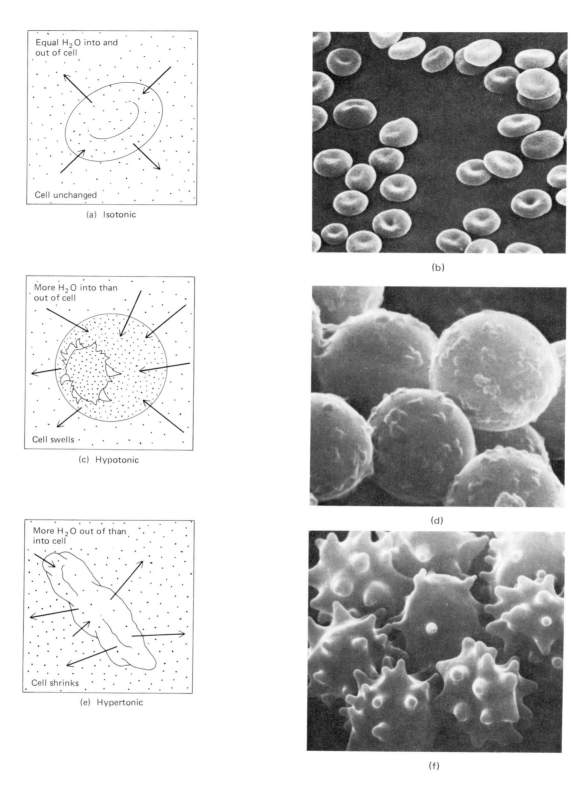

(a) Isotonic

Equal H₂O into and out of cell

Cell unchanged

(b)

More H₂O into than out of cell

Cell swells

(c) Hypotonic

(d)

More H₂O out of than into cell

Cell shrinks

(e) Hypertonic

(f)

Figure 5-1 A demonstration of osmotic pressure imbalances. The swelling or shrinkage of a red blood cell (or any cell) depends on the osmotic pressure of the solution in which it is placed. (a) A cell in an isotonic (isosmotic) solution—The osmotic pressure of the cell's external environment is equal to that of its interior. (b) A scanning micrograph of red blood cells in an isotonic solution. (c) A cell in a hypotonic (hypoosmotic) environment—The osmotic pressure is lower outside than inside the cell. Water molecules move into the cell faster than they can leave, diluting the cell's contents and causing it to swell. (d) A scanning micrograph showing swollen cells. (e) A cell in a hypertonic (hyperosmotic) solution—The osmotic pressure is greater outside than inside the cell; water molecules move out of the cell faster than they can enter, causing the cell to shrink. (f) A scanning micrograph showing crenation, one of the effects of a hypertonic environment. (All scanning micrographs from M. P. Sheetz, R. G. Painter, and S. J. Singer. *Journal of Cell Biology,* 70:193–203, 1976.)

Selective Membrane Permeability

Materials

The following materials should be provided for class use:

1. Yeast suspension, with Pasteur pipette and rubber bulb
2. 3 percent Congo red solution in dropper bottle
3. Test tubes
4. Water bath set at boiling temperature and equipped with test-tube racks
5. 1-ml pipettes and rubber bulbs
6. Pasteur pipettes and rubber bulbs
7. Glass slides and cover slips
8. Wax marking pencils or similar devices

Procedure

This procedure is to be performed by students in pairs.

1. Introduce about 1 ml of the yeast suspension into each of three test tubes. Use marking pencils to number these tubes 1, 2, and 3.
2. Add 3 drops of the Congo red solution to Tubes 1 and 2. Shake the tubes gently to mix the contents.
3. Place Tubes 1 and 3 in boiling water bath for 5 minutes. This treatment should kill the yeast cells. Tube 2 should contain living cells.
4. Add 3 drops of the Congo red solution to Tube 3, and shake gently to mix.
5. Allow the heated tubes to reach room temperature (25°C).
6. With separate Pasteur pipettes remove a sample from each tube, and prepare a temporary wet mount of each. (See Exercise 1.)
7. Examine each preparation first under low power and then under high power. Note which preparations contain stained cells.
8. Enter your findings and answer the questions in the results and observations section.

Structural Integrity of Cell Membranes

Materials

The following materials should be provided for class use:

1. Individual squeeze bottle containers of the following solutions—
   a. 70 percent alcohol
   b. Distilled water (H_2O)
   c. Physiological saline (0.85 percent sodium chloride)
   d. 2 percent sodium chloride (NaCl)
2. Sterile, disposable blood lancets
3. Alcohol prep pads (commercially packaged)
4. Sterile cotton squares
5. Petroleum jelly
6. Wooden applicator sticks
7. Glass depression slides and cover slips
8. Disposable Pasteur pipettes and rubber bulbs
9. Disinfectant solution for disposal of blood-containing materials

Procedure

This procedure should be performed by students in pairs. It is suggested that one laboratory partner serve as the blood donor. (Refer to Appendix C for alternative sources of blood.)

> **CAUTION**
> Working with blood or other body fluids will not pose any danger if the following precautions are observed:
>
> 1. Place any and all blood, or materials containing blood, into the disinfectant containers provided.
> 2. Dispose of any blood-containing materials as indicated by the instructor.
> 3. In cases of blood spillage, wipe the area with a disinfectant-soaked paper towel, and dispose of all materials as indicated by the instructor.
> 4. Always wash your hands after handling blood and associated materials.

1. Obtain three clean, dry glass vials, and label one for each solution, to be used as follows: distilled water, H_2O; physiological saline, *sal;* 2 percent sodium chloride, *NaCl.* See Figure 5-2 for steps to be followed in this procedure.
2. Introduce these solutions into their respective vials. The fluid levels should be about 8 mm from the bottom of the vial.
3. Select the finger that will be used, and wipe it with a cotton square that is moistened, but not saturated, with 70 percent alcohol. Do not use a finger that will be involved later in holding a pencil or other such activity.
4. Unwrap the blood lancet. Note that the lancet is to be used only once and then discarded as indicated by the instructor.
5. Hold the finger in a downward position and gently squeeze it.
6. Puncture the ball of the finger with the lancet quickly.
7. Discard the first 2 drops of blood with the aid of a sterile cotton square.

1. Obtaining specimen from alcohol-treated finger

2. Blood cell preparations

3. Hanging drop procedure

Cover slip

Blood cell specimen

Petroleum jelly (vaseline)

Depression slide (face down)

H₂O

Physiological saline

2% NaCl

Figure 5-2 Steps in the laboratory procedure for testing the structural integrity of cell membranes.

8. Keeping the finger in a downward position, place the bleeding site over the open end of the vial containing saline. Hold the vial firmly against the fingertip, and invert it so that the formed blood drop is washed off the finger into the solution.

9. Wipe the finger dry, and repeat Step 8 with each of the other solutions.

10. Examine the vials, and note whether each blood-containing solution is cloudy or clear.

11. With a wooden applicator stick put a *small* amount of petroleum jelly on each corner of three cover slips.

12. Transfer 1 drop of the suspension from the saline vial to the center of one of the prepared cover slips.

13. Place a depression slide face down over the cover slip so that the depression portion is directly over the drop of sample material. Press down gently on the slide so that it will stick firmly to the vaseline on the corners of the cover slip.

14. Turn the slide over in one quick motion so that the drop is suspended in the depression.

15. Examine this preparation first under low power and then under high power.

16. Sketch a number of cells, showing their general appearance, in the results and observations section. Note whether the cells are intact, are round, or show any unusual feature.

17. Repeat Steps 12–16 with the other solutions. (Avoid contamination with the pipette.)

18. Dispose of all blood-containing materials as indicated by your instructor.

19. Wash your hands.

20. Answer the questions in the results and observations section.

RESULTS AND OBSERVATIONS

Selective Membrane Permeability

1. Which of the preparations contained stained yeast cells? _____

2. Did living or dead yeast cells absorb the dye? _____

3. On the basis of this exercise, how are cell death and cell-membrane control functions related? _____
_____

Structural Integrity of Cell Membranes

1. Sketch representative fields.

Cells in distilled water

Cells in 0.85 percent saline

Cells in 2.0 percent sodium chloride

2. Which of the preparations used was hypotonic to the red blood cells? Explain. _____

_____

_____

3. Which of the preparations used was hypertonic to the red blood cells? Explain. _____

_____

_____

4. Which of the preparations used was isotonic to the red blood cells? Explain. _____

_____

_____

5. How would the different solutions used affect cells with cell walls. (Diagrams can be used.) _____

_____

6. Which of the preparations exhibited crenation?

_____

_____

LABORATORY REVIEW 5

NAME _____

LAB SECTION _____ DATE _____

SELECTIVE PERMEABILITY AND INTEGRITY OF CELL MEMBRANES

COMPLETION, FILL-IN, AND DEFINITION

1. a. Do all types of cells have cell membranes? _____

   b. What types of cells lack a cell membrane? _____

   c. What is the chemical composition of a cell membrane? _____

   _____

2. Briefly explain how exposure to the following factors would affect the functioning of a cell membrane:

   a. Heat _____

   b. Lipid solvents _____

3. Figure Q5-1 shows red blood cells suspended in a 2 percent salt solution.

   a. Describe the effect produced. _____

   b. Sketch the effect produced when blood cells are placed in a hypotonic solution.

Hypotonic solution

Figure Q5-1

4. Figure Q5-2 shows red blood cells suspended in an isotonic solution.

   a. Describe the effect produced. _____

   b. Sketch the effect produced when blood cells are placed in a hypertonic, not an isotonic solution.

Hypertonic solution

Figure Q5-2

5. Define or explain the following:

   a. Osmotic pressure _____

   _____

   b. Crenation _____

   c. Hemolysis _____

   d. Isotonic _____

6. In the system shown in Figure Q5-3, will water molecules pass from A to B or from B to A? Explain your answer. _____

   _____

   _____

   _____

   _____

Semipermeable membrane

A   B

Region of low
solute
concentration

Region of high
solute
concentration

Figure Q5-3

MULTIPLE CHOICE

Insert the correct answers in the spaces provided.

1. Which of the following terms is used to indicate the movement of molecules from a region of high concentration to regions of lower concentration, as a result of spontaneous molecular motions?
   a. diffusion
   b. osmosis
   c. equilibrium
   d. solute
   e. solvent

2. The most common substance in a solution, usually a liquid, is called which of the following?
   a. diffusion
   b. osmosis
   c. equilibrium
   d. solute
   e. solvent

3. Molecules other than those indicated in Question 2, which are dissolved in the liquid in lesser amounts than the liquid, are known as which of the following?
   a. diffusion
   b. osmosis
   c. equilibrium
   d. solute
   e. solvent

4. The diffusion of water across a selectively permeable membrane from an area of greater water concentration to an area of lesser water concentration is called what?
   a. diffusion
   b. osmosis
   c. equilibrium
   d. solute
   e. solvent

5. What is the term that refers to a solution that contains a lower concentration of solute particles (water moves across a membrane out of this solution)?
   a. isotonic
   b. hypertonic
   c. hypotonic
   d. osmotic pressure
   e. equilibrium

6. Which of the following terms refers to a solution with a higher solute concentration (water moves across a membrane into this solution)?
   a. isotonic
   b. hypertonic
   c. hypotonic
   d. osmotic pressure
   e. equilibrium

7. Which term refers to two solutions that have equal concentrations of solute particles?
   a. isotonic
   b. hypertonic
   c. hypotonic
   d. osmotic pressure
   e. equilibrium

ANSWERS

1. _____
2. _____
3. _____
4. _____
5. _____
6. _____
7. _____
8. _____
9. _____

8. Which of the following is the term for the potential pressure developed by a solution separated from pure water by a selectively permeable membrane?
    a. isotonic
    b. hypertonic
    c. hypotonic
    d. osmotic pressure
    e. equilibrium

9. Which of the following terms is used to indicate that cytoplasm is pushing out against the cell membrane?
    a. isotonic
    b. hypertonic
    c. hypotonic
    d. osmotic pressure
    e. equilibrium

Tissues

***After you have completed this exercise,
you should be able to***

1. Describe the properties of the four primary tissues of the human body
2. Identify by microscopic examination representatives of primary tissues of the body
3. List the functions of specific tissues
4. Give at least one body location of each type of tissue
5. Identify the component parts of selected tissues

A *tissue* is a group of similar cells that perform a similar function. The study of tissues is called *histology*.

Tissue slides are prepared by fixing part of an organ in a buffered solution of formaldehyde or other fixative, then dehydrating it with an organic solvent and embedding it in paraffin (or epoxyresins for electron microscopy). This embedded preparation is then sectioned with a *microtome* (a device that resembles a very precise meat slicer). All of the tissues shown in this exercise were cut to a thickness of 1 to 8 μm and then stained with appropriate dyes.

CLASSIFICATION OF TISSUES

The tissues in the human body usually are classified into four main types, based on their structure and function: *epithelial, connective, muscular,* and *nervous* tis-

sues. These four tissue types may be further subdivided into more specific categories (indicated in parentheses).

Epithelial tissue lines the inside or the outside of a body, an organ, or a part of an organ (simple, stratified, pseudostratified).

Connective tissue connects and supports other tissues in the body (fibrous, adipose, cartilage, bone, blood). Its cells produce a nonliving matrix which becomes located inside (intracellular) or outside (intercellular) the tissue cells.

Muscular tissue has the ability to contract (striated skeletal, striated cardiac, and smooth).

Nervous tissue (neurons) have the ability to conduct nerve impulses.

Most of the tissue types are discussed in the following parts of this exercise. Each example is accompanied by a photograph of that tissue, and below each photo is a space for your own drawing of any or all of the examples you have the opportunity to observe in the laboratory.[1] If drawings are made, they should be as accurate a representation as possible of the tissue slide you study. If you have no stained slide, you may draw the photograph. Use a hard-lead pencil. Stipple cytoplasmic detail, and label all parts. If you draw more than one cell, label only one. Make careful observations, examine the entire slide to make certain you have found the correct tissue for the drawing, and limit yourself to a small but representative area.

[1]Your instructor may assign some drawings and omit others, or ask for no drawings at all. Follow your instructor's directions.

EPITHELIAL TISSUE

Materials

The following prepared slides should be provided for class use:

Epithelial Tissue

1. Cheek cells (squamous)
2. Liver, kidney tubules, or thyroid gland (cuboidal
3. Human skin (stratified squamous)
4. Small intestine cross-section (columnar epithelium)
5. Lining of urinary bladder (stratified transitional epithelium)

Epithelial Tissue—Exocrine Glands

1. Large intestine (simple tubular glands)
2. Skin (sweat glands)
3. Pyloric region of the stomach (mucous glands)
4. Penile urethra (mucus-secreting glands)
5. Skin containing hair follicles (sebaceous glands)
6. Brunner's glands of duodenum
7. Pancreas (compound acinar glands)
8. Submandibular salivary gland (compound tubulo-acinar gland)

Simple Epithelium
Simple Squamous Epithelium

Squamous means "flat." These cells often join together at the surface of an organ and give the appearance of pavement tiles (Figure 6-1); hence it is often termed *pavement epithelium*. The cheek cells you studied in Exercise 2 were examples of isolated squamous cells but were actually part of a layered epithelium, to be discussed later in this exercise (stratified epithelium).

Plasma membrane

Cytoplasm

Nucleus

Figure 6-1 Squamous epithelium. (© Biophoto, Photo Researchers.)

Student Drawing

Location of simple squamous epithelium in the body:

1. Peritoneum (lining of the body cavity and its visceral organs)
2. Blood vessel lining (termed *endothelium*)
3. Pleural membranes that surround the lungs
4. Alveoli of lungs
5. Glomerulus of kidney
6. Lens of eye
7. Eardrum (tympanic membrane) layers

Simple Cuboidal Epithelium

These cells are cube shaped (Figure 6-2) or, if pinched together at one end—as they are in a duct with a small lumen, they are pyramidal (triangular).

— Simple cuboidal epithelium

— Lumen of tubule

— Nucleus of simple cuboidal cell

Figure 6-2 Cuboidal epithelium. (Courtesy of Ed Reschke.)

Student Drawing

Location of cuboidal epithelium in the body:

1. Lining of ducts of kidney tubules and most other tubules.
2. Germinal epithelium (outer layer) of ovary
3. Inner surfaces of the cornea and lens of the eye

Simple Columnar Epithelium

These cells are elongated (Figure 6-3). The shape of the nucleus usually conforms to the cell shape by becoming somewhat flattened, and it is located at the basal end of the cell (away from the free end). *Goblet cells* are columnar cells that produce mucus. The cells in Figure 6-3 have a great deal of mucus, which is released from the cells' free ends into the digestive tract.

Mucus-producing goblet cell

Nucleus of absorptive cell

Connective tissue

Simple columnar (nonciliated) epithelium

Figure 6-3 Columnar epithelium. (Courtesy of Lester Bergman and Associates.)

Student Drawing

Location of simple columnar epithelium in the body:

1. Lining of the digestive tract from stomach to anus
2. Gallbladder lining
3. Ducts of some glands

Stratified Epithelium

This type of epithelium consists of more than one layer of cells, often combining the three cell types we have just examined: squamous epithelium, cuboidal epithelium, and columnar epithelium. Here, however, all these cells are considered as *one* type of epithelium. The cheek cells previously studied are a part of this tissue.

Stratified Squamous Epithelium

The uppermost layer of this many-layered tissue consists of squamous cells. In Figure 6-4, the cells shown are of the cuboidal type, but they may range into the columnar variety in some tissues.

— Flattened squamous surface cell

— Stratified squamous epithelium

— Connective tissue

Figure 6-4 Stratified squamous epithelium (human cheek cells—430×). (Courtesy of Biophoto, SPL, Photo Researchers.)

Student Drawing

Location of stratified squamous epithelium in the body:

1. Epidermis of the skin
2. Lining of the mouth and esophagus
3. Lining of the vagina

Stratified Transitional Epithelium

This type of tissue is similar to stratified squamous except that the layer of cells at the free end is more cube-shaped than squamous (Figure 6-5).

Lumen of
urinary
bladder

Nucleus of
transitional
cell

Connective
tissue

Stratified transitional
epithelium

Figure 6-5 Stratified transitional epithelium. (Courtesy of Andrew J. Kuntzman.)

Student Drawing

Location of stratified transitional epithelium in the body:
lining of urinary bladder

Epithelium of Exocrine Glands

Glands usually are composed of epithelial cells that have grown inward from the free surface. They are classified structurally as follows and as shown in Figure 6-6.

Unicellular—a one-celled gland that produces mucus. An example has already been seen in the *goblet cell* of columnar epithelium.

Multicellular—glands composed of many cells.

Simple—has a single, nonbranching duct
1. Tubular
2. Branched tubular
3. Coiled tubular
4. Acinar
5. Branched acinar

Compound—has a branched duct
1. Tubular
2. Acinar

CONNECTIVE TISSUE

Materials

The following prepared slides should be provided for class use:

1. Trachea cross-section (hyaline cartilage)
2. External ear (elastic cartilage)
3. Long-bone cross-section (compact bone)
4. Subcutaneous tissue (adipose or fat cells)
5. Human blood smear (red and white blood cells)
6. Subcutaneous tissue (areolar tissue)
7. Bone marrow (hemopoietic tissue)
8. Tendon (dense connective tissue)
9. Liver (reticular connective tissue)
10. Human blood smear (formal blood elements)

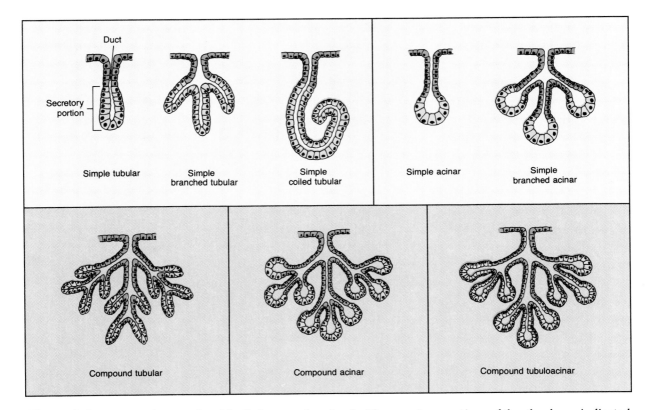

Figure 6-6 Structural types of multicellular exocrine glands. The secretory portions of the glands are indicated in purple; the blue areas represent the ducts of the glands. (From G. J. Tortora, *Principles of Human Anatomy*, 6th ed., New York: HarperCollins, 1992.)

This large group of diverse tissues has two determining characteristics:

1. The cells produce a nonliving intercellular matrix.
2. The cells are derived from embryonic mesenchyme cells.

There are some exceptions to the production of an intercellular matrix, but in a short classification such as this, exceptions are made for two of the connective tissue types. The special peculiarities of these tissues are noted as they are encountered in this exercise. The connective tissue types you study in this exercise are areolar, adipose, dense, elastic, reticular, cartilage, bone, and vascular (blood).

Connective Tissue Proper

The many tissues in this category are found in all parts of the body. They bind cells of an organ together, surround and protect major organs, and serve as important links between muscle and bone, as well as between bone and bone.

Areolar or Loose Connective Tissue

The living cells of this tissue (fibroblasts) produce both collagenous and elastic fibers, arranged loosely (Figure 6-7).

Collagenous fibers

Elastic fibers

Figure 6-7 Areolar connective tissue. (Courtesy of Biophoto, Photo Researchers.)

Student Drawing

Location of areolar connective tissue in the body:

1. Binding nerves, blood vessels, and body organs to adjacent tissue.
2. Subcutaneous layer of the skin

Adipose Tissue

These cells are specialized for *intracellular* (within the cells) fat storage (Figure 6-8). This is an exception to the usual intercellular matrix.

Fat–storage area of adipocyte

Nucleus of adipocyte

Blood vessel

Figure 6-8 Adipose tissue. (© Bruce Iverson.)

Student Drawing

Location of adipose tissue in the body:

1. Subcutaneous layer of the skin
2. Surrounding heart, kidneys, and major arteries
3. Yellow bone marrow
4. Various other fatty depots

Dense (Collagenous) Connective Tissue

The fibroblasts produce strong collagenous (sticky) fibers (Figure 6-9), which are often arranged in parallel fashion, forming tough capsules around organs and forming tendons, ligaments, and aponeuroses (a-po-noo-RO-sees).

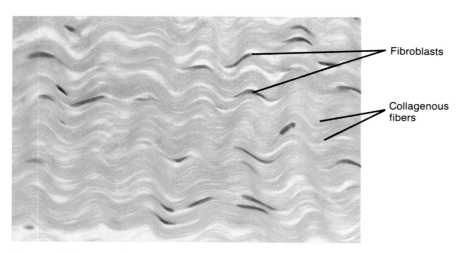

Figure 6-9 Dense (collagenous) connective tissue. (Courtesy of Andrew J. Kuntzman.)

Student Drawing

Location of dense connective tissue in the body:

1. Tendons
2. Ligaments
3. Aponeuroses
4. Capsule surrounding testes
5. Periosteum
6. Perichondrium
7. Fascia

Elastic Connective Tissue

The fibroblasts of this tissue produce both collagenous and elastic fibers (Figure 6-10). The latter lend elasticity to organs.

Figure 6-10 Elastic connective tissue. (Courtesy of Biophoto, Photo Researchers.)

Student Drawing

Location of elastic connective tissue in the body:

1. Lungs
2. Larynx, trachea, and bronchi
3. Vocal cords

Reticular Connective Tissue

These very delicate fibers, produced by the fibroblasts, tend to interlace, forming a weblike structure (Figure 6-11).

Nucleus of
reticular cell

Reticular
fiber

Figure 6-11 Reticular connective tissue. (Courtesy of Biophoto, Photo Researchers.)

Student Drawing

Location of reticular connective tissue in the body: the infrastructure of many organs, such as the liver and spleen.

Cartilage

This type of connective tissue is a skeletal tissue. Chondrocytes (cells in cartilage) produce a nonliving matrix, which surrounds and traps the cells within it. There are three types of cartilage: hyaline cartilage, fibrocartilage, and elastic cartilage. Only one, hyaline cartilage, is depicted here.

Hyaline Cartilage

The matrix surrounding the cells is clear because the fibers within it are so small and fine that they cannot be seen (Figure 6-12).

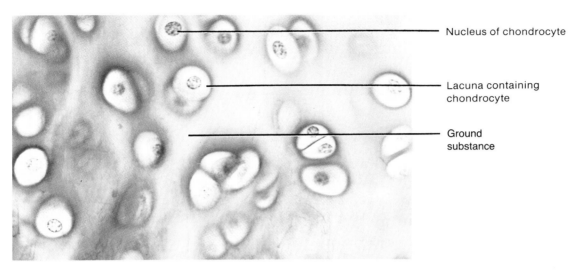

Nucleus of chondrocyte

Lacuna containing chondrocyte

Ground substance

Figure 6-12 Hyaline cartilage. (Courtesy of Biophoto, Photo Researchers.)

Student Drawing

Location of hyaline cartilage in the body:

1. Articular cartilages of long bones
2. Earlobe cartilages
3. Larynx, trachea, and bronchi
4. Costal (rib) cartilages

Bone

The *osteoblasts* (cells that produce the collagenous fibers in this skeletal tissue) become embedded in a hard matrix but retain connections with each other and the blood vessels by means of cytoplasmic extension (canaliculi) (Figure 6-13).

Figure 6-13 Bone. (Courtesy of Biophoto, Photo Researchers.)

Student Drawing

Location of bone in the body: the entire framework of bones that makes up the body

Vascular Tissue (Blood)

This tissue consists of formed elements (erythrocytes, leukocytes, and platelets) and a liquid plasma (Figure 6-14). A stained blood slide will show the pink, disc-shaped *enucleated* (no nucleus) red blood cells and the blue-stained white cells. Examine a blood slide, and pick out as many different white blood cells as you can. (Blood is studied in detail in a later exercise.) If you wish to identify all leukocytes, see Figure 39-2.

Thrombocyte (platelet)

Erythrocyte (red blood cell)

Leukocyte (white blood cell)

Plasma

Figure 6-14 Human blood cells: Photo showing erythrocytes and some leukocytes. (Courtesy of Lester Bergman and Associates.)

Student Drawing

Location of blood in the body: found circulating within arteries, veins, and capillaries of the body's closed circulatory system

MUSCULAR TISSUE

This highly specialized tissue contains cells that are able to shorten and thicken (contract) upon stimulation.

Classification of Muscle

Muscle is classified in three ways:

1. *Based on Location in the Body*
   a. *Skeletal*—attached to parts of the skeleton
   b. *Cardiac*—found only in the heart
   c. *Visceral*—found in many visceral organs
2. *Based on Function*
   a. *Voluntary*—skeletal muscle that is under the control of the will
   b. *Involuntary*—muscle that is not under the control of the will (For example, cardiac and visceral muscle are involuntary.)
3. *Based on Microscopic Appearance*
   a. *Striated Skeletal*—cross-striations visible under the light microscope (See text for explanations of this phenomenon.)
   b. *Striated Cardiac*—found only in the heart; cross-striations also visible here
   c. *Nonstriated or Smooth*—found in visceral organs; usually layered; no cross-striations present

Examine each of the aforementioned types of muscle for the given characteristics.

Striated Skeletal Muscle

In a good preparation, an observer can see that these muscle fibers are *multinucleate* (contain more than one nucleus). This is an unusual characteristic because most cells in the body are *uninucleate* (one nucleus). Seen within the cells are longitudinal muscle fibrils or *myofibrils* and dark-staining masses that appear to be lined up in a row transversely, giving the impression of a line. The myofibrils, which are often difficult to see, run longitudinally, and the dark-staining masses at right angles to them are the cross-striations. A highly specialized cell membrane, the *sarcolemma,* surrounds the cytoplasm, which is referred to as *sarcoplasm*. Find all parts labeled in Figure 6-15 and in the slide given you for laboratory study.

Nucleus of skeletal muscle fiber (cell)

Blood capillary

Striations

Sarcolemma

Skeletal muscle fiber (cell)

Figure 6-15 Striated skeletal muscle. (Courtesy of Andrew J. Kuntzman.)

Student Drawing

Striated Cardiac Muscle

The fibers in this type of muscle are all united, the cells are much smaller, and it is difficult to see the myofibrils (Figure 6-16). Notice the dark-staining bars (not seen in striated skeletal muscle) that appear at regular intervals. These bars are *intercalated discs*. Although both cardiac and skeletal muscle are striated, they differ in the following ways:

STRIATED SKELETAL

1. Multinucleate
2. Nuclei peripherally located

3. Muscle fibers do not branch
4. There are no intercalated discs

STRIATED CARDIAC

1. Uninucleate
2. Nucleus centrally located
3. Muscle fibers are branched, leaving prominent spaces between branches
4. Intercalated discs present

Figure 6-16 Cardiac muscle. (Courtesy of Ed Reschke.)

Student Drawing

Visceral or Smooth Muscle

These muscle cells are long and tapered and more difficult to see (Figure 6-17). They may be observed *in situ* (in place in the organ where they are located) or teased apart. A good source of smooth muscle is the digestive tract. Look for the myofibrils, the single nucleus, and the cell membrane (sarcolemma).

Student Drawing

Smooth muscle fiber (cell)

Nucleus of smooth muscle fiber (cell)

Sarcolemma

Figure 6-17 Smooth muscle *in situ.* (Courtesy of Andrew J. Kuntzman.)

NERVOUS TISSUE

Student Drawing

Materials

The following prepared slide should be provided for class use: Macerated ox nerve-cord tissue

Nervous Tissue

The unit of structure and function in nervous tissue is the *neuron.* A neuron is a nerve cell body and its processes. These processes are *dendrites* and *axons.* In a typical nerve cell smear from the spinal cord, you should be able to identify small triangular motor neuron cell bodies and the larger, many-sided association (*internuncial*) neurons. Figure 6-18 shows a single motor neuron, some of its processes, the nucleus, and a nucleolus. Make a drawing of several neurons, showing these parts clearly.

Figure 6-18 A neuron. (Courtesy of Biophoto Associates.)

LABORATORY REVIEW *6*

TISSUES

IDENTIFICATION, COMPLETION, AND DEFINITION

1. Identify the types of tissue shown in Figure Q6-1, and label a nucleus and cytoplasmic membrane of a representative cell in each micrograph.

a. _____

b. _____

c. _____

d. _____

e. _____

f. _____

Figure Q6-1

2. List the types and functions of the primary tissues in Table Q6-1.

Table Q6-1

Primary Tissue	Type and Function	
Connective	a.	
	b.	
	c.	
	d.	
	e.	
	f.	
Epithelium	g.	
	h.	
	i.	
	j.	
	k.	
Muscle	l.	
	m.	
	n.	
Nervous	o.	
	p.	

3. With the aid of your laboratory manual and your text, determine the locations in the body of the following types of tissue.

   a. Smooth muscle _____

   b. Elastic cartilage _____

   c. Compact bone _____

   d. Squamous epithelium _____

   e. Hyaline cartilage _____

   f. Adipose _____

   g. Cardiac muscle _____

   h. Columnar epithelium _____

   i. Hemopoietic _____

4. Place a check (√) in the appropriate space of Table Q6-2 to indicate which muscle type(s) exhibits the property listed.

Table Q6-2

	Cardiac	Skeletal	Visceral (Smooth)
a. Attached to parts of the skeleton			
b. Found only in the heart			
c. Found in many visceral organs			
d. No cross-striations present			
e. Shows cross-striations			
f. Has intercalated discs			
g. Multinucleated			
h. Uninucleated			
i. Tapered cells			
j. Voluntarily controlled			
k. Involuntarily controlled			

5. Define the word *tissue*. _____

   _____

6. What is the name for the study of tissues? _____

   _____

7. List below the four main types of tissue, and name one place in the body where each is found:

   TISSUE   **LOCATION IN BODY**

   a. _____   _____

   b. _____   _____

   c. _____   _____

   d. _____   _____

8. What is the shape of the cell found in the following tissues?

   TISSUE   **SHAPE OF CELL**

   a. Squamous epithelium   _____

   b. Cuboidal epithelium   _____

   c. Columnar epithelium   _____

9. What does the word *stratified* mean? _____

   _____

10. Why is the tissue that makes up the epidermis of the skin called "stratified epithelium"?

    _____

11. Connective tissues are made up of cells and a nonliving (usually intercellular) matrix. Describe the appearance of the following:

    a. A cartilage cell _____

    b. A bone cell _____

    c. A fat cell _____

    d. What is the matrix composed of in adipose tissue? _____

       _____

    e. What is the name of the tiny canals that connect bone cells? _____

    f. What is the technical name for a bone cell? _____

    g. What is the technical name of a cartilage cell? _____

12. What is the function of the neuron? _____

    _____

13. Name one organ where each of the following muscle-cell types is located:

    a. Cardiac muscle _____

    b. Visceral muscle _____

    c. Skeletal muscle _____

14. What is the name of the instrument that slices tissue preparations? _____

15. What is the technical name for a red blood cell? _____

16. What is the technical name for a white blood cell? _____

17. Name five different white blood cells: _____

    _____

An Introduction to the Organ and Systems Levels of Organization

A division of labor occurs early in the development of higher animals. As Exercise 6 demonstrated, such cells become specialized to perform specific functions and are arranged to form tissues, a process known as *histogenesis*. As the animal body continues to enlarge and develop further, the tissues become organized to produce organs. Later, the organs become clustered to form organ systems, as their functions become more complex.

The exercises of this section emphasize the terms used to specify body position and direction.

GENERAL REFERENCES

McMinn, R. M. H., and R. T. Hutchings. *Color Atlas of Human Anatomy,* 2nd ed. Chicago: New York Medical Publishers, 1988.
Melloni, J. L., et al. *Melloni's Illustrated Review of Human Anatomy.* Philadelphia: Lippincott, 1988.
Tortora, G. J., and S. R. Grabowski. *Principles of Anatomy and Physiology,* 7th ed. New York: HarperCollins, 1993.

SPECIFIC REFERENCES

Singer, C. *A Short History of Anatomy and Physiology from the Greeks to Harvey.* New York: Dover Publications, 1957.
Sutton, C. "A Magnetic Window into Bodily Functions." *New Scientist,* 111: 32, 1986.

Anatomical Orientation and Terminology

After completing this exercise, you should be able to

1. Define the anatomical terms used for location and position
2. List the fundamental body planes
3. Identify the major body cavities and their subdivisions
4. Apply proper medical terminology to regions of the body surface for the purpose of indicating location of internal organs
5. Use proper terminology to identify body parts

The purpose of this exercise is to acquaint the student with some of the more fundamental terms used in the study of anatomy. *Anatomy* is a science that describes how an animal is built. One must, therefore, learn the names of body parts, both *macroscopic* (those seen with the naked eye) and *microscopic* (those requiring the use of a microscope). In addition to body parts, there are terms of *location* and *position,* and terms describing *fundamental body planes, surface anatomy, general body regions,* and *body cavities.* A knowledge of these terms is necessary, in order to follow instructions in this lab guide and for comprehension of lecture material and textbook assignments.

Materials

Models of the human torso should be provided for class use for all the following exercises.

TERMS OF LOCATION AND POSITION

Terminology for the location and position of body parts of the two-legged human animal varies somewhat from that used for four-legged animals because in four-legged animals the head end is forward in locomotion and the belly surface is parallel to the ground. The differences in terminology are minor, but they can present some difficulty for the beginning student. There is no alternative but to learn all the variations because they are used (often interchangeably) by different authors. These terms are very important, and you should make them a part of your vocabulary as soon as possible (see Figure 7-1).

1. Terms referring to the head end and the rear end fall into three groups.
   Superior and Inferior
   a. *Superior*—the head, or upper end; used for humans but not for four-legged animals
   b. *Inferior*—the rear, or lower end, away from the head; used for humans, but not for four-legged animals
   Cranial and Caudal
   a. *Cranial*—the head end; used for all animals (including humans) that have a brain case or cranium
   b. *Caudal*—the rear or tail end; used for all animals that have a tail
   Anterior and Posterior
   a. *Anterior*—the part of the body that goes forward in locomotion (i.e., the front)
      i. The head end, as in four-legged animals
      ii. The front or belly side of humans

Figure 7-1 Anterior view of the human body, including terms of location, position, and body planes.

b. *Posterior*—the part of the body that is not forward in locomotion (i.e., the rear)
  i. The rear end of four-legged animals
  ii. The back side of the human
2. Terms referring to the back side and the belly side usually of a four-legged animal
  Dorsal and Ventral
  a. *Dorsal*—the back side of four-legged animals; sometimes also applied to the human back (i.e., the dorsal body cavity)
  b. *Ventral*—the belly side or underside of four-legged animals; sometimes also applied to the human belly side (i.e., the ventral body cavity)
  Anterior and Posterior
  a. *Anterior*—the front surface of human
  b. *Posterior*—the back surface of human
3. Terms referring to points on or away from the midline or center
  Medial and Lateral
  a. *Medial*—toward or on the midline of the body or an organ
  b. *Lateral*—either away from or to the side of the

midline of the body or an organ
Central and Peripheral
a. *Central*—the center of a spheroidal part or organ
b. *Peripheral*—along the outside of a spheroidal part or organ
Superficial and Deep
a. *Superficial* (external)—on the outside
b. *Deep* (internal)—on the inside
4. Terms describing appendages or parts of appendages, in reference to the point of attachment to the main axis of the body or to some specific point of reference inside the body
Proximal and Distal
a. *Proximal*—nearer to the point of attachment to the body axis or trunk in the case of extremities; nearer to the point of reference on an internal organ
b. *Distal*—farther from the point of attachment to the body axis or trunk in the case of extremities; farther from the point of reference on an internal organ

TERMS DESCRIBING BODY PLANES

Sections cut through a body or through an organ are very helpful in understanding anatomy. A section or plane is a surface passing through the body that divides it into two parts (see Figure 7-1).

1. *Transverse or cross-section*—a section at right angles to the longitudinal axis of the body or organ, dividing it into superior and inferior parts (in the human)
2. *Sagittal section*—parallel to the longitudinal axis of the body or organ, dividing it into right and left sides
   a. *Midsagittal*—a sagittal section through the midline, resulting in "equal" right and left sides
   b. *Parasagittal*—a sagittal section to the right or left of the midline, resulting in unequal right and left sides
3. *Coronal or frontal section*—a section parallel to the longitudinal axis of the body or organ, dividing it into dorsal and ventral (anterior and posterior in the case of humans) parts

BODY CAVITIES

The spaces within the body that contain the internal organs are divided into two principal cavities (Figure 7-2), both of which contain further subdivisions (Figure 7-3).

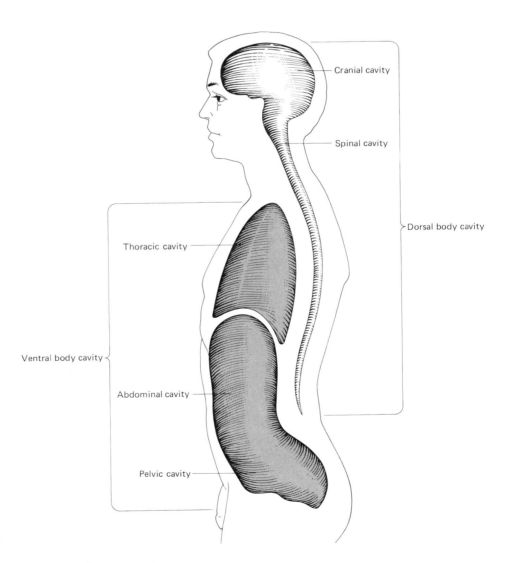

Figure 7-2 Major body cavities.

1. *Dorsal body cavity*—cavity that has its embryonic origin along the back side of the animal (Notice that the word *dorsal* is used and not *posterior,* for both humans and other animals.)
   a. *Cranial cavity*—cavity formed by the skull, which contains the brain
   b. *Spinal cavity*—cavity formed by the vertebrae, which contains the spinal cord
2. *Ventral body cavity*—cavity that is located in the belly region in both two-legged and four-legged animals
   a. *Thoracic cavity*—located above the diaphragm; the chest cavity (see Figure 7-3)
      i. *Pleural cavities*—cavities between the pleural membranes that surround the lung
      ii. *Pericardial cavity*—cavity between the pericardial membranes that surround the heart
      iii. *Mediastinal cavity*—the cavity between the lungs, containing the heart, trachea, esophagus, and thymus gland
   b. *Abdominal (abdominopelvic) cavity*—cavity located below the diaphragm, containing the viscera
      i. *Abdominal cavity proper*—contains stomach, spleen, liver, gallbladder, pancreas, small intestines, most of large intestine
      ii. *Pelvic cavity*—contains urinary bladder, sigmoid colon, rectum, internal reproductive organs

Note: The abdominal cavity, the pleural cavities, and the pericardial cavity are located between two layers of peritoneum: the outer layer is the *parietal peritoneum,* and the inner layer that surrounds the organs is the *visceral peritoneum*. A large cavity, called the abdominal cavity, occupies the space between the parietal and visceral peritoneum below the diaphragm. In the case of both pericardial and pleural cavities, however, the space is very small, and the two layers of peritoneum are closely applied to the respective organs, with only a small fluid-filled space between them.

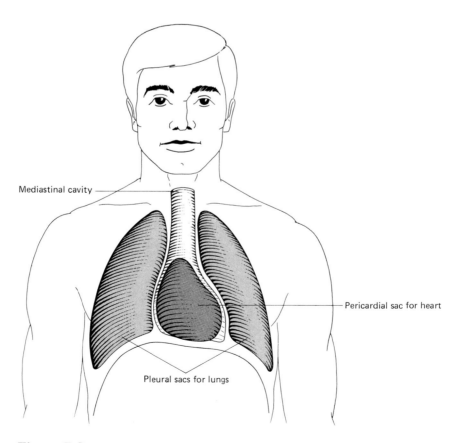

Mediastinal cavity

Pericardial sac for heart

Pleural sacs for lungs

Figure 7-3 Subdivision of thoracic cavity, anterior view.

SUBDIVISIONS OF THE ABDOMINAL REGION

The outer surface of the body is marked off into regions for the purpose of locating and specifying internal parts. The regions are shown in Figure 7-4.

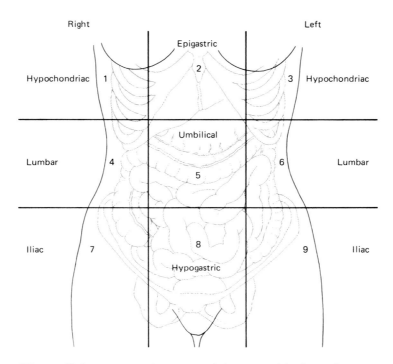

Figure 7-4 Anatomical regions of the ventral body surface.

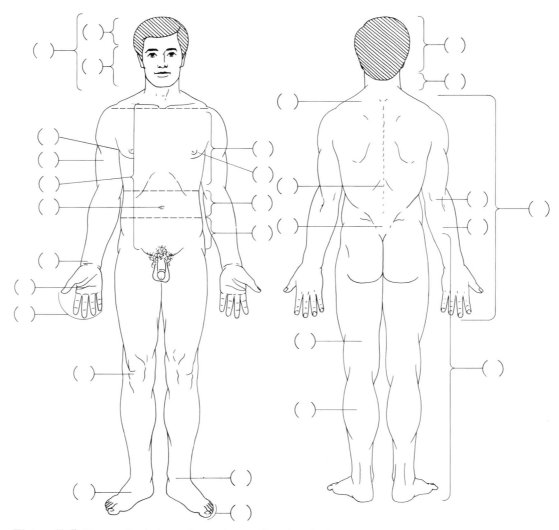

Figure 7-5 Human body (anterior and posterior views) showing major body regions

TERMS NAMING BODY PARTS

Identify the body parts on Figure 7-5, using the numerals referring to the common terms and some anatomical terms in the following list. These terms are used in future laboratory exercises, in which more detailed descriptions are given. Insert the numbers from the list on the diagrams of the human body. You may use your textbook.

Head (1)
  Skull, cranium (2)
  Skull, face (3)

Neck (4)

Trunk (5)
  Thorax (chest) (6)
  Breast (7)
  Axilla (armpit) (8)
  Shoulder (9)

Abdomen (10)
Pelvis (11)
Back (12)
Lumbar region (13)
Navel (14)

Upper extremity or arm (15)
  Upper arm (brachium) (16)
  Elbow (17)
  Forearm (antebrachium) (18)
  Wrist (19)
  Hand (20)
  Fingers (21)

Lower extremity or leg (22)
  Thigh (23)
  Knee (24)
  Lower leg or calf (25)
  Ankle (26)
  Foot (27)
  Toes (28)

LABORATORY REVIEW *7*

NAME _____

LAB SECTION _____ DATE _____

ANATOMICAL ORIENTATION AND TERMINOLOGY

COMPLETION, MATCHING, AND FILL-IN

1. Match the terms on the right with the definitions on the left.
   ( ) A longitudinal cut, dividing body into right and left sides
   ( ) The belly or underside
   ( ) Cavity containing the brain
   ( ) Transverse cut dividing the body into anterior and posterior
   ( ) Space between the membranes that surround the heart
   ( ) Term meaning on the outside of an organ
   ( ) Longitudinal cut dividing body into dorsal and ventral parts
   ( ) Either away from or to the side of the midregion
   ( ) Cavity between the lungs
   ( ) Rear end of an animal

2. In which specific body cavities would the following organs be found? Use your textbook or a chart to locate the organs.

   a. Brain _____

   b. Stomach _____

   c. Heart _____

   d. Thymus gland _____

   e. Spinal cord _____

   f. Pancreas _____

   g. Lungs _____

   h. Esophagus _____

   i. Uterus _____

   j. Liver _____

3. Fill in the correct term for location and position.

   a. The human head is _____ to the neck.

   b. The cat head is _____ to its neck.

   c. The human neck is _____ to the head.

   d. The cat neck is _____ to its head.

   e. The elbow is _____ to the wrist.

   f. The eye is _____ to the nose.

   g. The ankle is _____ to the knee.

   h. The mouth is _____ to the ears.

   i. The tail of the cat is _____ to the trunk.

   j. The vertebral column of the cat is _____ and _____ in position.

ANSWERS

1. Cranial cavity
2. Coronal section
3. Posterior
4. Lateral
5. Pericardial cavity
6. Sagittal section
7. Mediastinum
8. Ventral
9. Peripheral
10. Cross section

4. On Figure Q7-1, draw the following lines.
   a. A red line to indicate a transverse section
   b. A blue line to indicate a midsagittal section
   c. A green line to indicate a parasagittal section

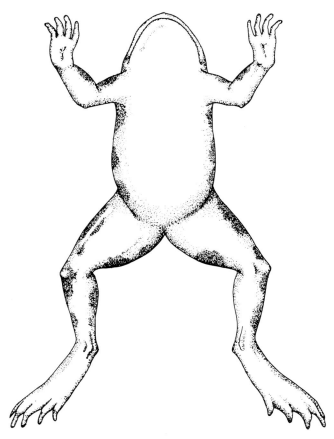

Figure Q7-1

5. To which of the five major parts of the body (head, neck, trunk, arm, leg) do the following belong?

   a. Navel _____

   b. Knee _____

   c. Eye _____

   d. Wrist _____

   e. Lumbar area _____

   f. Breast _____

   g. Ankle _____

   h. Chin _____

   i. Pelvis _____

   j. Thigh _____

The Integumentary System

*T*he skin and several specific derivatives of the skin, such as hair, nails, and various glands, make up the integument of the human body. The skin is only a few millimeters thick, but it is the body's largest organ. Within it, a variety of highly specialized cells are organized into intricate structures and subsystems, to perform many complex functions critical for an individual's survival. For example, the skin (1) provides protection against infection and injury; (2) protects against too little or too much heat loss; (3) serves as the chief means of obtaining vitamin D; (4) plays a vital role in fluid and electrolyte imbalance; and (5) functions as an indicator of changes in the body's external environment. In addition, this organ has been recently recognized to be an integral and active component of the body's immune system.

The organization, structure, and functions of the skin—together with those of its associated derivatives—are presented in this section. The experiments emphasize various properties of the skin, nails, and hair.

GENERAL REFERENCES

Edelson, T. L., and J. M. Fink. "The Immunologic Function of the Skin." *Scientific American,* 252: 46–53, 1985.

Goldsmith, L. A. (Ed.). *Biochemistry and Physiology of the Skin.* New York: Oxford University Press, 1983.

Mckie, R. *Clinical Dermatology: An Illustrated Textbook,* 3rd ed. New York: Oxford University Press, 1991.

Schaumann, B., and M. Alter. *Dermatoglyphics in Medical Disorders.* New York: Springer-Verlag, 1976.

Tortora, G. J., and S. R. Grabowski. *Principles of Anatomy and Physiology,* 7th ed. New York: HarperCollins, 1993.

SPECIFIC REFERENCES

Beutner, E. H., T. P. Chorzelski, and S. F. Bean (Eds.). *Immunopathology of the Skin.* New York: Wiley, 1979.

Koss, L. G. *Diagnostic Cytology and Its Histopathologic Bases,* 3rd ed. Philadelphia: Lippincott, 1979.

Pine, D. "Hair! From Personal Statement to Personal Problem." *FDA Consumer,* 25: 20–23, 1991.

Streilein, J. W. "Skin-Associated Lymphoid Tissue (SALT): Origins and Functions." *Journal of Investigative Dermatology,* 80: 12–16, 1983.

The Skin and Epidermal Derivatives

1. Describe the general structure and organization of skin
2. Identify the major parts of the epidermal derivatives, nails and hair
3. Perform a basic fingerprinting procedure
4. Identify basic digital patterns
5. Measure regional variations in skin temperature

The outer surface of the human body is covered with a protective layer of skin that includes such modified derivatives as nails and hair. The skin is composed of two distinct layers—the superficial layer, known as the *epidermis,* and the deeper layer, termed the *dermis* or *corium.* A subcutaneous layer of tissues is located below the dermis.

The dermis is composed of connective tissue interlaced with blood vessels, fat, glands, hair follicles, and nerves. Fibers from the dermis serve to attach the skin to the subcutaneous layer, which is firmly attached to underlying structures. The subcutaneous layer nourishes, supports, and cushions the skin. The general structure and organization of the skin are shown in Figure 8-1.

The thickness of the skin and of the individual layers varies in different parts of the body. For example, it is thickest on the back, palms, and soles. The thinner skin areas of the body occur on the tympanic (eardrum) membrane and over the eyelids.

In the areas where the epidermis is thickest, it is made up of five *strata* or regions. The outermost region, the *stratum corneum,* consists of numerous layers of dead, scalelike cells composed of proteins termed *ker-*

atins (Figure 8-2). Cells of this region are constantly being sloughed or lost by a process termed *exfoliation.* The deepest epidermal region, known as the *stratum basale* or *stratum germinativum,* consists of a single layer of columnar cells undergoing division. Newly formed cells are pushed upward into each successive layer and are eventually sloughed. The epidermis does not contain blood vessels and so depends on the dermis for nutrients.

Skin color depends on the presence of the yellow pigment *carotene* in the surface layer, on the varying concentrations of the brown pigment *melanin* throughout epidermal layers, and on the color of the circulating blood in the dermis. The occasional reddish flush appearance is influenced by the degree of oxygenation of the blood and the extent to which tiny blood vessels open up or close (peripheral blood flow). When the skin is exposed to heat, numerous small blood vessels open up, and more blood flows through the heated tissue. This blood is slightly warmer than usual because it picks up more heat from the warmed tissue. In contrast, the external cooling of the skin causes several arterioles and capillaries to constrict, which reduces the net amount of blood flowing through the tissue.

Of all the human body's organs, none is more exposed to disease and injury than is the skin. Diseases of several internal tissues and organs may be revealed by changes in the skin. These include both color changes or abnormal eruptions (such as rashes) and tumors (Color Plates 13 and 14). The skin is also subject to some disorders and diseases, such as warts and pimples (Color Plate 15).

The entire surface of the skin of humans, in contrast to that of most other mammals, is pitted and grooved by many intersecting lines that form geometric patterns characteristic of each specific area. No area is absolutely free of such markings. These lines, craters, and various microscopic features are stamped indelibly on the skin surface. The palm (palmar) and sole (plan-

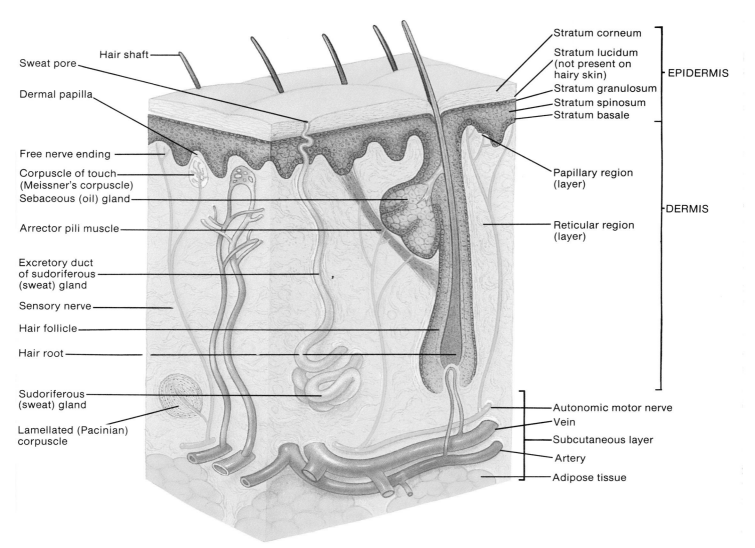

Sweat pore
Hair shaft
Dermal papilla
Free nerve ending
Corpuscle of touch (Meissner's corpuscle)
Sebaceous (oil) gland
Arrector pili muscle
Excretory duct of sudoriferous (sweat) gland
Sensory nerve
Hair follicle
Hair root
Sudoriferous (sweat) gland
Lamellated (Pacinian) corpuscle

Stratum corneum
Stratum lucidum (not present on hairy skin)
Stratum granulosum
Stratum spinosum
Stratum basale
EPIDERMIS
Papillary region (layer)
DERMIS
Reticular region (layer)
Autonomic motor nerve
Vein
Subcutaneous layer
Artery
Adipose tissue

Figure 8-1 Microscopic features of human skin, showing its most important components. The layers of skin also show a longitudinal section through a hair shaft. (From G. J. Tortora, *Principles of Human Anatomy,* 6th ed., New York: HarperCollins, 1992.)

Figure 8-2 A scanning micrograph of the skin surface showing the uniform appearance and size of stratum corneum cells. (From F. C. Tring and R. I. Jolly, *Acta Dermatovener (Stockholm),* 53: 343–346, 1973.)

tar) surfaces of an individual characteristically consist of alternating ridges and grooves, which are deeper, closer together, and more clearly seen than those elsewhere on the body. These markings, referred to as the true *dermatoglyphics* (from *derma,* "skin," and *glyphics,* "carvings"), are nearly indestructible and are different in every individual. The ridge patterns on fingertips are divided into three basic digital groups: *arches, loops,* and *whorls* (Figure 8-3). The simplest pattern found on the fingertips is the arch, and the most common is the loop. These markings increase the surface area of the body, which is an adaptation for a more efficient dissipation of heat. In addition, dermatoglyphics are obviously adapted to give the hand and foot a better grip.

Several rules, principles, and definitions have been proposed that allow reliable analyses of dermatoglyphics. In criminology investigation, the intricate details of ridge patterns serve as the bases for fingerprint identifications. In addition, relationships have been reported between unusual dermatoglyphic features and certain genetic abnormalities. Unusual ridge configurations have been shown to exist in individuals with certain chromosomal defects, in persons with single-gene disorders, and in some in whom the genetic basis of disorder is unclear.

Normally, every person has hair on most external areas of the skin surface except the palms and soles. Each individual hair is the product of the epidermis and develops from hair *follicles* (see Color Plates 16a and 16b). The free portion of a hair is referred to as the *shaft,* and the part within the follicle is the *root* (see Figure 8-4). A hair shaft consists of (1) an inner layer, the *medulla,* which is composed of loosely arranged,

soft, horny cells and several air spaces; (2) a layer surrounding the medulla, the *cortex,* which is composed of hard, closely placed, horny cells; and (3) an outer layer, the *cuticle,* which is composed of hard, tightly joined, horny cells (Figure 8-4). Individual hairs have one or several oil (sebaceous) glands, the ducts of which empty into the follicle. Smooth muscles, called *arrector pili,* are attached to the follicle below the sebaceous glands and can move the shaft in response to various emotions and temperatures. "Goose pimples" result from hair-shaft movements.

Adult nails consist of a plate of horny keratinized cells situated on a *nail bed* over the terminal *phalanges* (bones) of the fingers and toes (Figure 8-5). The visible portion of the nail *(nail body)* is bordered laterally and proximally by a fold of skin called the *nail fold.* A *nail groove* lies between this fold and the nail. The portion of the nail hidden under the nail fold is known as the *root.*

Other noticeable features include the *lunula,* a white semicircle at the base of each nail, and the cuticle or *eponychium,* which is the region where the base of the nail connects to the finger or toe. The distal end of the finger or toe under the free edge of the nail is the *hyponychium.* The characteristic pink color of the nail is due to the blood vessels that show through the semitransparent nail structure.

Nails also are subject to injury, disorders, and diseases. Discoloration of a nail not only represents a major part of the disease or disorder, but may also serve as a clue to a specific diagnosis. Causes of nail discoloration include injury (Color Plate 17), infections, malignancies, and the effects of some specific medications (Color

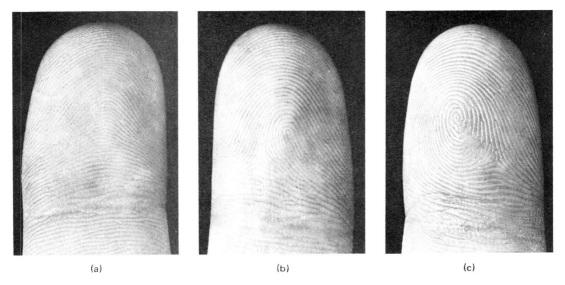

(a)   (b)   (c)

Figure 8-3 Dermatoglyphics. Photographs demonstrating the three basic digital patterns: (a) *arch,* (b) *loop,* and (c) *whorl.* (From J. R. Miller and J. Giroux. *Journal of Pediatrics,* 69: 302–312, 1966.)

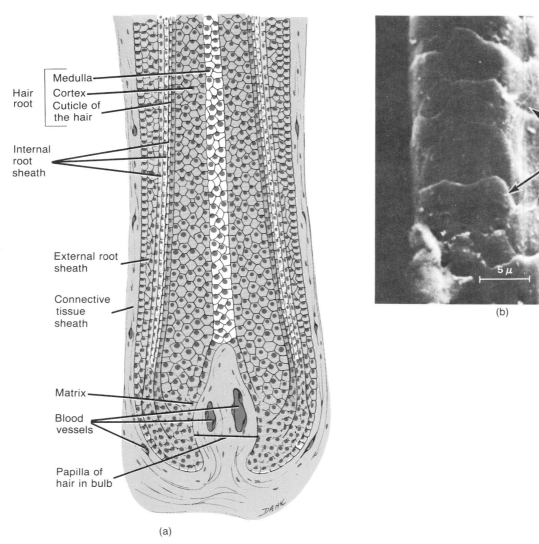

(a)

(b)

Figure 8-4 (a) The main parts of a hair root and associated structures. (From G. J. Tortora, *Principles of Human Anatomy,* 6th ed., New York: HarperCollins, 1992. © Leonard Dank.) (b) The surface appearance of hair. A scanning micrograph of the hair shaft showing the cuticular scales (*arrows*) on its surface. (From D. M. Singh, W. B. Green, R. A. Osborne, J. Upshur, and G. R. Hennigar, *Journal of Invest. Dermatology,* 63: 334–336, 1974.)

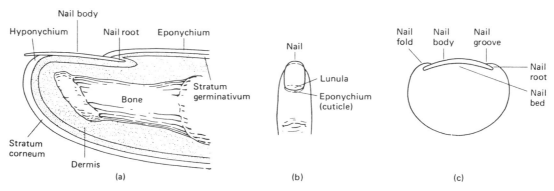

Figure 8-5 Structure of the nail: (a) longitudinal section through the finger showing important features of the nail; (b) top view of the finger; (c) cross-section of nail and nail bed.

Plate 18). Hemorrhaging, nail loosening or detachment, transverse pigmented lines, and white spots or lines are among the common signs of nail changes (Color Plate 19).

The procedures in this exercise emphasize the external structure of nails, the dermatoglyphics of fingertips, and the microscopic properties of skin and hair.

Examination of Fingers and Fingerprints

Materials

The following materials should be provided for class use:

1. Printer's ink
2. One ink roller
3. Two glass or metal slabs or ordinary stamp pads
4. Soap (cleanser and paper towels)

Procedure 1: External Structure of Nails

This procedure is to be performed by students in pairs.

1. Examine the fingers of your partner's right hand.
2. Locate all parts of the nail shown in Figure 8-5.

Procedure 2: Fingerprints

This procedure is to be performed by students in pairs.

1. Put a small amount of ink on an inking slab. Spread it with a roller into a thin, even film.
2. Place one edge of the thumb of your right hand against the inked slab. Allow your thumb to roll naturally across the slab to its opposite edge, so that the entire area to be printed is covered with ink.
3. Repeat the movement with the thumb on the space provided in the results and observations section. The sheet of paper on which the inked finger is pressed must rest on a firm surface.
4. One at a time, place the remaining fingers of your hand directly down on the inked slab, and press them directly onto the spaces provided in the results and observations section.
5. Repeat Steps 2–4 with your partner's right hand.
6. Compare the fingerprints obtained, and answer the questions in the results and observations section.

Microscopic Identification of the Skin and Hair

Materials

The following materials should be provided for class use:

1. Prepared slides of
   a. Human skin showing both epidermis and dermis
   b. Human scalp showing hair follicles
2. Glass slides and cover slips
3. Eyedroppers

Procedure

This procedure is to be performed by students individually.

1. Examine the prepared slides of human skin.
2. With the aid of Figure 8-1 and your textbook, identify the specific layers of the epidermis and the various glands typically found in the dermis.
3. Examine the preparation of human scalp, and identify the components of a hair follicle (see Color Plate 16).
4. Remove a hair from your scalp and one from your partner's scalp. Prepare a wet mount. (Refer to Exercise 1 for directions.)
5. Examine the hair preparations under both low power and high power.
6. Sketch the appearance of both hairs in the space provided in the results and observations section.
7. Look for any unusual surface features, such as scales and patterns.
8. Enter your findings in the results and observations section, and answer items in the laboratory review pertaining to this exercise.

RESULTS AND OBSERVATIONS

Fingerprints

1. Student's name: _____

Thumbprint   Index finger

Remaining fingerprints

2. Partner's name: _____

Thumbprint	Index finger

Remaining fingerprints

3. What type of fingerprint pattern did your fingers exhibit? (Refer to Figure 8-3.)

_____

4. What type of fingerprint pattern did your partner's fingers exhibit?

_____

5. Were the patterns the same for both individuals?

_____

6. Were there any unusual features to the patterns obtained? If so, circle them on the specific fingerprint, and describe them here.

_____

_____

_____

Microscopic Examination of the Hair

1. Hair sketch

Your hair

Partner's hair

2. Describe any unusual surface features.

_____

_____

_____

_____

LABORATORY REVIEW *8*

NAME _____

LAB SECTION _____ DATE _____

THE SKIN AND EPIDERMAL DERIVATIVES

COMPLETION, FILL-IN, AND DEFINITION

1. Identify the parts of the skin in Figure Q8-1.

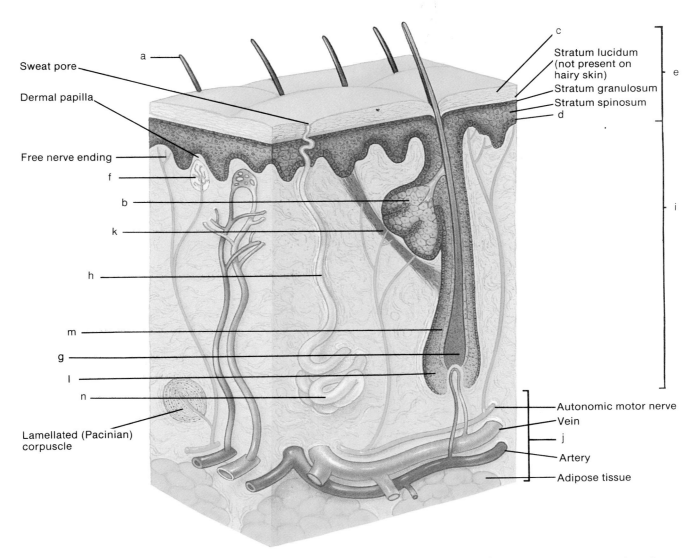

Figure Q8-1 A view of the skin and associated structure. (From G. J. Tortora and S. R. Grabowski, *Principles of Anatomy and Physiology,* 7th ed., New York: HarperCollins, 1993.)

a. _____ f. _____ k. _____

b. _____ g. _____ l. _____

c. _____ h. _____ m. _____

d. _____ i. _____ n. _____

e. _____ j. _____

2. Identify the parts of the nail in Figure Q8-2.

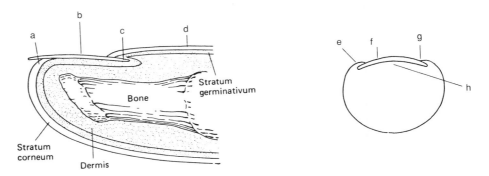

Figure Q8-2 Parts of the nail.

a. _____   e. _____

b. _____   f. _____

c. _____   g. _____

d. _____   h. _____

3. The fingerprint in Figure Q8-3a was found on a stolen object. The fingerprints of three suspects (b, c, and d) were taken and compared.

Figure Q8-3 Fingerprints.

a. On the basis of this evidence, which of the three individuals would you suspect?

_____

b. What are the digital patterns shown for each fingerprint?

a. _____

b. _____

c. _____

d. _____

4. Define or explain the following:

a. Palmar _____

b. Dermatoglyphics _____

c. Dermis _____

d. Sebaceous gland _____

e. Stratum basale _____

f. Stratum corneum _____

g. Eponychium _____

h. Lunula _____

MULTIPLE CHOICE

Insert the correct answers in the spaces provided.

1. The outermost layer of epidermis is which of the following?
   a. stratum spinosum
   b. stratum germinativum
   c. hypodermis
   d. stratum corneum
   e. stratum lucidum
2. The hair follicle is which of the following?
   a. an enlarged covering of skin
   b. a shaft of hair
   c. the hair root
   d. an enlarged sebaceous gland
   e. an eccrine gland
3. Sweat is produced by?
   a. sebaceous glands
   b. eccrine glands
   c. cerumenous glands
   d. endocrine glands
   e. papillary skin layer
4. Sebum is secreted by which of the following?
   a. sweat glands
   b. eccrine glands
   c. apocrine glands
   d. papillary skin layer
   e. sebaceous glands
5. Sebaceous glands produce which of the following?
   a. corneum lucidum
   b. sweat
   c. sebum
   d. water
   e. keratin
6. The attached visible part of a nail is known as the
   a. nail root
   b. nail fold
   c. nail bed
   d. nail proper
   e. nail body
7. The cuticle of a nail is known as the
   a. eponychium
   b. hyponychium
   c. matrix
   d. fold
   e. root
8. What are the fingerlike projections of the dermis containing capillaries and receptors called?
   a. pili
   b. polyps
   c. dermal papillae
   d. skin folds
   e. skin roots
9. The waterproofing property of human skin is associated with the presence of
   a. carotene
   b. keratin

ANSWERS	
1.	
2.	
3.	
4.	
5.	
6.	
7.	
8.	
9.	
10.	

c. blood vessels

d. roots

e. melanin

10. Which of the following choices show the proper sequence of the layering of the epidermis, moving from the free outer surface inward to the underlying tissues?

a. corneum, granulosum, lucidium, spinosum

b. corneum, lucidum, granulosum, spinosum, basale

c. spinosum, basale, granulosum, corneum

d. basale, spinosum, granulosum, corneum

e. none of the sequences listed are correct

The Support System

*I*n vertebrates, the skeleton is associated with three important body functions: support, protection, and locomotion. This internal framework contains both cartilage and bone. As indicated in Exercise 6, cartilage is one of several types of connective tissue. It consists of scattered cells embedded in a flexible, noncellular matrix which the cells secrete.

Vertebrate skeletons generally are divided into two components: the axial skeleton and the appendicular skeleton. The present section contains several exercises that consider the structure, arrangement, and functions of these skeletal components.

GENERAL REFERENCES

Caplan, A. I. "Cartilage." *Scientific American,* 251: 84–94, 1984.

Platzer, W. "Locomotor System." In W. Kahle, H. Leonhardt, and W. Platzer (Eds.), *Color Atlas and Textbook of Human Anatomy,* Vol. 1. Chicago: Year Book Medical Publishers, 1978.

Roose, C., and D. K. Clawson. *The Musculoskeletal System in Health and Disease.* New York: Harper & Row, 1980.

Tortora, G. J., and S. R. Grabowski. *Principles of Anatomy and Physiology,* 7th ed. New York: HarperCollins, 1993.

Vaughan, J. M. *The Physiology of Bone,* 3rd ed. New York: Oxford University Press, 1981.

SPECIFIC REFERENCES

Chestnut, C. H., III. "Osteoporosis and Its Treatment." *The New England Journal of Medicine,* 326: 406–407, 1992.

Nadel, E. R. "Physiological Adaptations to Aerobic Training." *American Scientist,* 73: 334–343, 1985.

Ott, S. M. "Bone Density in Adolescents." *The New England Journal of Medicine.* 325: 1646–1647, 1991.

Sharpe, W. D. "Age Changes in Human Bone: An Overview." *Bulletin of The New York Academy of Medicine,* 55: 757, 1979.

Sonstegard. D. A., L. S. Matthews, and H. Kaufer. "The Surgical Replacement of the Human Knee Joint." *Scientific American,* January 1978.

Strauss, R. H. (Ed.). *Sports Medicine and Physiology.* New York: Saunders, 1979.

Introduction to the Study of the Skeleton

After completing this exercise, you should be able to

1. Describe the histology of bone
2. List the parts of bone
3. Be familiar with the major subdivisions of the skeleton

Materials

The following materials should be provided for class use:

1. Mounted human skeleton
2. Long bone, split lengthwise
3. Prepared slides of bone, for demonstration
4. Microscopes
5. Charts and models of typical bones

MICROSCOPIC STRUCTURE OF BONE

Examine a demonstration slide showing a section through human compact bone that has been ground by hand. You will see that bone is made up of numerous *Haversian systems* (see Figure 9-1, and review the work done in Exercise 6). These systems are very active. They develop quickly, pushing other older Haversian systems out of the way. As a result you see many Haversian system fragments on the slide. These are *interstitial lamellae*. In the center of each system is the *Haversian canal*—containing blood vessels, lymph vessels, and nerves—and surrounding the canal are concentric rings of bone called *lamellae*. Trapped between the lamellae are the bone cells or *osteocytes*, with their long cytoplasmic extensions in the *canaliculi*.

MACROSCOPIC APPEARANCE OF BONE

If a long bone that has been split from its proximal to distal ends is available, study the parts. If no specimen is available, use Figure 9-2 to locate the following parts.

1. *Head or epiphysis* (ē-PIF-i-sis)—the end of the bone beyond the epiphyseal (ep´i-FIZ-ēal) line sometimes visible on the outside of the bone. A layer of compact bone covers the inner spongy bone. The *cancellous* (spongy) bone contains the *red bone marrow,* which is the site of blood-cell formation. A layer of *articular cartilage* covers the head, where it is part of a joint.
2. *Shaft* or *diaphysis* (dī-AF-i-sis)—the part of the long bone between the two heads. The bone shaft has an outer layer of compact bone and a hollow center, called the marrow or *medullary cavity,* which contains *yellow bone marrow* in the living specimen.
3. *Periosteum* (pair-ē-OS-tē-um)—a membrane surrounding the outside of a bone (silvery in a freshly killed or living specimen, but absent in the dried bone). This membrane is active in producing more bone on the outer surface.
4. *Endosteum* (end-OS-tē-um)—a membrane lining the medullary cavity. It is similar to the periosteum and produces bone on the inner surface. (It, too, cannot be seen in the dried specimen.)

MAJOR DIVISIONS OF THE SKELETON

The bones of the human skeleton are grouped into two major divisions, as shown in Figure 9-3.

1. *Axial skeleton* (*blue*)—the bones located along the midline axis of the body

Figure 9-1 Enlarged aspect of osteons (Haversian systems). (From G. J. Tortora and S. R. Grabowski, *Principles of Anatomy and Physiology,* 7th ed., New York: HarperCollins, 1993.)

2. *Appendicular skeleton* (yellow)—the bones that make up the appendages and the girdles

Each major division is further subdivided into regions. With the help of Figure 9-3, look over the mounted skeleton, and identify as many bones as you can. Be sure you have a clear idea of the bones in each subdivision before beginning detailed study in Exercises 10 through 13.

The subdivisions of the axial skeleton are

1. *Skull—cranial bones,* surrounding the brain, and *facial bones*
2. *Visceral skeleton*—derivatives of the embryonic visceral or branchial (gill) arches; the only bone in this skeletal division which can be seen on the laboratory skeleton is the Hyoid bone
3. *Vertebral column*—the vertebrae, series of bones in the middorsal region inferior to the skull

4. *Sternum*—commonly known as the "breast bone," located in the midventral or anterior thoracic region
5. *Ribs*—12 pairs of widely separated bones in the thoracic region

The subdivisions of the appendicular skeleton are

1. *Pectoral (shoulder) girdle*—two bones located in the shoulder that form the support for the articulation of the arm
2. *Upper extremity*—series of bones, beginning at the pectoral girdle, that make up the arm
3. *Pelvic girdle (os coxae)*—fused bones that form the support for articulation with the leg
4. *Lower extremity*—series of bones, beginning at the pelvic girdle, that make up the leg

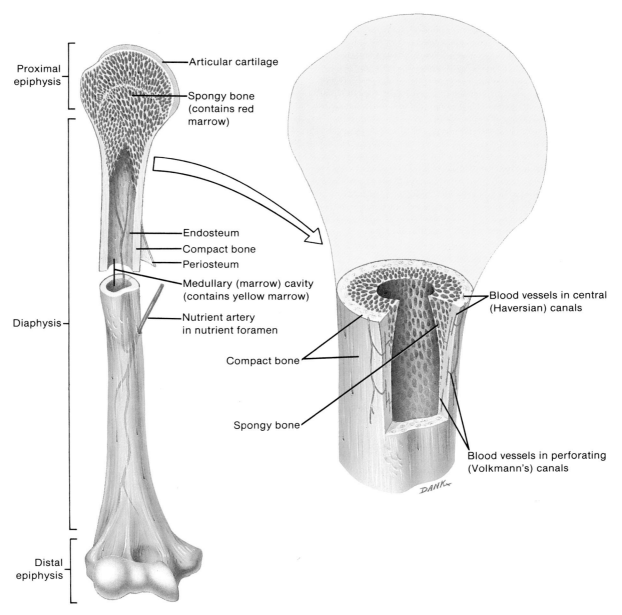

Figure 9-2 Longitudinal section through a long bone. (From G. J. Tortora, *Principles of Human Anatomy*, 6th ed., New York: HarperCollins, 1992. © Leonard Dank.)

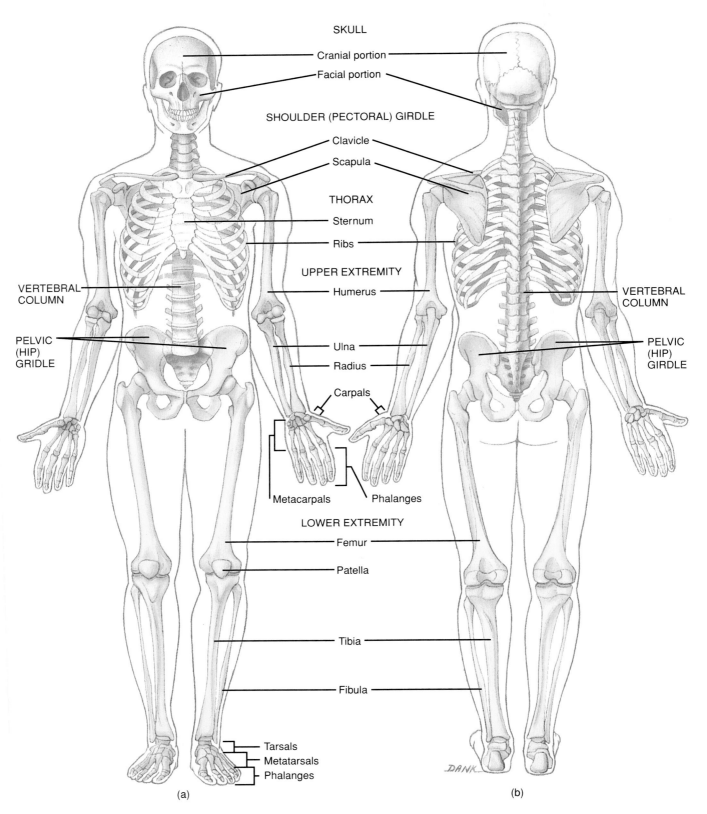

Figure 9-3 Division of the skeletal system (the axial skeleton is indicated in blue): (a) anterior view; (b) posterior view. (From G. J. Tortora and S. R. Grabowski,*Principles of Anatomy and Physiology,* 7th ed., New York: HarperCollins. 1993. © Leonard Dank.)

LABORATORY REVIEW *9*

NAME _____

LAB SECTION _____ DATE _____

INTRODUCTION TO THE STUDY OF THE SKELETON

COMPLETION

1. Name the divisions of the axial skeleton.

    a. _____

    b. _____

    c. _____

    d. _____

    e. _____

2. Name the divisions of the appendicular skeleton.

    a. _____

    b. _____

    c. _____

    d. _____

3. What is the key characteristic of any bone in the axial skeleton?

    _____

4. What is the key characteristic of any bone in the appendicular skeleton?

    _____

5. Pronounce the names of the bones described in this exercise at least two times or until you do not have any difficulty with them. Use this method throughout the laboratory exercises whenever large numbers of unfamiliar words are encountered.

The Skull (Part I)

**After completing this exercise,
you should be able to**

1. Identify all major bones of the human skull
2. Locate the major skull sutures and fontanels
3. Correlate the material in this exercise with the chapters in your textbook, with respect to origin and composition of bone
4. Pronounce all names correctly and fluently

Materials

The following materials should be provided for class use:

1. Mounted human skeleton
2. Human skull
3. Mounted disarticulated skull (Beauchene skull)
4. Disarticulated skull bones

The bones of the skull make up the head of an animal. They are subdivided into *cranial bones* and *facial bones*. Figures 10-1 through 10-6 provide various views and labels. Use them, along with the laboratory specimens, so that you will be able to identify the actual bones and not just pictures of bones.

THE CRANIAL BONES

Eight bones surround and protect the brain, forming the brain case or *cranium*. Be able to identify these bones on the skulls in the laboratory. Always learn the correct pronunciation and spelling.

1. *Frontal* (FRUN-tal)—large, single bone that forms the forehead
2. *Parietal* (pa-RĪ-i-tal)—paired bones that make up most of the top of the head, posterior to the frontal bone
3. *Occipital* (ok-SIP-i-tal)—large, single bone at the back of the head, posterior to the parietal bones
4. *Temporal* (TEM-por-al)—paired bones at the sides of the skull, inferior to the parietal bones
5. *Sphenoid* (SFĒ-noyd)—large, single bone that forms most of the floor of the cranium
6. *Ethmoid* (ETH-moyd)—single bone on the inner surface of the cranium, anterior to the sphenoid

THE FACIAL BONES

The remainder of the skull is made up of 14 bones, which together are the *face*.

1. *Nasal* (NĀ-zal)—small, paired bones that constitute the bridge of the nose (Other skeletal structures within the nose are either cartilage and not preserved on this skeleton or parts of other skull bones)
2. *Maxillary* (MAK-si-lair-ē)—large, paired bones of the upper jaw that bear teeth
3. *Lacrimal* (LAK-ri-mal)—small, paired bones located within the orbit of the eye and posterior to the uppermost portion of the maxillary bones
4. *Zygomatic* (zī-gō-MAT-ik)—paired bones that articulate with the maxillaries, commonly known as the *cheek bones*
5. *Palatine* (PAL-a-tīn)—paired bones located in the mouth cavity, posterior to the maxillaries and forming part of the hard palate

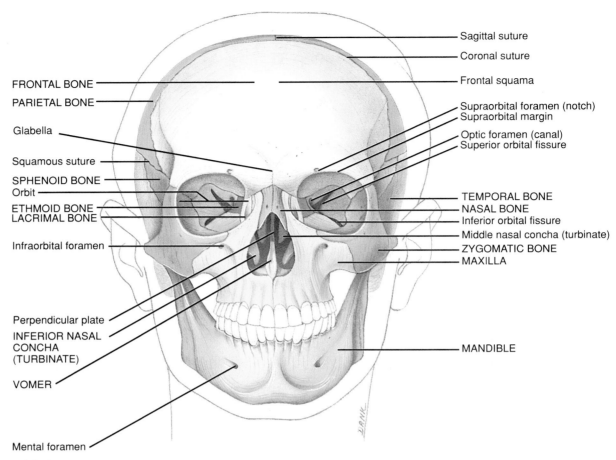

Figure 10-1 Skull, anterior view. (From G. J. Tortora and S. R. Grabowski, *Principles of Anatomy and Physiology,* 7th ed., New York: HarperCollins, 1993. © Leonard Dank.)

6. *Vomer* (VŌ-mer)—single, dagger-shaped bone located in the center of the nasal cavity superior to the maxillary bones.
7. *Inferior nasal concha* (KONG-ka)—paired, flat bones that form the lateral walls of the nasal cavity
8. *Mandible* (MAN-di-bul)—single bone that forms the lower jaw, and bears teeth.

THE SKULL SUTURES

Continue your examination of the skull. Bones come together at lines called "sutures" (SOO-churs). A *suture* is an immovable joint in which bones with serrated edges fit together, separated by a thin layer of fibrous connective tissue. Joints, or articulations, will be taken up in a later exercise, but it is appropriate to name the major skull sutures at this time.

1. *Coronal* (kō-RŌ-nal)—suture between the frontal and parietal bones
2. *Sagittal* (SAJ-i-tal)—suture extending from the coronal suture in the midline and separating the two parietal bones
3. *Squamosal* (skwa-MŌ-sal)—suture extending around the temporal bone and separating it from the parietal and sphenoid bones
4. *Lambdoidal* (lam-DOY-dal)—suture separating the occipital bone from the parietals

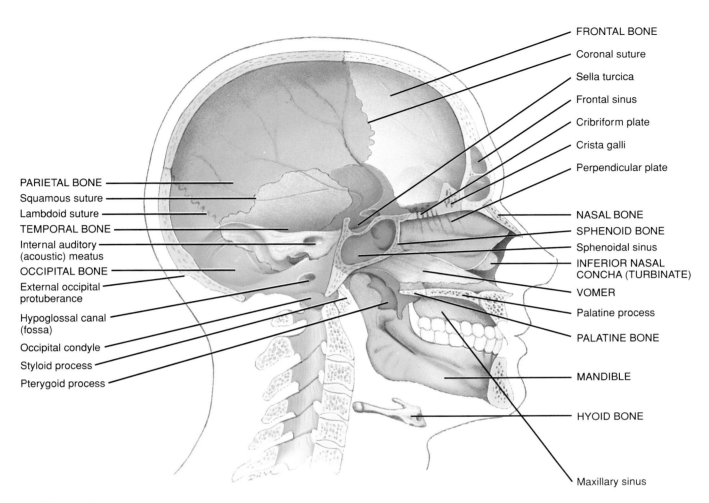

PARIETAL BONE
Squamous suture
Lambdoid suture
TEMPORAL BONE
Internal auditory (acoustic) meatus
OCCIPITAL BONE
External occipital protuberance
Hypoglossal canal (fossa)
Occipital condyle
Styloid process
Pterygoid process

FRONTAL BONE
Coronal suture
Sella turcica
Frontal sinus
Cribriform plate
Crista galli
Perpendicular plate
NASAL BONE
SPHENOID BONE
Sphenoidal sinus
INFERIOR NASAL CONCHA (TURBINATE)
VOMER
Palatine process
PALATINE BONE
MANDIBLE
HYOID BONE
Maxillary sinus

Figure 10-2 Skull, median view. (From G. J. Tortora and S. R. Grabowski, *Principles of Anatomy and Physiology,* 7th ed., New York: HarperCollins, 1993. © Leonard Dank.)

THE INFANT SKULL

Examine Figure 10-7 and other demonstration materials. Note the absence of teeth and the general fragility of the bones. It is obvious that the infant skull is still in the process of developing and maturing. Look for the following features:

1. Locate the suture separating the paired frontal bones. In the fully developed skull, the bones will have fused to form a single frontal bone, and the suture will have disappeared.

2. The *fontanels* or "soft spots" are places where the bone has not yet ossified over the membrane that gives rise to it. Locate the *anterior fontanel*, the *posterior fontanel*, and the smaller, paired *posterolateral* and *anterolateral fontanels*. Usually, all of the fontanels will have closed by age 1½ to 2 years.

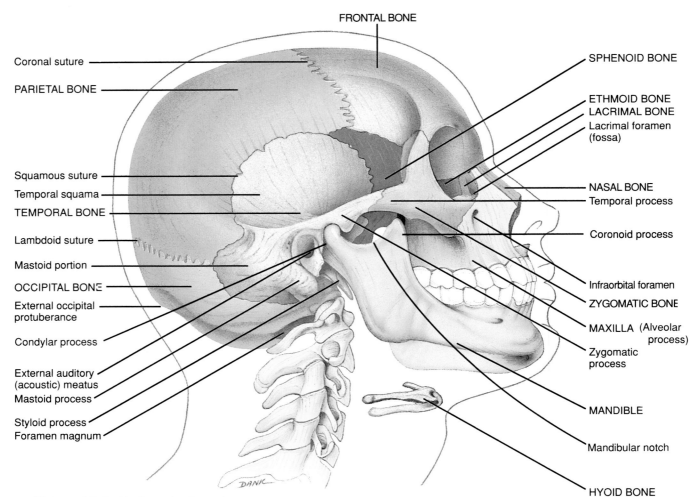

Figure 10-3 Skull, right lateral view. (From G. J. Tortora and S. R. Grabowski, *Principles of Anatomy and Physiology,* 7th ed., New York: HarperCollins, 1993. © Leonard Dank.)

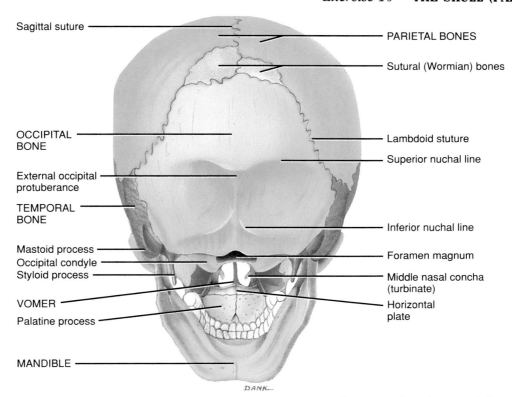

Sagittal suture

PARIETAL BONES

Sutural (Wormian) bones

OCCIPITAL
BONE

Lambdoid stuture

Superior nuchal line

External occipital
protuberance

TEMPORAL
BONE

Inferior nuchal line

Mastoid process

Occipital condyle

Styloid process

Foramen magnum

Middle nasal concha
(turbinate)

VOMER

Horizontal
plate

Palatine process

MANDIBLE

DANK

Figure 10-4 Skull, posterior view. (From G. J. Tortora and S. R. Grabowski, *Principles of Anatomy and Physiology,* 7th ed., New York: HarperCollins, 1993. © Leonard Dank.)

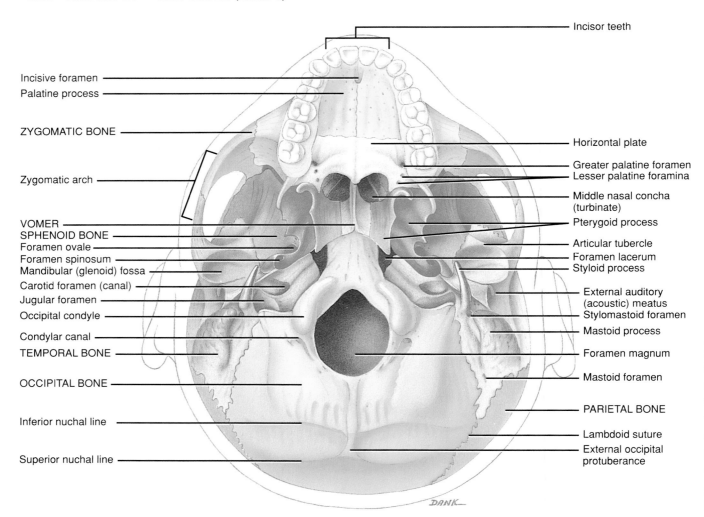

Incisor teeth

Incisive foramen

Palatine process

ZYGOMATIC BONE

Zygomatic arch

VOMER
SPHENOID BONE
Foramen ovale
Foramen spinosum
Mandibular (glenoid) fossa
Carotid foramen (canal)
Jugular foramen
Occipital condyle

Condylar canal
TEMPORAL BONE

OCCIPITAL BONE

Inferior nuchal line

Superior nuchal line

Horizontal plate
Greater palatine foramen
Lesser palatine foramina
Middle nasal concha (turbinate)
Pterygoid process
Articular tubercle
Foramen lacerum
Styloid process
External auditory (acoustic) meatus
Stylomastoid foramen
Mastoid process
Foramen magnum
Mastoid foramen
PARIETAL BONE
Lambdoid suture
External occipital protuberance

DANK

Figure 10-5 Skull in inferior view. (From G. J. Tortora and S. R. Grabowski, *Principles of Anatomy and Physiology,* 7th ed., New York: HarperCollins, 1993. © Leonard Dank.)

Figure 10-6 Beauchene skull.
1. Parietal
2. Frontal
3. Nasal
4. Lacrimal
5. Temporal
6. Sphenoid
7. Ethmoid
8. Vomer
9. Zygomatic
10. Maxillary
11. Mandible

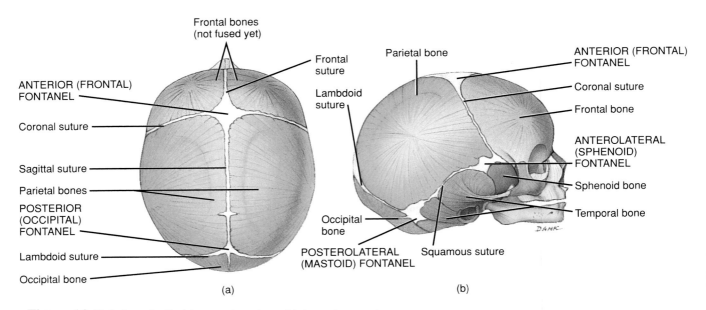

Figure 10-7 Infant skull: (a) superior view; (b) lateral view. (From G. J. Tortora and S. R. Grabowski, *Principles of Anatomy and Physiology,* 7th ed., New York: HarperCollins, 1993. © Leonard Dank.)

LABORATORY REVIEW *10*

THE SKULL (PART I)

1. Match the phrase on the left with the numbered choices on the right.

    a. _____ Total number of cranial bones

    b. _____ Dagger-shaped bone in the nasal cavity

    c. _____ Bone that forms the forehead

    d. _____ The cheek bone

    e. _____ Total number of facial bones

    f. _____ Bone that forms the bridge of the nose

    g. _____ Bone that forms the back of the skull

    h. _____ One bone that makes up the hard palate

    i. _____ Another bone that makes up the hard palate

    j. _____ Bone that makes up the lower jaw

ANSWERS

1. Lacrimal
2. 14
3. Mandible
4. Occipital
5. 20
6. Maxillary
7. Sphenoid
8. Parietal
9. 8
10. Palatine
11. Zygomatic
12. Inferior nasal concha
13. Vomer
14. Frontal
15. Nasal

2. Identify the bones in Figure Q10-1. You may color them if you wish. (If you do use color, be sure to use the same color for each bone in all the drawings.)

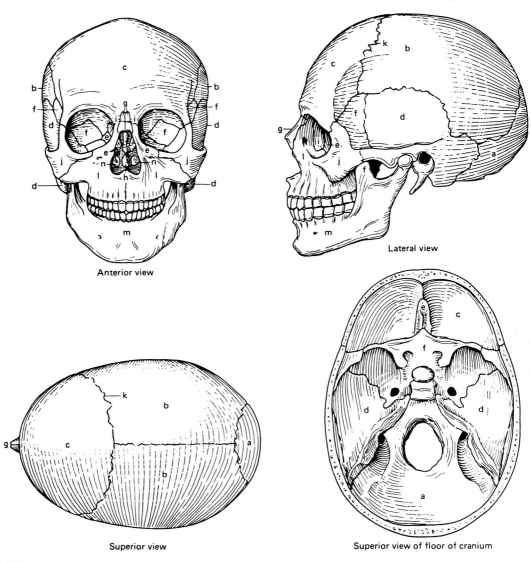

Anterior view

Lateral view

Superior view

Superior view of floor of cranium

Figure Q10-1

a. _____   h. _____

b. _____   i. _____

c. _____   j. _____

d. _____   k. _____

e. _____   l. _____

f. _____   m. _____

g. _____   n. _____

The Skull (Part II) and the Visceral Skeleton

After completing this exercise, you should be able to

1. Describe the special features of some skull bones
2. Correlate bone markings with their functions
3. Identify the bones of the visceral skeleton
4. Give the correct pronunciation and spelling of each bone and bone marking

Materials

The following materials should be provided for class use:

1. Mounted human skeleton
2. Human skull
3. Beauchene skull
4. Disarticulated skull bones
5. Demonstration of ear ossicles

SPECIAL FEATURES OF SKULL BONES

Special surface features of bones are referred to as *markings*. They include prominent or projecting parts, termed *processes* (for example, facets, crests, and spines); depressions, or *fossae;* openings, or *foramina;* grooves, or *sulci;* and enclosed cavities, or *sinuses*. Some of the more important markings of the skull bones are listed here. No attempt is made to name all markings on a bone. If greater detail is sought, the student should consult *Gray's Anatomy* or other similar texts.

Sphenoid Bone

Situated at the base of the skull and forming most of the base of the cranium, this very complex bone articulates with the temporal, frontal, occipital, and ethmoid bones. Locate the following on a separate sphenoid bone or on the Beauchene skull. The Beauchene skull is a mounted demonstration skull that has been "exploded" so that each bone can be seen readily and its suture lines accurately determined. Consult Figure 11-1.

1. *Sphenoidal sinus*—two large air cavities that drain into the nasal cavity
2. *Sella turcica* (SEL-a TUR-si-ka)—a centrally located depression on the superior surface, which contains the pituitary gland in the living organism
3. *Greater wing*—large, lateral, paired projections that can be seen on the outer surface of the skull
4. *Lesser wing*—small, paired projections superior and medial to the greater wings
5. *Pterygoid process* (TER-i-goyd)—large, paired projections on the inferior surface of the bone
6. *Optic foramen* (OP-tik fō-RĀ-men) or canal—paired holes through which the optic nerves pass
7. *Superior orbital fissure*—paired openings through which pass the cranial nerves III, IV, and parts of V and VI
8. *Foramen ovale* (fō-RĀ-men ō-VA-lē)—paired openings through which part of cranial nerve V passes
9. *Foramen spinosum* (spī-NŌ-sum)—paired openings through which medial meningeal vessels pass
10. *Foramen lacerum* (la-SER-um)—paired slits between the sphenoid and temporal bones, through which the internal carotid arteries pass
11. *Foramen rotundum* (rō-TUN-dum)—paired openings through which part of cranial nerve V passes

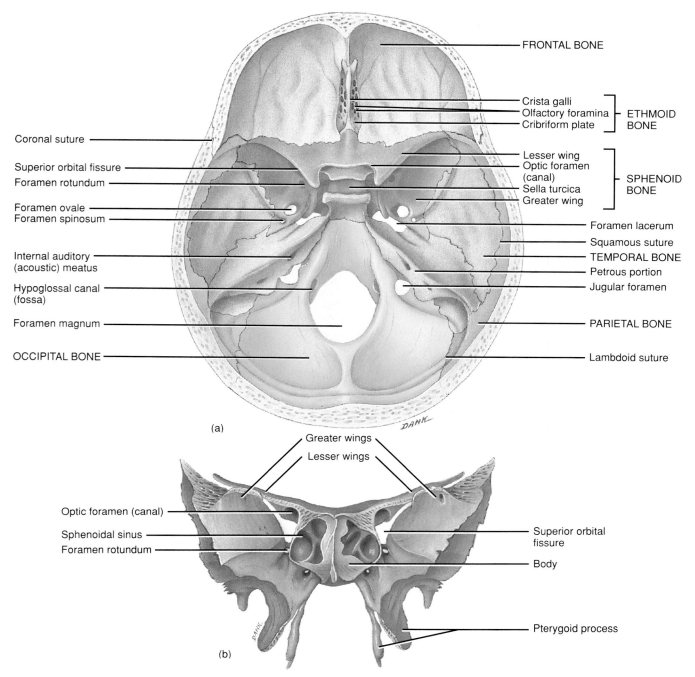

FRONTAL BONE

Crista galli
Olfactory foramina — ETHMOID BONE
Cribriform plate

Coronal suture

Superior orbital fissure
Foramen rotundum

Lesser wing
Optic foramen (canal)
Sella turcica — SPHENOID BONE
Greater wing

Foramen ovale
Foramen spinosum

Foramen lacerum
Squamous suture
TEMPORAL BONE

Internal auditory (acoustic) meatus

Petrous portion

Hypoglossal canal (fossa)

Jugular foramen

Foramen magnum

PARIETAL BONE

OCCIPITAL BONE

Lambdoid suture

(a)

DANK

Greater wings
Lesser wings

Optic foramen (canal)

Sphenoidal sinus
Foramen rotundum

Superior orbital fissure

Body

Pterygoid process

(b)

DANK

Figure 11-1 Sphenoid bone: (a) viewed in the floor of the cranium from above; (b) anterior view. (From G. J. Tortora and S. R. Grabowski, *Principles of Anatomy and Physiology,* 7th ed., New York: HarperCollins, 1993.)

Ethmoid Bone

The ethmoid bone is a large but very light spongy bone with a honeycombed structure (see Figures 10-1, 10-2, 10-6, and 11-2). It is the main support of the nasal cavity and forms part of the cranial floor and orbits. The orbit is the eyeball socket. It is made up of several bones.

1. *Perpendicular plate*—forms nasal septum, with the vomer bone
2. *Cribriform plate* (KRIB-ri-form)—porous plate that

forms part of the floor of the cranium and contains openings for the olfactory nerve

3. *Crista galli* (KRIS-ta GAL-lē)—triangular process projecting upward from the cribriform plate, which forms part of the floor of the cranium
4. *Superior nasal conchae* (KONG-kē)—paired lateral folds that create air flow for efficient circulation of inhaled air
5. *Middle nasal conchae*—paired lateral folds below the superior nasal conchae, with the same function as the superior nasal conchae

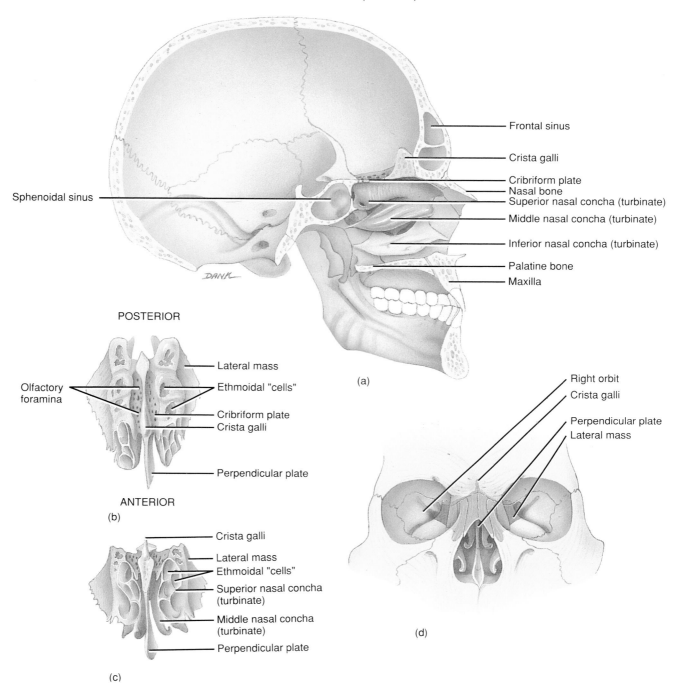

Figure 11-2 Ethmoid bone. (From G. J. Tortora and S. R. Grabowski, *Principles of Anatomy and Physiology*, 7th ed., New York: HarperCollins, 1993.)

Maxillary Bone

The paired maxillary bones form the upper jaw and most of the hard palate. They also form part of the orbits and the nasal cavity. (See Figures 10-1, 10-2, 10-3, 10-5, 10-6, and 11-2.)

1. *Maxillary sinus*—air cavity emptying into the nasal cavity (Figure 10-2)
2. *Alveolar process* (al-VĒ-ō-lar)—part of the bone that contains sockets for the teeth (Figure 10-3)

3. *Infraorbital foramen* (in′fra-OR-bi-tal)—paired holes through which the infraorbital nerves and arteries pass (Figure 10-3)

Temporal Bone

These paired bones form the inferior lateral portion of the cranium and are often referred to as the "temple bones." (See Figures 11-1, 11-2, and 11-3).

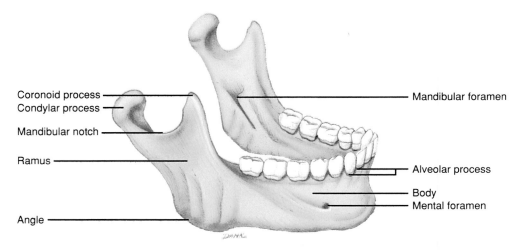

Figure 11-3 Mandible: right lateral view. (From G. J. Tortora and S. R. Grabowski, *Principles of Anatomy and Physiology,* 7th ed., New York: HarperCollins, 1993.)

1. *Zygomatic process*—projection that articulates with the temporal process of the zygomatic bone and forms most of the zygomatic arch
2. *Mandibular fossa* (man-DIB-yoo-ler)—depression into which the lower jaw inserts (Figure 10-5)
3. *Mastoid process* (MAS-toyd)—portion of the bone located behind the earlobe that contains mastoid air cells and serves as the point of attachment for several neck muscles (Figures 10-2, 10-3, 10-4, 10-5)
4. *External auditory meatus* (AW-di-tor-ē mē-Ā-tus)— canal that leads toward the middle ear deep within the temporal bone (Figure 10-3)
5. *Styloid process* (STĪ-loyd)—long, downward-projecting process to which muscles of the tongue and neck are attached (Figures 10-2, 10-3, 10-4, 10-5)

Frontal Bone

The frontal bone is a bone of both the cranium and the face. It forms the forehead and parts of the orbit and the cranial floor. (See Figure 11-2; see also Figures 10-1, 10-2, 10-3, 10-6, 10-7, and 11-1.)

1. *Glabella* (gla-BELL-a)—flattened area above the nose (Figure 10-1)
2. *Supraorbital foramen*—paired openings through which supraorbital nerve and supraorbital artery pass (Figure 10-1)
3. *Frontal sinus*—air cavity that communicates with the nasal cavity (Figure 10-2, 11-2)

Mandible

The mandible is the large bone that forms the entire lower jaw and contains all the teeth of the lower jaw (Figures 10-1, 10-2, 10-3, 10-4, 10-6, 10-7, 11-3)

1. *Condylar process* (KON-di-lar)—projection that articulates with the mandibular fossa of the temporal bone (Figures 10-3, 11-3)
2. *Coronoid process* (KOR-ō-noyd)—projection to which temporal muscle attaches (Figures 10-3, 11-3)
3. *Mandibular notch*—depression between the coronoid and condylar processes (Figures 10-3, 11-3)
4. *Mental foramen*—opening through which mental nerve and mental blood vessels pass (Figures 10-1, 11-3)
5. *Alveolar process* (al-VĒ-ō-lar)—part of the bone that contains sockets for the teeth (Figure 11-3)

THE VISCERAL SKELETON

These and other important skeletal elements (some bone and some cartilage) arise from the embryonic gill arch (visceral arch) region:

Hyoid Bone

The hyoid (HĪ-oyd) is a U-shaped bone suspended from the styloid process of the temporal bone by ligaments. It serves as point of origin (attachment) for some of the muscles of the tongue. Observe it on the mounted skeleton just inside the mandibular area (Figures 10-2, 10-3).

Ear Ossicles

Three pairs of tiny bones span the air cavity of the middle ear and serve to transmit sound waves to the inner ear. They are not preserved on the laboratory specimens, but you may have a special preparation to study. Diagrams can be found in your textbook in the chapter on the special senses or in Exercise 3 of this manual. The names of the bones are *malleus* (MAL-ē-us), meaning "hammer"; *incus* (ING-kus), meaning "anvil"; and *stapes* (STĀ-pēz), meaning "stirrup."

Cartilages of the Larynx

These highly specialized skeletal elements of a visceral organ cannot be seen on the preserved skeleton. They can be observed when the dissection of the cat is made. There are nine cartilages, and they aid in the production of sound (see Figure 11-4). Three are single, and three are paired:

Thyroid cartilage (1)
Epiglottis (1)
Cricoid (1)
Arytenoid (2)
Corniculate (2)
Cuneiform (2)

Figure 11-4 Larynx: (a) diagram of anterior view; (b) diagram of posterior view. (From G. J. Tortora, *Principles of Human Anatomy*, 6th ed., New York: HarperCollins, 1992.)

LABORATORY REVIEW *11*

THE SKULL (PART II) AND THE VISCERAL SKELETON

MATCHING AND COMPLETION

1. Match the terms on the left with the bones listed on the right. This question reviews the markings of the bones of the skull.

    a. _____ Sella turcica

    b. _____ Mastoid process

    c. _____ Zygomatic process

    d. _____ Mandibular notch

    e. _____ Cribriform plate

    f. _____ Greater wing

    g. _____ Mental foramen

    h. _____ Middle nasal concha

    i. _____ Maxillary sinus

    j. _____ Pterygoid processes

    k. _____ Foramen spinosum

    l. _____ Supraorbital foramen

    m. _____ Infraorbital foramen

    n. _____ Perpendicular plate

    o. _____ Condylar process

    p. _____ Foramen lacerum

    q. _____ Glabella

    r. _____ Coronoid process

    s. _____ Styloid process

    t. _____ Superior nasal concha

    u. _____ Foramen ovale

    v. _____ Crista galli

    w. _____ Sphenoidal sinus

    x. _____ Optic foramen

    y. _____ External auditory meatus

2. Answer the following questions briefly.

    a. Why is the "visceral skeleton" so named? _____

    _____

    _____

ANSWERS

1. Frontal
2. Parietal
3. Occipital
4. Temporal
5. Sphenoid
6. Ethmoid
7. Nasal
8. Maxillary
9. Lacrimal
10. Zygomatic
11. Palatine
12. Vomer
13. Inferior nasal concha
14. Mandible

b. What is the function of the laryngeal cartilages? _____

_____

_____

c. What is the function of the middle-ear ossicles? _____

_____

_____

d. Which skeletal muscle originates on the hyoid bone, and what is the function of this muscle? _____

_____

_____

e. Distinguish between the terms *visceral skeleton* and *viscera*. _____

_____

The Vertebral Column, Sternum, and Ribs

*After completing this exercise,
you should be able to*

1. List the major subdivisions of the vertebral column
2. Describe the special features of an individual vertebra
3. Correlate markings on a single vertebra with its functions
4. Identify the bones of the sternum
5. Divide ribs into groups, according to their type of articulation

Materials

The following materials should be provided for class use:

1. Mounted human skeleton
2. Disarticulated skeleton
3. Strung vertebral columns
4. Individual vertebrae

THE VERTEBRAL COLUMN

The bones of the vertebral column surround and protect the spinal cord in the living organism. They are named for their location along the body axis. Study them on the laboratory specimens, and compare them with Figure 12-1.

Continue your observations by locating the parts of a vertebra shown in Figure 12-2. This happens to be a *thoracic vertebra,* but it is possible to find the same parts on any typical vertebra.

1. *Spinous process*—posterior projection that serves as point of attachment for muscles (It can be felt on your own body in the midback region.)
2. *Vertebral foramen*—central canal through which the spinal cord passes
3. *Body* or *centrum*—large circular area anterior to the vertebral foramen, which forms the joints at the intervertebral discs
4. *Transverse processes*—lateral projections that serve as points of attachment for muscles
5. *Articular processes*—smooth surfaces that articulate with other vertebrae
   a. *Superior articular processes*—pair of smooth articular surfaces facing toward the spinous process
   b. *Inferior articular processes*—pair of smooth articular surfaces facing away from the spinous process
6. *Vertebral arch*—arch over the vertebral foramen, between the centrum and the spinous process
   a. *Pedicles*—lateral portions of arch
   b. *Laminae* or *roof plates*—posterior portion of arch

Cervical Vertebrae

The seven vertebrae in the neck region are the *cervical vertebrae* (see Figure 12-3). They differ from other vertebrae, in that each transverse process contains a lateral opening termed the *transverse foramen.* The first two vertebrae are highly specialized to form the type of joint that enables you to turn your head (to pivot the head on its axis). The first cervical vertebra is the *atlas,* and it supports the head. The second is the *axis.* The centrum, or body of the axis, is the peglike *odontoid process,* or *dens.* It projects into the ring of the atlas, which lacks a centrum, and forms a pivot, which enables us to turn our heads. Blood vessels pass through the *transverse foramina* of the cervical group.

(a)

(b)

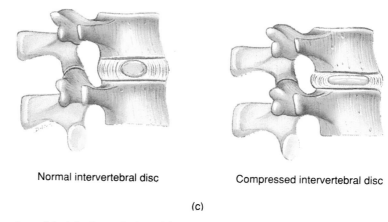

Normal intervertebral disc   Compressed intervertebral disc

(c)

Figure 12-1 Vertebral column: (a) anterior view; (b) right lateral view; (c) intervertebral disc in its normal position (left) and under compression (right). The relative size of the intervertebral disc has been enlarged for emphasis. An apparent window has been cut in the annulus fibrosus so that the nucleus pulposus can be seen. (From G. J. Tortora and S. R. Grabowski, *Principles of Anatomy and Physiology,* 7th ed., New York: HarperCollins, 1993.)

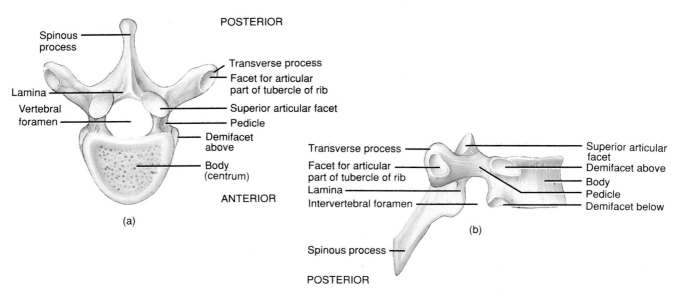

Figure 12-2 Vertebra, as illustrated by a thoracic vertebra: (a) superior view; (b) right lateral view. (From G. J. Tortora and S. R. Grabowski, *Principles of Anatomy and Physiology,* 7th ed., New York: HarperCollins, 1993.)

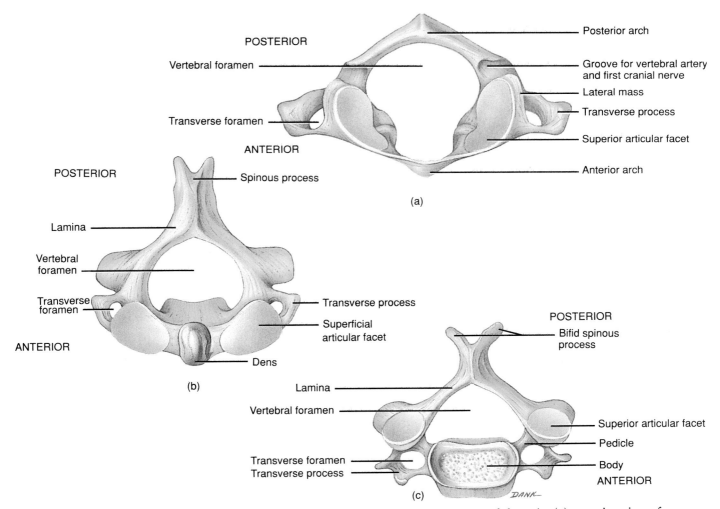

Figure 12-3 Cervical vertebrae: (a) superior view of the atlas; (b) superior view of the axis; (c) superior view of a typical cervical vertebra. (From G. J. Tortora and S. R. Grabowski, *Principles of Anatomy and Physiology,* 7th ed., New York: HarperCollins, 1993.)

Thoracic Vertebrae

The 12 thoracic vertebrae are characterized by long spinous processes. Each vertebra articulates with a pair of ribs. On Figure 12-2, locate the four smooth facets (two pairs) on each centrum that are the points of articulation with the head of a rib. These smooth surfaces, termed *costal demifacets,* are found on all thoracic vertebrae. In addition, there are paired *facets* on the transverse processes of all but the last two thoracic vertebrae for articulation with the *tubercle* of the rib. (See Figure 12-4.)

Ribs

The ribs are divided into two groups (Figure 12-4). The first seven pairs of ribs articulate with the sternum by means of their own *costal cartilages.* They are *true ribs.*

The last five pairs of ribs do not articulate directly with the sternum by means of their own costal cartilages, but indirectly, if at all, through the cartilages of the seventh true rib. These are *false ribs.* The cartilages of the eighth, ninth, and tenth false ribs join each other and then articulate with the cartilage of the seventh true rib.

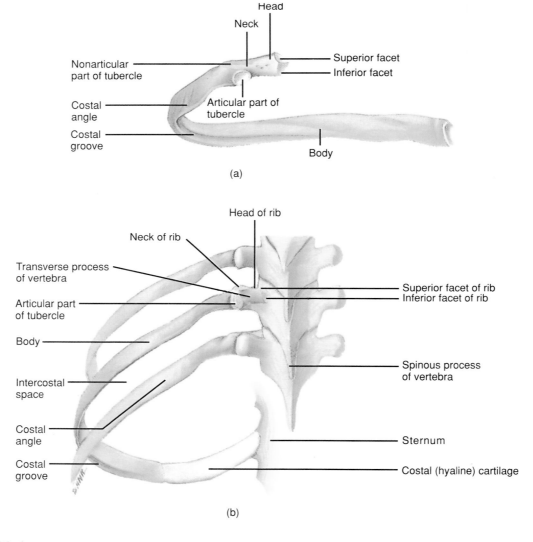

Head
Neck
Nonarticular part of tubercle
Superior facet
Inferior facet
Costal angle
Articular part of tubercle
Costal groove
Body

(a)

Head of rib
Neck of rib
Transverse process of vertebra
Articular part of tubercle
Body
Superior facet of rib
Inferior facet of rib
Intercostal space
Spinous process of vertebra
Costal angle
Costal groove
Sternum
Costal (hyaline) cartilage

(b)

Figure 12-4 Typical rib: (a) a left rib viewed from below and behind; (b) a left rib articulated with a thoracic vertebra and the sternum. (From G. J. Tortora and S. R. Grabowski, *Principles of Anatomy and Physiology,* 7th ed., New York: HarperCollins, 1993.)

POSTERIOR

ANTERIOR

Spinous process

Lamina

Superior articular facet

Transverse process

Pedicle

Vertebral foramen

Body

(a)

POSTERIOR

ANTERIOR

Superior articular process

Transverse process

Lamina

Spinous process

Body

Intervertebral foramen

Inferior articular facet

(b)

POSTERIOR

ANTERIOR

Intervertebral foramen

Intervertebral disc

Superior articular process

Transverse process

Spinous process

Body

Inferior articular facet

DANK

(c)

Figure 12-5 Lumbar vertebrae: (a) superior view; (b) right lateral view; (c) diagram of lumbar vertebrae articulated in right lateral view. (From G. J. Tortora and S. R. Grabowski, *Principles of Anatomy and Physiology*, 7th ed., New York: HarperCollins, 1993.)

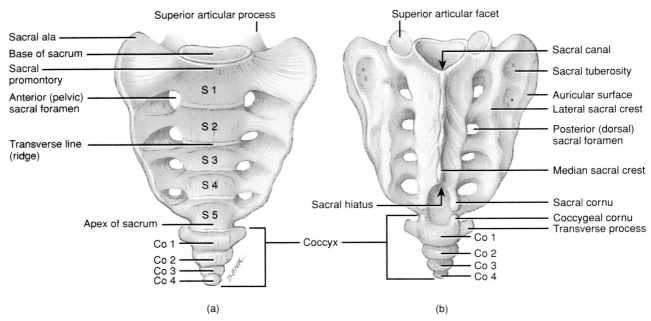

Superior articular process

Sacral ala

Base of sacrum

Sacral promontory

Anterior (pelvic) sacral foramen

Transverse line (ridge)

S 1

S 2

S 3

S 4

S 5

Apex of sacrum

Co 1
Co 2
Co 3
Co 4

Coccyx

(a)

Superior articular facet

Sacral canal

Sacral tuberosity

Auricular surface

Lateral sacral crest

Posterior (dorsal) sacral foramen

Median sacral crest

Sacral hiatus

Sacral cornu

Coccygeal cornu

Transverse process

Co 1
Co 2
Co 3
Co 4

Coccyx

(b)

Figure 12-6 Sacrum and coccyx: (a) anterior view; (b) posterior view. (From G. J. Tortora and S. R. Grabowski, *Principles of Anatomy and Physiology,* 7th ed., New York: HarperCollins, 1993.)

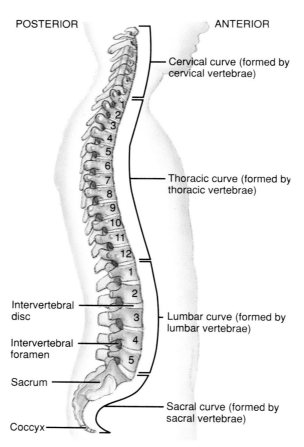

POSTERIOR   ANTERIOR

Cervical curve (formed by cervical vertebrae)

Thoracic curve (formed by thoracic vertebrae)

Intervertebral disc

Intervertebral foramen

Lumbar curve (formed by lumbar vertebrae)

Sacrum

Sacral curve (formed by sacral vertebrae)

Coccyx

Figure 12-7 Vertebral column in right lateral view. (From G. J. Tortora and S. R. Grabowski, *Principles of Anatomy and Physiology,* 7th ed., New York: HarperCollins, 1993.)

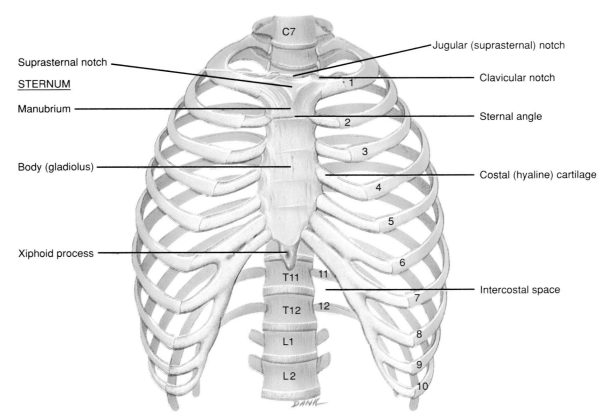

Figure 12-8 Sternum and ribs: anterior view. (From G. J. Tortora and S. R. Grabowski, *Principles of Anatomy and Physiology,* 7th ed., New York: HarperCollins, 1993.)

The last two pairs of the false rib group do not have costal cartilages and do not attach to the sternum at all. They are *floating ribs.*

Lumbar Vertebrae

The five lumbar vertebrae (Figure 12-5) are generally larger and heavier than the other vertebrae and are located in an area of the back that bears a considerable amount of the body's weight (see Figure 12-2). They have no costal demifacets on their centra, and their spines, which project backward almost horizontally, are not as long and pointed as those of the thoracic group (see Figure 12-2).

Sacral Vertebrae (Sacrum) and Coccyx

In this region of the vertebral column, five vertebrae have fused (for added strength) to form one large triangular bone termed the *sacrum* (Figures 12-1 and 12-6). It supports the pelvic girdle at the *sacroiliac* joint. The coccyx is a vestigial (nonfunctional) part of the vertebral column that represents the tail (Figure 12-6). There may be three, four, or five of these small bones, and they may or may not be fused.

Spinal Curvatures

The spinal column exhibits four normal curves. When viewed from the side, two are concave, and two are convex (see Figure 12-7):

Cervical curve (convex)

Thoracic curve (concave)

Lumbar curve (convex)

Sacral curve (concave)

A convex curve, when viewed from the side, has its convexity anteriorly. A concave curve when viewed from the side, has its concavity posteriorly.

THE STERNUM AND RIBS

Sternum

Identify the three bones that make up the sternum (Figure 12-8):

1. *Manubrium* (ma-NOO-brē-um)—most superior in position

2. *Body* or *gladiolus* (glad-́ē-Ō-lus)—middle and longest bone

3. *Xiphoid process* (ZĪ-foyd)—most inferior and smallest

With the aid of the diagram, locate the *jugular notch* and the *clavicular notch* on the manubrium.

LABORATORY REVIEW *12*

THE VERTEBRAL COLUMN, STERNUM, AND RIBS

COMPLETION

a. _____ Name the part of a vertebra that forms the point of articulation for a rib.

b. _____ How many cervical vertebrae does the human have?

c. _____ What is the special name given to the first cervical vertebra?

d. _____ What is the special name given to the second cervical vertebra?

e. _____ Which bone is formed from five fused sacral vertebrae?

f. _____ Which part of the human vertebral column is considered vestigial?

g. _____ What is the name for the cavity in a vertebra through which the spinal cord passes?

h. _____ Which vertebrae are the largest and heaviest?

i. _____ On which part of the vertebral column does walking place the most stress?

j. _____ What is the function of the transverse processes?

k. _____ Do spinal nerves exit through vertebrae or between them?

l. _____ Name the superior portion of the sternum.

m. _____ Name the middle portion of the sternum.

n. _____ Name the inferior portion of the sternum.

o. _____ Which part of a vertebra is most posterior in position?

2. What criterion makes a rib a *true rib?* _____

_____

3. Name and state the location of the spinal curvatures.

a. _____

b. _____

c. _____

d. _____

ANSWERS

1. Vertebral column

2. Ribs

3. Sternum

4. Match the terms on the right with the part of the skeleton (listed on the left) in which they belong.

a. _____ Transverse process

b. _____ Tubercle

c. _____ Atlas

d. _____ Gladiolus

e. _____ Centrum

f. _____ Manubrium

g. _____ Inferior articular process

h. _____ Spinous process

i. _____ Xiphoid process

j. _____ Jugular notch

k. _____ Coccyx

The Appendicular Skeleton

After completing this exercise, you should be able to

1. List the major bones and markings of the appendicular skeleton
2. Give the correct pronunciation and spelling of all major bones and markings
3. Identify all parts on laboratory specimens

Materials

The following materials should be provided for class use:

1. Mounted human skeleton
2. Disarticulated skeleton
3. Individual pelvis and scapula bones
4. Articulated hand-wrist and foot-ankle bones

SUBDIVISIONS OF THE APPENDICULAR SKELETON

The bones of the appendicular skeleton form the *appendages* (arms and legs) and the *girdles* (pectoral and pelvic). The *girdles* are bony structures to which the arms and legs attach and join the *torso*. The bones are grouped into four subdivisions.

1. *Upper extremity* or *arm* (Figure 13-1)
   a. *Humerus* (HYOO-mer-us)—bone of the upper arm
   b. *Radius* (RĀ-dē-us)—lateral bone of the lower arm or forearm
   c. *Ulna* (UL-na)—medial bone of the lower arm or forearm
   d. *Carpals* (KAR-puls)—bones of the wrist
   e. *Metacarpals*—bones of the hand
   f. *Phalanges* (fa-LAN-jēz; singular is *phalanx*, FĀ-lanks)—bones of the fingers
2. *Pectoral girdle* (Figure 13-1)
   a. *Scapula* (SKAP-yoo-la)—posterior component of girdle, commonly known as the shoulder blade
   b. *Clavicle* (KLAV-i-kul)—anterior component, commonly known as the collarbone
3. *Lower extremity* or *leg* (Figure 13-2)
   a. *Femur* (FĒ-mer)—bone of the thigh
   b. *Patella* (pa-TEL-a)—sesamoid bone (which develops in a tendon) that forms the kneecap
   c. *Tibia* (TIB-ē-a)—medial bone of the lower leg
   d. *Fibula* (FIB-yoo-la)—lateral bone of the lower leg
   e. *Tarsals* (TAR-sals)—bones of the ankle
   f. *Metatarsals*—bones of the foot
   g. *Phalanges*—bones of the toes
4. *Pelvic girdle* (Figures 13-2 and 13-3)
   a. *Coxal bone*—commonly known as the hipbone, made of three fused bones
      i. *Ilium* (ILL-ē-um)
      ii. *Ischium* (IS-kē-um)
      iii. *Pubis* (PYOO-bis)

BONES OF THE UPPER EXTREMITY

Having learned the names of the bones of the appendicular skeleton, the student should now examine each bone for some of the more important markings. We consider the bones of the upper extremity first.

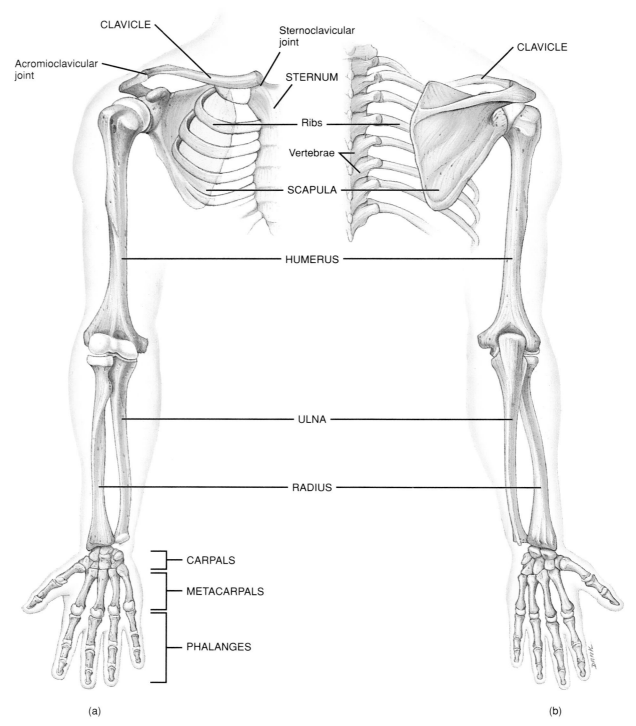

Figure 13-1 Right pectoral (shoulder) girdle and upper extremity: (a) anterior view; (b) posterior view. (From G. J. Tortora and S. R. Grabowski, *Principles of Anatomy and Physiology,* 7th ed., New York: HarperCollins, 1993.)

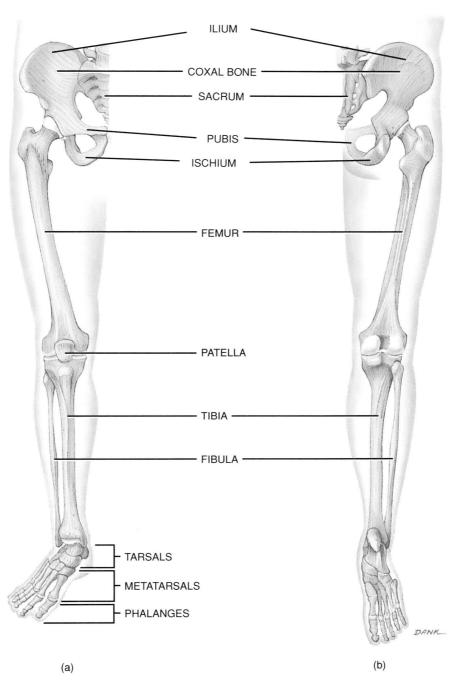

Figure 13-2 Right pelvic girdle and lower extremity: (a) anterior view; (b) posterior view. (From G. J. Tortora and S. R. Grabowski, *Principles of Anatomy and Physiology,* 7th ed., New York: HarperCollins, 1993.)

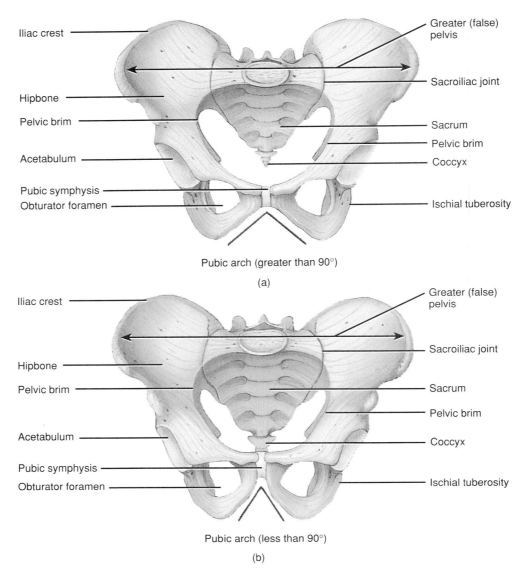

Iliac crest

Greater (false) pelvis

Hipbone

Sacroiliac joint

Pelvic brim

Sacrum

Pelvic brim

Acetabulum

Coccyx

Pubic symphysis

Obturator foramen

Ischial tuberosity

Pubic arch (greater than 90°)

(a)

Iliac crest

Greater (false) pelvis

Hipbone

Sacroiliac joint

Pelvic brim

Sacrum

Pelvic brim

Acetabulum

Coccyx

Pubic symphysis

Obturator foramen

Ischial tuberosity

Pubic arch (less than 90°)

(b)

Figure 13-3 Pelvis, anterior views: (a) female pelvis; (b) male pelvis. (From G. J. Tortora and S. R. Grabowski, *Principles of Anatomy and Physiology,* 7th ed., New York: HarperCollins, 1993.)

Humerus

The longest bone of the arm, the *humerus,* articulates with the pectoral girdle (at the scapula) proximally and with the radius and ulna distally at the elbow. Locate the following parts on a laboratory specimen (Figure 13-4):

1. *Greater tubercle* (TOO-ber-kul)—lateral projection distal to the neck of the humerus
2. *Lesser tubercle*—anterior projection distal to the neck of the humerus
3. *Intertubercular sulcus* or *bicipital groove* (in′-ter-too-BER-kyoo-lar SUL-kus) (bī-SIP-i-tal)—furrow between the two tubercles

4. *Deltoid tuberosity* (too-ber-OS-i-tē)—roughened area in the middle of the shaft
5. *Capitulum* (ka-PIT-yoo-lum)—knoblike process articulating with the radius
6. *Radial fossa*—depression for the head of the radius
7. *Trochlea* (TRŌK-lē-a)—articular surface for the ulna
8. *Coronoid fossa*—anterior depression for the ulna when forearm is flexed
9. *Olecranon fossa* (ō-LEK-ra-non)—posterior depression for the olecranon process of the ulna when forearm is extended
10. *Medial epicondyle* (ep′-i-KON-dīl)—roughened projection on the medial distal surface
11. *Lateral epicondyle*—roughened projection on the lateral distal surface

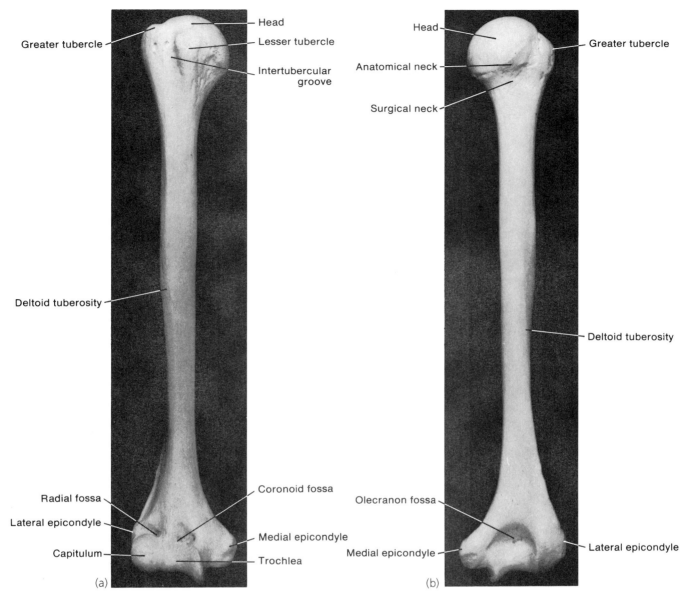

Figure 13-4 The humerus: (a) anterior view; (b) posterior view. (Photographs courtesy of Lenny Patti.)

Radius and Ulna

The *radius* is the lateral bone of the forearm. It forms a radial or pivotal joint with the ulna at the proximal ends of each bone (Figure 13-5):

Carpals, Metacarpals, and Phalanges

The *carpus* or wrist consists of eight small bones closely held together by ligaments. They are arranged in two rows, each with four bones. Using Figure 13-6 and a human skeleton or separate hand–wrist preparation, locate all the carpal bones.

The four bones of the proximal row, from medial to lateral, are

1. *Pisiform* (PĪ-si-form)
2. *Triquetral* (trī-KWĒ-tral) or *triangular*
3. *Lunate* (LOO-nāt)
4. *Scaphoid* (SKAF-oyd) or *navicular* (na-VIK-yoo-lar)

The four bones of the distal row, from medial to lateral, are

1. *Hamate* (HAM-āt)
2. *Capitate* (KAP-i-tāt)
3. *Trapezoid* (TRAP-i-zoyd) or *lesser multangular* (mul-TANG-gyoo-lar)
4. *Trapezium* (tra-PĒ-zē-um) or *greater multangular*

The *metacarpus,* the region between the wrist and the fingers, consists of five bones that form the palm of the

Figure 13-5 Right radius and ulna: (a) anterior view; (b) posterior view; (c) right elbow, lateral view. (From G. J. Tortora, *Principles of Human Anatomy,* 3rd, 4th, 5th eds., New York: HarperCollins, 1983, 1986, 1989.)

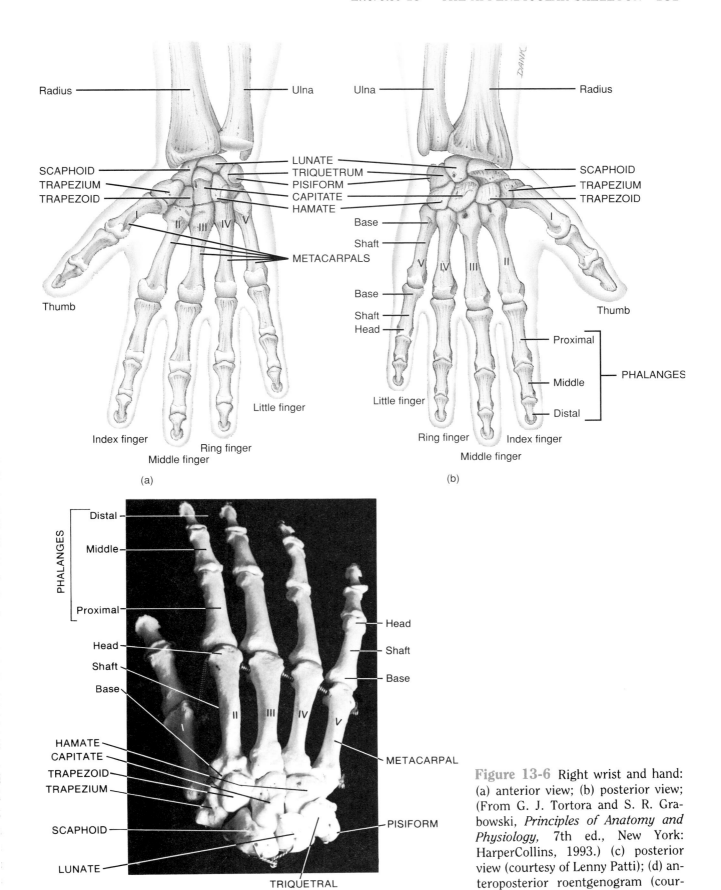

Figure 13-6 Right wrist and hand: (a) anterior view; (b) posterior view; (From G. J. Tortora and S. R. Grabowski, *Principles of Anatomy and Physiology*, 7th ed., New York: HarperCollins, 1993.) (c) posterior view (courtesy of Lenny Patti); (d) anteroposterior roentgenogram (courtesy of Daniel Sorrentino.)

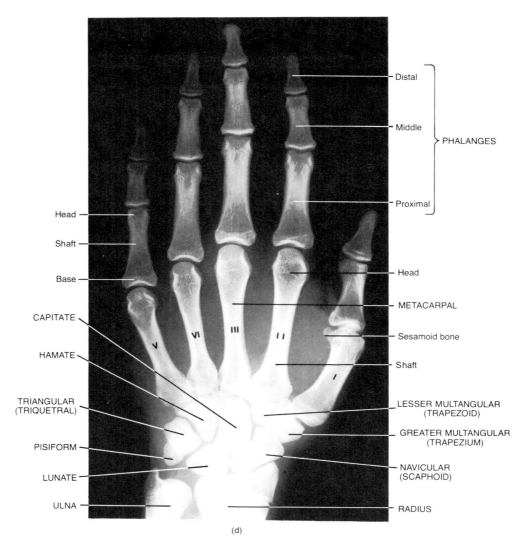

(d)

Figure 13-6 *(Continued)*

hand. They are numbered from I to V, beginning with the metacarpal that articulates with the thumb. Assign the proper numbers to the metacarpals on your lab specimens.

The *phalanges* are the bones of the digits, or fingers. There are 14 in each hand, and 3 in each finger except the thumb, which has only 2.

BONES OF THE PECTORAL GIRDLE

The pectoral girdle is the point of attachment of the arm to the torso (Figures 13-7 and 13-8). It is made up of only two bones: the clavicle and the scapula. The anterior bone is the *clavicle,* a long, slender, horizontally placed bone. It lies just above the first rib and articulates with the sternum and the scapula. The pos-

terior bone is the *scapula,* an essentially flat triangular bone with many markings for muscle attachment. Identify the following parts on an individual bone.

1. *Spine*—large, sharp ridge that runs transversely across the posterior surface
2. *Acromion* (a-KRŌ-mē-on)—large projection at the end of the spine that articulates with the clavicle
3. *Coracoid process*—(KOR-a-koyd)—projection on the anterior surface
4. *Glenoid fossa* or *cavity*—depression below the acromion and coracoid process that articulates with the proximal head of the humerus to form the shoulder joint
5. *Supraspinous fossa* (soo'-pra-spī-NOUS)—depression above the spine on the posterior surface, point of attachment for supraspinatus muscles
6. *Infraspinous fossa* (in'-fra-spī-NOUS)—large con-

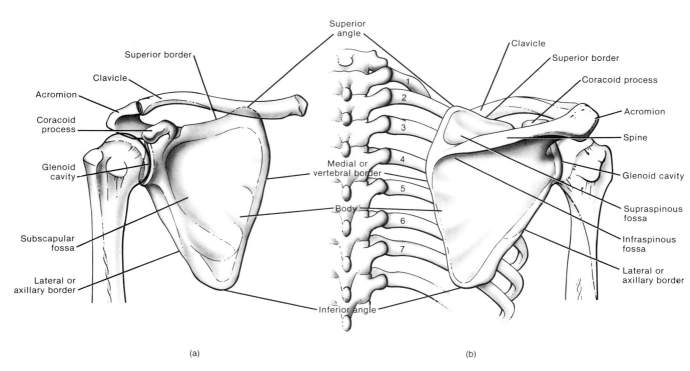

Figure 13-7 Right scapula: (a) anterior view; (b) posterior view. (From G. J. Tortora, *Principles of Human Anatomy,* 3rd, 4th, 5th eds., New York: HarperCollins, 1983, 1986, 1989.)

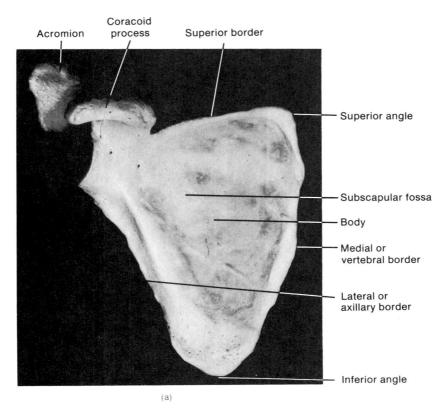

Figure 13-8 Right scapula: (a) anterior view; (b) posterior view; (c) lateral view. (Courtesy of Lenny Patti.)

(b)

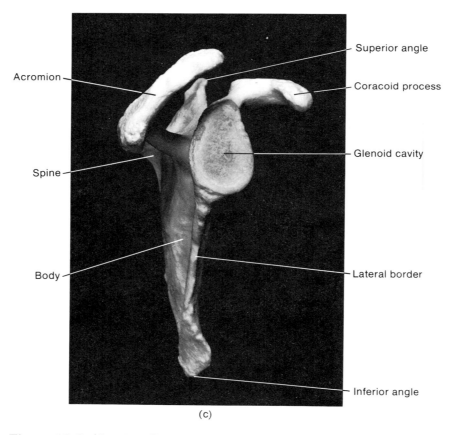

(c)

Figure 13-8 *(Continued)*

cavity below the spine on the posterior surface, point of attachment for infraspinatus muscles

7. *Subscapular fossa* (sub-SKAP-yoo-lar)—slight concavity occupying most of the anterior surface, point of attachment for shoulder muscles
8. *Vertebral* or *medial border*—thin edge of the body near the vertebral column
9. *Axillary* or *lateral border*—thickened edge of the body near the arm
10. *Superior border*—superior edge of the body
11. *Inferior angle*—angle formed where medial and lateral borders meet, most inferior part of the bone
12. *Superior angle*—angle formed where superior and medial borders meet

BONES OF THE LOWER EXTREMITY

Review the names of the bones of the leg, and then study their important markings.

Femur

The *femur,* or thigh bone, is the longest and heaviest bone in the body. Using Figure 13-9 and a laboratory specimen, locate the following parts.

1. *Head*—rounded proximal end that articulates with the hip socket in the pelvic girdle
2. *Neck*—constricted area distal to the head

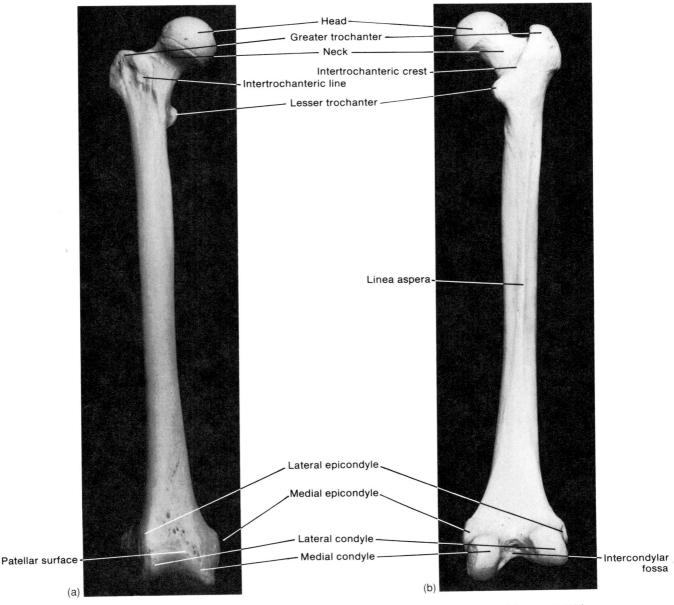

Head
Greater trochanter
Neck
Intertrochanteric crest
Intertrochanteric line
Lesser trochanter

Linea aspera

Lateral epicondyle
Medial epicondyle
Lateral condyle
Medial condyle
Patellar surface
Intercondylar fossa

(a)   (b)

Figure 13-9 The femur: (a) anterior view; (b) posterior view. (Photographs courtesy of Lenny Patti.)

3. *Greater trochanter* (trō-KAN-ter)—large, roughened lateral projection at the proximal end

4. *Lesser trochanter*—smaller, roughened medial projection

5. *Linea aspera* (LIN-ē-a AS-per-a)—roughened vertical ridge on the shaft

6. *Medial condyle* (KON-dīl)—projection at the distal end for articulation with the tibia

7. *Lateral condyle*—projection at the distal end for articulation with the tibia

8. *Intercondylar fossa*—depression between the medial and lateral condyles

9. *Medial epicondyle*—projection superior to the medial condyle

10. *Lateral epicondyle*—projection superior to the lateral condyle

Tibia and Fibula

The *tibia* is the larger of the two bones of the lower leg. It is located medially. Identify the following parts, using Figure 13-10 and the laboratory specimen.

1. *Lateral condyle*—lateral projection at the proximal end that articulates with the femur

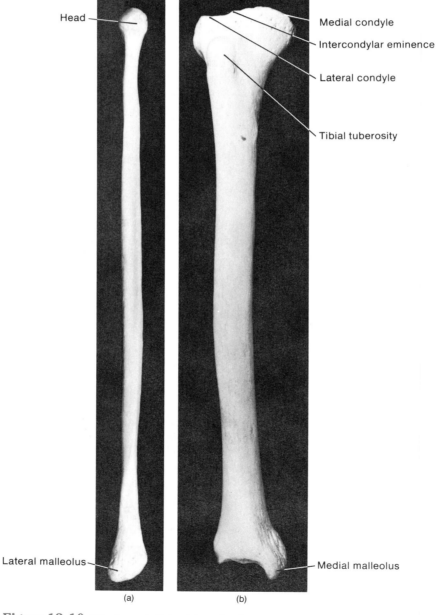

Head

Medial condyle

Intercondylar eminence

Lateral condyle

Tibial tuberosity

Lateral malleolus

Medial malleolus

(a)   (b)

Figure 13-10 Fibula and tibia: (a) Right fibula, anterior view; (b) right tibia, anterior view. (Photographs courtesy of Lenny Patti.)

2. *Medial condyle*—medial projection at the proximal end that articulates with the femur
3. *Intercondylar eminence* (in′-ter-KON-di-lar)—small projection between the condyles
4. *Tibial tuberosity*—roughened projection on the anterior surface just below the head
5. *Medial malleolus* (mal-LĒ-ō-lus)—medial prominence at the distal end
6. *Fibular notch*—lateral surface on the lateral condyle that articulates with the fibula

The *fibula* is the lateral bone of the lower leg. It is much smaller than the tibia. Locate the following parts.

1. *Head*—proximal end that articulates with the tibia
2. *Lateral malleolus*—long projection at the distal end, which articulates with the talus of the ankle

Tarsals, Metatarsals, and Phalanges

The *tarsus,* or ankle, consists of seven bones. Using Figure 13-11 and the laboratory skeleton, locate the following tarsal bones.

1. *Calcaneus* (kal-KĀ-ne-us)—largest of the tarsals, forming the heel

2. *Cuboid* (KYOO-boyd)
3. *Navicular*
4. *First (medial) cuneiform* (kyoo-NĒ-i-form)
5. *Second (intermediate) cuneiform*
6. *Third (lateral) cuneiform*
7. *Talus*

The *metatarsals,* the bones of the foot, are numbered from I to V (Figure 13-11), starting with I, which articulates with the big toe.

The *phalanges* of the foot have the same number and arrangement as those of the hand. There are 14 toe bones, 2 in the big toe, and 3 each in the other digits (Figure 13-11).

BONES OF THE PELVIC GIRDLE

The pelvic girdle serves as the point of attachment of the leg to the main axis of the body. It is made up of a total of six fused bones, three on each side (see Figure 13-12). The three bones on the right and left sides form the *coxal bones* or *os coxae* (formerly termed the *innominate bones*). It is not possible to see the lines of

Figure 13-11 Right foot and ankle: (a) superior view; (b) medial view; (From G. J. Tortora and N. P. Anagnostakos, *Principles of Anatomy and Physiology,* 3rd, 4th, 5th, 6th eds., New York: HarperCollins, 1981, 1984, 1987, 1990.)

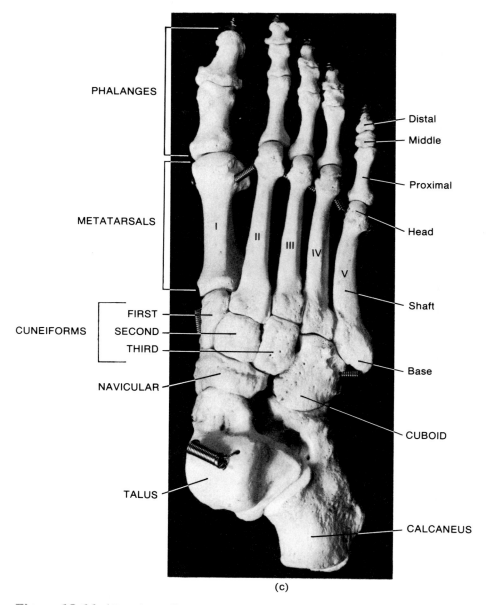

PHALANGES

— Distal

— Middle

METATARSALS

— Proximal

— Head

— Shaft

CUNEIFORMS

FIRST —

SECOND —

THIRD —

NAVICULAR —

— Base

— CUBOID

TALUS —

— CALCANEUS

(c)

Figure 13-11 *(Continued)* (c) superior view. (Courtesy of Lenny Patti.)

fusion of these bones, but the approximate suture lines are shown in Figure 13-12. Try to locate the approximate lines in your laboratory specimen.

The pelvic girdles of the two sexes are shaped differently, with adaptations related to childbearing (Figure 13-3). The female pelvis is broader and shallower than the male, and the bones are lighter weight. If you have both male and female skeletons in the lab, compare them.

Study the bones that make up one coxal bone and their markings. The *ilium* is the largest of the three bones and the most superior in position. Locate the following parts (see Figures 13-3 and 13-12).

1. *Iliac crest*—convex superior border
2. *Anterior superior iliac spine*
3. *Anterior inferior iliac spine*
4. *Posterior inferior iliac spine*
5. *Greater sciatic notch* (sī-AT-ik)—concavity below the posterior inferior iliac spine
6. *Iliac fossa*—medial concavity
7. *Iliac tuberosity*—roughened area medial to the iliac fossa, point of attachment of the sacroiliac ligament
8. *Auricular surface* (o-RIK-yoo-lar)—point of articulation with the sacrum

Figure 13-12 Right coxal bone: (a) lateral view (fusion lines are shown but are probably not seen on your specimen); (b) medial view. (From G. J. Tortora and S. R. Grabowski, *Principles of Anatomy and Physiology,* 7th ed., New York: HarperCollins, 1993.)

The *ischium* is the inferior and posterior part of the coxal bone. It is the part of the pelvic girdle you sit on. Locate the following parts.

1. *Ischial tuberosity*—roughened area at the most posterior and inferior point
2. *Ischial spine*—sharp projection above the ischial tuberosity
3. *Lesser sciatic notch*—concavity below the ischial spine

The *pubis* is the most anterior part of the coxal bone. At the midline, where right and left pubic bones meet,

a joint is formed, which is called the *pubic symphysis* (symphysis pubis). It is discussed in the exercise on articulations.

Other markings on the pelvic girdle are

1. *Acetabulum* (as'-e-TAB-yoo-lum)—fossa formed by the ilium, ischium, and pubis, which articulates with the proximal head of the femur; commonly called the hip socket
2. *Obturator foramen* (OB-tyoo-rā'-ter)—large oval hole surrounded by parts of the ischium and the pubis

LABORATORY REVIEW *13*

THE APPENDICULAR SKELETON

ANSWERS

1. Upper arm
2. Lower arm
3. Wrist
4. Hand
5. Fingers
6. Pectoral girdle
7. Upper leg
8. Lower leg
9. Ankle
10. Foot
11. Toes
12. Pelvic girdle

1. Match the bones on the left with the major areas to which they belong.

   a. _____ Metatarsal

   b. _____ Radius

   c. _____ Clavicle

   d. _____ Phalanx

   e. _____ Tibia

   f. _____ Humerus

   g. _____ Carpal

   h. _____ Fibula

   i. _____ Femur

   j. _____ Ilium

   k. _____ Scapula

   l. _____ Tarsal

   m. _____ Ischium

   n. _____ Ulna

   o. _____ Metacarpal

   p. _____ Pubis

2. Match the markings on the left with the bones on which they appear.

1. Humerus
2. Radius
3. Ulna
4. Scapula

   a. _____ Ulnar notch

   b. _____ Coracoid process

   c. _____ Semilunar notch

   d. _____ Glenoid fossa

   e. _____ Deltoid tuberosity

   f. _____ Coronoid process

   g. _____ Acromion process

   h. _____ Radial fossa

   i. _____ Subscapular fossa

   j. _____ Olecranon fossa

   k. _____ Olecranon process

   l. _____ Inferior angle

   m. _____ Radial tuberosity

   n. _____ Spine

   o. _____ Coronoid fossa

   p. _____ Supraspinous fossa

ANSWERS

q. _____ Radial notch

r. _____ Intertubercular sulcus

s. _____ Infraspinous fossa

t. _____ Lesser tubercle

3. Write the names of the eight carpal bones. Try to do it from memory.

   a. _____

   b. _____

   c. _____

   d. _____

   e. _____

   f. _____

   g. _____

   h. _____

4. Match the markings on the right with the bones on which they appear.

1. Femur

2. Tibia

3. Fibula

4. Pelvic girdle

   a. _____ Ischial tuberosity

   b. _____ Linea aspera

   c. _____ Lateral malleolus

   d. _____ Iliac crest

   e. _____ Lateral epicondyle

   f. _____ Acetabulum

   g. _____ Intercondylar eminence

   h. _____ Greater trochanter

   i. _____ Greater sciatic notch

   j. _____ Medial epicondyle

   k. _____ Obturator foramen

   l. _____ Medial malleolus

   m. _____ Fibular notch

   n. _____ Lesser trochanter

5. Name the seven bones of the ankle. Try to do it from memory.

   a. _____

   b. _____

   c. _____

   d. _____

   e. _____

   f. _____

   g. _____

Articulations

After completing this exercise, you should be able to

1. List the terminology associated with joints
2. Classify joints
3. Demonstrate joint movements on your own body

Materials

The following materials should be provided for class use:

1. Mounted human skeleton
2. Disarticulated vertebral column

An *articulation,* or *joint,* is a place where two or more skeletal elements come together. *Arthrology* is the study of joints.

In the laboratory, joint movements can be demonstrated by referring to the mounted human skeletons or by making specific movements on your own body. Often, the latter is preferable, because your bones are not wired together, and you do not have a metal rod running the length of your vertebral column. The knowledge you have gained in the study of the skeleton will now be applied to good advantage. In order to study the joints intelligently, a familiarity with the names and parts of bones is essential.

It is a common misconception that a joint is always movable. As a matter of fact, most joints are only slightly movable (often, the movement is impossible to detect) or totally immovable.

There are many ways in which joints are classified, but we restrict the classification systems to two: a *functional classification,* based on the degree of movement at a joint, and a *structural classification,* based on the anatomical features of the joint area.

FUNCTIONAL CLASSIFICATION

This classification divides joints into three categories: synarthroses, amphiarthroses, and diarthroses.

A *synarthrosis* (sin-ar-THRŌ-sis) is an immovable joint. There are four general types of synarthroses.

1. *Suture*—bones joined by a thin layer of dense connective tissue, found only in the skull
   Example: Coronal suture between the frontal bone and the parietal bones
2. *Synchondrosis* (sin-kon-DRŌ-sis)—two bones separated by hyaline cartilage
   Example: Epiphyseal plate region of a child's bone, with epiphysis and diaphysis separated by cartilage
3. *Synostosis* (sin-os-TŌ-sis)—two bones separated by another plate of bone (ossification of connective tissue or cartilage between two bones)
   Example: a. End of a long bone after growth has stopped, when the cartilaginous epiphyseal plate turns to bone
   b. Joint between right and left frontal bones, after they have fused into one bone
4. *Gomphosis*—a cone-shaped peg that fits into a socket
   Example: Roots of teeth in the alveolar processes of skull

An *amphiarthrosis* (am-fē-ar-THRO-sis) is a slightly movable joint (see Figure 14-1). The degree of movement is almost imperceptible. The two types of amphiarthroses are

1. *Symphysis* (SIM-fi-sis)—two bones separated by white fibrocartilage (Figure 14-1)
   Example: a. *Pubic symphysis* between the two pubic bones of the pelvic girdle

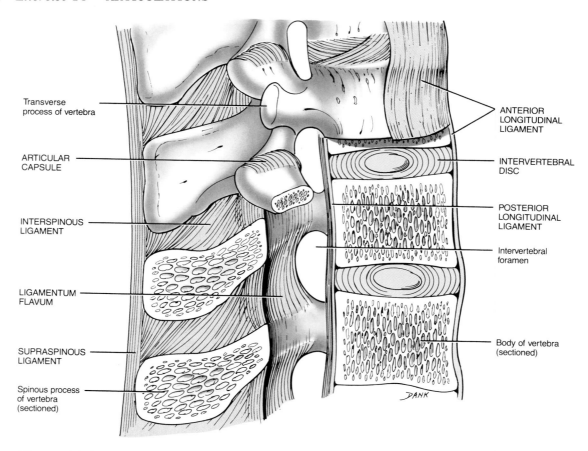

Figure 14-1 A slightly movable joint (amphiarthrosis). The intervertebral discs of the spinal column separate the centra of adjacent vertebrae. (From G. J. Tortora, *Principles of Human Anatomy*, 6th ed., New York: HarperCollins, 1992. © Leonard Dank.)

b. *Intervertebral discs* between the centra of adjacent vertebrae

2. *Syndesmosis* (sin-́dez-MŌ-sis)—two bones separated by relatively thick, dense fibrous connective tissue
   Example: Distal ends of tibia and fibula

A *diarthrosis* (dī-́ar-THRŌ-sis) is a freely movable joint (Figure 14-2). These joints are characterized by the presence of a space between the bones involved. The space is termed the *synovial* or *joint cavity*. This separation allows great freedom of movement. Six types of diarthrotic joints, as shown in Figure 14-3, are as follows:

1. *Hinge joint*—convex surface of one bone fits into concave surface of another in such a way that movement in only one plane (monaxial) is allowed (see Figure 14-3b)
   Example: a. *Elbow* between the distal end of the humerus and the proximal end of the ulna
   b. *Knee* between the distal end of the femur and the proximal end of the tibia

c. *Interphalangeal joints* between adjacent phalanges of fingers or toes

2. *Condyloid* (KON-di-loyd) *joint* or *ellipsoidal* (ē-lip-SOYD-al) *joint*—spherical head of one bone fits into an elliptical cavity, allowing movement in two planes (biaxial) (see Figure 14-3d)
   Example: a. Between the distal end of the metatarsals 2 to 5 and the proximal end of the first phalanx
   b. Between the atlas of the vertebral column and the occipital condyles of the skull
   c. Between radius and ulna and adjacent carpals (wrist joint)

3. *Ball and socket joint*—rounded head of a bone fits into a capsule of another, allowing movement in three planes (triaxial) (see Figure 14-3f)
   Example: a. *Hip* between the proximal end of the femur and the acetabulum
   b. *Shoulder* between the proximal end of the humerus and the glenoid fossa (the most freely movable joint in the body)

4. *Pivotal* or *rotary joint*—one bone turns on an-

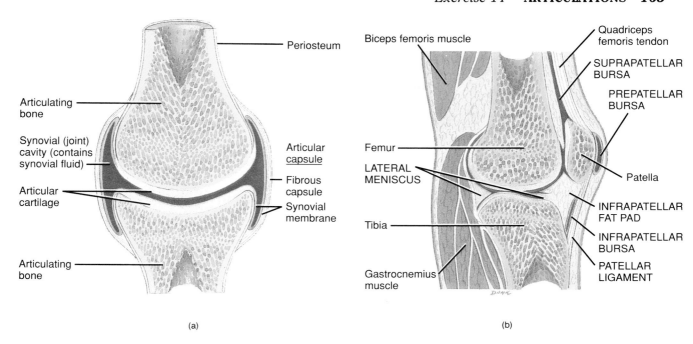

Figure 14-2 Synovial joint: (a) generalized synovial joint in frontal section; (b) sagittal section of the knee joint, showing articular discs (menisci) and bursae. (From G. J. Tortora and S. R. Grabowski, *Principles of Anatomy and Physiology,* 7th ed., New York: HarperCollins, 1993.)

other, allowing monaxial movement (see Figure 14-3c)

Example: a. Between the atlas and the axis
b. Between the proximal and distal ends of the radius and the ulna

5. *Plane* or *gliding joint*—flat articulating surfaces move side to side or forward and back on one another, allowing highly restricted biaxial movement (see Figure 14-3a)

Example: a. Between adjacent carpal bones of the wrist
b. Between adjacent tarsal bones of the ankle
c. Between superior and inferior articular surfaces of the vertebrae
d. Between costal demifacets of thoracic vertebrae and ribs

6. *Saddle joint*—concave surface of one bone fits over a convex surface of another, allowing biaxial movement (much like the ellipsoidal joint) (see Figure 14-3e)

Example: Between the first metacarpal (thumb) and the greater multangular of the wrist

STRUCTURAL CLASSIFICATION

On the basis of structure, joints may be classified according to the presence or absence of a joint cavity and the kind of connective tissue between the bones. There are three structural classifications.

1. *Fibrous joint*—bones held together by fibrous connective tissue; no joint cavity; may be synarthrotic or amphiarthrotic
2. *Cartilaginous joint*—bones are held together by cartilage; no joint cavity; may be synarthrotic or amphiarthrotic
3. *Synovial joint*—bones not bound together; separated by a synovial cavity; diarthrotic

Examples of a Fibrous Joint

Suture—immovable (review skull sutures); syndesmosis—slightly movable (review distal ends of tibia and fibula); and gomphosis—immovable (review teeth in bony sockets).

Examples of a Cartilaginous Joint

Synchondrosis—immovable (Figure 14-4); symphysis—slightly movable.

Examples of Synovial Joints

Hinge, condyloid (ellipsoidal), ball and socket pivotal (rotary), plane (gliding), and saddle joints have been shown in Figures 14-2 and 14-3, and Figure 14-5 shows a photograph of a partially dissected diarthrotic synovial joint.

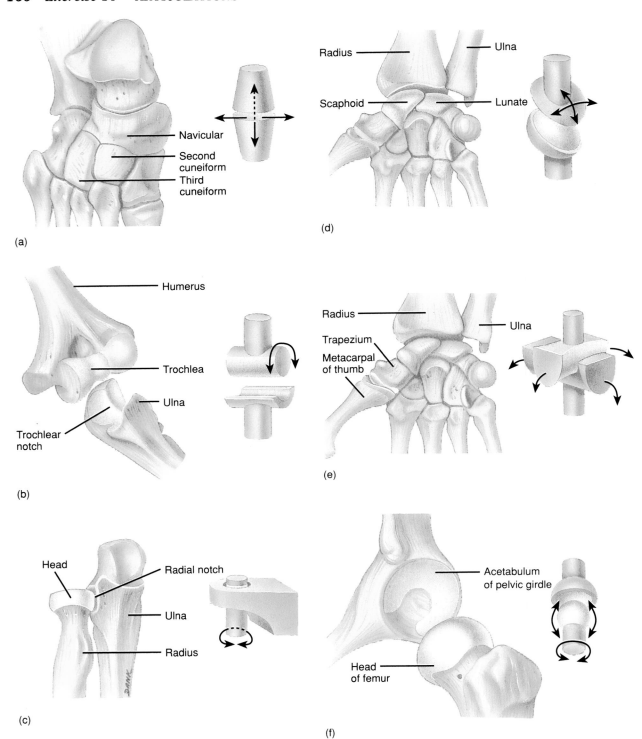

(a)

(b)

(c)

(d)

(e)

(f)

Figure 14-3 Simplified representations of types of synovial joints. (From G. J. Tortora and S. R. Grabowski, *Principles of Anatomy and Physiology,* 7th ed., New York: HarperCollins, 1993.)

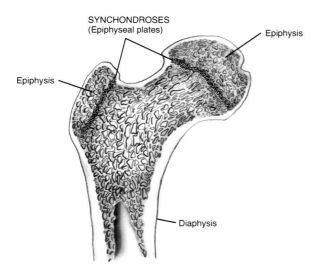

SYNCHONDROSES
(Epiphyseal plates)

Epiphysis

Epiphysis

Diaphysis

Figure 14-4 An immovable joint *(synchondrosis)*. The epiphyseal plates (cartilage) separate the heads of this young femur from the shaft. (From G. J. Tortora and N. P. Anagnostakos, *Principles of Anatomy and Physiology*, 3rd, 4th, 5th, 6th eds., New York: HarperCollins, 1981, 1984, 1987, 1990.)

JOINT MOVEMENT

The types of movement allowed at freely movable joints are many and varied. There are always factors that prevent dislocation or restrict movement, such as (a) large muscle masses that limit bending, (b) tight tendons and ligaments, and (c) muscle tension. However, most movements can be clearly defined, and the student should try to demonstrate them on his or her own body. Try to make the following movements, using Figure 14-6 as a guide.

Types of Movements

1. *Plane or gliding*—Make a simple sliding movement from side to side between adjacent carpals of the wrist.
2. *Flexion*—Decrease the angle between two bones: Fold the elbow.
3. *Extension*—Increase the angle between two bones: Straighten the elbow (arm).
4. *Dorsiflexion*—Bend the foot at the ankle region: With your foot on the ground, bring your toes upward.
5. *Hyperextension*—Straighten one part on another beyond its normal position: Bend the head backward.
6. *Plantar flexion*—Extend the foot on the ankle as far as possible: Stand on your toes.
7. *Abduction*—Move an appendage away from the

Figure 14-5 A freely movable joint (diarthrosis). A photograph of the shoulder joint partially dissected to expose the glenoid fossa between the humerus and scapula.

midline: Move your arm or leg out and away from the body; spread your fingers.
8. *Adduction*—Move an appendage toward the midline: Bring your arm or leg back to the midline from an abducted position.
9. *Circumduction*—While the proximal end of the appendage remains stationary, make a circle or arc with the distal end: Extend the arm, and make the arc with your hand.
10. *Rotation*—Turn one bone on another: Turn the atlas on the axis (turn the head); rotate the radius on the ulna to turn the palm of the hand upward (*supination*) or downward (*pronation*); turn the leg inward or outward by rotating at the hip.

Axes of Movement

Further, joints may be classified as

1. *Monaxial*—movement in one plane only
2. *Biaxial*—movement in two planes
3. *Triaxial*—a wide range of movements

Examples of the aforementioned joints are

Monaxial

Hinge joint—movement is flexion and extension.
Pivotal joint—movement is primarily rotation.

Flexion
of lower leg

Extension
of lower arm

Dorsiflexion
of foot

Hyperextension
of head

Plantar flexion
of foot

Abduction of thigh

Adduction of
thigh

Circumduction
arm

Pronation   Supination
(Rotation)

Figure 14-6 Joint movements.

Biaxial

Condyloid or ellipsoidal—movement is side-to-side and back and forth.

Saddle—a modified condyloid joint with movement side-to-side and back and forth.

Triaxial

Ball and socket—movement is flexion–extension, abduction–adduction, and rotation

LABORATORY REVIEW *14*

ARTICULATIONS

NAME _____

LAB SECTION _____ DATE _____

ANSWERS

1. Fibrous
2. Cartilaginous
3. Synovial

MATCHING AND COMPLETION

1. Match the functional classifications of joints on the left with the structural classifications to which they belong.

   a. _____ Suture

   b. _____ Synchondroses

   c. _____ Synostoses

   d. _____ Symphyses

   e. _____ Syndesmoses

   f. _____ Diarthroses

2. Give an example of a specific joint for each functional classification listed in Question 1.

   a. _____

   b. _____

   c. _____

   d. _____

   e. _____

   f. _____

3. List the types of diarthrotic joints, and give an example of each.

   a. _____

   b. _____

   c. _____

   d. _____

   e. _____

   f. _____

4. Match the joints on the left with the term describing its action.

   a. _____ Shoulder

   b. _____ Between atlas and axis

   c. _____ Pubic symphysis

   d. _____ Between adjacent skull bones

   e. _____ Between radius and ulna

   f. _____ Between articular surfaces of vertebrae

   g. _____ Knee

   h. _____ Hip

   i. _____ Child's epiphyseal plate region in the long bone

   j. _____ Elbow

1. Monaxial
2. Biaxial
3. Triaxial
4. Partially movable
5. Immovable

The Muscle System

*T*he human body has more than 600 skeletal muscles that account for approximately 40–45 percent of an individual's body weight. These muscles are used to move and manipulate body parts.

This section contains exercises on the dissection of fetal pig muscles and their comparison with human muscles. The muscles are divided into groups so that a logical progression can be made to dissect the superficial muscles and the more important deep muscles. No attempt is made to cover all the fetal pig's muscles. These muscles should be compared with human muscles, as well as with the text drawings and with charts and models.

With each dissection sequence, there are the following study aides: (1) a black-and-white photo of the specific muscle group; (2) a line drawing corresponding to the black-and-white photo; (3) a table summarizing the origins, insertions, and actions of pig muscles; (4) a table summarizing the origins, insertions, and actions of corresponding human muscles.

Because of their importance in body movements, skeletal muscles have long been the subject of studies concerned with their chemical composition, ultrastructure, and mechanism of contraction. Exercises in this section also present several aspects of muscle physiology, including the microscopic and submicroscopic properties of skeletal muscle fibers, the biochemistry of muscular contraction, the demonstration and measurement of the degree of muscular contraction (using a laboratory animal), isotonic and isometric muscle contractions, muscle fatigue, and electromyography.

GENERAL REFERENCES

Berger, R. A. *Applied Exercise Physiology.* Philadelphia: Lea & Febiger, 1982.

Eisenberg, E., and L. E. Greene. "The Relation of Muscle Biochemistry to Muscle Physiology." *Annual Review of Physiology,* 42: 293, 1980.

Kapit, W., R. I. Macey, and E. Meisami. *The Physiology Coloring Book.* New York: Harper & Row, 1987.

Tortora, G. J., and S. R. Grabowski. *Principles of Anatomy and Physiology,* 7th ed. New York: HarperCollins, 1993.

Wilson, F. C. *Musculoskeletal System: Basic Processes and Disorders,* 2nd ed. Philadelphia: Lippincott, 1983.

SPECIFIC REFERENCES

Basmajian, J. V. "Electromyography—Dynamic Gross Anatomy: A Review." *American Journal of Anatomy,* 159: 245–260, 1980.

Edwards, R. H. T. "Human Muscle Function and Fatigue." In *Human Muscle Fatigue: Physiological Mechanisms.* Ciba Foundation Symposium, Vol. 82, pp. 1–18. London: Pittman Medical, 1981.

Hamilton, W. G. "Surgical Anatomy of the Foot and Ankle." *Clinical Symposia,* 37(3), 1985.

Lampe, E. W. "Surgical Anatomy of the Hand." *Clinical Symposia,* 40(3), 1988.

LeVeau, B. *William's & Lissner's Biomechanics of Human Motion,* 3rd ed. Philadelphia: Saunders, 1991.

Nadel, E. R. "Physiological Adaptations to Aerobic Training." *American Scientist,* 73: 334–343, 1985.

Shepard, R. J. *Physiology and Biochemistry of Exercise.* New York: Praeger Publishers, 1982.

Veikko, A. K. III. "The Physiology of Marathon Running." *Science Progress,* 70: 109–127, 1986.

External Anatomy, Removal of the Skin, and Muscle Review

After completing this exercise, you should be able to

1. Remove the skin of the fetal pig in one piece
2. Prepare the specimen for future dissection by removing all fat, fascia, and hair from the muscle surface, so that muscle outlines and fiber direction can be seen
3. Compare fetal-pig muscles with those of the human

Materials

The following materials should be provided for class use:

1. Preserved fetal pigs
2. Dissecting tools
   a. Heavy-duty scissors
   b. Scalpel
   c. Probe
   d. Two dissecting needles
   e. Forceps
3. String
4. Rubber bands
5. Name tags

INTRODUCTION

This portion of the manual contains exercises on the dissection of fetal-pig muscles and their comparison to human muscles. The muscles are divided into groups so that a logical progression can be made from dissecting the superficial muscles to dissecting the more important deep muscles. No attempt has been made to cover all of the pig muscles. The fetal-pig muscles can be compared to human muscles, using the drawings that follow each group of dissections.

For each dissection sequence, there are three study aides: (1) a black-and-white photo of the specific muscle group for dissection, (2) a line drawing corresponding to the black-and-white photo, and (3) tables summarizing the origins, insertions, and actions of both fetal-pig and corresponding human muscles.

The ideal way to make a dissection of the muscles is for each student to work alone on a specimen. However, two or more students may share an animal. If this is the case, it is to be understood that each student does an equal amount of work and accepts responsibility for the dissection. If laboratory grades are given on these exercises, they must be shared by those responsible for the specimen.

For the safety of the student, it is recommended that each student use a heavy-duty, nonrazor scalpel and a heavy-duty pair of scissors for cutting cartilage. Lightweight scissors (cuticle scissors) and a razor scalpel may be used for the more delicate internal organs. Always point scissors away from you when cutting tissue, and be careful to keep fingers away from the scalpel blade.

EXTERNAL ANATOMY OF THE FETAL PIG

Examine the surface of the fetal pig, and review some of the terms of location and position studied in Exercise 7. Figures 15-1 and 15-2 are diagrams of pig anatomy. The pig is a four-legged animal, and the terms *anterior* and *posterior* or *cranial* and *caudal* are used for front end and rear end. The belly side is *ventral,* and the back side is *dorsal.* The terms *lateral* and *medial*

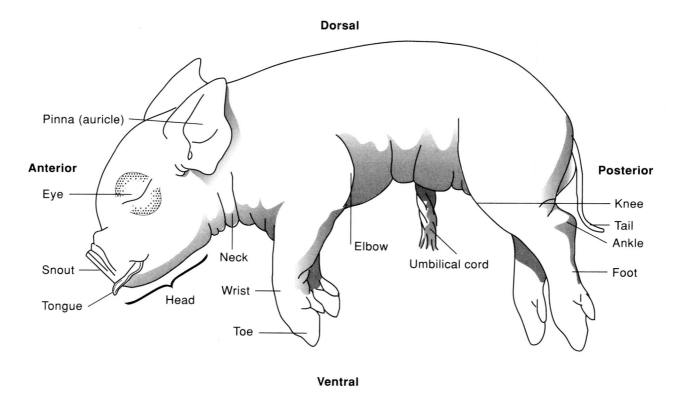

Figure 15-1 External anatomy of the fetal pig (lateral view).

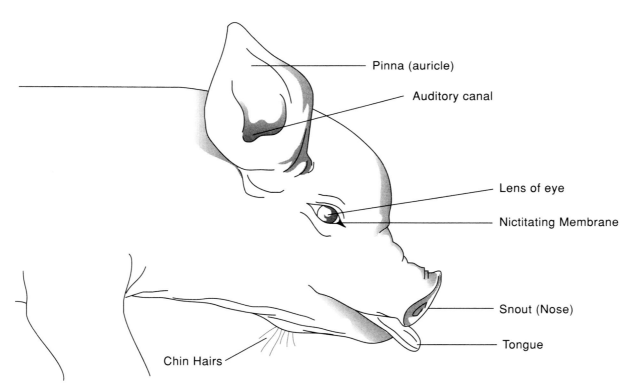

Figure 15-2 Fetal-pig head (lateral view).

are the same in both pig and human. Also review body planes. *Transverse* and *sagittal* planes are important in your future dissections.

External Features

The body is divided into head, neck, and trunk. There are well-developed arms, legs, and a small tail. The head has laterally placed ears and eyes and a medially placed mouth and snout (nose).

Examine the ears. The outer flap, which is quite large proportionately in the pig, is the *auricle* or *pinna.* Look carefully near the base of the auricle, and locate the entrance to the *external auditory canal (meatus).* Sound waves pass down this canal to the middle ear, which is located beyond the *tympanic membrane* (eardrum). The pig has muscles that move the pinna.

The eyes are closed. Two eyelids (upper and lower) meet at about the center of the eye. Take your scissors, and cut forward and backward at the corners of the eye, so that the eyeball can be seen. In the corner of the eye (anterior in the pig and medial in the human) is the *nictitating membrane.* In the living animal, this mem-

brane can be extended over the entire eyeball of the pig, to protect the eye and to clean it of foreign particles. Humans have a remnant of this membrane, termed the *plica semilunaris,* but it is *vestigial* (nonfunctional).

The mouth of the pig is located medially and is bounded by two rather thick lips. The tongue usually protrudes from the mouth, which is closed.

The snout is prominent and has paired *external nares* (nostrils). Pigs use this broad, protruding organ to *root,* a process not unlike shoveling. Habitual rooting causes muscles of the neck to become very well developed and gives the pig the appearance of having no neck at all.

The arms and legs are well developed. Look at the distal ends of each appendage. Note that the pig walks on its phalanges. It is classified as an even-toed ungulate (order Artiodactyla). The horse is an odd-toed ungulate (order Perissodactyla) and actually walks on the nail of its one toe. Animals whose digits touch the ground only in walking are referred to as *digitigrade.* In the human, the entire sole touches the ground in walking. This type is called *plantigrade.*

On the ventral surface of the body is an umbilical

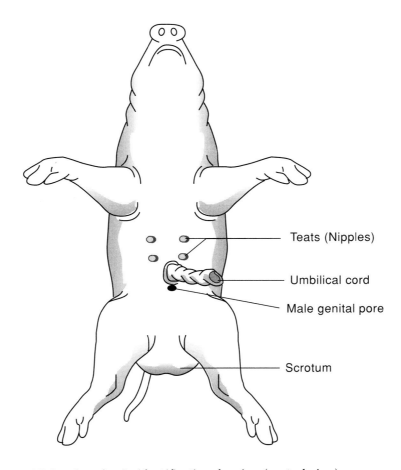

Figure 15-3   A male pig identification drawing (ventral view).

cord, which has been tied off with string. In the male pig, there is a conspicuous opening in the center of a small mound posterior to the umbilical cord. This is the urogenital opening for the *penis,* which can be felt beneath the skin (see Figure 15-3). A male also has paired testes located beneath the skin posteriorly between the legs. This skin is the *scrotum* (Figure 15-3). If your specimen is a female, the urogenital opening is located ventral to the anus (Figure 15-4). This opening can be seen in the center of a mound from which a papilla arises (Figure 15-4).

Examine the cut end of the *umbilical cord* (or make a fresh cut in it). Locate two *umbilical arteries* and one *umbilical vein.* You may also see another vessel, called the *allantois,* one of the extraembryonic membranes during fetal life (Figure 15-5).

REMOVAL OF THE SKIN

Read this entire section carefully before you begin. Removing the skin in one piece is not difficult, but care

must be taken to ensure that no muscles are damaged or destroyed in the process. In the fetal pig, many of the superficial muscles are not well developed. Pigs that measure about 35 cm are best for this study. You will find masses of heavy connective tissue and stringy undeveloped fat subcutaneously. This tissue should be left covering the muscles as you proceed with the skinning. It can be removed later with each separate muscle dissection. Figures 15-6 and 15-7 show the dissection lines you should follow, listed in numerical order.

To begin, place the pig face down on a dissecting tray. Using Figure 15-6 as a guide, insert the scissors in the midline just anterior to the tail, and cut forward, separating skin from muscle and fascia. Use your fingers or the dull end of the scalpel, to ensure that no muscle is left attached to the skin. Extend your first incision line to a point just above the neck. Now separate as much skin as possible to the right and the left of that incision line.

When you have loosened all the skin on the back side of the trunk and neck, begin a second incision line, extending down each limb, to the ankles and wrists. Cut a complete circle around each appendage. Next, cut

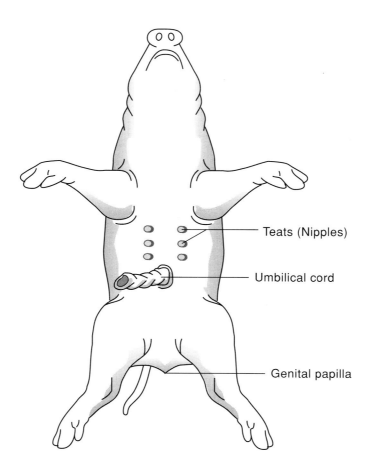

Figure 15-4 A female pig identification drawing (ventral view).

Teats (Nipples)

Umbilical cord

Genital papilla

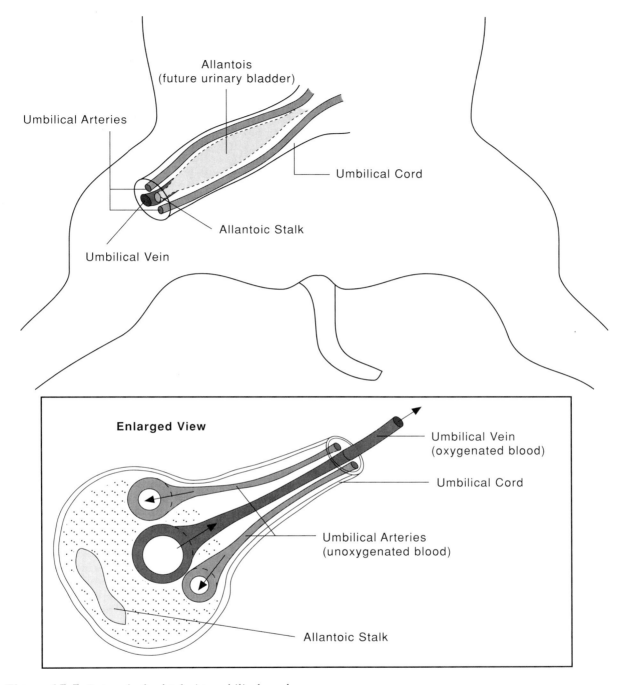

Figure 15-5 Cut end of a fetal-pig umbilical cord.

around the neck from dorsal to ventral (incision line 3). At this point, all skin should be loosened except on the ventral side.

Turn the fetal pig onto its back, and remove the skin from the ventral surface, from anterior to posterior. Press the blunt end of the scalpel against the muscle, and hold the skin with the other hand, to be certain that no muscle is removed with the skin. Pull the skin down carefully until you reach the umbilical cord. It will be necessary to cut around this region. When you reach the umbilical cord, make a U-shaped cut around it, and finish pulling off the skin from the ventral surface of the body and from the legs (Figure 15-7).

Keep the skin so that you can wrap the pig in it at the end of the laboratory period. This will ensure that no loose parts fall off and no damage will be done to

Figure 15-6 Skin removal (dorsal view) showing incision lines.

the muscle dissections. Fill out a name tag for your pig, and tie it to one of the legs. The string should be long enough to reach outside the pail in which the specimens are stored.

OVERVIEW OF MUSCLES AND MUSCLE TISSUE

Although muscular tissue can be divided into three categories on the basis of its location, the cells in muscle tissue all have one thing in common. The fibrils within them shorten and thicken and thus effect movement. Review the histology of the following muscles in Exercise 6.

1. *Visceral muscle*—found in visceral organs, largely involuntary action
2. *Cardiac muscle*—found only in the heart, involuntary
3. *Skeletal muscle*—associated with skeletal parts, voluntary (*Note:* Skeletal muscle is the subject for the study of muscles in this unit.)

Figure 15-7 Skin removal (ventral view) showing incision lines.

The Structure of a Muscle

Each muscle has three main parts:

1. *Origin*—end attached to the part of the skeleton that does not move when the muscle contracts
2. *Insertion*—end attached to the part of the skeleton that moves when the muscle contracts
3. *Body* or *belly*—center of the muscle between the origin and the insertion

Most muscles are subdivided into bundles or *fasciculae.* This is an important concept to keep in mind because there is often a tendency for the student to try to subdivide a muscle into several parts simply because it seems possible to do so. The entire extent, size, and shape of a muscle must be determined *before* beginning to dissect it, and its origin and insertion should be located on the skeleton. Locate each origin and insertion as you go along. This manual provides photo-

Table 15-1

Muscle Action	Description
Flexion	*Flexor muscle* bends a joint or reduces the angle between two bones.
Extension	*Extensor muscle* straightens a joint or increases the angle between two bones.
Abduction	*Abductor muscle* carries an appendage or a part away from the midline.
Adduction	*Adductor muscle* brings an appendage or a part back to the midline.
Elevation	*Levator muscle* raises a part.
Depression	*Depressor muscle* lowers a part.
Rotation	*Rotator muscle* turns one part on another.

graphs, line drawings, and tables indicating the origin, insertion, and action of each muscle, so that careful study of each muscle can be made *before* the dissection is begun.

Dissection

To *dissect* a muscle does not mean to cut into it or to cut it out, but merely to separate it cleanly from the adjacent muscles between its origin and its insertion. If deeper muscles are to be viewed and dissected, it will be necessary to transect a muscle. To *transect* a muscle, find the midpoint between its origin and its insertion, and cut it in half (or as close to half as possible in the case of odd-shaped muscles), perpendicular to the direction of the fibers. Fold each half back to its origin or insertion. You will then be able to observe the underlying muscles. Never transect a muscle unless you are specifically instructed to do so in this manual or by your instructor.

Each muscle is surrounded by connective tissue—which is called *fascia*. Extensions of fascia around and into the muscle are differentiated into *epimysium, perimysium,* and *endomysium,* but because these dissection exercises do not separate muscles into their bundles, only the term *fascia* is used in this lab book.

Most muscles do not attach directly to the bone, but rather by extensions of fascia. Connective tissue that binds a muscle to a bone is a *tendon.* The name *aponeurosis* (ap'-ō-nyoo-RŌ-sis) is often applied to a broad, flat tendon. Connective tissue that connects one bone with another bone is a *ligament.*

The connective tissue that binds muscles together may be very tough, requiring scissors or a scalpel to cut it, or it may be so soft and fragile that it can be broken by running a finger between the muscles. Do whatever separating you can with your fingers, resorting to scissors or scalpel only when necessary. The older the fetal pig, the better developed the muscles will be.

The fascia that covers many of the superficial muscles of the fetal pig is thick and gelatinous. Great care must be taken to remove it without injuring the muscles beneath. Remove the fascia from each muscle group just before dissecting that group.

Instructions for dissection are given for one side of the pig only. Corresponding photographs and line drawings in each exercise usually show the left side. Immediately following this dissection exercise, there are diagrams and photographs of human musculature (Figures 15-8; 15-9; 15-10; and 15-11). Figures 15-8 and 15-9 are drawings of human musculature in ventral and dorsal views. Figures 15-10 and 15-11 are photos which depict the superficial human muscles which have been dissected by Dr. Yokochi of Japan after very careful skinning procedure. Refer to these and other diagrams throughout all your work on human muscles, and compare the pig muscles you have dissected with their human counterparts. The laboratory review sections at the end of each exercise attempt to call attention to the functions of your own skeletal muscles. Always try to apply the knowledge gained from the dissections to the function of your own muscles.

Figure 15-8 Principal superficial muscles of the human (ventral view). (From G. J. Tortora and S. R. Grabowski, *Principles of Anatomy and Physiology,* 7th ed., New York: HarperCollins, 1993. © Leonard Dank.)

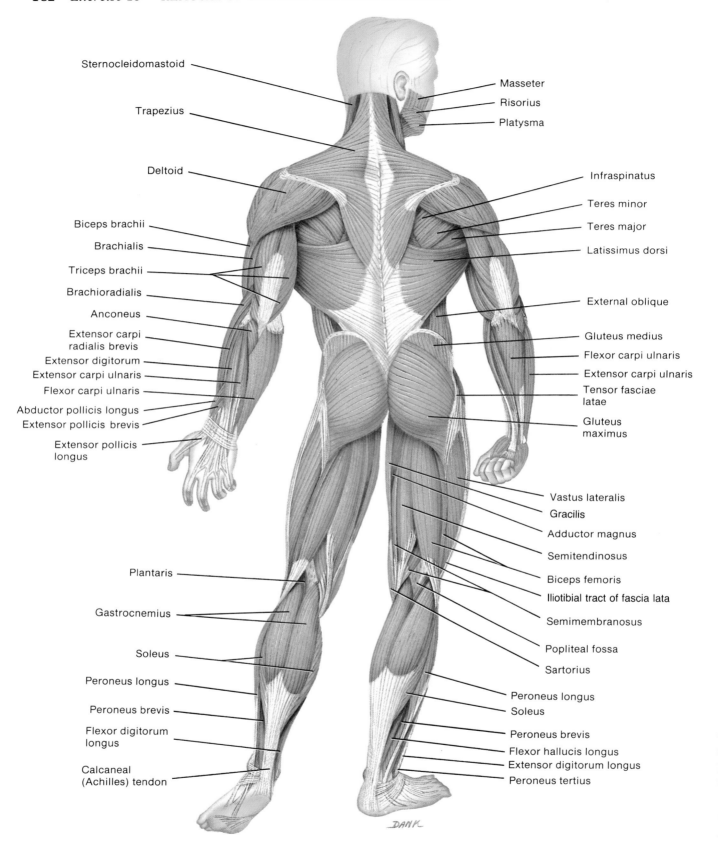

Sternocleidomastoid

Trapezius

Deltoid

Biceps brachii

Brachialis

Triceps brachii

Brachioradialis

Anconeus

Extensor carpi radialis brevis

Extensor digitorum

Extensor carpi ulnaris

Flexor carpi ulnaris

Abductor pollicis longus

Extensor pollicis brevis

Extensor pollicis longus

Plantaris

Gastrocnemius

Soleus

Peroneus longus

Peroneus brevis

Flexor digitorum longus

Calcaneal (Achilles) tendon

Masseter

Risorius

Platysma

Infraspinatus

Teres minor

Teres major

Latissimus dorsi

External oblique

Gluteus medius

Flexor carpi ulnaris

Extensor carpi ulnaris

Tensor fasciae latae

Gluteus maximus

Vastus lateralis

Gracilis

Adductor magnus

Semitendinosus

Biceps femoris

Iliotibial tract of fascia lata

Semimembranosus

Popliteal fossa

Sartorius

Peroneus longus

Soleus

Peroneus brevis

Flexor hallucis longus

Extensor digitorum longus

Peroneus tertius

DANK

Figure 15-9 Principal superficial muscles of the human (dorsal view). (From G. J. Tortora and S. R. Grabowski, *Principles of Anatomy and Physiology,* 7th ed., New York: HarperCollins, 1993. © Leonard Dank.)

Figure 15-10 Superficial muscles—anterior aspect: 1. Frontal belly of occipitofrontalis; 2. procerus; 3. orbicularis oculi; 4. zygomaticus major, minor, levator anguli oris, levator labii superioris alaeque nasi; 5. depressor labii inferioris; 6. risorius, depressor anguli oris; 7. platysma; 8. deltoid; 9. sternocleidomastoid; 10. sternohyoid; 11. pectoralis major; 12. lateral head of triceps; 13. long head of biceps; 14. short head of biceps; 15. brachialis; 16. latissimus dorsi; 17. serratus anterior; 18. pronator teres; 19. brachioradialis; 20. flexor carpi radialis; 21. external abdominal oblique; 22. rectus abdominis; 23. abductor pollicis longus; 24. flexor digitorum superficialis; 25. palmaris longus; 26. palmaris brevis; 27. thenar eminence; 28. abductor digiti minimi; 29. tensor fasciae latae; 30. iliopsoas; 31. pectineus; 32. adductor longus; 33. rectus femoris; 34. vastus medialis; 35. sartorius; 36. gracilis; 37. gastrocnemius; 38. soleus; 39. flexor digitorum longus; 40. tibialis anterior; 41. tendo calcaneus; 42. ext. hallucis longus; 43. abductor hallucis aponeurosis of external oblique (rectus sheath). (From Yokochi/Rohen. *Photographic Anatomy of the Human Body,* 2nd ed. Baltimore: University Park Press, and Tokyo: Igaku-Shoin Ltd., 1978).

Figure 15-11 Superficial muscles, posterior aspect: 1. temporalis; 2. occipital belly of occipitofrontalis; 3. sternocleidomastoideus; 4. trapezius; 5. deltoid; 6. infraspinatus; 7. teres major; 8. lateral head of triceps; 9. long head of triceps; 10. medial head of triceps; 11. brachialis; 12. rhomboideus major; 13. anconeus; 14. latissimus dorsi; 15. flexor carpi ulnaris; 16. extensor carpi ulnaris; 17. flexor digitorum superficialis; 18. serratus posterior inferior; 19. extensor digitorum (communis); 20. ext. abdominal oblique; 21. lumbar triangle; 22. gluteus medius; 23. dorsal interosseous; 24. gluteus maximus; 25. iliotibial tract of fascia lata; 26. gracilis; 27. biceps femoris (long head); 28. semitendinosus; 29. semimembranosus; 30. gastrocnemius; 31. soleus; 32. peroneus longus; 33. tendo calcaneus; 34. peroneus brevis *ext. digiti minimi **abductor digiti minimi ***lumbodorsal fascia. (From Yokochi/Rohen. *Photographic Anatomy of the Human Body,* 2nd ed. Baltimore: University Park Press, and Tokyo: Igaku-Shoin Ltd., 1978.)

LABORATORY REVIEW *15*

NAME _____

LAB SECTION _____ DATE _____

EXTERNAL ANATOMY, REMOVAL OF THE SKIN, AND MUSCLE REVIEW

COMPLETION AND DEFINITION

1. For each of the following terms, state what it is, where it is located, and what function, if any, it has.

   a. Pinna _____
   _____
   _____

   b. Umbilical Cord _____
   _____
   _____

   c. Fascia _____
   _____
   _____

2. How is the origin of a muscle different from the insertion? _____
   _____
   _____

3. What is the difference between dissecting and transecting? _____
   _____
   _____

4. Under what circumstances would it be necessary to transect a muscle? _____
   _____

5. Define these terms:

   a. Tendon _____
   _____

   b. Ligament _____
   _____

   c. Aponeurosis _____
   _____

Muscles of the Back and Shoulder

After completing this exercise, you should be able to

1. Dissect the muscles of the back of the fetal pig
2. Identify comparable human muscles

Materials

The following materials should be provided for class use:

1. Preserved fetal pigs
2. Dissecting tools (scissors, scalpel, probe, dissecting needles, and forceps)

The following fetal-pig muscles are dissected in this exercise:

SUPERFICIAL BACK AND SHOULDER MUSCLES

Spinotrapezius
Acromiotrapezius
Clavotrapezius
Clavodeltoid (clavobrachialis) } Brachiocephalic
Acromiodeltoid
Spinodeltoid
Latissimus dorsi

MUSCLES ASSOCIATED WITH THE SCAPULA

Rhomboideus major
Rhomboideus minor

Rhomboideus capitis
Supraspinatus
Infraspinatus
Teres major
Anterior deep pectoral

DEEP BACK MUSCLES

Splenius
Spinalis dorsi
Longissimus dorsi
Iliocostalis
Serratus dorsalis (anterior part)
Serratus dorsalis (posterior part)
Multifidus spinae
Serratus ventralis
Subscapularis
Rhomboideus (transected)

Figure 16-1 (a) Superficial back muscles of fetal pig (photograph).

SUPERFICIAL BACK AND SHOULDER MUSCLES

Instructions for Dissection

Begin your dissection in the upper back region (see Figure 16-1a) between the two arms. Make sure that the skin has been removed from the neck as far back as the ears. Determine the midline *raphe* (RĀ-fē), and locate the *trapezius* group of muscles, which have their origins on this raphe. Humans have only one trapezius muscle, but the pig has three, as shown in Figure 16-1b.

Start with the most posterior of the group, the *spinotrapezius*. This is a triangular muscle that lies over the latissimus dorsi at its posterior end and joins the *acromiotrapezius* anteriorly. Dissect it completely, but be careful not to loosen either its origin or its insertion.

Next, dissect the acromiotrapezius and then the *clavotrapezius*. The trapezius group of muscles inserts on the spine of the scapula. For the pig, the word *clavotrapezius* is misleading because *clavo* means clavicle and the pig has no clavicle. In animals that *do* have clavicles, this part of the trapezius inserts on the clavicle. Its distal part goes on to insert on the humerus as one of the shoulder muscles, the clavodeltoid or cla-

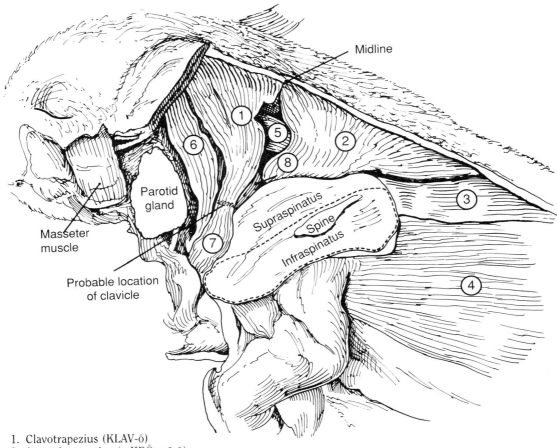

1. Clavotrapezius (KLAV-ō)
2. Acromiotrapezius (a-KRŌ-mē-ō)
3. Spinotrapezius (spī-nō-tra-PĒ-zē-us)
4. Latissimus dorsi (la-TISS-i-mus DOR-sī)
5. Rhomboideus capitis
6. Sternomastoid
7. Clavodeltoid or clavobrachialis (klav-ō-brak-ē-AL-is)
8. Acromiodeltoid

Figure 16-1 (b) Superficial back muscles of fetal pig (line drawing).

vobrachialis. The distal part of the muscle is present in the pig and is labeled here as the *clavodeltoid*. (Together, the clavotrapezius and the clavodeltoid are often called the "brachiocephalic.")

Posterior to the spinotrapezius is the large *latissimus dorsi*. Free it between the origin and the insertion, if possible. If the muscle is too thin to dissect, just observe it. It originates by a broad, tough aponeurosis along the midline of the back, and it inserts on the humerus.

In the pig, the *deltoid* group of muscles (shoulder muscles) is represented by three muscles, whereas in the human, there is only one deltoid muscle. The afore-

mentioned *clavodeltoid* extends down the arm from the clavicular region at the proximal end of the humerus. Posterior to it is a long, slender *acromiodeltoid*, which runs down over the anterior surface of the scapula to the proximal end of the humerus. The most prominent of the deltoid group is the *spinodeltoid*, which occupies a large part of the scapula from the spine to the proximal end of the humerus. The spinodeltoid has been removed in the photographs, to show muscles beneath it. It is outlined in dotted lines.

Figure 16-2 (a) Superficial back muscles associated with the scapula of fetal pig (photograph).

MUSCLES ASSOCIATED WITH THE SCAPULA

Instructions for Dissection

For this dissection, it is necessary to transect the trapezius group of muscles, the latissimus dorsi, and the spinodeltoid (Figures 16-2 and 16-2b). Begin with the spinotrapezius. It is triangular in shape, so that the transection line will have to be angled to divide the muscle into two separate parts. The acromiotrapezius and the clavotrapezius should be transected mid-way between origin and insertion and their cut ends pulled back. The latissimus dorsi is also irregularly shaped. Transect it near the scapula, where it is thickest. Pull the scapula away from the body, and remove excess fascia.

Observe the *rhomboideus* group of muscles. Counterintuitively, the *rhomboideus minor* is larger than the *rhomboideus major*. The *rhomboideus capitis* is the part of the group that originates on the skull.

On the lateral surface of the scapula, locate the *supraspinatus* (above the spine), the *infraspinatus* (below the spine), and the *teres major* (below the infraspinatus). Also observe the *deep anterior pectoral* muscle, which originates on the ventral midline. It is observed in more detail in another exercise. Be sure that the scapular spine is clear. The *subscapularis* lies on the underside of the scapula, beneath another conspicuous muscle, the *serratus ventralis*. Both serratus ventralis and subscapularis are shown in a later exercise in ventral view.

1. Rhomboideus capitis
2. Rhomboideus minor
3. Rhomboideus major
4. Trapezius group (reflected)
5. Infra spinatus
6. Supraspinatus
7. Spinodeltoid (reflected) (dotted lines indicate its position over scapula).
8. Scapular spine
9. Anterior deep pectoral
10. Teres major
11. Clarodeltoid

Figure 16-2 (b) Superficial back muscles associated with the scapula of fetal pig (line drawing).

Figure 16-3 (a) Deep back muscles of fetal pig (photograph).

DEEP BACK MUSCLES

Instructions for Dissection

To observe the deep back muscles, it is necessary to transect the rhomboideus major and rhomboideus minor (Figures 16-3a and 16-3b). Do *not* transect the rhomboideus capitis.

Anterior to the rhomboids, near the head, is a broad flat muscle, the *splenius*. Observe its extent, but do not dissect it. Posterior to the splenius and medial in position is the *spinalis dorsi*. Lateral to the spinalis dorsi is a very long muscle, the *longissimus dorsi,* which divides into three distinct bundles at the posterior end of the trunk. Observe both ends of the muscle. Running parallel and medial to the longissimus is the multifidus spinae. Covering the anterior end of the longissimus dorsi is a delicate fascia, which forms the origin of the *serratus dorsalis* muscle located lateral to the longissimus. Tear this fascia, and pull the serratus dorsalis downward to expose the *iliocostalis*. The subscapularis and the serratus ventralis are shown again in Figure 17-1 from a ventral view. See Table 16-1 for a summary of both the superficial and the deep muscles of the fetal pig.

1. Splenius (SPLĒ-nē-us)
2. Spinalis dorsi (spī-NAL-is)
3. Longissimus (lon-JISS-i-mus) dorsi (posterior part)
4. Iliocostalis (ill'-ē-ō-kos-TAL-is)
5. Serratus dorsalis (anterior part)
6. Serratus dorsalis (posterior part)
7. Serratus ventralis (ser-RĀ-tus ven-TRAL-is)
8. External oblique
9. Spinotrapezius (transected)
10. Latissimus dorsi (transected)
11. Rhomboideus capitis
12. Multifidus spinae
13. Subscapularis
14. Rhomboids (transected)

Figure 16-3 (b) Deep back muscles of fetal pig (line drawing).

Table 16-1   Superficial Back Muscles and Muscles Associated with the Scapula of the Fetal Pig

Muscle	Origin	Insertion	Action
Spinotrapezius	Spines of first ten thoracic vertebrae	Scapula	Moves scapula dorsally and posteriorly
Acromiotrapezius	Spines of cervical vertebrae	Spine of scapula	Pulls scapula toward middorsal line
Clavotrapezius	Occipital bone of skull and transverse processes of anterior cervical vertebrae	Proximal end of clavobrachialis (in absence of clavicle in pig)	Moves forelimb anteriorly
Clavodeltoid	Proximal end of clavotrapezius (in absence of a clavicle in the pig)	Distal end of humerus	Moves forelimb anteriorly
Acromiodeltoid	Scapula	Proximal end of humerus	Raises humerus
Spinodeltoid	Spine of scapula	Proximal end of humerus	Raises humerus
Latissimus dorsi	Lumbodorsal fascia of the last few thoracic vertebrae and the lumbar vertebrae	Proximal end of humerus by a tendon	Moves forelimb dorsally and posteriorly
Rhomboideus major	Spines of thoracic vertebrae	Scapula	Moves scapula dorsally
Rhomboideus minor	Spines of cervical and thoracic vertebrae	Scapula	Moves scapula dorsally
Rhomboideus capitis	Occipital bone of skull	Scapula	Moves scapula dorsally
Supraspinatus	Scapula cranial to spine	Proximal end of humerus	Protracts the arm
Infraspinatus	Lateral surface of scapula caudal to spine	Proximal end of humerus	Rotates arm outward
Teres major	Caudal border of scapula	Proximal end of humerus	Aids in retracting arm
Subscapularis	Medial surface of scapula	Proximal end of humerus	Adducts arm
Spinalis dorsi	Middle thoracic vertebrae	Anterior thoracic vertebrae and caudal vertebrae	Extends vertebral column
Longissimus dorsi	Spines of lumbar and thoracic vertebrae; sacrum and ilium	Transverse processes of lumbar and thoracic vertebrae; last four cervical vertebrae; lateral surfaces of ribs	Extends back and neck; aids in exhalation
Iliocostalis	Angles of posterior thoracic ribs by slips	On ribs cranial to origins of slips	Draws ribs together
Serratus dorsalis	Dorsal fascia covering spinalis dorsi muscle	Dorsal surface of ribs	Cranial slips—inhalation; caudal slips—exhalation
Serratus ventralis	Cervical vertebrae; lateral surfaces of fourth to eighth ribs	Dorsomedial surface of scapula	Suspends and raises thorax

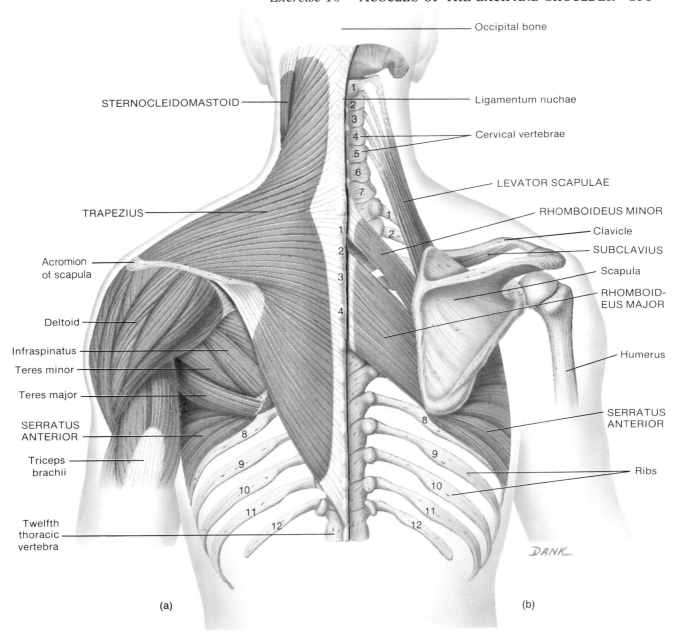

Occipital bone

STERNOCLEIDOMASTOID

Ligamentum nuchae

Cervical vertebrae

LEVATOR SCAPULAE

RHOMBOIDEUS MINOR

TRAPEZIUS

Clavicle

SUBCLAVIUS

Acromion of scapula

Scapula

RHOMBOID-EUS MAJOR

Deltoid

Infraspinatus

Teres minor

Teres major

Humerus

SERRATUS ANTERIOR

SERRATUS ANTERIOR

Triceps brachii

Ribs

Twelfth thoracic vertebra

DANK

(a)

(b)

Figure 16-4 Human back muscles: (a) posterior superficial view; (b) posterior deep view. (From G. J. Tortora and S. R. Grabowski, *Principles of Anatomy and Physiology,* 7th ed., HarperCollins, 1993. © Leonard Dank.)

COMPARABLE HUMAN MUSCLES

The muscles associated with the scapula in the human, which are comparable to those you dissected in the fetal pig, are the *trapezius* and the *rhomboids (major, minor, and capitis)*. In addition, there are the *serratus anterior* and the *pectoralis minor,* which you will dissect in a later exercise. Locate as many of these muscles on your own body as possible and make movements that demonstrate their actions. (See Laboratory Review.)

The human *supraspinatus, infraspinatus, subscapularis,* and *teres minor* form what is called a *musculotendinous cuff* over the head of the humerus. They reinforce the joint while allowing a great deal of movement. Make as many shoulder-joint movements as you can, and you will see why it is called the most freely movable joint in the body. The subscapularis is described later.

The deep muscles of the human back stabilize the bones of the vertebral column. Many of them are short and span only the distance between two vertebrae. Others are relatively long. They are postural muscles and not only maintain the upright posture, but also influence the curvatures of the spinal column. They extend, flex, and rotate parts of the vertebral column.

The location of these postural muscles is between the angle of the ribs and the vertebral spines. Refer to Tables 16-2 and 16-3 for the origin, insertion, and action of human back muscles.

Table 16-2   Superficial Back Muscles of the Human

Muscle	Origin	Insertion	Action
Trapezius	Occipital bone and cervical and thoracic vertebrae	Spine and acromion process of scapula	Raises clavicle; adducts, elevates, and depresses scapula
Deltoid	Clavicle and spine and acromion process of scapula	Humerus	Abducts arm
Levator scapulae	Cervical vertebrae	Vertebral border of scapula	Raises scapula
Latissimus dorsi	Spines of thoracic and lumbar vertebrae, crests of sacrum and ilium, last four ribs	Intertubercular groove of humerus	Extends, adducts, and rotates arm medially; pulls shoulder down and back
Rhomboideus major	Spines of second through fifth thoracic vertebrae	Vertebral border of scapula	Moves scapula back and up and rotates it downward
Rhomboideus minor	Spines of seventh cervical and first thoracic vertebrae	Superior angle of scapula	Adducts scapula
Supraspinatus	Supraspinous fossa of scapula	Greater tubercle of humerus	Adducts arm
Infraspinatus	Infraspinous fossa of scapula	Greater tubercle of humerus	Rotates humerus laterally
Teres major	Inferior angle of scapula	Proximal end of humerus	Extends humerus; adducts and rotates arm

Exercises involving some of the human back muscles which can be compared with those dissected in the fetal pig are given below:

1. *Trapezius* (Figure 16-5)

   The human trapezius muscle elevates and depresses the clavicle and the scapula, adducts the scapula, and extends the head. Perform each of the actions of the trapezius muscle on your own body. Shrug your left shoulder with your right hand on first the clavicle, then the scapula:

   a. Describe the movement of the clavicle. _____

   _____

   b. Describe the movement of the scapula. _____

   _____

   c. What is the name of the muscle action in Question (a)? _____

   _____

Figure 16-5 Trapezius (posterior view). (From P. J. Donnelly and G. A. Wistreich, *Exploring the Human Body*, New York: Harper & Row, 1989.)

Table 16-3 Deep Back Muscles of the Human

Muscle	Origin	Insertion	Action
Splenius capitis	Transverse processes of last four cervical vertebrae	Occipital bone and mastoid process	Flexes head; rotates face
Semispinalis capitis	Articular process of the fourth through the sixth cervical vertebrae and transverse processes of the seventh cervical vertebra and the first six thoracic vertebrae	Occipital bone	Extends head; rotates face
Spinalis thoracis	Spines of upper lumbar and lower thoracic vertebrae	Spines of upper thoracic vertebrae	Extends vertebral column
Longissimus capitis	Transverse processes of last four cervical vertebrae	Mastoid process	Extends head, rotates face
Longissimus cervicis	Transverse processes of fourth and fifth thoracic vertebrae	Transverse processes of second through sixth cervical vertebrae	Extends cervical region of vertebral column
Longissimus thoracis	Transverse processes of lumbar vertebrae	Transverse processes of thoracic and upper lumbar vertebrae and ninth and tenth ribs	Extends thoracic region of vertebral column
Iliocostalis cervicis and iliocostalis lumborum	Iliac crest and all ribs	All ribs and transverse processes of fourth through sixth cervical vertebrae	Extends back

What is the name of the muscle action in Question (b)? _____

_____

d. Move the left shoulder toward the midline and back several times. What effect does this have on the scapula? _____

_____

e. Bow your head, then slowly raise it. What effect is the trapezius muscle having on the head?

_____

_____

2. *Rhomboideus major* (Figure 16-6)
The rhomboideus major adducts the scapula and rotates the scapula upward, slightly outward, and downward.

Contract the rhomboideus muscle by trying to pull the scapula closer to the midline. Have your

Figure 16-6 Rhomboideus (posterior view). (T1 and T12 refer to the first and twelfth thoracic vertebrae, respectively.) (From P. J. Donnelly and G. A. Wistreich, *Exploring the Human Body,* New York: Harper & Row, 1989.)

lab partner lie face down with arms behind the back, palms up. Ask your partner to raise his or her arm while you place your hand between the medial border of the scapula and the midline of the body (over the rhomboideus major). Observe the origins and insertions of the muscle. What movement can you

detect in the scapula? _____

_____

3. *Levator scapulae* (Figure 16-7)
This muscle elevates the scapula and bends the neck laterally when the scapula is fixed.

   Shrug the shoulders to elevate the scapulae. Next, try to keep the scapulae fixed while contracting the levator scapulae on the left side of the body only. Do you feel any movement of the neck? Are there any significant differences in the distance the shoulder can be raised? Make observations and com-

   ments: _____

   _____

   _____

   _____

4. *Latissimus dorsi* (Figure 16-8)
This muscle extends, adducts, and rotates the arm medially. It also draws the shoulder downward and backward. It is a large triangular muscle covering the lower half of the back. Along with the trapezius, rhomboideus major, and levator scapulae muscles, the latissimus dorsi connects the arm to the vertebral column.

   With elbow fixed, arm at your side, turn the arm medially in the glenoid fossa and extend the arm slowly. Draw the shoulder downward and backward. Where on the back do you feel muscle

   contraction? _____

   _____

5. *Deltoid* (Figure 16-9)
The deltoid muscle has an anterior and a posterior portion. If the entire muscle contracts, the arm is abducted. The anterior portion alone flexes the arm, and the posterior portion alone extends the arm.

   Perform the following three actions:
   a. Hold your arm at your side, elbow straight, and move your arm away from your body (abduction). Return your arm to its resting position at the side of your body.
   b. Move your arm forward (contracting the anterior fibers).
   c. Move your arm backward (contracting the posterior fibers). This is the action you would make if you were lifting, then rolling a bowling ball.

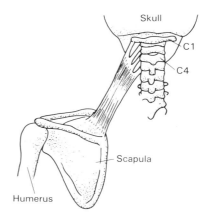

Figure 16-7 Levator scapulae (posterior view). (From P. J. Donnelly and G. A. Wistreich, *Exploring the Human Body*, New York: Harper & Row, 1989.)

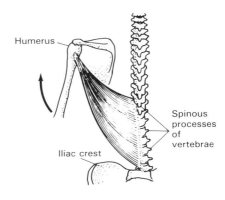

Figure 16-8 Latissimus dorsi (posterior view). (From P. J. Donnelly and G. A. Wistreich, *Exploring the Human Body*, New York: Harper & Row, 1989.)

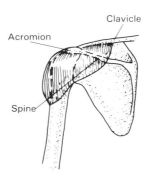

Figure 16-9 Deltoid (posterior view). (From P. J. Donnelly and G. A. Wistreich, *Exploring the Human Body*, New York: Harper & Row, 1989.)

Name other activities that involve the use of the deltoid muscle: _____

_____

_____

6. *Supraspinatus* (Figure 16-10)
This muscle abducts the arm, along with the deltoid muscle, and slightly rotates the arm.

Move your arm away from your body laterally. With your elbow fixed, rotate your arm laterally. Describe how you can palpate contraction in the supraspinous region of the scapula. _____

_____

7. *Infraspinatus* (Figure 16-11)
The infraspinatus rotates the arm laterally and both abducts and adducts the humerus. The upper part of the muscle abducts, and the lower part adducts.

With your arm at the side of the body, and the elbow fixed, rotate the arm laterally. Describe how you can palpate contraction in the infraspinous region of the scapula. _____

_____

8. *Teres minor* (Figure 16-12)
Together with the subscapularis, supraspinatus, and infraspinatus, this muscle forms the *rotator cuff* of the shoulder. By itself, it rotates the arm laterally. Which sports actitivities might aggravate this group of muscles? _____

_____

9. *Subscapularis* (Figure 16-13)
The subscapularis (a rotator-cuff muscle) rotates the arm medially. It is often injured in sports activities.

Name some everyday activities that involve the rotator cuff muscles. _____

_____

_____

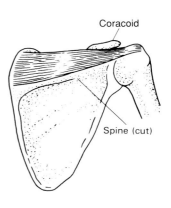

Figure 16-10 Supraspinatus (posterior view). (From P. J. Donnelly and G. A. Wistreich, *Exploring the Human Body,* New York: Harper & Row, 1989.)

Figure 16-11 Infraspinatus (posterior view). (From P. J. Donnelly and G. A. Wistreich, *Exploring the Human Body,* New York: Harper & Row, 1989.)

Figure 16-12 Teres minor (posterior view). (From P. J. Donnelly and G. A. Wistreich, *Exploring the Human Body,* New York: Harper & Row, 1989.)

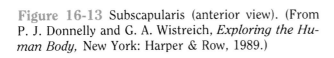

Figure 16-13 Subscapularis (anterior view). (From P. J. Donnelly and G. A. Wistreich, *Exploring the Human Body,* New York: Harper & Row, 1989.)

LABORATORY REVIEW *16*

MUSCLES OF THE BACK AND SHOULDER

IDENTIFICATION

1. Identify the muscles in Figure Q16-1 by matching the letters in the figures with the names in the following list. You may color the diagram if you wish. Use your text as a guide.

_____ Spinalis thoracis

_____ Iliocostalis thoracis

_____ Semispinalis capitis

_____ Iliocostalis cervicis

_____ Splenius capitis

_____ Longissimus thoracis

_____ Longissimus cervicis

_____ Longissimus capitis

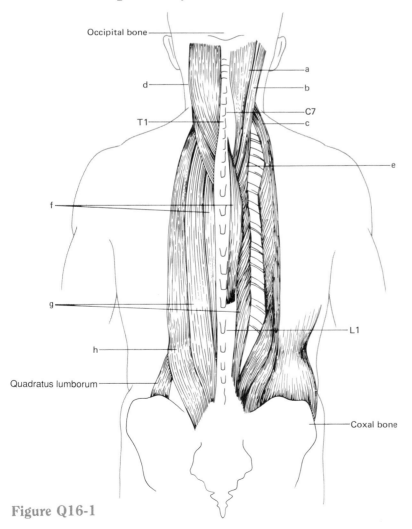

Figure Q16-1

2. Name the muscles indicated in Figure Q16-2. Color the diagram if you wish.

a. _____

b. _____

c. _____

d. _____

e. _____

f. _____

g. _____

h. _____

i. _____

j. _____

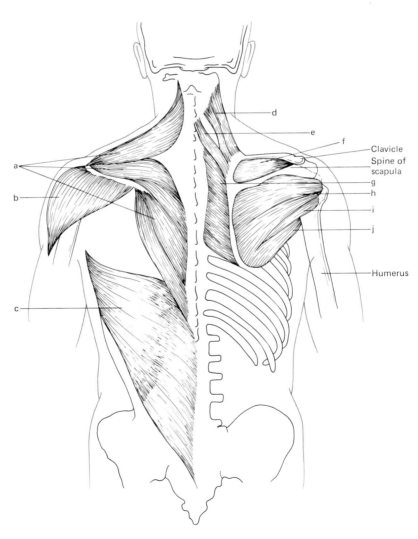

Figure Q16-2

3. Figure Q16-3 is a posterior view of the human skeleton. Draw in and label the
   following muscles. Review origins and insertions in Table 16-2.
   a. Trapezius
   b. Deltoid
   c. Latissimus dorsi
   d. Levator scapulae
   e. Rhomboideus major
   f. Rhomboideus minor
   g. Supraspinatus
   h. Infraspinatus

Figure Q16-3

4. Figure Q16-4 is a posterior view of the human skeleton. Draw in and label the following muscles on side A or B.
   a. Semispinalis capitis
   b. Longissimus capitis
   c. Iliocostalis cervicis
   d. Splenius capitis

Draw in and label the following muscles on side B.
   e. Longissimus cervicis
   f. Spinalis thoracis
   g. Longissimus thoracis
   h. Iliocostalis thoracis

Side A   Side B

Figure Q16-4

Muscles of the Chest, Ventral Scapula, Upper Arm (Medial Surface), Abdomen, and Ribs

After completing this exercise, you should be able to

1. Dissect the pectoral and abdominal muscles of the fetal pig
2. Identify comparable human muscles

Materials

The following materials should be provided for class use:

1. Preserved fetal pigs
2. Dissecting tools

The fetal-pig muscles dissected in this exercise are listed here:

PECTORAL MUSCLES

Anterior superficial pectoral, pectoralis major
Posterior superficial pectoral, pectoralis minor
Anterior deep pectoral
Posterior deep pectoral

SCAPULAR MUSCLES

Subscapularis

Serratus ventralis
Coracobrachialis
Teres major

UPPER ARM MUSCLES

Biceps brachii
Triceps brachii

ABDOMINAL AND INTERCOSTAL MUSCLES

Rectus abdominis
External oblique
Internal oblique
Transversus abdominis
External intercostal
Internal intercostal

Figure 17-1 (a) The pectoral scapular and medial arm muscles of the fetal pig (photograph).

SUPERFICIAL PECTORALS

Instructions for Dissection

Turn the pig on its back, as shown in Figure 17-1, and find the general outline of the pectoral group of muscles on both sides. The most anterior muscle is the *anterior superficial pectoral.* Posterior to it is the *posterior superficial pectoral.* Transect these muscles on the left side (see Figure 17-1). You can now see the *anterior deep pectoral*, which moves over the shoulder to insert on the scapula. The *posterior deep pectoral* is posterior to the anterior deep pectoral and quite conspicuously separated from it. The pectoral muscles adduct the arm.

Forming the undersurface of the scapula is the large *subscapularis,* and below it is the *teres major.* A prominent muscle that connects the shoulder and the scapula is the *coracobrachialis.* On the medial surface of the upper arm is the spindle-shaped *biceps brachii,* and forming both lateral and medial borders of the arm is the *triceps brachii long head;* the latter muscle is observed from the lateral surface in Exercise 19.

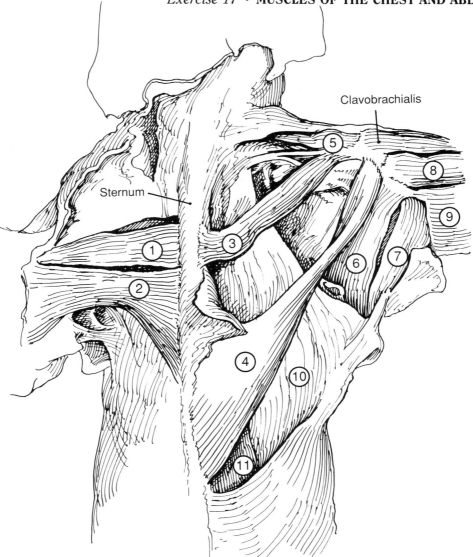

1. Anterior superficial pectoral or pectoralis major
2. Posterior superficial pectoral or pectoralis minor
3. Anterior deep pectoral
4. Posterior deep pectoral
5. Coraco brachialis
6. Subscapularis
7. Teres major
8. Biceps brachii
9. Triceps brachii
10. Serratus ventralis
11. Rectus abdominis

Figure 17-1 (b) The pectoral, scapular and medial arm muscles of fetal pig (line drawing).

Figure 17-2 (a) The abdominal and intercostal muscles of the fetal pig (photograph).

ABDOMINAL MUSCLES

Instructions for Dissection

Three thin layers of muscle form the abdominal wall. With the aid of Figure 17-2 and Table 17-1, locate the *external oblique* muscle. Insert the tip of small dissecting scissors midway between the origin (on the costal cartilages) and the insertion (on the ventral midline) of the muscle. Taking care not to cut too deeply, extend the transection in an arc from the pectoral region to the groin. Carefully separate with a probe the external oblique from the muscle beneath it, the *internal oblique.* When you have exposed all of the internal oblique, fold back the cut ends of the external oblique.

Now find the midpoint of the internal oblique. Plan ahead just how you are going to make the next incision, which will separate the internal oblique from the *transversus abdominis,* the third layer of abdominal muscle. Just as you did with the external oblique, insert the tip of the scissors midway between the origin and the insertion of the internal oblique, and cut a tiny hole in it and then extend it for about 1 to 2 inches. Do not cut too deeply, or you will puncture the *parietal peritoneum,* which lines the abdominal cavity. At this point, fold back the cut ends of the internal oblique.

1. External oblique (reflected)
2. Internal oblique (reflected)
3. Transversus abdominis
4. Rectus abdominis
5. Posterior superficial pectoral
6. External intercostal
7. Internal intercostal

Figure 17-2 (b) The abdominal and intercostal muscles of the fetal pig (line drawing).

INTERCOSTAL MUSCLES

Find a place between two ribs, and use small, sharp scissors to cut through the external layer of muscle, to expose the second layer of muscle, the fibers of which run obliquely in a different direction. The outer layer is the *external intercostal* muscle, and the inner layer is the *internal intercostal* muscle. Below the internal intercostal layer is the *parietal pleura,* which lines the thoracic cavity. Finally, in the midline between the *external oblique* and the *posterior superficial pectoral* muscle, locate a portion of the *rectus abdominis.* The pig rectus abdominis does not exhibit the numerous compartments that are so obvious in the human.

Table 17-1 Pectoral and Abdominal Muscles of the Fetal Pig

Muscle	Origin	Insertion	Action
Anterior superficial pectoral (pectoralis major)	Ventral sternum	Proximal end of humerus	Adducts forearm
Posterior superficial pectoral (pectoralis minor)	Ventral sternum	Proximal end of humerus	Adducts forearm
Anterior deep pectoral	Anterior end of sternum	Aponeurosis covering supraspinatus	Adducts and retracts forelimb
Posterior deep pectoral	Ventral sternum and costal cartilages (fourth to ninth ribs)	Proximal end of humerus	Adducts and retracts forelimb
External oblique	Lateral surfaces of caudal ribs and lumbodorsal fascia	On linea alba by an aponeurosis	Compresses abdomen
Internal oblique	Lumbodorsal fascia	On linea alba	Compresses abdomen
Transversus abdominis	Lumbodorsal fascia	On linea alba	Compresses abdomen
Rectus abdominis	Pubic symphysis	Sternum	Compresses abdomen
Subscapularis	Medial surface of scapula	Proximal end of humerus	Adducts arm
Teres major	Caudal border of scapula	Proximal end of humerus	Aids in retracting arm
Biceps brachii	Ventral border of scapula	Proximal end of humerus, radius and ulna	Flexes lower arm
Triceps brachii	Proximal humerus	Olecranon process of ulna	Extends lower arm
External intercostal	Caudal border of one rib dorsoventrally	Cranial border of next rib	Inhalation
Internal intercostal	Caudal border of one rib dorsoventrally	Cranial border of next rib	Exhalation

COMPARABLE HUMAN MUSCLES

The human pectoral group consists of a *pectoralis major* and a *pectoralis minor* (see Figures 17-3 and 17-4). The function of the pectoral muscles in the human is the same as in the fetal pig—that is, to adduct the arm, and to perform the other actions listed in Table 17-2.

The abdominal group of human muscles is also similar to those in the pig. The *transversus abdominis* lies over the parietal peritoneum. The middle layer is the *internal oblique,* and the outer layer is the *external oblique.* The *rectus abdominis* arises in segments connected by flat tendons and is enclosed in a sheath formed by the aponeuroses of the other three abdominal muscles. The segmentation of the rectus abdominis is seldom seen in the pig. The abdominal muscles compress the contents of the abdominal cavity and also contribute to flexion and rotation of the torso. In the next section, try to demonstrate as many of these muscles as possible on your own body.

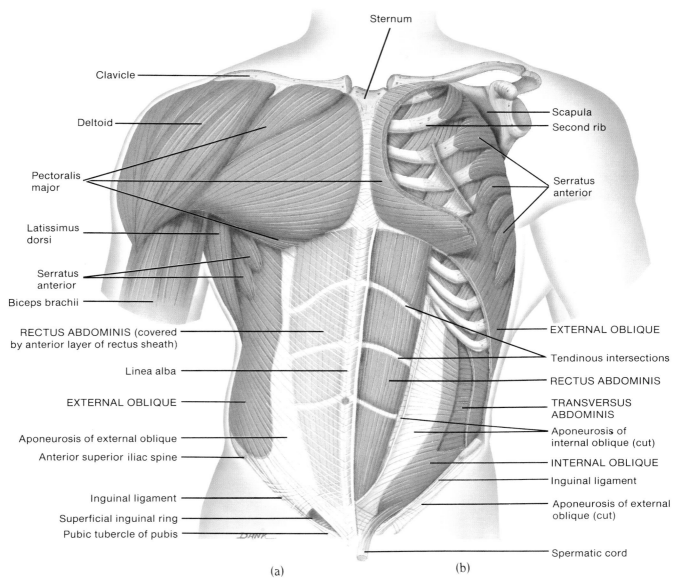

Figure 17-3 Muscles of the anterior abdominal wall: (a) superficial view; (b) deep view. (From Tortora, G. J., and S. R. Grabowski. *Principles of Anatomy and Physiology,* 7th ed., New York: HarperCollins, 1993.)

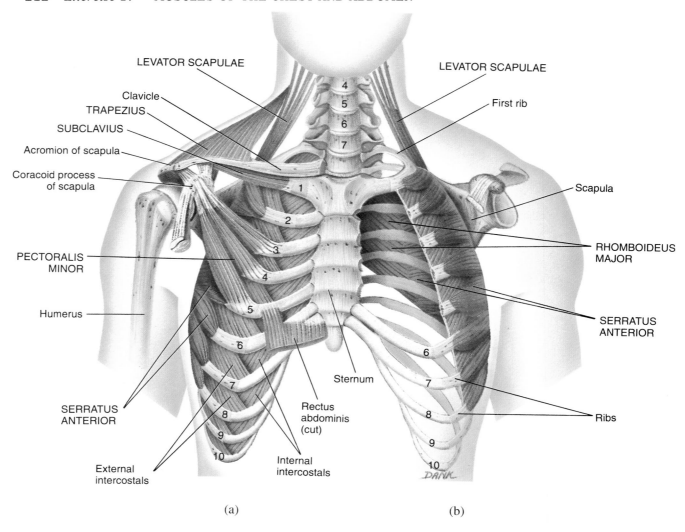

Figure 17-4 The human pectoral muscle group: (a) anterior deep view; (b) anterior deeper view. (From G. J. Tortora and S. R. Grabowski, *Principles of Anatomy and Physiology*, 7th ed., New York: HarperCollins, 1993.)

HUMAN CHEST MUSCLES

Study the actions of the following with respect to your own body.

1. *Pectoralis major* (Figure 17-5)
   This muscle adducts, flexes, and rotates the arm medially. It occupies most of the chest area above the diaphragm.

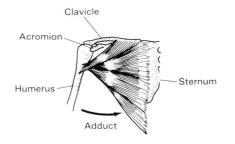

Figure 17-5 Pectoralis major (anterior view). (From P. J. Donnelly and G. A. Wistreich, *Exploring the Human Body*, New York: Harper & Row, 1989.)

Table 17-2 Pectoral Abdominal and Intercostal Muscles of the Human

Muscle	Origin	Insertion	Action
Pectoralis major	Clavicle, sternum, and cartilage of second through sixth ribs	Greater tubercle of humerus	Flexes, adducts, and rotates arm medially
Pectoralis minor	Third through fifth ribs	Coracoid process of scapula	Depresses scapula; rotates shoulder upward
External oblique	Lower eight ribs	Iliac crest and linea alba	Compresses abdomen
Internal oblique	Iliac crest, inguinal ligament, and thoracolumbar fascia	Costal cartilage of last three or four ribs	Compresses abdomen
Transversus abdominis	Iliac crest, inguinal ligament, lumbar fascia, and cartilage of last six ribs	Xiphoid process, linea alba, and pubis	Compresses abdomen
Rectus abdominis	Pubic crest and symphysis pubis	Costal cartilage of fifth to seventh ribs and xiphoid process	Flexes vertebral column
External intercostal	Inferior border of rib above	Superior border of rib below	Raise ribs at inhalation, increasing size or thoracic cavity
Internal intercostal	Superior border of rib below	Inferior border of rib above	Draws adjacent ribs together and lower ribs at exhalation

a. With your right hand on the left chest, near the arm, hold your left arm straight out in front of you. Then move the arm as far medially as possible. Do you feel the contraction of the pectoralis major muscle? _____

_____

b. With your right hand in the same place, flex and rotate the left arm. Describe the contraction of the pectoralis major. Is it greater or less than during adduction? _____

_____

2. *Pectoralis minor* (Figure 17-6)

This muscle depresses the scapula and rotates the shoulder joint anteriorly. It also elevates the third through fifth ribs during forced inhalation when scapulae are fixed.

a. To demonstrate the pectoralis minor, keep the shoulders in a fixed position, place your hands over the superior rib cage, and force inhale. Describe the movement you feel in the rib cage.

_____

b. Relax the left shoulder, then lower it. Slowly rotate the shoulder anteriorly without shrug-

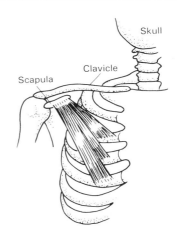

Figure 17-6 Pectoralis minor (anterior view). (From P. J. Donnelly and G. A. Wistreich, *Exploring the Human Body,* New York: Harper & Row, 1989.)

ging it. Describe how you can feel the muscle contract. _____

The abdominal muscles compress the abdomen and flex the vertebral column.

1. *External oblique* (Figure 17-7), *internal oblique* (Figure 17-8), and *transversus abdominus* (Figure 17-9)

Figure 17-7 External oblique. (From P. J. Donnelly and G. A. Wistreich, *Exploring the Human Body,* New York: Harper & Row, 1989.)

Figure 17-8 Internal oblique. (From P. J. Donnelly and G. A. Wistreich, *Exploring the Human Body,* New York: Harper & Row, 1989.)

Figure 17-9 Transversus abdominis. (From P. J. Donnelly and G. A. Wistreich, *Exploring the Human Body,* New York: Harper & Row, 1989.)

These thin sheets of muscle, the fibers of which run in different directions from one another, compress the abdomen and bend the vertebral column laterally.

a. Contract this group of muscles in order to pull in your abdomen. Try to do it without holding your breath. Can you hold this position without holding your breath? _____

_____

b. Contract the same three muscles on the left side, only to the extent that you notice a slight bending of your torso to the left. Repeat this on the right side. Can you do this without holding your breath or using your shoulder muscles? _____

_____

2. *Rectus abdominis* (Figure 17-10)
The rectus abdominis plays a role in compressing the abdomen, along with the muscles above it, but it also flexes the vertebral column.

a. To demonstrate the action of this muscle, first compress your abdomen, then contract your rectus abdominis. Can you feel a slight bending forward of your torso? _____

_____

b. Lie flat on your back, with hands behind your head. Raise your head and shoulders about 2 inches. Can you feel your rectus abdominus contracting? _____

_____

Figure 17-10 Right rectus abdominis (anterior view). (From P. J. Donnelly and G. A. Wistreich, *Exploring the Human Body,* New York: Harper & Row, 1989.)

MUSCLES OF THE CHEST, VENTRAL SCAPULA, UPPER ARM (MEDIAL SURFACE), ABDOMEN, AND RIBS

IDENTIFICATION

1. Identify the muscles in Figure Q17-1 by matching the letters in the figure with the names listed after the figure.

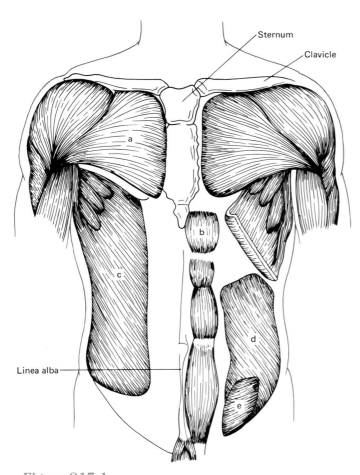

Figure Q17-1

_____ Internal oblique

_____ Transversus abdominis

_____ Pectoralis major

_____ External oblique

_____ Rectus abdominis

2. Figure Q17-2 is an anterior view of the skeleton. Refer to Table 17-2. Draw in and label the following muscles on Side A.
    a. Pectoralis major
    b. External oblique
    Draw in and label the following muscle on Side B.
    c. Rectus abdominis

Side A Side B

Figure Q17-2

Muscles of the Neck, Jaw, Face and the Salivary Glands

After completing this exercise, you should be able to

1. Dissect the muscles of the neck and jaw of the fetal pig
2. Locate and identify the salivary glands of the fetal pig
3. Identify comparable human muscles

Materials

The following materials should be provided for class use:

1. Preserved fetal pigs
2. Dissecting tools
3. Small trays
4. String
5. Name tags

The muscles and salivary glands dissected in this exercise are listed here:

MASSETER MUSCLE AND SALIVARY GLANDS

Masseter muscle

Parotid gland

Submandibular gland

Sublingual gland

Buccal glands

NECK AND JAW MUSCLES

Sternomastoid

Cleidomastoid

Digastric

Mylohyoid

Sternohyoid

Sternothyroid

Cricothyroid

Figure 18-1 (a) The masseter muscle and salivary glands of the fetal pig (photograph).

MASSETER AND SALIVARY GLANDS

Instructions for Dissection

Turn the pig onto its right side. Remove the skin from its neck and face. Remove the skin carefully. The large central muscle mass with fibers running dorsoventrally is the *masseter.* If you have removed the skin carefully, you will see four salivary glands surrounding this muscle. These glands all have ducts that empty into the mouth cavity.

The large triangular mass at the angle of the jaw is the *parotid gland.* It is located below the ear and runs ventrad along the side of the shoulder. It then makes a sharp angle and continues anteriad below the masseter muscle. Its duct, the *parotid duct* or *duct of Stensen* can be seen below and in front of the masseter muscle.

Gently pull the parotid gland posteriorly, and you will see a compact oval glandular mass, the *submandibular gland.* Its duct extends anteriorly and runs past another smaller glandular mass, the *sublingual gland.* The ducts from the sublingual and submandibular glands run forward, parallel to one another, and then empty into the oral cavity ventral to the tongue. Anterior to the masseter muscle are the *buccal glands,* located above the mandible over the lips. They are numerous and small.

1. Parotid gland
2. Submandibular gland
3. Sublingual gland
4. Buccal glands
5. Masseter (MASS-e-ter)
6. Parotid duct
7. Branches of cranial duct
8. Oral cavity

Figure 18-1 (b) The masseter muscle and the salivary glands of the fetal pig (line drawing).

Figure 18-2 (a) The neck and jaw muscles of the fetal pig (photograph).

NECK AND JAW MUSCLES

Instructions for Dissection

Turn the pig onto its back. Dissect the long V-shaped muscle between the sternum and the mastoid process of the temporal bone. This is the *sternomastoid*. The *cleidomastoid* arises from the shoulder region (*cleido* means "clavicle"). The pig has no clavicle, so the muscle arises from the fascia of the clavodeltoid–clavotrapezius complex, discussed in Exercise 16. This complex is in the region where a scapula would be located if the pig had one. It inserts on the mastoid process next to the sternomastoid. In the human, the cleidomastoid and the sternomastoid muscles are joined and are given one name, the sternocleidomastoid.

In the middle of the neck, between the two sternomastoids, are the *sternohyoids*. Separate them with a probe, and note the trachea lying beneath them. Transect the left sternohyoid. Lying beneath it is the *sternothyroid* muscle. Lying on the ventral surface of the larynx is the *cricothyroid* muscle. Along the inner border of the mandible are the *digastric* muscles, and between them, with fibers running horizontally, are the *mylohyoids*. (See Table 18-1 for each muscle's origin, insertion, and action.)

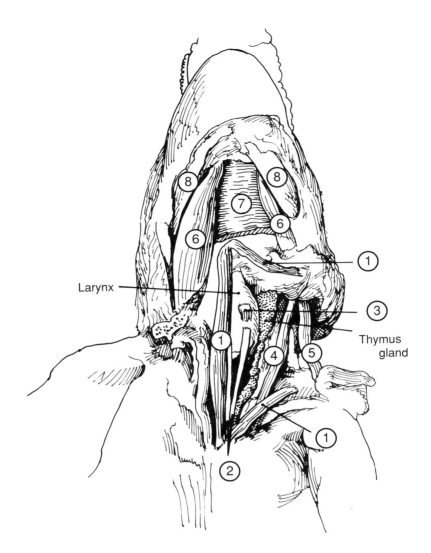

1. Sternohyoid (reflected)
2. Sternothyroid (ster'-nō-THĪ-royd)
3. Cricothyroid (kri-kō-THĪ-royd)
4. Sternomastoid
5. Cleidomastoid (klī-dō-MAS-toyd)
6. Digastric
7. Mylohyoid
8. Mandible bone

Figure 18-2 (b) The neck and jaw muscles of the fetal pig (line drawing).

Table 18-1   Face, Neck, and Jaw Muscles of the Fetal Pig

Muscle	Origin	Insertion	Action
Masseter	Zygomatic arch	Lateral surface of mandible	Raises mandible
Sternomastoid	Anterior end of sternum	Mastoid process	Flexes head and neck
Cleidomastoid	Region of clavicle; fascia of clavobrachialis	Mastoid process	Pulls head laterally and craniad
Digastric	Occipital bone and mastoid process	Ventral border of mandible	Depresses mandible
Mylohyoid	Cricoid cartilage of larynx	Hyoid bone	Pulls hyoid forward; raises floor of mouth
Sternohyoid	Anterior sternum and first costal cartilage	Hyoid bone	Moves hyoid posteriorly
Sternothyroid	Anterior sternum	Thyroid cartilage	Moves larynx posteriorly
Cricothyroid	Cricoid cartilage of larynx	Thyroid cartilage of larynx	Tenses larynx

COMPARABLE HUMAN MUSCLES

The largest muscle of the neck in the human is the *sternocleidomastoid*. In the fetal pig, this muscle is divided into two distinct muscles, the sternomastoid and the cleidomastoid. In the human, its position between the clavicle, sternum, and mastoid process of the skull gives it complex capability of movement. If only the right or the left sternocleidomastoid contracts, it turns the head to the opposite side. If both right and left sternocleidomastoids contract, the head is flexed on the chest.

Many of the deep muscles of the human neck are considered to be in the chest and the deep back on the pig. These are the *scalenus* muscles, which turn the vertebral column to one side when they contract, and the *semispinalis, longissimus,* and *splenius* muscles, which extend the head and neck. They are found in the *posterior triangle* of the human neck.

The muscles associated with the hyoid bone are grouped into the *suprahyoid* muscles and the *infrahyoid* muscles, depending on whether they are found above or below the hyoid bone, respectively. The suprahyoid muscles lie in the floor of the mouth, and the infrahyoid muscles are found lower in the neck, around the larynx and trachea. The following drawings (Figures 18-3 through 18-6) and tables (Tables 18-2 through 18-5) include the human facial muscles of the lower jaw and of the tongue and larynx, and the muscles that move the head.

The neck is a complicated area. The muscles in Figure 18-7 are neck and jaw muscles, shown in side view. The large *sternocleidomastoid* forms the *anterior triangle* of the neck. Using the hyoid bone as a point of reference, the muscles of the anterior triangle are divided into two groups: the *suprahyoid* muscles and the *infrahyoid* muscles. The muscles of the *posterior triangle* are those located posterior to the sternocleidomastoid. Muscles included in these groups are listed in the caption for Figure 18-7. Color the muscles, and compare their functions and locations with their counterparts in the pig. An anterior view of the muscles of the neck is shown in Figures 18-8 and 18-9. Color the muscles. Note their relationship to the larynx and the trachea.

Figure 18-3 Muscles that move the lower jaw: (a) superficial right lateral view; (b) deep right lateral view. (From G. J. Tortora and S. R. Grabowski, *Principles of Anatomy and Physiology*, 7th ed., New York: HarperCollins, 1993. © Leonard Dank.)

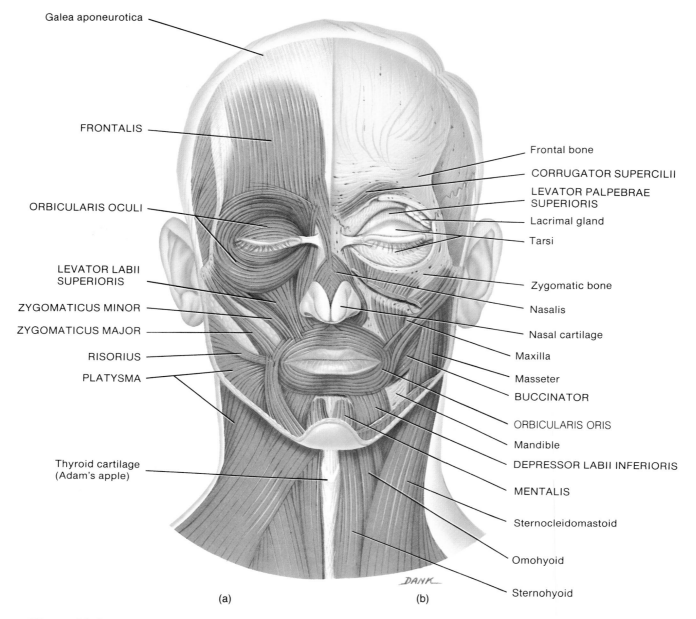

Galea aponeurotica

FRONTALIS

ORBICULARIS OCULI

LEVATOR LABII
SUPERIORIS

ZYGOMATICUS MINOR

ZYGOMATICUS MAJOR

RISORIUS

PLATYSMA

Thyroid cartilage
(Adam's apple)

Frontal bone

CORRUGATOR SUPERCILII

LEVATOR PALPEBRAE
SUPERIORIS

Lacrimal gland

Tarsi

Zygomatic bone

Nasalis

Nasal cartilage

Maxilla

Masseter

BUCCINATOR

ORBICULARIS ORIS

Mandible

DEPRESSOR LABII INFERIORIS

MENTALIS

Sternocleidomastoid

Omohyoid

Sternohyoid

(a)   (b)

Figure 18-4 Muscles of facial expression: (a) anterior superficial view; (b) anterior deep view. (From G. J. Tortora and S. R. Grabowski, *Principles of Anatomy and Physiology,* 7th ed., New York: HarperCollins, 1993. © Leonard Dank.)

Figure 18-5 Muscles of the scalp, face, and neck (lateral view). (From G. J. Tortora and S. R. Grabowski, *Principles of Anatomy and Physiology,* 7th ed., New York: HarperCollins, 1993. © Leonard Dank.)

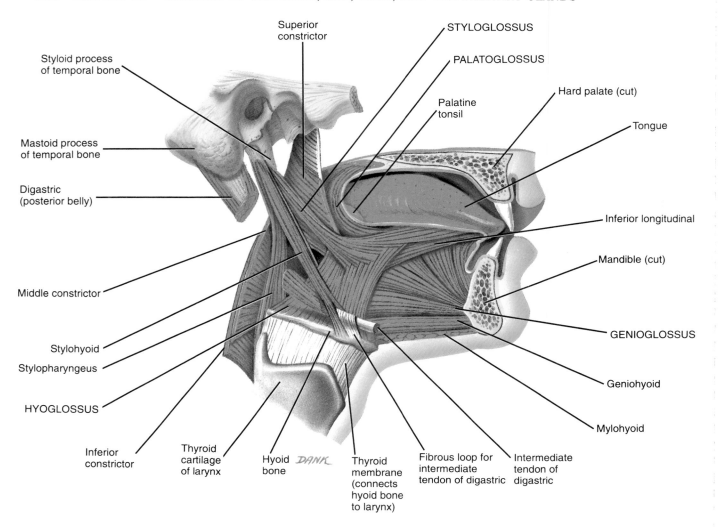

Figure 18-6 Muscles that move the tongue (viewed from the right side). (From G. J. Tortora and S. R. Grabowski, *Principles of Anatomy and Physiology,* 7th ed., New York: HarperCollins, 1993. © Leonard Dank.)

Table 18-2   Muscles That Move the Lower Jaw

Muscle	Origin	Insertion	Action	Innervation
Masseter (*maseter* = chewer)	Maxilla and zygomatic arch	Angle and ramus of mandible	Elevates mandible, as in closing the mouth, and protracts (protrudes) mandible	Mandibular branch of trigeminal nerve (V)
Temporalis (*tempora* = temples)	Temporal bone	Coronoid process of mandible	Elevates and retracts mandible	Temporal nerve from mandibular division of trigeminal nerve (V)
Medial pterygoid (*medial* = closer to midline; *pterygoid* = like a wing)	Medial surface of lateral pterygoid plate of sphenoid; maxilla	Angle and ramus of mandible	Elevates and protracts mandible and moves mandible from side to side	Mandibular branch of trigeminal nerve (V)
Lateral pterygoid (*lateral* = farther from midline)	Greater wing and lateral surface of lateral pterygoid plate of sphenoid	Condyle of mandible; temporomandibular articulation	Protracts mandible, opens mouth, and moves mandible from side to side	Mandibular branch of trigeminal nerve (V)

Table 18-3   Muscles of Facial Expression

Muscle	Origin	Insertion	Action	Innervation
Epicranius (*epi* = over; *crani* = skull)	This muscle is divisible into two portions: the frontalis over the frontal bone and the occipitalis over the occipital bone. The two muscles are united by a strong aponeurosis, the galea aponeurotica, which covers the superior and lateral surfaces of the skull.			
Frontalis (*front* = forehead)	Galea aponeurotica	Skin superior to supraorbital line	Draws scalp forward, raises eyebrows, and wrinkles forehead horizontally	Facial nerve (VII)
Occipitalis (*occipito* = base of skull)	Occipital bone and mastoid process of temporal bone	Galea aponeurotica	Draws scalp backward	Facial nerve (VII)
Orbicularis oris (*orb* = circular; *or* = mouth)	Muscle fibers surrounding opening of mouth	Skin at corner of mouth	Closes lips, compresses lips against teeth, protrudes lips, and shapes lips during speech	Facial nerve (VII)
Zygomaticus major (*zygomatic* = cheek bone; *major* = greater)	Zygomatic bone	Skin at angle of mouth and orbicularis oris	Draws angle of mouth upward and outward, as in smiling or laughing	Facial nerve (VII)

Table 18-3 *(Continued)*

Muscle	Origin	Insertion	Action	Innervation
Levator labii superioris (*levator* = elevates; *labii* = lip; *superioris* = upper)	Superior to infraorbital foramen of maxilla	Skin at angle of mouth and orbicularis oris	Elevates upper lip	Facial nerve (VII)
Depressor labii inferioris (*depressor* = depresses; *inferioris* = lower)	Mandible	Skin of lower lip	Depresses lower lip	Facial nerve (VII)
Buccinator (*bucc* = cheek)	Alveolar processes of maxilla and mandible and pterygomandibular raphe (fibrous band extending from the pterygoid hamulus to the mandible)	Orbicularis oris	Major cheek muscle; compresses cheek, as in blowing air out of mouth, and causes the cheeks to cave in, producing the action of sucking	Facial nerve (VII)
Mentalis (*mentum* = chin)	Mandible	Skin of chin	Elevates and protrudes lower lip and pulls skin of chin up, as in pouting	Facial nerve (VII)
Platysma (*platy* = flat, broad)	Fascia over deltoid and pectoralis major muscles	Mandible, muscles around angle of mouth, and skin of lower face	Draws outer part of lower lip downward and backward, as in pouting; depresses mandible	Facial nerve (VII)
Risorius (*risor* = laughter)	Fascia over parotid (salivary) gland	Skin at angle of mouth	Draws angle of mouth laterally, as in tenseness	Facial nerve (VII)
Orbicularis oculi (*ocul* = eye)	Medial wall of orbit	Circular path around orbit	Closes eye	Facial nerve (VII)
Corrugator supercilii (*corrugo* = to wrinkle; *supercilium* = eyebrow)	Medial end of superciliary arch of frontal bone	Skin of the eyebrow	Draws eyebrow downward as in frowning	Facial nerve (VII)
Levator palpebrae superioris (*palpebrae* = eyelids)	Roof of orbit (lesser wing of sphenoid bone)	Skin of the upper eyelid	Elevates the upper eyelid	Oculomotor nerve (III)

Table 18-4   Muscles of the Larynx

Muscle	Origin	Insertion	Action	Innervation
EXTRINSIC Omohyoid (*omo* = relationship to the shoulder; *hyoedies* = U-shaped, pertaining to the hyoid bone)	Superior border of scapula and superior transverse ligament	Body of hyoid bone	Depresses hyoid bone	Branches of ansa cervicalis nerve (C1–C3)
Sternohyoid (*sterno* = sternum)	Medial end of clavicle and manubrium of sternum	Body of hyoid bone	Depresses hyoid bone	Branches of ansa cervicalis nerve (C1–C3)
Sternothyroid (*thyro* = thyroid gland)	Manubrium of sternum	Thyroid cartilage of larynx	Depresses thyroid cartilage	Branches of ansa cervicalis nerve (C1–C3)
Thyrohyoid	Thyroid cartilage of larynx	Greater cornu of hyoid bone	Elevates thyroid cartilage and depresses hyoid bone	Cervical nerves C1–C2 and descending hypoglossal nerve (XIII)
Stylopharyngeus Palatopharyngeus Inferior constrictor Middle constrictor				
INTRINSIC Cricothyroid (*crico* = cricoid cartilage of larynx)	Anterior and lateral portion of cricoid cartilage of larynx	Anterior border of inferior cornu of thyroid cartilage of larynx and posterior part of inferior border of lamina of thyroid cartilage	Produces tension and elongation of vocal folds	External laryngeal branch of vagus nerve (X)

Table 18-5   Muscles That Move the Tongue

Muscle	Origin	Insertion	Action	Innervation
Genioglossus (*geneion* = chin; *glossus* = tongue)	Mandible	Undersurface of tongue and hyoid bone	Depresses and thrusts tongue forward (protraction)	Hypoglossal nerve (XII)
Styloglossus (*stylo* = stake or pole)	Styloid process of temporal bone	Side and undersurface of tongue	Elevates tongue and draws it backward (retraction)	Hypoglossal nerve (XII)
Stylohyoid (*hyoeides* = U-shaped; pertaining to hyoid bone)	Styloid process of temporal bone	Body of hyoid bone	Elevates and retracts tongue	Facial nerve (VII)
Hyoglossus	Body of hyoid bone	Side of tongue	Depresses tongue and draws down its sides	Hypoglossal nerve (XII)

Anterior triangle muscle
 1. Sternocleidomastoid

Suprahyoid muscles
 2. Stylohyoid
 3. Digastric
 4. Mylohyoid
 5. Hyoglossus

Infrahyoid muscles
 6. Sternohyoid
 7. Omohyoid
 8. Thyrohyoid
 9. Sternothyroid

Posterior triangle muscles
10. Splenius capitis
11. Levator scapulae
12. Scalenus anterior
13. Scalenus medius
14. Scalenus posterior

Figure 18-7 Muscles of the neck and jaw (left lateral view).

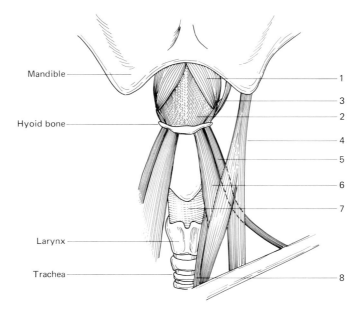

1. Digastric
2. Mylohyoid
3. Stylohyoid
4. Sternocleidomastoid
5. Omohyoid
6. Sternohyoid
7. Cricothyroid
8. Sternothyroid

Figure 18-8 Muscles of the neck (anterior view).

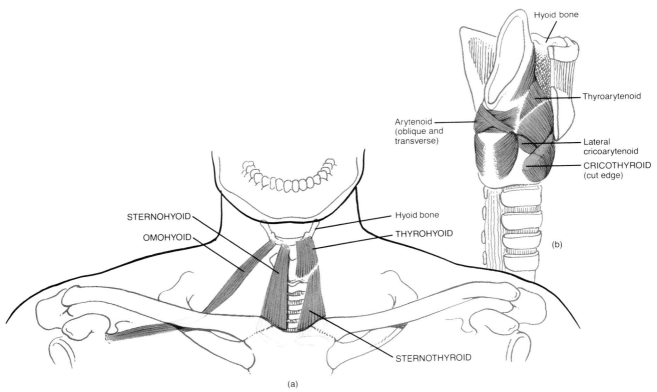

Figure 18-9 Muscles of the larynx: (a) anterior view; (b) right lateral view. (From G. J. Tortora, *Principles of Human Anatomy,* 3rd, 4th, 5th eds., New York: HarperCollins, 1983, 1986, 1989.)

HUMAN NECK, JAW, AND FACE MUSCLES

The muscles discussed here are some of the more important facial muscles of the human. These were not dissected in the fetal pig, but because they are so important as a means of silent communication in the human, they are studied here.

1. *Epicranius or occipitofrontalis* (Figure 18-10.)
   The epicranius is divisible into two parts: (1) the *frontalis,* which covers the frontal bone, and (2) the *occipitalis,* which covers the superior and lateral surfaces of the skull. The two muscles are united by an aponeurosis (galea aponeurotica) that covers the superior and lateral surfaces of the skull. They raise the eyebrows and wrinkle the skin of the forehead horizontally.
   a. Look in a mirror. Wrinkle your forehead. What do you observe? How many horizontal lines appear? Are there differences among your classmates?

Observations: _____

_____

_____

_____

b. Comparisons: _____

_____

_____

c. When you raise your eyebrows or wrinkle your forehead, can you feel movement over the hairy portion of the skull? Does your hair move?

_____

2. *Orbicularis oculi* (Figure 18-10)
   This circular muscle surrounds the opening of the eye. It closes the eye.
   Close your eye, to feel the contraction of this muscle. Have your lab partner do the same, so that you can observe the varying degrees of contracture and closure of the eye. What are your observations?

_____

_____

1. Depressor supercilli
2. Orbicularis oculi
3. Levator labii superioris alaeque nasi
4. Levator labii superioris
5. Zygomaticus minor
6. Levator anguli oris
7. Zygomaticus major
8. Risorius
9. Platysma
10. Depressor anguli oris
11. Depressor labii inferioris
12. Frontal belly of occipito-frontalis
13. Corrugator supercilli
14. Procerus
15. Transverse part of nasalis (compressor naris)
16. Orbicularis oris
17. Buccinator
18. Orbicularis oris (incisivus labii inferioris)
19. Mentalis

Figure 18-10 Muscles of facial expression. (From Yokochi/Rohen. *Photographic Anatomy of the Human Body,* 2nd ed. Baltimore: University Park Press, and Tokyo: Igaku-Shoin Ltd., 1978.)

3. *Zygomaticus major* (Figure 18-10)
   This muscle draws the angle of the mouth upward and outward, as in smiling or laughing.

   Smile broadly, using muscles on both sides of your face. Make as many facial expressions as you can involving this muscle, and identify these expressions. Use a mirror or your lab partner. Describe these expressions. _____

   _____

   _____

4. *Platysma* (Figure 18-10)
   The platysma draws the outer part of the lower lip downward and backward, as in pouting or crying. It also depresses (lowers) the mandible. This muscle was probably removed in the fetal pig dissection because it sticks so closely to the skin.
   a. Demonstrate the platysma by pouting for your lab partner. With your fingers at the corners of your mouth, feel the contraction.
   b. Make other facial expressions similar to pouting, using this muscle. Describe these expressions.

   _____

   _____

5. *Orbicularis oris* (Figure 18-10)
   The orbicularis oris produces many facial expressions. Perform each of the four actions listed next, and describe in your own words the resulting facial expressions. Perhaps you can associate these expressions with a particular mood.

   a. Close the lips. _____

   _____

   b. Compress the lips against the teeth. _____

   _____

   c. Protrude the lips. _____

   _____

   d. Prepare your mouth to say "oh." _____

   _____

   e. Prepare your mouth to kiss. _____

   _____

6. *Buccinator* (Figure 18-10)
   The buccinator compresses the cheek, as in blowing air out of the mouth. It causes the cheeks to cave in as in sucking, and it is used a great deal of the time to manipulate the lips.

   Make both blowing and sucking movements with your mouth, and describe the configuration of your cheeks. _____

   _____

   _____

Two of the three muscles that move the lower jaw and head (masseter and temporalis) are discussed next. They are used in eating and chewing.

1. *Masseter* (Figure 18-11)
   The masseter raises the mandible as the mouth is closed. It causes the protrusion (protraction) of the mandible. It is used in speech, as the mandible is moved up and down, forward and back, to achieve not only a particular sound, but also particular facial posturing that often accompanies these sounds.
   a. Demonstrate the muscle by creating a phrase or sentence in which the masseter is involved with both elevation and protrusion of the mandible. Try slowly reading this aloud, and describe

      the movement of the mandible. _____

      _____

      _____

   b. The masseter is also used while eating. Make chewing movements with your mouth closed. Place your fingers at the angles of your jaw to feel the contraction.

Figure 18-11 Masseter muscle. (From P. J. Donnelly and G. A. Wistreich, *Exploring the Human Body*, New York: Harper & Row, 1989.)

2. *Temporalis* (Figure 18-12)
   The temporalis elevates and retracts the mandible. Pull your lower jaw back to demonstrate it. Keep your mouth closed, and let your teeth be your guide as to how much movement can be achieved. Place your fingers on your temples. Describe what you

   feel. _____

   _____

Figure 18-12 Temporal muscle. (From P. J. Donnelly and G. A. Wistreich, *Exploring the Human Body,* New York: Harper & Row, 1989.)

3. *Sternocleidomastoid* (Figure 18-13)
   The large muscle in the human neck is the *sternocleidomastoid.* It is complicated and has two origins, one on the manubrium and the other on the clavicle. By contracting both muscles of the pair together, you can flex the cervical vertebrae, draw the head forward, and elevate the chin. By contracting one muscle of the pair only, you can rotate your face toward the side opposite the contracting muscle.

   a. To demonstrate both muscles of the pair, stand straight, eyes looking forward. Bend your head forward, and feel the muscle contraction below your ears. If your muscles are strong, another test for this muscle is to be able to raise your head while lying in a supine position. Do not assist this action by raising your shoulders.

   b. To demonstrate only one muscle at a time, contract only the left sternocleidomastoid. Turn your head to the right, and look up. Now contract only the right muscle. Turn your head to the left, and look up. Do this exercise while looking in a mirror, or make observations on your lab

   partner. _____

   _____

   _____

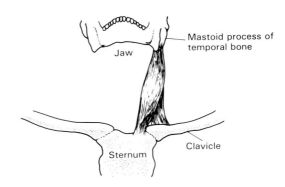

Figure 18-13 Sternocleidomastoid (anterior view). (From P. J. Donnelly and G. A. Wistreich, *Exploring the Human Body,* New York: Harper & Row, 1989.)

LABORATORY REVIEW *18*

MUSCLES OF THE NECK, JAW, FACE, AND THE SALIVARY GLANDS

COMPLETION

1. Which human muscles perform the following functions?

   a. _____ Compress lips against teeth

   b. _____ Depress lower lip

   c. _____ Close the eye

   d. _____ Raise the eyebrows

   e. _____ Elevate upper lip

   f. _____ Elevate mandible to close the mouth

   g. _____ Move mandible from side to side

   h. _____ Draw tongue backward

   i. _____ Draw soft palate down on the tongue

   j. _____ Depress thyroid cartilage of larynx

   k. _____ Elongate the vocal folds

   l. _____ Thrust tongue forward

2. Name the human salivary glands.

   a. _____

   b. _____

   c. _____

3. What are the functions of saliva?

   _____

   _____

   _____

Muscles of the Arm

After completing this exercise, you should be able to

1. Dissect the muscles of the arm and deep chest of the fetal pig
2. Identify comparable human muscles

Materials

The following materials should be provided for class use:

1. Preserved fetal pigs
2. Dissecting tools (scissors, scalpel, probe, dissecting needles, and forceps)

The muscles dissected in this exercise are listed here:

UPPER ARM, LATERAL VIEW

Triceps brachii long head

Triceps brachii lateral head

Triceps brachii medial head

Brachialis

LOWER ARM, LATERAL VIEW

Brachioradialis

Extensor carpi radialis

Extensor digitorum communis

Extensor digitorum lateralis

Extensor carpi ulnaris

LOWER ARM, MEDIAL VIEW

Brachioradialis

Flexor carpi radialis

Flexor digitorum profundis

Flexor digitorum superficialis

Flexor carpi ulnaris

Figure 19-1 (a) Superficial upper- and lower-arm muscles (lateral surface) of fetal pig (photograph).

SUPERFICIAL UPPER-ARM MUSCLES

Instructions for Dissection

Most of the lateral surface is covered by the *triceps brachii* muscle (see Figure 19-1). As the name implies, this muscle has three heads of origin, although it only has one point of insertion on the ulna bone. The large mass over the posterior surface is the *triceps long head;* lying next to it on the lateral surface is the *triceps lateral*

head. Free both of these muscles, then transect the triceps lateral head to see the *triceps medial head* lying along the humerus bone. Next to the triceps medial head is the *brachialis,* which continues into the lower arm as the *brachioradialis.*

Turn to the lateral surface of the lower arm. Begin with the *brachioradialis.* Lying next to it are the *extensor carpi radialis, extensor carpi radialis, extensor digitorum commuvis, extensor digitorum lateralis,* and *extensor carpi ulnaris,* in that order. See Table 19-1 for muscle origins, insertions, and actions.

1. Triceps brachii—lateral head
2. Triceps brachii—long head
3. Triceps brachii—medial head
4. Brachialis
5. Brachioradialis
6. Extensor carpi radialis
7. Extensor digitorum communis
8. Extensor digitorum lateralis (longus)
9. Extensor carpi ulnaris

Figure 19-1 (b) Superficial upper- and lower-arm muscles (lateral surface) of fetal pig (line drawing).

Figure 19-2 (a) Superficial lower-arm muscles (medial surface) of fetal pig (photograph).

LOWER-ARM MUSCLES

Instructions for Dissection

On the medial surface of the lower arm (see Figure 19-2), locate the *brachioradialis* on the lateral edge. Next to it are the *flexor carpi radialis, flexor digitorum profundis, flexor digitorum superficialis* and the *flexor carpi ulnaris,* in that order.

Review related muscles that were dissected in Exercise 17. On the medial surface of the upper arm, locate again the *biceps brachii,* a flexor of the lower arm. At the proximal end of the humerus, locate the *coracobrachialis,* which connects the arm to the scapula. The serratus ventralis and the subscapularis can also be seen in this dissection.

COMPARABLE HUMAN MUSCLES

Compare the fetal pig and human muscles of the upper and lower arms, using the photos and diagrams (Figures 19-1 and 19-2 and 19-3 through 19-4) for both animals, as well as Tables 19-1, 19-2, and 19-3. You will find that there are very few differences between them anatomically.

1. Brachio radialis
2. Flexor carpi radialis
3. Flexor digitorum profundis
4. Flexor digitorum superficialis
5. Flexor carpi ulnaris

Figure 19-2 (b) Superficial lower-arm muscles medial surface) of fetal pig (line drawing).

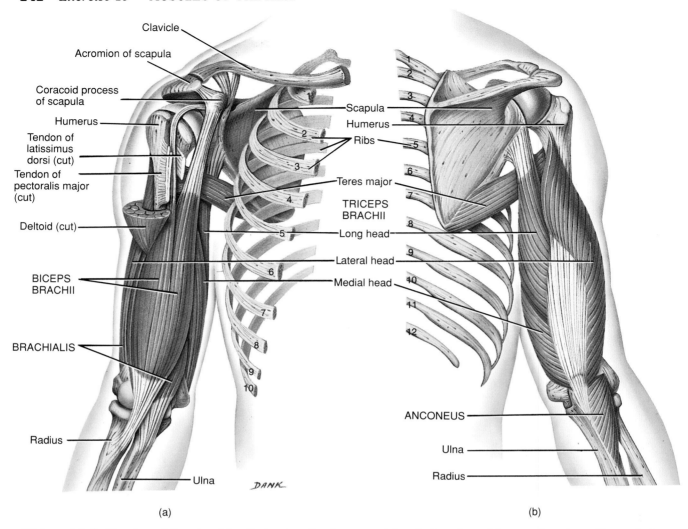

Figure 19-3 Muscles of the shoulder that move the forearm: (a) anterior view; (b) posterior view. (From G. J. Tortora and S. R. Grabowski, *Principles of Anatomy and Physiology,* 7th ed., New York: HarperCollins, 1993 © Leonard Dank.)

Biceps brachii
Brachialis
Brachial artery
Median nerve
Medial epicondyle of humerus
Tendon of biceps brachii
PRONATOR TERES

PL
FCR
PT
FDS
FCU

BRACHIORADIALIS
SUPINATOR
PALMARIS LONGUS
FLEXOR CARPI RADIALIS
FLEXOR CARPI ULNARIS
FLEXOR DIGITORUM PROFUNDUS
PRONATOR TERES (cut)
FLEXOR DIGITORUM SUPERFICIALIS
FLEXOR POLLICIS LONGUS
ABDUCTOR POLLICIS LONGUS
PRONATOR QUADRATUS

Ulna

Flexor retinaculum
Metacarpals
Tendon of flexor
digitorum superficialis
Tendon of flexor
digitorum profundus

*Key to abbreviations for
cut muscles in (b)

PL = PALMARIS LONGUS
PT = PRONATOR TERES
FCR = FLEXOR CARPI RADIALIS
FDS = FLEXOR DIGITORUM
       SUPERFICIALIS
FCU = FLEXOR CARPI ULNARIS

(a)

(b)

DANK

Figure 19-4 Muscles that move the wrist and fingers: (a) anterior superfial view; (b) anterior deep view. (From G. J. Tortora and S. R. Grabowski, *Principles of Anatomy and Physiology,* 7th ed., New York: HarperCollins, 1993. © Leonard Dank.)

Table 19-1 Muscles of the Arm and Deep Pectoral Region

Muscle	Origin	Insertion	Action
Triceps brachii—long head	Posterior border of scapula	Olecranon process of ulna	Extends lower arm
Triceps brachii—lateral head	Proximal end of humerus	Olecranon process of ulna	Extends lower arm
Triceps brachii—medial head	Proximal one-third of humerus	Olecranon process of ulna	Extends lower arm
Brachialis	Proximal one-third of humerus	Proximal ends of radius and ulna	Flexes lower arm
Biceps brachii	Ventral part of scapula above glenoid fossa	Proximal ends of radius and ulna	Flexes lower arm
Coracobrachialis	Ventral scapula above glenoid fossa	Middle one-third of humerus	Adducts arm and flexes shoulder
Serratus ventralis	Cervical vertebrae and lateral surfaces of fourth to eighth ribs	Dorsal medial border of scapula	Raises thorax
Splenius	Thoracic vertebral spines 3, 4, and 5	Occipital bone and temporal bone	Raises head and neck
Brachioradialis	Humerus	Lower part of radius	Rotates the forearm
Extensor carpi radialis	Distal end of humerus	Third metacarpal	Extend wrist and flexes forearm
Extensor digitorum communis	Distal humerus	On the digits	Extends digits
Extensor digitorum lateralis	Distal end of humerus	On the digits	Extends digits
Extensor carpi ulnaris	Lateral epicondyle of humerus	Fifth metacarpal	Extends wrist
Flexor carpi radialis	Distal end of humerus	Third metacarpal	Flexes wrist and extends forearm
Flexor digitorum profundis	Distal tibia and fibula	All digits	Flexes digits
Flexor digitorum superficialis	Distal tibia	All digits	Flexes digits
Flexor carpi ulnaris	Distal end of humerus	Lateral surface of wrist	Flexes wrist; extends lower arm

Table 19-2   Arm Muscles of the Human

Muscles	Origin	Insertion	Action
Triceps brachii			
Long head	Infraglenoid tuberosity of scapula	Olecranon process of ulna	Extends forearm
Lateral head	Above radial groove of humerus	Olecranon process of ulna	Extends forearm
Medial head	Below radial groove of humerus	Olecranon process of ulna	Extends forearm
Brachialis	Anterior surface of humerus	Tuberosity and coronoid process of ulna	Flexes forearm
Biceps brachii			
Long head	Near glenoid fossa of scapula	Radial tuberosity	Flexes and supinates forearm
Short head	Coracoid process of scapula	Radial tuberosity	Flexes and supinates forearm
Brachioradialis	Supracondyloid ridge of humerus	Above styloid process of radius	Flexes forearm
Flexor carpi radialis	Medial epicondyle of humerus	Second and third metacarpals	Flexes and abducts wrist
Flexor carpi ulnaris	Medial epicondyle of humerus and upper dorsal border of ulna	Pisiform, hamate, and fifth metacarpal	Flexes and adducts wrist
Extensor carpi radialis longus	Lateral epicondyle of humerus	Second metacarpal	Extends and abducts wrist
Extensor carpi ulnaris	Lateral epicondyle of humerus and dorsal border of ulna	Fifth metacarpal	Extends and adducts wrist
Flexor digitorum profundus	Body of ulna	Bases of distal phalanges	Flexes all distal phalanges
Flexor digitorum superficialis	Medial epicondyle of humerus, coronoid process of ulna, and oblique line of radius	Middle phalanges	Flexes middle phalanges of each finger
Extensor digitorum	Lateral epicondyle of humerus	Middle and distal phalanges	Extends phalanges
Extensor indicis	Dorsal surface of ulna	Tendon of extensor digitorum of index finger	Extends index finger
Supinator	Lateral epicondyle of humerus and ridge on ulna	Oblique line of radius	Supinates forearm
Pronator teres	Medial epicondyle of humerus and coronoid process of ulna	Lateral surface of radius	Pronates forearm
Palmaris longus	Medial epicondyle of humerus	Radial tuberosity	Flexes and supinates forearm
Coracobrachialis	Coracoid process of scapula	Middle of medial border of humerus	Flexes and adducts arm

Some arm muscles easily experimented on are listed:

1. *Triceps brachii* (Figure 19-5)
The triceps has three heads of origin in the human, just as it does in the fetal pig. It extends the forearm.

    With the elbow flexed, hand touching the shoulder, slowly straighten (extend) the elbow joint. Feel the triceps contract. (a) Name two sports activities during which the triceps brachii is used extensively, and (b) describe the specific phase of the activity during which it is contracted.

a. _____

_____

b. _____

_____

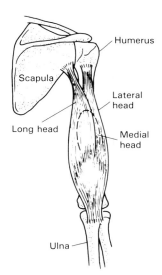

Figure 19-5 Human triceps brachii (posterior view). (From P. J. Donnelly and G. A. Wistreich, *Exploring the Human Body,* New York: Harper & Row, 1989.)

2. *Biceps brachii* (Figure 19-6)
The biceps has two distinct heads of origin on the scapula, by tendons that pass over the proximal head of the humerus. It flexes the forearm and supinates the forearm.

    Demonstrate the action of this muscle by flexing the elbow joint. Name some specific activities that you perform regularly, which require the use

of the biceps. _____

_____

_____

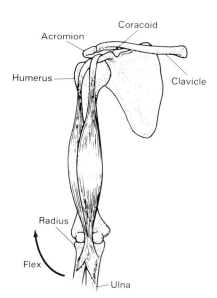

Figure 19-6 Biceps brachii (anterior view). (From P. J. Donnelly and G. A. Wistreich, *Exploring the Human Body,* New York: Harper & Row, 1989.)

3. *Supinator* (Figure 19-7)

The supinator turns the forearm and hand so that the palm faces upward.

To demonstrate its action, move your arm (with the hand relaxed) away from your body in front of you. Slowly turn the palm of your hand upward (supination). Place your fingers on the forearm below the elbow, to feel the action. Name two specific

activities in which this muscle is used. _____

_____

_____

Figure 19-7 Supinator (posterior view). (From P. J. Donnelly and G. A. Wistreich, *Exploring the Human Body,* New York: Harper & Row, 1989.)

4. *Pronator teres* (Figure 19-8)

This muscle pronates (turns downward) the forearm and hand. It has two heads of origin, one on the humerus and the other on the ulna. Its fibers pass obliquely over the forearm.

With your arm outstretched, in the palm-up position, turn your palm down. Make the same motion with your arm at your side. What specific action does a baseball player make when contract-

ing this muscle? _____

_____

Figure 19-8 Pronator teres (anterior view). (From P. J. Donnelly and G. A. Wistreich, *Exploring the Human Body,* New York: Harper & Row, 1989.)

Some of the muscles that move the wrist and fingers, which perform some of the most intricate movements of which the human is capable, are as follows:

1. *Flexor carpi radialis* (Figure 19-9)
   This flexor muscle bends and abducts the wrist and hand.

   With your arm in front of you, palm in the pronated position, flex your hand as far as possible. Then turn your wrist laterally (abduction). Name an activity in which you would move your hand in

   this way. _____

   _____

Figure 19-9 Flexor carpi radialis (posterior view). (From P. J. Donnelly and G. A. Wistreich, *Exploring the Human Body,* New York: Harper & Row, 1989.)

2. *Flexor carpi ulnaris* (Figure 19-10)
   This muscle also has two heads of origin, one on the humerus, the other on the ulna.

   To demonstrate this muscle, hold your arm in front of you, palm down. Flex your hand, and turn your hand inward (adduction). Name an activity in which the hand and wrist would be moved in this

   way. _____

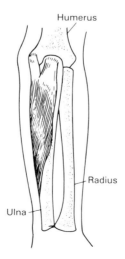

Figure 19-10 Flexor carpi ulnaris (posterior view). (From P. J. Donnelly and G. A. Wistreich, *Exploring the Human Body,* New York: Harper & Row, 1989.)

3. *Palmaris longus* (Figure 19-11)
   This is a slender muscle with a long tendon of insertion. It frequently sends a small tendon to the thumb muscles (not shown in Figure 19-11).

   To demonstrate it, flex the wrist with the fingers curled as if you were about to strike piano keys. Feel the palmer aponeurosis tense.

   Observations: _____

   _____

   _____

   _____

Figure 19-11 Palmaris longus (anterior view). (From P. J. Donnelly and G. A. Wistreich, *Exploring the Human Body*, New York: Harper & Row, 1989.)

4. *Flexor digitorum superficialis* (Figure 19-12)
   This flexor muscle bends the phalanges and the hand.

   Demonstrate this muscle's action by flexing fingers two through five while keeping the interphalangeal joints extended. What purposeful activity could be accomplished with this muscle? _____

   _____

Figure 19-12 Flexor digitorum superficialis (anterior view). (From P. J. Donnelly and G. A. Wistreich, *Exploring the Human Body*, New York: Harper & Row, 1989.)

5. *Flexor digitorum profundis* (Figure 19-13)
   The tendon of this muscle runs deep to the tendon of the flexor digitorum superficialis. It flexes the distal phalanges of each finger, and by continued action, it flexes all phalanges (two through five) and the hand.

   To demonstrate it, make the hand into a fist.

   Observations: _____

   _____

   _____

   _____

Figure 19-13 Flexor digitorum profundis (anterior view). (From P. J. Donnelly and G. A. Wistreich, *Exploring the Human Body,* New York: Harper & Row, 1989.)

6. *Extensor carpi radialis longus* (Figure 19-14)
   This muscle extends and abducts the wrist and hand.

   To demonstrate, hold your hand out, palm down. Bend the hand backward (hyperextension). Relax the hand and turn it away from the body (abduction). In what specific activities does this muscle play a part? _____

   _____

Figure 19-14 Extensor carpi radialis longus (posterior view). (From P. J. Donnelly and G. A. Wistreich, *Exploring the Human Body,* New York: Harper & Row, 1989.)

7. *Extensor carpi ulnaris* (Figure 19-15)
This extensor straightens and adducts the wrist and hand.

   To demonstrate, hold your hand in front of you. Bend the hand backward (hyperextension). Relax again. Then move the hand and wrist inward (toward the midline). Describe an activity during

which this muscle is used. _____

_____

Figure 19-15 Extensor carpi ulnaris (posterior view). (From P. J. Donnelly and G. A. Wistreich, *Exploring the Human Body,* New York: Harper & Row, 1989.)

LABORATORY REVIEW *19*

MUSCLES OF THE ARM

COMPLETION

1. Complete Table Q19-1. Note how differences in location affect the action of the muscles. If the actions are similar, note that, too.

Table Q19-1

Muscle	Location	Action
Coracobrachialis (pig)		
Coracobrachialis (human)		
Serratus ventralis (pig)		
Serratus anterior (human)		
Intercostal muscles (pig)		
Intercostal muscles (human)		

2. Study Figure Q19-1, showing the deep chest and neck muscles of the human. Compare them, as to size and location, with the muscles just dissected in the fetal pig. Color this diagram if you wish.

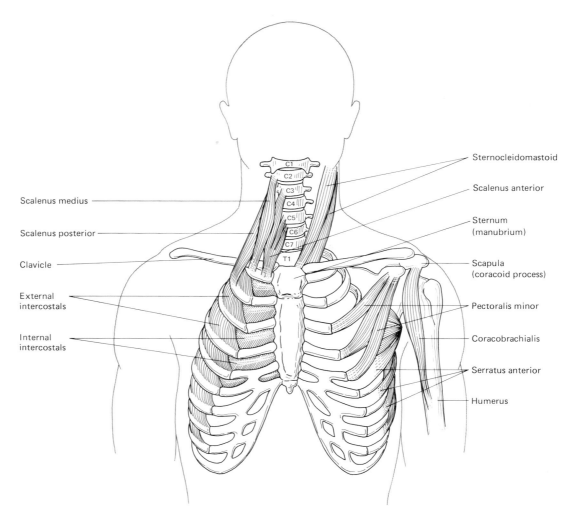

Figure Q19-1 (From G. J. Tortora, *Principles of Human Anatomy,* 3rd, 4th, 5th eds., New York: HarperCollins, 1983, 1986, 1989.)

3. Fill in the blanks.

a. _____   Name a muscle that flexes the forearm.

b. _____   Name a muscle that extends and adducts the wrist.

c. _____   Name a muscle that flexes and abducts the wrist.

d. _____   Name a muscle that extends the forearm.

e. _____   Name a muscle that extends a phalanx.

f. _____   Name another muscle that extends a phalanx.

g. _____   Name a muscle that pronates the forearm.

h. _____   Name a muscle that supinates the forearm.

i. _____   What is the position of the hand in pronation?

j. _____   What is the position of the hand in supination?

k. _____   Name a muscle that flexes the forearm.

l. _____   Name another muscle that flexes the forearm.

4. Study Figure Q19-2, showing the human arm muscles. Relate each muscle to one dissected in the fetal pig. Color the drawing for clarity.

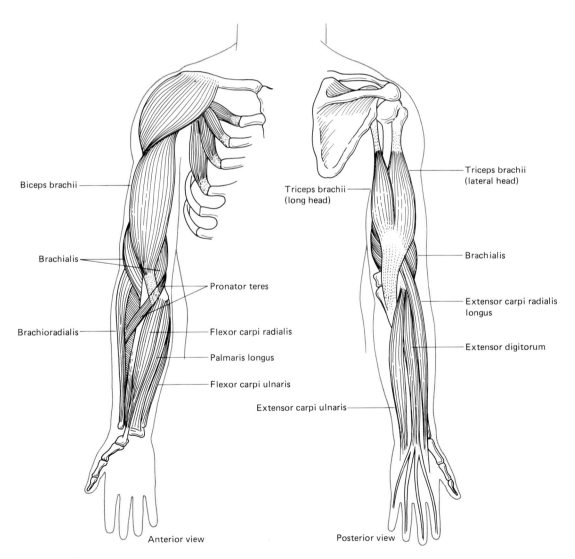

Biceps brachii

Brachialis

Brachioradialis

Pronator teres

Flexor carpi radialis

Palmaris longus

Flexor carpi ulnaris

Anterior view

Triceps brachii (long head)

Triceps brachii (lateral head)

Brachialis

Extensor carpi radialis longus

Extensor digitorum

Extensor carpi ulnaris

Posterior view

Figure Q19-2

Muscles of the Hip, Thigh, and Lower Leg

After completing this exercise, you should be able to

1. Dissect the muscles of the hip, thigh, and lower leg of the fetal pig
2. Identify comparable human muscles

Materials

The following materials should be provided for class use:

1. Preserved fetal pigs
2. Dissecting tools (scissors, scalpel, probe, dissecting needles, and forceps)

The muscles dissected in this exercise are listed here:

SUPERFICIAL THIGH MUSCLES, MEDIAL SURFACE

Sartorius

Gracilis

DEEP THIGH MUSCLES, MEDIAL SURFACE

Quadriceps femoris
  Vastus lateralis
  Vastus medialis
  Rectus femoris
  Vastus intermedius

Iliacus (iliopsoas)

Pectineus

Adductor

Semimembranosus

Semitendinosus

SUPERFICIAL HIP, THIGH, AND LOWER-LEG MUSCLES, LATERAL SURFACE

Gluteus medius

Gluteus maximus

Tensor fasciae latae

Biceps femoris

Semitendinosus

Gastrocnemius

Extensor digitorum

Peroneus longus

Peroneus tertius

SUPERFICIAL LOWER-LEG MUSCLES, MEDIAL SURFACE

Flexor digitorum longus

Tibialis anterior

Gastrocnemius

Soleus

Tibial Bone

Figure 20-1 (a) The superficial thigh muscles of the fetal pig, medial surface (photograph).

1. Sartorius (sar-TOR-ē-us)
2. Gracilis (gra-SIL-is)
3. Femoral triangle
4. Linea alba
5. External oblique

Figure 20-1 (b) The superficial thigh muscles, medial surface (line drawing).

SUPERFICIAL THIGH MUSCLES, MEDIAL SURFACE

Instructions for Dissection

Almost the entire medial surface of the thigh (see Figure 20-1) is covered by two ribbonlike muscles separated proximally by the area of the *femoral triangle*. In this triangle are blood vessels and a nerve. The blood vessels are the femoral artery (injected red on your specimen) and the femoral vein (injected blue). The large nerve running parallel to the blood vessels is the femoral nerve.

The thin, wide muscle located laterally is the *sartorius*. Free it, and transect it, pulling the cut ends as far back as possible. The thin, wide muscle located medially is the *gracilis*. Transect the gracilis. The muscles lying in the femoral triangle are considered in the next section. (See Table 20-1 for the origins, insertions, and actions of these muscles.)

Table 20-1   Muscles of the Hindlimb of the Fetal Pig

Muscle	Origin	Insertion	Action
Gluteus medius	Aponeurosis of the latissimus dorsi; lateral pelvis	Greater trochanter of femur	Extends hip; abducts thigh
Gluteus maximus	Last sacral and first caudal vertebrae	Cranial margin of biceps femoris and on fascia lata	Abducts thigh
Tensor fasciae latae	Iliac crest	Patella	Tightens fascia lata; flexes hip; extends lower leg
Biceps femoris	Ischium and sacrum	Distal end of femur; proximal tibia	Abducts thigh; flexes lower leg
Semitendinosus	First and second caudal vertebrae; ischial tuberosity	Proximal tibia and calcaneus	Flexes lower leg
Peroneus longus	Proximal tibia and fibula	Metatarsals	Flexes ankle
Extensor digitorum	Proximal ends of tibia and fibula	By tendons on each digit	Extends digits
Peroneus tertius	Distal femur	Third metatarsal; tarsal bones	Flexes ankle when knee is fixed
Soleus	Proximal fibula	Calcaneus	Extends foot
Sartorius	Pelvis; iliac fascia	Proximal tibia	Flexes hip; abducts leg
Gracilis	Ventral surface of pubis	Proximal one-third of tibia	Adducts thigh
Vastus intermedius	Craniolateral surface of femur	Patella	Extends lower leg
Vastus lateralis	Lateral surface of femur	Patella	Extends lower leg
Vastus medialis	Proximal two-thirds of femur	Patella	Extends lower leg
Rectus femoris	Ilium above acetabulum	Patella	Extends lower leg; flexes hip
Iliacus (iliopsoas)	Ventral surface of ilium; wing of sacrum	Lesser trochanter of femur	Flexes hip; rotates thigh outward
Adductor	Ventral surfaces of pubis and ischium	Distal end of femur	Adducts leg; extends hip
Semimembranosus	Ischial tuberosity	Proximal tibia; distal femur	Extends hip; adducts leg
Flexor digitorum longus	Proximal ends of tibia and fibula	Digits	Flexes digits
Tibialis anterior	Proximal end of tibia	Second tarsal and metatarsal	Flexes ankle
Gastrocnemius	Distal femur	Calcaneus (by tendon of Achilles)	Extends foot

Figure 20-2 (a) The deep thigh muscles of the fetal pig, medial view (photograph).

DEEP THIGH MUSCLES, MEDIAL SURFACE

Instructions for Dissection

If not already done, now transect the sartorius and gracilis muscles (see Figure 20-1). Lay their ends back as far as possible. Begin your observations at the lateral border of the thigh.

Covering about one-half of the area of the thigh is a large muscle, the *quadriceps femoris.* The four muscles that make it up are considered separately in the pig, as they are in the human. The quadriceps separates into three distinct bundles superficially (see Figure 20-2): (1) a large lateral muscle, the *vastus lateralis;* (2) a middle muscle, the *rectus femoris;* and (3) a medial muscle, the *vastus medialis.* The fourth muscle of the group, the *vastus intermedius,* may be seen by tran-

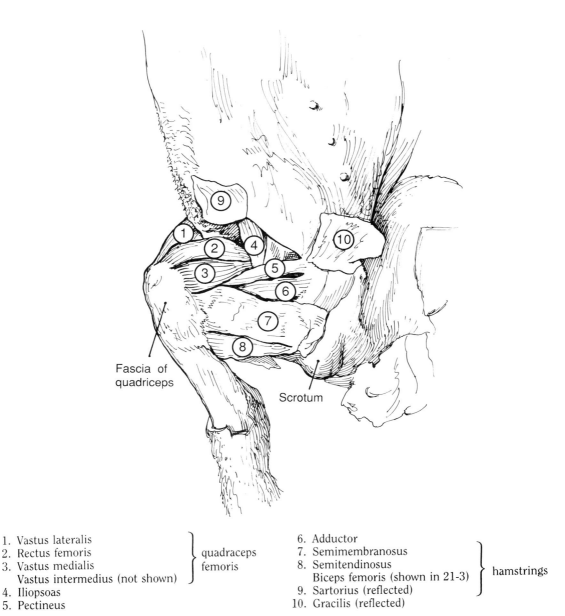

Fascia of
quadriceps

Scrotum

1. Vastus lateralis
2. Rectus femoris   } quadraceps
3. Vastus medialis   femoris
    Vastus intermedius (not shown)
4. Iliopsoas
5. Pectineus

6. Adductor
7. Semimembranosus
8. Semitendinosus   } hamstrings
    Biceps femoris (shown in 21-3)
9. Sartorius (reflected)
10. Gracilis (reflected)

Figure 20-2 (b) The deep thigh muscles of the fetal pig,
medial view (line drawing).

secting the rectus femoris and then pulling its cut ends
back to its origin and its insertion. The vastus intermedius partially surrounds the femur bone. It is not
shown in Figure 20-2.

Cut away the femoral artery, femoral vein, and femoral nerve so that you are better able to see the group
of muscles beneath them. The most lateral of this muscle group is the iliopsoas (iliacus). It can be seen pen-

trating the body wall. Medial to the iliopsoas is the
pectineus, medial to the pectineus is the *adductor* muscle. Forming the medial surface of the thigh is the large
semimembranosus. Along the edge of the medial surface, you may see the *semitendinosus,* which can be
seen more clearly from the dorsal surface. (See Table
20-1 for more information on these muscles.)

Figure 20-3 (a) The superficial thigh, hip, and leg muscles of the fetal pig, lateral view (photograph).

SUPERFICIAL HIP, THIGH, AND LOWER-LEG MUSCLES, LATERAL SURFACE

Instructions for Dissection

Turn the pig over so that the lateral surface is facing you. The hip region is covered with heavy fascia beneath which are the gluteal muscles. Remove this heavy fascia by cutting off as much as is necessary to expose the *gluteus medius* and posterior to it the *gluteus maximus*. Note that the gluteus maximus is the smaller of the two, contrary to the condition in the human. It is not well developed in the pig. Lying within the same mass of fascia that you have just removed, and below the gluteus medius is the *tensor fasciae latae*. The tensor is an irregularly shaped muscle that not only originates in a heavy fascia that covers the gluteus medius, but also inserts by a long broad fascia (the *fascia lata*), almost completely covering the vastus lateralis. Loosen the fascia lata and transect it so that you can see the entire vastus lateralis. The large *biceps femoris* covers most of the lateral surface of the thigh. The *semiten-*

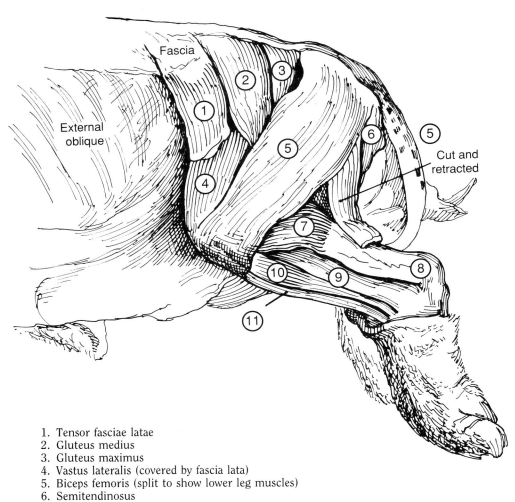

1. Tensor fasciae latae
2. Gluteus medius
3. Gluteus maximus
4. Vastus lateralis (covered by fascia lata)
5. Biceps femoris (split to show lower leg muscles)
6. Semitendinosus
7. Gastrocnemius
8. Tendon of Achilles
9. Flexor digitorum
10. Peroneus longus
11. Peroneus tertius

Figure 20-3 (b) The superficial thigh, hip, and leg muscles of the fetal pig, lateral view (line drawing).

dinosus forms the posterior border of the thigh. Dissect the semitendinosus completely from the semimembranosus and the biceps femoris.

In the lower leg (lateral view), find the large *gastrocnemius,* which is partially hidden by the *biceps femoris.* Pull the biceps femoris upward, or cut into it (see Figure 20-3a) to see clearly the muscles of the lower leg. The *gastrocnemius* forms the posterior border of the leg. It is large and ends in the tendon of Achilles. Identify the muscles in order from that point as the *flexor digitorum, peroneus longus,* and *peroneus tertius.* Very closely applied to the *gastrocnemius* is the *soleus.* It has the same origin and insertion as the gastrocnemius, but can be cleanly separated from it.

Figure 20-4 (a) The fetal pig's muscles of the lower leg, medial surface (photograph).

SUPERFICIAL LOWER-LEG MUSCLES, MEDIAL VIEW

Instructions for Dissection

In the lower leg, the large, prominent *gastrocnemius* can be seen on both the medial and lateral surfaces (see Figure 20-4). It has two distinct heads of origin, but the two muscle masses unite to form the single very tough *tendon of Achilles* or *calcaneal tendon,* which inserts on the calcaneus bone.

The soleus can also be seen on this surface. Lying between the soleus and the tibial bone is the *flexor digitorum longus.* This muscle has two heads of origin. On the anterior surface of the tibial bone is the *tibialis anterior* muscle.

1. Tibialis anterior
2. Tibial bone
3. Flexor digitorum longus
4. Soleus
5. Gastrocnemius
6. Tendon of Achilles

Figure 20-4 (b) Muscles of the lower leg, medial surface (line drawing).

Twelfth rib
Quadratus lumborum
Iliac crest
ILIACUS
Anterior superior iliac spine
TENSOR FASCIAE LATAE
SARTORIUS
RECTUS FEMORIS (cut)
VASTUS LATERALIS
VASTUS INTERMEDIUS
VASTUS MEDIALIS
Iliotibial tract
RECTUS FEMORIS (cut)
Section of fascia lata
Tendon of quadriceps femoris
Patellar ligament

Psoas minor
PSOAS MAJOR
Sacrum
Inguinal ligament
Pubic tubercle
PECTINEUS
ADDUCTOR LONGUS
GRACILIS
ADDUCTOR MAGNUS
Patella

(a)

Figure 20-5 Human muscles that act on the leg: (a) anterior superficial view; (b) anterior deep view.

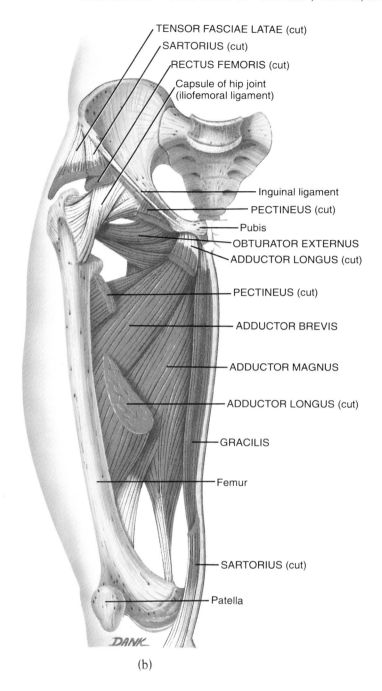

TENSOR FASCIAE LATAE (cut)
SARTORIUS (cut)
RECTUS FEMORIS (cut)
Capsule of hip joint
(iliofemoral ligament)

Inguinal ligament
PECTINEUS (cut)
Pubis
OBTURATOR EXTERNUS
ADDUCTOR LONGUS (cut)
PECTINEUS (cut)
ADDUCTOR BREVIS
ADDUCTOR MAGNUS
ADDUCTOR LONGUS (cut)
GRACILIS
Femur
SARTORIUS (cut)
Patella

DANK

(b)

COMPARABLE HUMAN MUSCLES

The gluteus muscles in the human are divided into three groups. Externally, there is the large posterior *gluteus maximus,* and above and lateral to it is the gluteus medius. The smaller *gluteus minimus* lies beneath the other two (Figure 20-5).

The muscles of the posterior thigh are equally effective at both hip and knee joints as extensors and flexors. The tendons of these muscles can be felt at the back of the knee when the knee is partly flexed. In addition to the *biceps femoris,* there are the *semimembranosus* and the *semitendinosus.* These muscles, called the "hamstrings," enable you to stand in a relaxed posture.

The muscles of the anterior thigh extend the knee joint. They are the quadriceps group: the vastus lateralis, rectus femoris, vastus medialis, and vastus intermedius. Flexors of the hip joint are the *iliopsoas, rectus*

Iliac crest

Gluteus maximus (cut)

Sacrum

Coccyx

Obturator internus

Ischial tuberosity

Sciatic nerve

GRACILIS

SARTORIUS

Gluteus medius (cut)

Gluteus minimus

Piriformis

Superior gemellus

Greater trochanter

Inferior gemellus

Obturator externus

Quadratus femoris

Gluteus maximus (cut)

Femur

Adductor magnus

SEMITENDINOSUS

BICEPS FEMORIS

SEMIMEMBRANOSUS

Vastus lateralis

Femur deep to popliteal fossa

Plantaris

Gastrocnemius

Tendon of biceps femoris

(c)

Figure 20-5 *(Continued)* (c) posterior superficial view; (d) posterior deeper view. (From G. J. Tortora and S. R. Grabowski, *Principles of Anatomy and Physiology,* 7th ed., New York: HarperCollins, 1993. © Leonard Dank.)

Iliac crest

GLUTEUS MEDIUS
(cut)

GLUTEUS MINIMUS

GLUTEUS MAXIMUS
(cut)

Sacrum

PIRIFORMIS

SUPERIOR GEMELLUS

Greater trochanter of femur

OBTURATOR
INTERNUS

INFERIOR GEMELLUS

Ischial tuberosity

OBTURATOR EXTERNUS

QUADRATUS FEMORIS

Sciatic nerve

ADDUCTOR
MAGNUS

Femur

Common peroneal nerve
in popliteal fossa

Tibial nerve in
popliteal fossa

DANK

(d)

femoris, tensor fasciae latae, pectineus, gracilis, and—to some extent—the *sartorius.*

There is a great deal of similarity between the fetal pig and the human in these groups of muscles. Study as many human hip and thigh muscles as you can. Refer to Table 20-2 for the muscles' origins, insertions, and actions.

The muscles of the anterior lower leg (Figure 20-6) are dorsiflexors of the ankle and extensors of the toes. They are the *tibialis anterior, extensor digitorum longus,* and *peroneus tertius.*

On the lateral surface of the leg are the peroneal muscles, which evert the foot. They may also protect against excessive inversion. Specifically, they are the *peroneus longus* and the *peroneus brevis.*

The *gastrocnemius* muscle covers the posterior surface of the lower leg and has two heads. It ends in the calcaneal tendon (Achilles tendon) and inserts on the calcaneus bone. It plantar flexes (extends) the foot. Beneath it is the *soleus,* which inserts on the calcaneus and also plantar flexes the foot. Stand on your toes and feel these muscles contract.

Muscles of this region are similar in the fetal pig and the human. Learn as many human muscles as you can, and check all origins, insertions, and actions with Table 20-3. Demonstrate the action of as many muscles

Table 20-2 Hip and Thigh Muscles of the Human

Muscle	Origin	Insertion	Action
Gluteus maximus	Iliac crest, sacrum, and coccyx	Iliotibial tract of fascia lata and gluteal tuberosity of femur	Extends and rotates femur laterally
Gluteus medius	Ilium	Greater trochanter of femur	Abducts and rotates femur laterally
Gluteus minimus	Ilium	Greater trochanter of femur	Abducts and rotates femur medially
Tensor fasciae latae	Iliac crest	Iliotibial tract of fascia lata	Flexes and abducts femur
Adductor longus	Pubic crest and symphysis	Linea aspera of femur	Adducts, rotates, and flexes femur
Adductor magnus (femoris)	Pubis and ischium	Linea aspera of femur	Adducts, flexes, and extends femur
Psoas major	Transverse processes and centra of lumbar vertebrae	Lesser trochanter of femur	Flexes and rotates femur laterally; flexes vertebral column
Iliacus	Iliac fossa	Tendon of psoas major	Flexes and rotates femur laterally; flexes vertebral column
Pectineus	Fascia of pubis	Pectineal line of femur	Flexes, adducts, and rotates thigh laterally
Biceps femoris	Ischial tuberosity and linea aspera of femur	Head of fibula and lateral condyle of tibia	Flexes knee; extends thigh
Semitendinosus	Ischial tuberosity	Proximal end of tibia	Flexes knee; extends thigh
Semimembranosus	Ischial tuberosity	Medial condyle of tibia	Flexes knee; extends thigh
Gracilis	Pubic symphysis and pubic arch	Medial surface of tibia	Flexes knee; adducts thigh
Sartorius	Anterior superior spine of ilium	Medial surface of tibia	Flexes knee; flexes thigh and rotates it laterally
Quadriceps femoris Rectus femoris	Anterior inferior spine of ilium	Upper border of patella	Extends knee; flexes thigh
Vastus lateralis	Greater trochanter and linea aspera of femur	Patella and tibial tuberosity	Extends knee
Vastus medialis	Linea aspera of femur	Patella and tibial tuberosity	Extends knee
Vastus intermedius	Anterior and lateral surface of femur	Patella and tibial tuberosity	Extends knee

Gracilis

Sartorius

Biceps femoris

Semitendinosus

Semimembranosus

Femur

Popliteal fossa

PLANTARIS

GASTROCNEMIUS (cut)

Tendon of biceps femoris (cut)

Tibia

POPLITEUS

GASTROCNEMIUS

SOLEUS (cut)

Fibula

TIBIALIS POSTERIOR

SOLEUS

PERONEUS LONGUS

FLEXOR DIGITORUM LONGUS

FLEXOR HALLUCIS LONGUS

Tibia

PERONEUS BREVIS

Tendon of TIBIALIS POSTERIOR

Fibula

Calcaneal (Achilles) tendon (cut)

DANK

(a)

(b)

Figure 20-6 Human muscles that move the foot and toes: (a) superficial posterior view; (b) deep posterior view.

Quadriceps femoris

Tendon of quadriceps femoris

Fascia lata

Biceps femoris

Patella

PLANTARIS

Head of fibula

Patellar ligament

Tibia

TIBIALIS ANTERIOR

GASTROCNEMIUS

PERONEUS LONGUS

SOLEUS

EXTENSOR DIGITORUM LONGUS

FLEXOR DIGITORUM LONGUS

PERONEUS BREVIS

PERONEUS TERTIUS

EXTENSOR HALLUCIS LONGUS

Calcaneal (Achilles) tendon

Fibula

Extensor digitorum brevis

Metatarsals

Superior extensor retinaculum

Inferior extensor retinaculum

(c)

(d)   DANK

Figure 20-6 *(Continued)* (c) superficial anterior view; (d) right lateral superficial view. (From G. J. Tortora and S. R. Grabowski, *Principles of Anatomy and Physiology,* 7th ed., New York: HarperCollins, 1993. © Leonard Dank.)

Table 20-3 Superficial Lower-Leg Muscles of the Human

Muscle	Origin	Insertion	Action
Gastrocnemius	Lateral and medial condyles of femur and capsule of knee	Calcaneus by Achilles tendon	Plantar flexes (extends) foot
Soleus	Head of fibula and medial border of tibia	Calcaneus by Achilles tendon	Plantar flexes foot
Peroneus longus	Head and body of fibula and lateral condyle of tibia	First metatarsal and first cuneiform	Plantar flexes and everts foot
Peroneus brevis	Body of fibula	Fifth metatarsal	Plantar flexes and everts foot
Tibialis anterior	Lateral condyle and body of tibia	First metatarsal and first cuneiform	Flexes foot dorsally and inverts it
Tibialis posterior	Membrane between tibia and fibula	Second, third, and fourth metatarsals; navicular; third cuneiform; cuboid	Plantar flexes and inverts foot
Flexor digitorum longus	Tibia	Distal phalanges of four outer toes	Flexes toes; plantar flexes and inverts foot
Extensor digitorum longus	Lateral condyle of tibia and anterior surface of fibula	Middle and distal phalanges of four outer toes	Extends toes, flexes foot dorsally and everts it

of the leg as you can. Try to see or feel the contraction as you make a specific movement.

Hip and Thigh Muscles

Two large groups of muscles move the leg. They are the quadriceps femoris group and the hamstrings group. The quadriceps group ("quads") flexes the thigh and extends the leg. The hamstring group flexes the leg and extends the thigh. In addition there are abductors, adductors, and rotators. Perform the movements indicated in the following discussion.

The Quadriceps Femoris

1. *Rectus femoris* (Figure 20-7) extends the leg and flexes the thigh.

Figure 20-7 Rectus femoris (anterior view). (From P. J. Donnelly and G. A. Wistreich, *Exploring the Human Body,* New York: Harper & Row, 1989.)

2. *Vastus medialis* (Figure 20-8) extends the leg.

Figure 20-8 Vastus medialis (anterior view). (From P. J. Donnelly and G. A. Wistreich, *Exploring the Human Body,* New York: Harper & Row, 1989.)

3. *Vastus lateralis* (Figure 20-9) extends the leg.

Figure 20-9 Vastus lateralis (anterior view). (From P. J. Donnelly and G. A. Wistreich, *Exporting the Human Body,* New York: Harper & Row, 1989.)

4. *Vastus intermedius* (Figure 20-10) extends the leg.

To demonstrate leg extension, sit on a chair. Slowly raise your legs off the floor until the upper and lower legs form a straight line (knee extended). Place your hands on the anterior thigh to feel the contraction of the quads. These powerful extensors are indispensable to walking.

Figure 20-10 Vastus intermedius (anterior view). (From P. J. Donnelly and G. A. Wistreich, *Exploring the Human Body,* New York: Harper & Row, 1989.)

The Hamstrings

1. *Biceps femoris* (Figure 20-11) flexes the leg at the hip and extends the thigh.

Figure 20-11 Biceps femoris (posterior view). (From P. J. Donnelly and G. A. Wistreich, *Exploring the Human Body,* New York: Harper & Row, 1989.)

2. *Semitendinosus* (Figure 20-12) flexes the leg at the hip and extends the thigh.

Figure 20-12 Semitendinosus (posterior view). (From P. J. Donnelly and G. A. Wistreich, *Exploring the Human Body,* New York: Harper & Row, 1989.)

3. *Semimembranosus* (Figure 20-13) flexes the leg at the hip and extends the thigh.

The hamstring group of muscles, along with the quads, is used in walking, running, and jumping. To demonstrate the hamstrings, flex your leg, and place your hand on the posterior surface of the thigh above your knee. Feel the powerful contraction. To test for the strength of your hamstrings, lie down in a supine position. Have the instructor or another student hold one leg down while you raise the other.

Figure 20-13 Semimembranosus (posterior view). (From P. J. Donnelly and G. A. Wistreich, *Exploring the Human Body,* New York: Harper & Row, 1989.)

Abductors, Adductors, and Rotators

1. *Sartorius* (Figure 20-14)
   This muscle enables you to rotate your leg laterally so that you can sit in a cross-legged (tailor) fashion. It originates on the ilium of the pelvic girdle and inserts on the tibia, running across the leg from lateral to medial. Sit with your legs crossed, and feel the muscle contract. It flexes the leg, rotates the leg laterally, and flexes the thigh.

   a. Name one synergist of the sartorius. _____

   b. Name one antagonist of the sartorius. _____

Figure 20-14 Sartorius (anterior view). (From P. J. Donnelly and G. A. Wistreich, *Exploring the Human Body,* New York: Harper & Row, 1989.)

2. *Tensor fasciae latae* (Figure 20-15)
   The tensor flexes the thigh, rotates the thigh medially, and abducts the thigh. It also tenses the fascia latae.

   Name other muscles in the leg that perform the same functions as these listed here:

   Flexes thigh _____

   Rotates thigh medially _____

   Abducts thigh _____

Figure 20-15 Tensor fasciae latae (lateral view). (From P. J. Donnelly and G. A. Wistreich, *Exploring the Human Body,* New York: Harper & Row, 1989.)

3. *Adductor magnus (femoris)* (Figure 20-16)
   This is the largest and most powerful muscle of the adductor group. Its anterior part adducts and flexes the thigh and also rotates the thigh medially. The posterior part extends the thigh and rotates it laterally.

   a. Flex the thigh of one leg a short distance. Simultaneously adduct it and rotate it medially. This action is controlled by the anterior portion

      of which muscle? _____

   b. Straighten the thigh and rotate it laterally. This is accomplished by the posterior portion of which

      muscle? _____

Figure 20-16 Adductor magnus (anterior view). (From P. J. Donnelly and G. A. Wistreich, *Exploring the Human Body,* New York: Harper & Row, 1989.)

4. *Psoas major* (Figure 20-17)
   The psoas flexes and rotates the thigh laterally and flexes the trunk on the thigh. It also bends the trunk laterally. It originates in the lumbar region of the spine and inserts on the lesser trochanter of the femur.

   To demonstrate the muscle, lie on your back. Have a student or the instructor place a hand on the left iliac crest for stabilization. Slightly abduct the right thigh, then flex it and rotate it laterally. Are you able to feel the contraction between the

   sartorius and the adductor tendons? _____

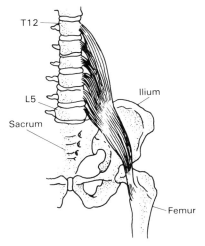

Figure 20-17 Psoas major (anterior view). (From P. J. Donnelly and G. A. Wistreich, *Exploring the Human Body,* New York: Harper & Row, 1989.)

5. *Iliacus* (Figure 20-18)

This muscle flexes the thigh and rotates it laterally. It also flexes the vertebral column. Its origin is high on the pelvic girdle, and it inserts on the lesser trochanter of the femur.

To demonstrate it, sit on the edge of a table, with your hands grasping the table edge. Flex your hips, one leg at a time. Can you feel its contraction?

Figure 20-18 Iliacus (anterior view). (From P. J. Donnelly and G. A. Wistreich, *Exploring the Human Body*, New York: Harper & Row, 1989.)

6. *Gluteus maximus* (Figure 20-19)

This is the largest of the gluteus group of muscles. It is very thick and forms the human buttocks. The pair of muscles is instrumental in maintaining the upright posture and is in a state of contraction when you are standing.

To demonstrate the muscles, rotate one leg at a time laterally while in a standing position. Next, bend over a table, elbows on the table and feet on the floor. Laterally extend one leg at a time. Feel the contraction, and compare it with the intensity of contraction in the standing position.

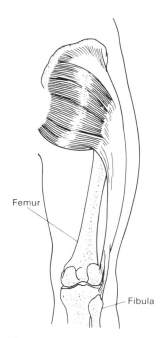

Figure 20-19 Gluteus maximus (posterior view). (From P. J. Donnelly and G. A. Wistreich, *Exploring the Human Body*, New York: Harper & Row, 1989.)

Muscles of the Lower Leg

Some of the muscles that move the foot, ankle, and toes are the following:

1. *Gastrocnemius* (Figure 20-20)
   This is the largest muscle of the lower leg. It plantar flexes the foot and flexes the knee.
   a. To demonstrate this action, stand on your tiptoes. Feel the contraction.
   b. In a sitting position, with both feet flat on the floor, raise your heels, keeping the ball of each foot on the floor. With your hand, feel the contraction of the entire gastrocnemius muscle.

Figure 20-20 Gastrocnemius (posterior view). (From P. J. Donnelly and G. A. Wistreich, *Exploring the Human Body,* New York: Harper & Row, 1989.)

8. *Peroneus longus* (Figure 20-21)
   This muscle plantar flexes the foot and abducts and everts the foot.

   To demonstrate this action, in a sitting position, raise your right foot, and turn it (evert) laterally. Place your fingers on the lateral surface of your leg, to feel the muscle contraction.

Plantar insertion

Figure 20-21 Peroneus longus (lateral view). (From P. J. Donnelly and G. A. Wistreich, *Exploring the Human Body,* New York: Harper & Row, 1989.)

9. *Tibialis anterior* (Figure 20-22)
The tibialis dorsiflexes and inverts the foot.

    To demonstrate this action, from a sitting position, raise your right leg slightly, and point your toes upward. Then turn your foot medially (inversion). To feel the contraction, place your fingers on the anterior surface of your leg.

Figure 20-22 Tibialis anterior (anterior view). (From P. J. Donnelly and G. A. Wistreich, *Exploring the Human Body,* New York: Harper & Row, 1989.)

10. *Flexor digitorum longus* (Figure 20-23)
This flexor muscle bends the toes (second through fifth). It also plantar flexes the foot and inverts the foot.

    To demonstrate this action, from a sitting position, raise your right foot, and flex the second through fifth toes. Then turn your foot inward (inversion) and plantar flex it.

Figure 20-23 Flexor digitorum longus (posterior view). (From P. J. Donnelly and G. A. Wistreich, *Exploring the Human Body,* New York: Harper & Row, 1989.)

11. *Extensor digitorum longus* (Figure 20-24)
This muscle extends the toes, dorsiflexes the foot, and everts the foot. The body of the muscle cannot be felt, but the tendons that insert on the dorsal surface of the phalanges of the second through fifth toes can be seen and felt.

    To demonstrate this action, from a sitting position, raise your right foot, and straighten (extend) those toes. Then move the foot toward you (dorsiflexion), and turn it outward (eversion). If you cannot see the tendons, try to spread your toes.

Figure 20-24 Extensor digitorum longus (anterior view). (From P. J. Donnelly and G. A. Wistreich, *Exploring the Human Body,* New York: Harper & Row, 1989.)

LABORATORY REVIEW *20*

MUSCLES OF THE HIP, THIGH, AND LOWER LEG

1. The gluteus muscles and the tensor fasciae latae are shown diagrammatically in Figure Q20-1. Color them, and review their origins and insertions.

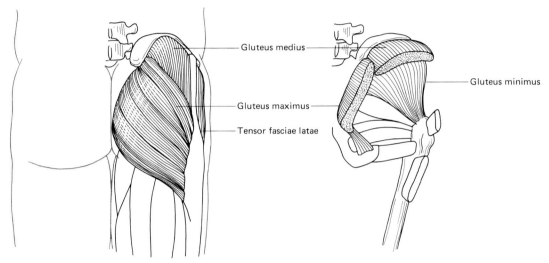

Gluteus medius

Gluteus minimus

Gluteus maximus

Tensor fasciae latae

Posterior view (superficial muscles)　　　　　　Posterior view (deep muscles)

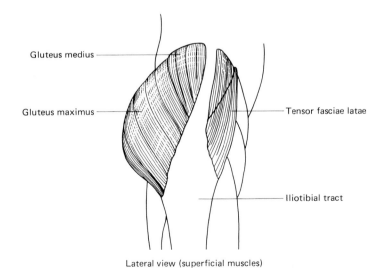

Gluteus medius

Gluteus maximus

Tensor fasciae latae

Iliotibial tract

Lateral view (superficial muscles)

Figure Q20-1

2. Label the thigh muscles on Figure Q20-2.

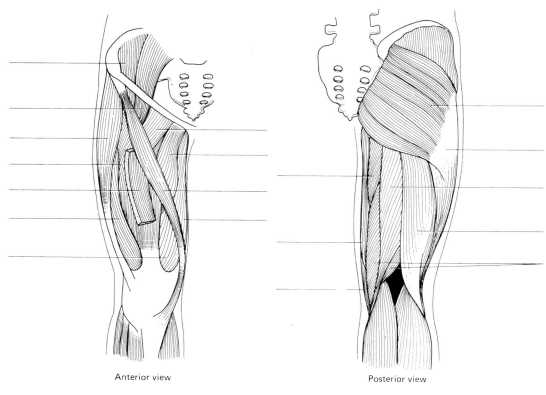

Anterior view

Posterior view

Figure Q20-2

3. Color the muscles of the lower leg in Figure Q20-3. Review the origin, insertion, and action of each muscle. Demonstrate as many of them as possible on your own leg.

Rectus femoris

Vastus lateralis

Biceps femoris

Iliotibial tract

Patella

Gastrocnemius

Tibialis anterior

Extensor digitorum longus

Soleus

Peroneus longus

Peroneus brevis

Peroneus tertius

Calcaneus

Extensor hallucis longus

Extensor digitorum brevis

V metatarsal

Figure Q20-3

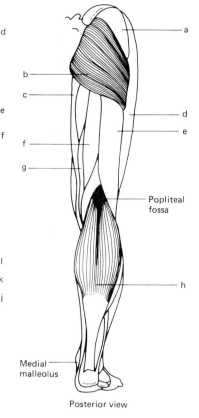

Figure Q20-4 From P. J. Donnelly, *Laboratory Manual for Human Anatomy,* revised ed., New York: Harper & Row, 1988.)

ANTERIOR VIEW

a. _____   f. _____   k. _____

b. _____   g. _____   l. _____

c. _____   h. _____   m. _____

d. _____   i. _____   n. _____

e. _____   j. _____   o. _____

LATERAL VIEW

a. _____   e. _____   i. _____

b. _____   f. _____   j. _____

c. _____   g. _____   k. _____

d. _____   h. _____   l. _____

POSTERIOR VIEW

a. _____   d. _____   g. _____

b. _____   e. _____   h. _____

c. _____   f. _____

Muscle Ultrastructure and Chemistry

After completing this exercise, you should be able to

1. Distinguish—on the basis of microscopic examination—among smooth, skeletal, and cardiac muscle
2. List the four protein components in muscle, and give the functions of each in muscle contraction
3. Interpret the striated pattern of skeletal muscle fibers in an electron micrograph
4. Describe the effects of temperature, adenosine triphosphate (ATP), potassium chloride (KCl), magnesium chloride ($MgCl_2$), and pH on muscular contraction

Skeletal muscle consists of thousands of elongated, cylindrical cells termed *myofibers* or muscle fibers. Each individual muscle fiber is covered by a plasma membrane (or *sarcolemma*) that surrounds the muscle cell's cytoplasm, known as *sarcoplasm*. Many nuclei are located within the sarcoplasm and close to the sarcolemma (Color Plates 20 and 21).

An electron microscopic view of skeletal muscle fibers reveals that they contain threadlike, cylindrical structures termed *myofibrils*. These myofibrils run longitudinally within the muscle fiber and consist of two types of smaller structures—thin and thick *myofilaments*. Alternating thick and thin myofilaments are stacked parallel to one another in units termed *sarcomeres*. (Figure 21-1).

The sarcomere, which measures 2–3 μm in length, is bounded at both ends by a narrow zone of dense material termed a *Z line*. Along the length of a muscle cell, there are other conspicuous bands, zones, or striations that are repeated regularly. These include the rel-

atively clear, broad, *I* (iostropic) *band*, which is bisected by the Z line; and the large, dark, dense area termed the *A* (anisotropic) *band* in the center of the sarcomere (Figure 21-1). The muscle's striated appearance is due to the combination of the alternating dark A bands and the lighter I bands.

Thin myofilaments contain three proteins, the most abundant of which is *actin*. Actin gives thin myofilaments their characteristic shape. The two other proteins are *tropomyosin* and *troponin*. Together, these molecules are referred to as the *tropomyosin–troponin complex,* and they are involved in the regulation of muscle contractions.

Thick myofilaments overlap the free ends of the thin myofilaments. *Myosin* is the major protein found with thick myofilaments. In intact muscle, one end of the thin filaments is attached to the Z line. These thin filaments extend perpendicularly from both of its surfaces. Transmission electron micrographs show that sarcomeres consist of a set of thick filaments arranged in the spaces between the thin filaments projecting from the Z lines. The two types of filament are arranged in parallel groups and overlap for part of their lengths (Figures 21-1(c) and 21-2). The spacing of the filaments is such that the two sets can slide past one another without restriction. This sliding motion is responsible for the shortening reaction during muscle contraction. All events associated with the movement of myofilaments are known as the *sliding-filament theory* (Figure 21-2).

A stimulus must be applied to skeletal muscle for it to contract. Such a stimulus is delivered by a *neuron* or nerve cell, specifically known as a *motor neuron*. A long threadlike process of the neuron, termed an *axon*, branches into terminals that come into close association with a muscle fiber's sarcolemma (Color Plate 22). The axon terminal of the motor neuron, together with the portion of the sarcolemma involved, form a *motor end plate* (Figure 21-3).

Figure 21-1 Structural and functional components of striated skeletal muscle: (a) This composite diagram shows a general relationship of skeletal muscle to bone, the components of a muscle fiber, and a sarcomere (after Shepro, D., F. Belamarich, and C. Levy. *Human Anatomy and Physiology.* New York: Holt, Rinehart and Winston, 1974). (b) Electron micrograph of a cylindrical sarcomere. (c) The arrangement and relationship of muscle fiber components. (From G. J. Tortora and S. R. Grabowski, *Principles of Anatomy and Physiology,* 7th ed., New York: HarperCollins, 1993.)

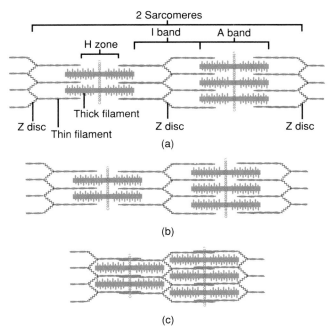

Figure 21-2 Selected positions of sarcomeres showing the sliding-filament theory of muscle contraction in action: (a) the relaxed state; (b) contracting; (c) fully or maximally contracted. Note that the lengths of both types of myofilaments do not change. (From G. J. Tortora and S. R. Grabowski, *Principles of Anatomy and Physiology,* 7th ed., New York: HarperCollins, 1993.)

The energy for muscle contraction is provided by the hydrolysis of ATP. In addition to the four proteins mentioned earlier, an ATP muscle contraction requires the presence of calcium ions. Calcium provides a means of regulation.

This exercise considers the microscopic properties of muscle and the particular effects that selected chemical and physical factors have on muscle contraction.

MICROSCOPIC FEATURES OF SKELETAL MUSCLE

Materials

1. The following prepared slides should be provided for class use:
   a. Teased skeletal muscle fibers
   b. Longitudinal section of skeletal muscle
2. Prepared slides stained to differentiate motor end plates should be provided for a class demonstration.

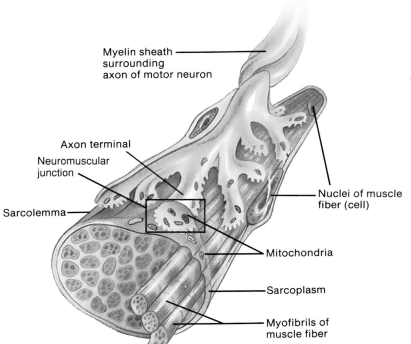

Figure 21-3 A diagrammatic view of a motor end plate. (From G. J. Tortora and S. R. Grabowski, *Principles of Anatomy and Physiology,* 7th ed., New York: HarperCollins, 1993.)

Procedure

This procedure is to be performed by students individually.

1. Examine a prepared slide of muscle under both low- and high-power objectives. (Refer to Color Plates 5 and 20).

2. Select a single fiber, and identify the following components: *muscle cell nuclei, sarcolemma* (a thin membrane that covers each fiber), *A bands* (repeating dark zones, perpendicular to the long axis of the muscle fiber), and *I bands* (repeating light zones, perpendicular to the long axis of the muscle fiber).

3. Sketch a representative field, showing the components listed in Step 2 in the subsequent results and observations section.

4. Examine a prepared slide of a longitudinal section of skeletal muscle. Identify the following components: *muscle cell nuclei, endomysium* (connective tissue layer surrounding each muscle fiber and external to the sarcolemma), and *local areas of contraction* (darker areas along the length of a muscle fiber).

5. Sketch a representative field showing the components listed in Step 4 in the results and observations section.

6. Examine the demonstration slide showing motor end plates. Locate the following: a *motor end plate, nerve fiber,* and *nerve fiber branches*. Note that each muscle fiber has only one end plate (refer to Color Plate 22).

7. Sketch and label a representative field in the results and observations section.

8. Answer the questions pertaining to this procedure in the results and observations section.

THE CHEMISTRY OF MUSCLE CONTRACTION

Materials

The following items should be provided for class use:

1. Rabbit psoas muscle strips in 50 percent glycerol (obtainable from Carolina Biological Supply Company)

2. The following prepared solutions in containers with droppers:
   a. Adenosine triphosphate (ATP)
   b. Magnesium chloride ($MgCl_2$)
   c. Potassium chloride (KCl)
   d. Three separate glycerol preparations maintained at 4°C, 25°C (room temperature), and 37°C
   e. Three separate Ringer's preparations, with pH of 5, 7, and 8, maintained at 37°C

3. Microscope slides
4. Forceps
5. Millimeter rulers
6. Microscopes
7. Water bath maintained at 37°C
8. Ice bath or refrigerator to maintain appropriate solutions at 4°C
9. Eye droppers
10. Stopwatch or other timing device

Procedure 1: Effects of Temperature, Adenosine Triphosphate (ATP), Potassium Chloride (KCl), and Magnesium Chloride ($MgCl_2$) on Muscle Contraction

This procedure should be performed by students in groups of four.

1. On a glass slide, place two rabbit psoas muscle fibers in a small amount of the glycerol preparation kept at 25°C (room temperature). Examine them under low power. Are striations and nuclei evident?

2. Measure the length of each fiber in millimeters. Enter your findings in the results and observations section.

3. Remove the glycerol, and replace it with the glycerol preparation kept at 37°C. Wait 30 seconds. Measure the length of the fibers, and enter your findings. If the length is different from the room-temperature measurement at 25°C, calculate the percentage of contraction by dividing the contraction distance by the fiber length at room temperature. Enter your findings in the results and observations section.

4. Next, replace the glycerol with the preparation kept at 4°C. Wait 30 seconds. Measure the length of the fibers, and enter your findings. If this length is different from the length at 25°C calculate the percentage of contraction, as described in Step 3, and enter your findings.

5. Replace the glycerol with the preparation kept at 25°C. Wait 30 seconds, and repeat Steps 3 and 4, using the 4°C and 37°C glycerol solutions for 60 seconds.

6. Now replace the glycerol with the preparation kept at 25°C. Wait 30 seconds, and measure the length of the fibers again. This measurement will be used as a baseline value for calculations of the effects of

chemicals added to muscle preparation in the following steps.

7. Replace the glycerol with the 4°C preparation, and after 30 seconds cover the fibers with the KCl solution. Wait 30 seconds, and measure the length of the fibers. If there is a difference, calculate the percentage of contraction, as described in Step 3. Enter your findings in the results and observations section.

8. Repeat Step 7, using the following preparations separately, first with the 4°C and then with the 37°C glycerol preparations. Enter your findings in Table 21-1.

    a. $MgCl_2$
    b. ATP
    c. A mixture of KCl, $MgCl_2$, and ATP

9. Repeat Steps 7 and 8, exposing the psoas muscle to the different preparations for 60 seconds each.

Procedure 2: Effects of pH on Muscle Contraction

This procedure should be performed by students in groups of four.

1. Place two rabbit psoas muscle fibers in a small amount of glycerol (37°C) on a glass slide.

2. Measure the length of each fiber in millimeters. Enter your findings in the results and observations section.

3. Replace the glycerol with the pH 7 Ringer's solution, wait 30 seconds, measure the length of the fibers, and enter your findings in Table 21-2.

4. Add to the fibers a mixture of KCl, $MgCl_2$, and ATP. Wait 30 seconds, measure the length of the fibers, and enter your findings.

5. Repeat Steps 3 and 4, using the Ringer's solution with pH 5. If there is a difference in contraction distance from that obtained with the pH 7 solution, calculate the percentage of contraction by dividing the contraction distance obtained at pH 5 by the fiber length at pH 7.

6. Repeat Step 5 with the Ringer's solution with pH 8.

RESULTS AND OBSERVATIONS

Microscopic Features of Muscle

1. Sketch and label representative microscopic fields of the muscle specimens provided.

Teased skeletal
muscle fibers

Longitudinal
section

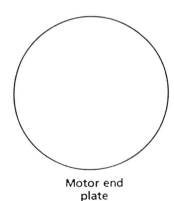

Motor end
plate

2. List three areas in the human body where smooth muscle characteristically is found.

    a. _____

    b. _____

    c. _____

3. Is striated muscle multinucleated? _____

Effects of Temperature, ATP, KCl, MgCl$_2$, and pH on Muscle Contraction

1. Enter your findings in the appropriate column of Table 21-1. Note that this table combines the results obtained from Procedure 1 in this part of the exercise.

2. Enter your findings obtained from Procedure 2 in Table 21-2.

3. Was muscular contraction affected by temperature? In what way? _____

_____

Table 21-1

Muscle Fiber Condition	Length at 30 Seconds (in mm)	Percentage of Contraction (if applicable)	Length at 60 Seconds (in mm)	Percentage of Contraction (if applicable)
Glycerol (25°C)				
Glycerol (4°C)				
ATP and glycerol (4°C)				
KCl and glycerol (4°C)				
MgCl$_2$ and glycerol (4°C)				
ATP, KCl, MgCl$_2$, and glycerol (4°C)				
Glycerol (37°C)				
ATP and glycerol (37°C)				
ATP and glycerol (4°C)				
KCl and glycerol (37°C)				
KCl and glycerol (4°C)				
MgCl$_2$ and glycerol (37°C)				
MgCl$_2$ and glycerol (4°C)				
ATP, KCl, MgCl$_2$, and glycerol (37°C)				
ATP, KCl, MgCl$_2$, and glycerol (4°C)				

Table 21-2

Muscle Fiber Condition	Length at 30 Seconds (in mm)	Percentage of Contraction (if applicable)
Glycerol (37°C)		
Ringer's solution (pH 7)		
Ringer's solution (pH 7) and ATP, KCl, MgCl$_2$		
Ringer's solution (pH 5)		
Ringer's solution (pH 5) and ATP, KCl, MgCl$_2$		
Ringer's solution (pH 8)		
Ringer's solution (pH 8) and ATP, KCl, MgCl$_2$		

4. Was muscular contraction affected by pH? In what way? _____

_____

5. Was muscular contraction affected by the absence of ATP? In what way? _____

_____

6. What is the role of calcium in muscle contraction?

_____

_____

LABORATORY REVIEW *21*

MUSCLE ULTRASTRUCTURE AND CHEMISTRY

COMPLETION AND DEFINITION

1. Draw a simple diagram to show the relationships among the components of a skeletal muscle, including muscle fibers, myofibrils, and sarcomere.

2. Draw a simple diagram to show the A bands, I bands, H zones, and Z lines within a sarcomere. (Include the overlapping of myosin and actin filaments.)

3. List four major proteins involved in muscle contraction.

   a. _____   c. _____

   b. _____   d. _____

ANSWERS

4. Briefly describe the events taking place in a muscle contraction. Attach a separate sheet, if diagrams are used. _____

_____

_____

5. Define or explain:

a. Sarcomere _____

_____

b. Myofibril _____

_____

c. Motor end plate _____

_____

d. Skeletal muscle _____

_____

e. Actin _____

_____

f. Myosin _____

_____

MULTIPLE CHOICE

Insert the correct answers in the spaces provided.

1. The ability of muscle tissue to receive and respond to stimuli is referred to as which of the following?
   a. tetanus
   b. contraction
   c. extension
   d. excitability
   e. relaxation

2. The sections of a muscle fiber separated by *two* lines are known as?
   a. sarcolemma
   b. sarcomere
   c. synapse
   d. T tubule
   e. myofilament

3. Which of the following is the plasma membrane surrounding a muscle fiber?
   a. sarcolemma
   b. sarcomere
   c. synapse
   d. T tubule
   e. myofilament

4. Which of the following tissues is striated and voluntary?
   a. smooth muscle
   b. cardiac muscle
   c. skeletal muscle
   d. visceral muscle
   e. heart muscle

1. _____

2. _____

3. _____

4. _____

5. _____

6. _____

7. _____

8. _____

9. _____

10. _____

5. The sarcomere portion consisting of thin myofilaments only is which of the following?
   a. H band
   b. Z line
   c. A band
   d. I band
   e. S zone

6. What is the connective tissue layer that surrounds muscle fiber bundles known as?
   a. myomysium
   b. Z band
   c. ectomysium
   d. perimysium
   e. endomysium

7. What muscle tissue is known to be nonstriated and involuntary?
   a. skeletal muscle
   b. cardiac muscle
   c. visceral muscle
   d. smooth muscle

8. Which of the following is known as the sarcomere portion where thin and thick myofilaments overlap?
   a. H band
   b. Z line
   c. A band
   b. I band
   e. S zone

9. Which of the following is known as the sarcomere portion in a sarcomere containing thick myofilaments only?
   a. H band
   b. Z line
   c. A band
   d. I band
   e. S zone

10. Which of the following is the muscle tissue known to contain intercalated discs?
    a. smooth muscle
    b. cardiac muscle
    c. skeletal muscle
    d. visceral muscle
    e. striated muscle

Introduction to Skeletal Muscle Contraction, Using a Laboratory Animal

After completing this exercise, you should be able to

1. Define or explain *myofibril, sarcoplasm, stimulus,* and the *all-or-none response*
2. Describe the events associated with a single muscle twitch and tetany
3. Prepare a laboratory animal for the demonstration of skeletal muscle contraction
4. Use appropriate equipment and stimuli to demonstrate a single muscle twitch, summation and tetanic contraction, determination of work, and muscle fatigue

Muscle is the most common tissue in the body of most vertebrates. As described in Exercise 21, it consists of *elongated cells* or *fibers* specialized for contraction. The contractile components within the muscle fiber are the *myofibrils.* These fibrils run lengthwise through the unspecialized cytoplasm of muscles or the *sarcoplasm.* Muscle bundles, termed *fascicles,* are formed through the joining of fibers to other parallel muscle cells by intertwining connective tissue. The fascicles, in turn, are organized into larger units referred to as *muscle organs.*

The ability of muscle to respond to stimulus is termed *irritability.* The response takes the form of a contraction. The stimulus is normally an impulse transmitted to the muscle cell membrane by the axon, or nerve fiber, of a neuron. Each fiber supplies from a few up to more than a hundred individual muscle cells. As described in Exercise 21, the portion of the membrane directly under the end of the axon is termed a *myoneural* or *motor end plate* (Color Plate 22). The impulse

causes changes within the myofibril and shortening of the sarcomere and the entire muscle cell.

In response to a single stimulus, individual muscle fibers either contract completely or do not contract at all (according to the *all-or-none-law*). Muscular contraction depends on how frequently nerve impulses are received by the muscle fibers and how many fibers are stimulated. The number of individual fibers contracting also determines the strength of a muscle contraction.

Skeletal muscle contracts only when it has been stimulated adequately. It does not contract spontaneously. In this exercise, isolated muscles from a frog are used to demonstrate how a whole muscle contracts and how changes in a stimulus can modify muscular contractions. Muscle used for such study is isolated and attached to appropriate equipment in such a way as to record contractions. An electrical stimulus either to the muscle itself or to an associated nerve produces a recorded response.

A single stimulus produces a single brief contraction known as a *muscle twitch.* An examination of a recorded response (Figure 22-1) reveals three principal events: (1) the *latent period,* which is the period between the stimulus and the beginning of contraction; (2) the *contraction period;* and (3) the *relaxation period.* The extent of the twitch depends on the stimulus applied. The stimulus must reach some *threshold level* before the muscle responds at all. Increased stimulus increases to the event of contraction until some maximum is obtained. As the stimulus is increased, more fibers are activated until all of them are contracting in response to the stimulus. Once a muscle is stimulated enough to contract, it briefly loses its ability to respond to additional stimuli. This is known as the *refractory period.*

If a second stimulus is applied after the refractory period, but before the contraction is over, the muscle will respond to both, and the additive effect of both stimuli will produce a stronger contraction. This is

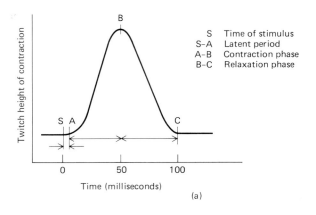

S	Time of stimulus
S–A	Latent period
A–B	Contraction phase
B–C	Relaxation phase

(a)

(b)

(c)

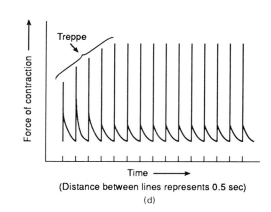

Figure 22-1 Muscle contraction responses: (a) muscle twitch; (b) summation of contraction; (c) tetanic contraction; (d) treppe.

known as a *summation of twitches* or *wave summation* (Figure 22-1b).

The application of two successive stimuli of maximum intensity causes two separate maximal contractions. If successive stimuli are applied so quickly that the muscle can relax only partially between stimuli, it remains in a state of contraction. This summation and fusion of individual twitches is known as *incomplete tetanus* (Figure 22-1c). *Complete tetanus* is the condition in which stimuli are applied so quickly that the muscle does not relax at all. Tetany is an important action because most muscular contractions are short-term tetanic contractions, and are thus smooth sustained contractions. Skeletal muscle contractions are all tetanic in nature, even the blinking of an eye.

The condition known as *treppe* occurs when a skeletal muscle contracts more forcefully in response to the same strength of stimulus after it has contracted several times (Figure 22-1d). It can be demonstrated by stimulating an isolated muscle with a series of stimuli at the same intensity and frequency, but not at a rate fast enough to produce tetanus (Figure 22-1c).

SPECIMEN PREPARATION

A laboratory animal is used in this exercise, to demonstrate some fundamental physiological principles. To ensure the success of the experiments, care must be exercised to avoid unnecessary injury, irritation, or pain to the laboratory animal. Any of these conditions will render the animal less capable of "normal" responses.

In this exercise, the living frog is prepared by a procedure known as *pithing*. Refer to Appendix B for the "pithing procedure."

Materials

The following materials should be provided per two students or for instructor's demonstration:

1. One normal, healthy single-pithed frog
2. Dissecting tools (needle, scalpel, probe, scissors)
3. Three sterile gauze pads (2 × 2 inch)
4. Frog Ringer's solution
5. Thread

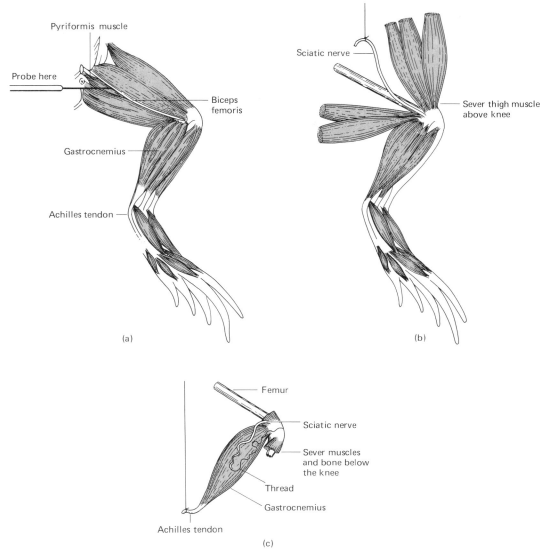

Figure 22-2 Diagrammatic representation of the musculature of the dorsal surface of a frog's leg: (a) musculature of dorsal surface; (b) relationship of sciatic nerve and gastrocnemius muscle; (c) appearance of muscle preparation.

Procedure: Preparation of Muscle

This procedure should be performed by students in pairs.

1. Remove the skin from one leg of the prepared frog by first cutting the skin as high as possible around the thigh. Roll the loosened skin back a short distance. Grasp it with one hand and the frog's thigh with the other. Peel the skin completely off the leg by pulling it quickly.
2. Use Figure 22-2a to locate the *gastrocnemius muscle*. Keep the frog's tissues moist with Ringer's solution during the remainder of the exercise.
3. Carefully insert a probe to separate the thigh muscles and to expose the sciatic nerve (Figure 22-2b).
4. Tie a piece of thread about 10 cm long (4 inches) around the sciatic nerve, and cut the nerve above the tie.
5. With the thread, gently ease the nerve from its location, and lay it on the surface of the gastrocnemius muscle.
6. Cut the thigh muscles first near their origins on the pelvic girdle and then just above the knee (Figure 22-2b).
7. Use the probe again to separate and free the gastrocnemius muscle from the other muscle of the frog's leg (Figure 22-2c).
8. Locate the Achilles tendon (Figure 22-2c). Tie a piece of thread about 15 cm long (6 inches) around

its lowest end. Cut the tendon at the point between the thread and the animal's heel bone.

9. Lift and move the gastrocnemius muscle to one side. Cut the other muscles of the frog's lower leg and the bone (tibiofibula) just below the knee.

10. Cut the femur bone to about 2.5 cm (1 inch) (Figure 22-2c). Your instructor will demonstrate the procedure for attaching the nerve–muscle preparation to the recording system. Be certain to keep this material and the other leg of the frog moist with Ringer's solution.

MUSCLE ACTIVITY MEASUREMENTS

Materials

The following materials should be provided for class use:

1. Stimulating apparatus
2. Physiograph or related recording system
3. Weight pans
4. Weights (5- and 10-g units)
5. Millimeter ruler
6. Litmus paper
7. Examples of muscle contraction tracings or records
8. Frog Ringer's solution
9. Sterile gauze pads (2 × 2 inch)

The procedures in this part of the exercise should be performed by students in groups of four. One individual should operate the equipment used. A sufficient number of tracings should be made for all members of the group.

Procedure 1: The Single Muscle Twitch

To cause a muscle twitch, it is necessary to produce a *threshold stimulus*—that is, the lowest voltage of electrical stimulation that produces a contraction. As the voltage is increased, more and more motor units are activated. The lowest voltage that activates all such units and produces a maximum contraction is called the *maximal stimulus*.

A stimulator is used to regulate the strength of the stimulus (voltage) and the frequency of stimuli (number of pulses or stimuli per second). The duration of each stimulus is measured in microseconds (μsec). The muscle stimulated to contract raises the writing device so that a trace of the contraction is made.

1. Set up the stimulating apparatus and recording system, as directed by your instructor. Refer to Appendix A for details of physiograph operation. Examine the stimulator, and note the position of its components.

2. Set the voltage dial of the stimulator at its lowest setting and the duration at 10 μsec.

3. Start the recording system at a slow speed.

4. Observe the recording for the presence of contractions.

5. Slowly increase the voltage by moving the voltage dial until contractions are observed.

6. Mark the response with the voltage needed to produce it. This is the threshold stimulus. Make four recordings of this response so that each member of the group can have one. Enter this value, and attach a representative tracing to the results and observations section. Refer to Figure 22-1a as a guide for labeling purposes.

7. Turn off the system. Moisten the muscle preparation with Ringer's solution.

8. Set the voltage dial at the threshold value. Turn on the recording system and the stimulating system.

9. Increase the voltage by one increment (step value), and stimulate the muscle.

10. Continue to stimulate the muscle and increase the voltage by one voltage increment at a time until the muscle contractions do not increase further. The voltage at this point is the *maximal stimulus,* and the effect produced is known as the *maximal response.*

11. Mark the record, and enter the value for the maximal stimulus in the results and observations section.

Procedure 2: Summation, Incomplete Tetanus, and Complete Tetanus

1. Use the same muscle preparation, stimulating apparatus, and recording system.

2. Set the voltage dial at the maximal stimulus value.

3. Stimulate the muscle at a frequency of 1 pulse per second (pps).

4. Record the effect for 15 seconds.

5. Increase the frequency to 2 pps for 15 seconds and then to 4 pps for 15 seconds.

6. Allow the muscle preparation to rest for 2 minutes. During this time, moisten the preparation again.

7. Continue to double the frequency, with a rest break following, until the muscle preparation goes into complete tetany. Make sufficient records for all members of the group.

8. Label the record, and attach it to the results and observations section. Be certain to indicate the following events on the record (refer to Figure 22-1).
   a. Point of maximal stimulus
   b. Wave summation
   c. Incomplete tetanus
   d. Complete tetanus

Procedure 3: Determination of Work Performed

1. Use the same muscle preparation, stimulating apparatus, and recording system.
2. Attach a weight pan or comparable device to the muscle preparation. Follow the directions of your instructor, and make the necessary adjustments.
3. Determine the threshold stimulus. This is the value for unloaded contraction. Enter your findings in the results and observations section.
4. Add a 5-g load to the weight pan, and determine the threshold stimulus. Enter your findings in the results and observations section.
5. Increase the voltage substantially above the threshold stimulus. Obtain a single muscle twitch. Mark the starting point of the stimulus application. Make enough recordings for each member of the group. Attach a labeled record in the results and observations section.
6. Continue to add 5-g loads until the muscle preparation can no longer contract. Mark the record with the total weight lifted. Measure the height or distance the weight was lifted. Enter your findings, and attach a label record in the results and observations section.
7. Calculate the work performed by the muscle preparation, using the following formula:

$$\frac{\text{Work}}{\text{performed}} = \text{weight of load} \times \text{distance lifted.}$$

Work can be recorded in units of gram-millimeter.
8. Turn off the stimulating apparatus and the recording system, and moisten the muscle preparation with Ringer's solution.

Procedure 4: Muscle Fatigue

Muscle fatigue represents a loss of *irritability* (ability to respond to stimuli) by muscle. It is caused by the accumulation of metabolic products such as acid phosphates, carbon dioxide, and lactic acid.

1. Use the same muscle preparation, stimulating apparatus, recording system, and weight pan.
2. Add a 10-g load to the weight pain, and determine the threshold stimulus. Record your finding in the results and observations section.
3. Increase the stimulus to 20 volts above the threshold value. Stimulate the muscle preparation about 15 times at a rate of 2 pps. Repeat this step until the recording line begins to slope downward toward the baseline. Mark this point. It represents the onset of fatigue. Continue to stimulate the muscle at the same rate. Determine the time it takes for the muscle preparation to reach the bottom of the record. This event is complete fatigue.

4. Turn off the stimulating apparatus and the recording system.
5. Remove the muscle tissue preparation and cut it into two parts. Press one cut end on a piece of red litmus paper and the other cut end on a piece of blue litmus paper. What happened? Enter and explain your findings in the results and observations section.

RESULTS AND OBSERVATIONS

The Single Muscle Twitch

1. Threshold voltage: _____
2. Maximal stimulus: _____
3. Attach a labeled representation of the single muscle twitch.

4. Distinguish between maximal stimulus and maximal response. _____

_____

_____

Summation, Incomplete Tetanus, and Tetanus

1. Attach a labeled representation of summation, incomplete tetanus, and tetanus.

2. Do the heights of individual contractions differ in summation? _____

Determination of Work Performed (Gram-Millimeter)

1. Threshold stimulus in unloaded contraction: ____

2. Threshold stimulus with first 5-g load: _____

3. Attach a labeled representation of a single twitch with first 5-g load and intense voltage.

4. Attach a labeled representation of maximum work performed.

5. Total weight (load) with which no further contraction of the muscle preparation occurred:

_____

6. Work performed by the muscle preparation:

Work performed = load × distance

____ = g × ____ mm

Muscle Fatigue

1. Threshold stimulus with 10-g load: _____

2. Threshold stimulus with increased voltage: ____

3. Time for complete fatigue to occur: _____

4. What were the results of litmus test? Explain what happened. _____

_____

_____

NAME _____

LAB SECTION _____ DATE _____

INTRODUCTION TO SKELETAL MUSCLE CONTRACTION, USING A LABORATORY ANIMAL

COMPLETION AND DEFINITION

1. Draw and label a representation (in the form of a graph) of the events of a single muscle twitch.

Height of contraction

Time (milliseconds)

2. Explain or define:

a. Tetanus (tetany) _____

_____

b. Threshold stimulus _____

_____

c. Summation _____

_____

d. Foramen magnum _____

_____

e. Maximal stimulus _____

_____

3. Distinguish between single pithing and double pithing. _____

_____

4. What substances are responsible for muscle fatigue? _____

_____

ANSWERS

1. _____

2. _____

3. _____

4. _____

5. _____

6. _____

7. _____

8. _____

9. _____

10. _____

Provide the correct answers in the spaces.

1. Which of the following is the weakest stimulus from a neuron that can still initiate a contraction?
   a. threshold stimulus
   b. subliminal stimulus
   c. tetanus
   d. twitch
   e. muscle tone

2. What is the rapid, jerky response to a single stimulus called?
   a. threshold stimulus
   b. subliminal stimulus
   c. tetanus
   d. twitch
   e. muscle tone

3. Which of the following is a stimulus that cannot initiate a stimulus?
   a. threshold stimulus
   b. subliminal stimulus
   c. tetanus
   d. twitch
   e. muscle tone

4. Which of the following factors are known to decrease the strength of muscle contractions?
   a. fatigue
   b. lack of nutrients
   c. lack of oxygen
   d. choices *a, b,* and *c*
   e. choices *b* and *c* only

5. Once a threshold stimulus is applied, individual muscle fibers of a motor unit will contract to their fullest extent or will not contract at all, provided conditions remain constant. This characteristic best describes which of the following?
   a. treppe
   b. wave summation
   c. sliding-filament theory
   d. tetanus
   e. all-or-none principle

6. What is the condition in which a skeletal muscle contracts more forcefully to the same strength of stimulus after it has contracted several times?
   a. treppe
   b. wave summation
   c. sliding filament theory
   d. tetanus
   e. all-or-none principle

7. What is the brief period between the application of a stimulus and the beginning of a contraction called?
   a. refractory period
   b. wave summation
   c. latent period
   d. relaxation period
   e. contraction period

8. Which of the following best describes the period during which two stimuli are applied, one immediately after the other, causing the muscle to respond to the first stimulus but not to the second?

   a. refractory period
   b. wave-summation period
   c. latent period
   d. relaxation period
   e. contraction period

9. Which of the following best describes the situation when two stimuli are applied, and the second is delayed until the refractory period is over, causing the skeletal muscle to respond to both stimuli?
   a. refractory period
   b. wave summation
   c. latent period
   d. relaxation period
   e. contraction period

10. A state of sustained muscle contraction is known as which of the following?
   a. relaxation period
   b. tetanus
   c. treppe
   d. threshold
   e. latent period

Human Muscle Contraction

After completing this exercise, you should be able to

1. Distinguish between *muscle tone* and *muscle contractions* that are associated with body movements
2. Differentiate between *isotonic* and *isometric* muscle contraction
3. Perform simple experiments to demonstrate muscle fatigue and to measure work done

The only action a muscle is capable of is contraction. Some other force, including gravity, must be exerted to return the muscle to its uncontracted or relaxed state. For example, the biceps (a flexor) bends the arm, and the triceps (an extensor) must contract to pull the forearm back into place and also to stretch the biceps muscle (Figure 23-1). The force involved in this case is generated by the sliding filaments (Exercise 21) and is exerted parallel to the muscle fibers.

When an individual uses the biceps muscle—for example, to lift a heavy object—the force exerted by the contracting muscle on the object is referred to as *muscle tension*. The force exerted on the muscle by the object is termed the *load*. In order for a heavy object or load to be moved, muscle tension must be greater than the load, because these two factors are opposing forces. With *isotonic* (equal tension) contractions, the muscle fibers shorten, but the tension remains the same (Figure 23-2a). A contraction of this type takes place when an individual uses the biceps to lift heavy weights or other objects. With *isometric* (equal length) contraction, another type of muscle activity, a muscle's length remains essentially the same (does not shorten), but the tension changes (Figure 23-2b). Contractions of this type occur when a muscle supports a load in a fixed position; for example, contraction of the quadriceps muscles of the leg when an individual is in a standing position.

Figure 23-1 Muscle contraction and the relaxed state. The movement of flexing the forearm involves contraction of the biceps and simultaneous relaxation of the triceps. To return the arm to the straight position the triceps contracts, extending the forearm, while the biceps relaxes and yields to the action of the triceps. The muscle that performs the action is the *agonist;* the muscle that is in opposition is the *antagonist.*

Figure 23-2 Muscle contractions: (a) isotonic; (b) isometric.

Muscle tone (tonus) is a reflex stretch response to motor-nerve impulses originating in the spinal cord. These impulses stimulate small numbers of different muscle fibers to contract at different times, resulting in a partial but sustained contraction of the entire muscle. Several muscles involved with body posture are associated with this muscle property. When muscles are not being stretched, as in the case of postural muscles during sleep or unconsciousness, the body collapses. The loss of muscle tone is due to an interference with the conduction of nerve impulses. It may result from certain disease states, such as poliomyelitis, or from accidents involving destruction to motor-nerve fibers. Muscles with less than normal tone are referred to as *flaccid* (FLAK-sid).

Some interesting aspects of muscle contractions can be revealed by *electromyography,* the procedure used to record the action potentials (electric discharges) from muscles. The records are termed *electromyograms.* When a nerve impulse arrives at the muscle fibers of a *motor unit* (a neuron and the group of muscle cells it stimulates), a reversal of the electrical charge on the muscle fiber membranes occurs. Such a reversal is the action potential that spreads throughout the muscle fibers to cause contraction. Electromyograms are recorded with the aid either of a needle or wire electrodes inserted directly into muscles or of surface electrodes placed on the skin. The action potentials from individual motor units are picked up by the needle or wire electrodes, while surface electrodes detect integrated action potentials from whole muscles. In some situations, electromyograms can be used to demonstrate which muscles are responsible for or control particular movements. For example, by placing surface electrodes over the muscles of the shin and calf of a test subject and then having the individual respond to a command such as "lean backward," "sit," or "stand," an electromyogram would show which muscles contract during specific movements and to what extent they contract (Figure 23-3).

This exercise deals with various aspects of muscle activities, including isometrics.

Materials

The following materials should be provided per four students to perform Procedures 1–4:

1. One 5- or 10-lb barbell
2. One cloth tape measure (1 meter length)
3. One relatively soft tennis or rubber ball
4. Stopwatch or other timing device
5. An ergograph or similarly constructed device to measure work performed
6. Weight pan or similar container to hold gram weights
7. Gram weights
   a. 500 g (2)
   b. 250 g (2)
   c. 100 g (5)
   d. 50 g (2)
   e. 10 g (10)
8. 1-meter stick

Procedure 1: Muscle Tone (Biceps, Triceps, Gastrocnemius, and Masseter Muscles)

This procedure should be performed by students in pairs.

1. Take turns in carrying out the following tasks. Note the degree of contraction of the biceps and triceps in each case by feeling them for *tightening* and/or

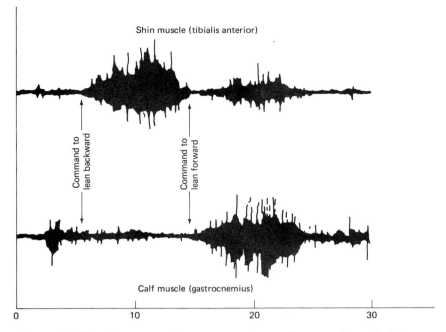

Figure 23-3 A diagrammatic representation and comparison of electromyograms obtained from human muscles of the shin and calf, during the execution of specific verbal commands. The test subject first was told to lean backward, and then to lean forward. The involvement of a specific muscle and the extent of muscle contractions following the command can be seen. (Adapted from Witherspoon, J. D., R. Flowers, M. Isaacson, and S. Keslensky. "Electromyograms." *Muscle Contraction* (audiovisual tape). Houston: Nacro Bio-Systems, Inc., 1973.)

hardening. Enter your findings, and answer the questions in the results and observations section.

a. Sit in a comfortable position, with your forearms supported on a level surface and with palms down.

b. Sit in the same position, but with palms up, and note any differences.

c. Sit in a comfortable position. Grasp the barbell, and lift it with palms up. This is known as a weight curl. (A book may be substituted for the weight.)

d. Stand erect, and allow both arms to hang freely at the sides of your body.

e. Stand erect and hold the barbell with one hand at the side of your body.

2. Take turns in carrying out the following tasks. Note the degree of contraction of the gastrocnemius muscle. Enter your findings, and answer the questions in the results and observations section.

a. Lie down on your abdomen (prone position).

b. Sit on the edge of a chair so that both feet touch the floor.

c. Sit on the edge of a table so that both feet are dangling and not touching the floor.

d. Stand erect, and lock your knees for 5 seconds.

e. Stand erect with legs relaxed.

f. Stand on your toes with your arms extended overhead, and reach toward the ceiling.

3. Locate the masseter muscle in your partner's face. Place your hands on it. Have your partner first relax and then tighten this muscle. Describe what happens in each case in the results and observations section.

Procedure 2: Distinguishing Between Isotonic and Isometric Muscle Contractions

This procedure should be performed by students in pairs.

1. Use a tape measure to measure the distance around the midline (circumference) of the biceps and triceps muscles of both arms while performing each of the following tasks. Enter your findings, and answer the questions in the results and observations section.

a. Extend both arms directly to the front of the body.

b. Flex both arms with palms facing upward, much like a weight-lifting exercise, 10 times in succession, then remeasure.

c. Extend the right arm while grasping a barbell or weight substitute.

d. Repeat Step (c) with the left arm.

e. Flex the right arm while grasping a barbell, 10 times in succession, then remeasure.

f. Repeat Step (e) with the left arm.

2. With a tape measure, determine the circumference of the midline calf region of both legs, under the condition of each of the following tasks. Enter your findings, and answer the questions in the results and observations section.

a. Lie down on your abdomen (prone position), with gastrocnemius relaxed.

b. Repeat Step (a), but contract the gastrocnemius muscles.

c. Stand erect.

d. Stand on your toes, and extend your arms upward.

e. Stand on your toes, with your arms at the side.

f. Sit on a table, so that your legs are dangling and do not touch the floor.

g. Repeat Step (f), but try to move your feet forward and upward with someone pushing against them (try to prevent any movement backward or downward).

3. Change places with your partner, and repeat the steps and tasks of this procedure.

Procedure 3: Determining Muscle Fatigue

This procedure should be performed by students in groups of four.

1. Hold a tennis ball (or soft rubber ball) in your right hand. As you squeeze the ball as rapidly as possible, have your laboratory partners count the number of complete contractions during each 10-second interval for a total period of 3 minutes nonstop. Enter the count in the results and observations section.

2. Continue to time while you squeeze the ball until fatigue is complete. Record the time. This is the end of Trial 1.

3. Wait 1 minute, and repeat Steps 1 and 2. This is the end of Trial 2.

4. Wait 5 minutes and repeat Steps 1 and 2. This is the end of Trial 3.

5. Repeat all steps with your left hand.

6. Now have each of your laboratory partners repeat Steps 1–4. Enter your findings, and answer the questions in the results and observations section.

Procedure 4: Determination of Work Performed

This procedure should be performed by students in groups of four.

1. Assemble an ergograph similar to the one shown in Figure 23-4. Tie one end of the cord to a weight pan or comparable container. Have one partner hold the other end of the cord until the experiment is ready to proceed.

2. Put the ergograph on a level surface, and sit next to it so that you can comfortably place your left forearm upon its surface. Your fingers should be pointing toward the pulley system. Strap the forearm to the board.

3. Tie the free end of the cord to the middle phalanx of the index finger of the hand.

4. Have a laboratory partner hold a meter stick next to the weight pan and note the level at which the pan is located.

5. Have a laboratory partner add weights, in 10-g increments, to the pan. Lift the weight pan by raising your index finger. Continue to add weights until the weight pan cannot be lifted more than 5 mm. In carrying out this task, your hand must not be lifted from the ergograph's surface, nor should the wrist be flexed. Enter the maximum number of grams lifted and the distance (in cm) the weight was lifted in the results and observations section.

6. Repeat Steps 3–5 with the middle finger and again with four fingers together, not including the thumb.

7. Repeat this entire procedure with your laboratory partners.

8. Calculate the work performed (weight lifted in grams × distance lifted in centimeters). Enter your calculations, and answer the questions in the results and observations section.

Materials

The following materials should be provided for class use, to produce electromyograms (Procedure 5):

1. Skin electrodes (8 mm or 12.5 mm diameters)

2. Electrode washers (8 mm or 12.5 mm diameters) or micropore tape for attaching electrodes to the skin

3. Skin cleanser to prepare the skin

4. Electrode gel

5. Kimwipes® or paper towels

6. Sterile, disposable alcohol prep-pads

7. Scissors

8. Physiograph

Procedure 5: Electromyograms from Leg Muscles

This procedure should be performed by students in pairs. Your instructor will have a physiograph or similar equipment ready for class use. Refer to Appendix A for a general review of the component parts of a physiograph.

Figure 23-4 An ergograph: This device can be used to determine work performed by skeletal muscle; it consists of a board, pulley, cord, weight pan, and restraining straps.

1. Obtain four of the skin electrodes, and—using the individual alcohol prep-pads—wipe them and allow them to dry on a clean surface.
2. Turn on the power switch of the physiograph, and allow the instrument to warm up for at least 10 minutes.
3. Using a paper towel and the skin cleaner provided, wipe the skin over the tibialis anterior (shin) and gastrocnemius (calf) muscles (See Figure 23-5.)
4. Apply electrode gel onto the sterilized electrode surfaces. Use a Kimwipe® to smooth the gel evenly so that it fills the well (depression) between the metal surfaces and surrounding plastic or similar type of ring of the electrode.
5. Cut four 1-inch micropore tape strips, or remove the paper coverings over the adhesive surfaces of the electrode washers.
6. Use the adhesive strips to apply two skin electrodes to the tibialis anterior muscle and two skin electrodes to the gastrocnemius muscle in the pattern shown in Figure 23-5. Make certain that the electrodes are placed parallel to one another and that one of the recording electrodes is placed over the proximal portion of the muscle and the second electrode is positioned over the distal portion.

Figure 23-5 Placement of biopotential skin electrodes for electromyograms (recordings) from the tibialis anterior muscle of the shin and the gastrocnemius muscle of the calf. The two muscles may contract simultaneously, as when standing. At other times, one muscle contracts while the other relaxes. This occurs when the toe end of the foot is raised.

7. Check to see that the electrodes are adhering securely to the skin. Press on them.
8. Make the following settings on two modules to be used for the electromyographs.
   a. Time constant setting to 0.03
   b. Sensitivity setting to between 20 and 100
   c. Gain knob setting to ×100
   d. Chart speed to 1.0 cm/sec
9. Plug the electrode into the appropriate modules, and turn on the physiograph to record.
10. Establish a baseline recording by having your partner relax in the standing position.
11. Next, issue the following commands to your partner to execute in the order given. Mark the electromyogram with each command. Refer to Figure 23-3.
    a. Stand straight with both legs together
    b. Sit (on the chair provided)
    c. Stand again
    d. Lean forward from the standing position
    e. Bend the knees slightly from the standing position.

    Note the degree of contraction for each muscle by examining the electromyogram. Enter your findings, and answer the questions in the results and observations section.

12. Change the chart speed to 0.5 cm/sec, and repeat Steps 10 and 11.
13. Change places with your laboratory partner, and repeat Steps 1–12.
14. Remove your electromyogram, and attach it to the space provided in the results and observations section.

RESULTS AND OBSERVATIONS

Muscle Tone

1. Enter your findings on the biceps and triceps and those of your laboratory partner in Table 23-1.
2. Enter your findings on the gastrocnemius and those of your laboratory partner in Table 23-2.
3. Describe the effects observed with the masseter muscle. _____
4. In which of the tasks performed was muscle tonus evident? _____

Table 23-1

Task	Degree of Contraction* of Biceps and Triceps	
	Your Results	Partner's Results
Forearms supported with palms down		
Forearms supported with palms up		
Weight curl		
Standing with arms at side		
Standing with one hand grasping weight		

*Determine by increases in size or firmness.

Table 23-2

Task	Degree of Contraction* of Gastrocnemius	
	Your Results	Partner's Results
Body in prone position		
Sitting with feet touching the floor		
Sitting with legs dangling over an edge		
Standing with knees locked		
Standing in a relaxed manner		
Standing erect on tiptoes		

*Determine by increases in size or firmness.

Distinguishing Between Isotonic and Isometric Muscle Contractions

1. In Table 23-3, enter both your and your laboratory partner's findings on the biceps.
2. In Table 23-4, enter both your and your laboratory partner's findings on the gastrocnemius.
3. Were any differences found between circumference measurements of the right and left arms? If so, explain these differences. _____

4. Were any differences found between circumference measurements of the right and left gastrocnemius muscles? If so, explain. _____

5. What variables influence the reading of measurements? _____

Table 23-3

Task	Measurements			
	Your Results		Partner's Results	
	Right Arm	*Left Arm*	*Right Arm*	*Left Arm*
Body arms extended				
Both arms flexed with palms up				
Right arm extended with weight				
Left arm extended with weight				
Right arm flexed with weight				
Left arm flexed with weight				

Table 23-4

Task	Measurements			
	Your Results		Partner's Results	
	Right Leg	*Left Leg*	*Right Leg*	*Left Leg*
Body in prone position with gastrocnemius relaxed				
Body in prone position with gastrocnemius contracted				
Standing erect				
Standing on toes with arms extended				
Standing on toes with arms at side				
Sitting with legs dangling				
Sitting with force applied to dangling legs				

Determining Muscle Fatigue

1. Enter your findings in Table 23-5.
2. Indicate the time required for complete fatigue to occur.

   Student 1: _____

   Student 2: _____

   Student 3: _____

   Student 4: _____

3. Were all subjects able to squeeze the ball for 3 minutes? If not, why not? _____

   _____

4. Explain any differences noted in the results obtained with the right hand or the left hand. _____

   _____

   _____

Determination of Work Performed

1. Enter both your measurements on the ergograph and those of your laboratory partners in Table 23-6.
2. Which of the fingers performed the most work?

   _____

3. Did your results differ from those of your partners?

   _____

4. How can you account for any differences in the work performed? Explain your answer by considering the muscles involved. _____

Table 23-5

Student	Trial	Number of Contractions During 10-sec Intervals															
		10		20		30		40		50		60		120		180	
		R*	L	R	L	R	L	R	L	R	L	R	L	R	L	R	L
	1																
	2																
	3																
	1																
	2																
	3																
	1																
	2																
	3																
	1																
	2																
	3																

*R = right hand; L = left hand.

Table 23-6

Student	Weight Lifted (in g)			Distance Lifted (in cm)			Work Performed*		
	IF†	MF	FF	IF	MF	FF	IF	MF	FF

*Work performed = weight lifted × distance lifted.
†IF = index finger; MF = middle finger; FF = four fingers.

Electromyograms from Leg Muscles

1. Place the appropriate portion of the electromyogram, or sketch the results in the space provided here.

2. Enter your findings in Table 23-7. Indicate which muscle(s) contracted during the execution of the commands listed. Use either a " + " or a " − ."

Table 23-7

	Muscles Involved	
Command	Tibialis Anterior (shin)	Gastrocnemius (calf)
Sit		
Stand		
Lean forward		
Knee bend		
Extend toe		

LABORATORY REVIEW *23*

NAME _____

LAB SECTION _____ DATE _____

HUMAN MUSCLE CONTRACTION

ANSWERS

COMPLETION AND DEFINITION

1. Distinguish between isotonic and isometric contractions.

   _____

   _____

2. What is muscle tension? _____

   _____

3. Lifting a weight exerts a force on a muscle. What is this force called? _____

   _____

4. Define the following:

   a. Flaccid _____

   b. Muscle fatigue _____

   c. Muscle tone _____

   d. Electromyogram _____

   _____

5. Distinguish between flexor and tensor muscles. _____

   _____

MULTIPLE CHOICE

Provide the correct answers in the spaces.

1. What is the force exerted by a contracting muscle used to lift a heavy object?
   a. load
   b. muscle tension
   c. tonus
   d. flaccid
   e. potential

2. What is the force exerted by the object being lifted on the muscle in Question 1?
   a. load
   b. muscle tension
   c. tonus
   d. flaccid
   e. potential

3. In order for a heavy object or load to be moved, muscle tension must have what quantitative relationship to the load?
   a. less than
   b. greater than
   c. same as

1. _____
2. _____
3. _____
4. _____
5. _____
6. _____
7. _____
8. _____
9. _____
10. _____

4. What type of contraction takes place when an individual uses the biceps to lift a heavy weight?
   a. isometric
   b. equal length
   c. isotonic
   d. impulse
   e. flaccid

5. What is a reflex stretch response to motor-nerve impulses originating in the spinal cord?
   a. a full contraction
   b. muscle tension
   c. muscle tone
   d. relaxed state
   e. flaccid

6. How do you describe a muscle with less than normal tone?
   a. in a full contraction
   b. showing muscle tension
   c. relaxed
   d. tonus
   e. flaccid

7. What is the actual record of action potentials from muscles called?
   a. an electrode
   b. a contraction
   c. electromyograph
   d. an electromyogram
   e. isometrics

8. In the case of an isometric contraction, what happens to a muscle's length?
   a. shortens
   b. remains the same
   c. lengthens

9. In the case of an isometric contraction, what happens to the tension on the muscle?
   a. changes
   b. does not change

10. With isotonic contractions, what happens to muscle fibers?
    a. shorten
    b. lengthen
    c. do not change

SECTION VI

Nervous System Coordination and Control

*T*he nervous and endocrine systems regulate the activities and functions of the body. These two systems, however, operate at different rates. The nervous system acts rapidly, while the endocrine system operates more slowly. This section considers the nervous system.

The nervous system is divided into the central nervous system, the peripheral nervous system, and the autonomic nervous system.

The *central nervous system (CNS)* includes the brain and the spinal cord. It contains the centers where connections are made between nerve fibers bringing information from receptor (sense) organs and nerve fibers carrying information to the effector organs (muscles and glands).

The *peripheral nervous system (PNS)* consists of the nerves connecting the peripheral parts of the body with the central nervous system. Twelve pairs of nerves arise from the brain and are called the cranial nerves; 31 pairs of nerves arise from the spinal cord and are called the spinal nerves.

The *autonomic nervous system (ANS)* is composed of nerves, nerve trunks with ganglia, and special ganglia that govern the operation of smooth muscle, cardiac muscle, and glands, which—for the most part—are not under voluntary control.

The exercises in this section consider the general structure and organization of the nervous system, as well as selected activities and functions associated with the components of the system.

GENERAL REFERENCES

Brown, A. G. *Nerve Cells and Nervous Systems: An Introduction to Neuroscience.* New York: Springer-Verlag, 1991.

Geschwind, N. "Specializations of the Human Brain." *Scientific American,* 241: 180–199, 1979.

Guyton, A. C. *Basic Neuroscience: Anatomy and Physiology,* 2nd ed. Philadelphia: Saunders, 1991.

Iversen, L. L. "The Chemistry of the Brain." *Scientific American,* 241: 134–149, 1979.

Netter, F. H. *The CIBA Collection of Medical Illustrations: Vol. 1, The Nervous System.* Summit, NJ: CIBA Pharmaceutical Products, 1953.

Noback, C. R., N. L. Strominger, and R. J. Damarst. *The Human Nervous System: Introduction and Review.* Philadelphia: Lea & Febiger, 1991.

Peters, A., S. L. Palay, and H. de F. Webster. *The Fine Structure of the Nervous System: Neurons and Their Supporting Cells,* 3rd ed. New York: Oxford University Press, 1991.

Roberts, P. A. *Neuroanatomy (Oklahoma Notes),* 2nd ed. New York: Springer-Verlag, 1991.

Tortora, G. J., and S. R. Grabowski. *Principles of Anatomy and Physiology,* 6th ed. New York: HarperCollins, 1993.

SPECIFIC REFERENCES

Aoki, C., and P. Siekevitz. "Plasticity in Brain Development." *Scientific American,* 259: 56–64, 1988.

Bennett, D. A., and D. A. Evans. "Alzheimer's Disease." *Disease-a-Month.* 38: 1–64, 1992.

Dunant, Y., and M. Israël. "The Release of Acetylcholine." *Scientific American,* 252: 58–66, 1985.

Houston, J. P. *Fundamentals of Learning and Memory,* 2nd ed. New York: Academic Press, 1981.

Stevens, C. F. "The Neuron." *Scientific American,* 241: 54–65, 1979.

Williams, M. *Brain Damage, Behavior and the Mind.* New York: Wiley, 1979.

Winter, P. M., and J. N. Miller. "Anesthesiology." *Scientific American,* 252: 124–131, 1985.

Anatomy of the Nervous System

After completing this exercise, you should be able to

1. Name the basic parts of the mammalian brain
2. Locate the cranial and spinal nerves
3. Dissect the mammalian brain

Materials

The following materials should be provided for class use:

1. Sheep brains (preserved)
2. Bone saw
3. Bone forceps
4. Model of human nervous system
5. Charts of human nervous system

The nervous system is divided into the central nervous system, the peripheral nervous system, and the autonomic nervous system.

The central nervous system (CNS) includes the brain and the spinal cord. It contains the centers where connections are made between nerve fibers bringing information from receptor (sense) organs, and nerve fibers carrying information to the effector organs (muscles and glands).

The peripheral nervous system consists of the nerves connecting the peripheral parts of the body with the central nervous system. Twelve pairs of nerves arise from the brain and are called the cranial nerves; 31 pairs of nerves arise from the spinal cord and are called the spinal nerves.

The autonomic nervous system is composed of nerves, nerve trunks with ganglia, and special ganglia that govern the operation of smooth muscle, cardiac muscle, and glands which are, for the most part, not under voluntary control.

SHEEP BRAIN, DORSAL SURFACE

For detailed study, it is best to use a sheep's brain (either fresh or preserved). It is large enough to see the parts clearly and is easily obtained. If a sheep's brain is not available and you wish to use the pig's brain, there are instructions for its dissection at the end of this exercise.

Covering the brain are three membranes or *meninges*. Usually, only the outermost and toughest membrane, the *dura mater,* is visible. The middle membrane is the *arachnoid,* and the innermost pigmented membrane is the *pia mater*.

In a dorsal view (Figure 24-1), the brain reveals a large anterior portion, the *cerebrum* or the *cerebral hemispheres*. Posterior to the cerebrum is the *cerebellum*. The *medulla oblongata* (Figure 24-2) is located anterior to the spinal cord. If the cerebellum is gently drawn back (Figure 24-1), some of the midsection of the dorsal brain is revealed. One can see four distinct spherical bodies, the *corpora quadrigemina,* which make up the dorsal midbrain, and which can be seen more clearly when the brain is cut in half. Much of the dorsal surface of the brain that is visible in lower animals is obscured by the drastic folding of both cerebral and cerebellar tissue. Observe both dorsal and ventral views, and locate the following:

IN A DORSAL VIEW (Figure 24-1)

1. Cerebral hemispheres
2. Midbrain (corpora quadrigemina)
   a. Superior colliculi (see Figure 24-6)
   b. Inferior colliculi (see Figure 24-6)
3. Cerebellum
4. Medulla oblongata

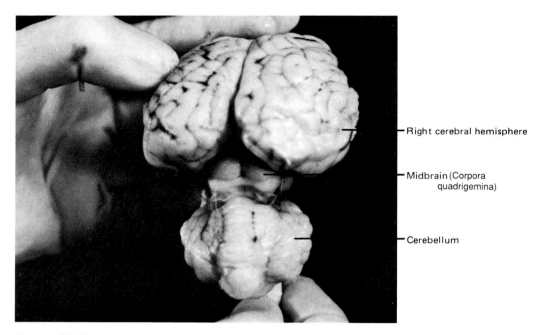

Right cerebral hemisphere

Midbrain (Corpora quadrigemina)

Cerebellum

Figure 24-1 Sheep brain (dorsal view). Separate the cerebrum and cerebellum to see the midbrain.

IN A VENTRAL VIEW (Figure 24-2)

1. Olfactory bulb
2. Cerebral hemispheres
3. Infundibulum and pituitary gland (if present)
4. Mammillary bodies
5. Cerebral peduncle
6. Pons varolii
7. Medulla oblongata

SHEEP BRAIN, VENTRAL SURFACE

Using Figures 24-2, 24-3, and your own specimen, compare the sheep brain with the human brain (Figure 24-4).

Once again, locate the cerebral hemispheres, cerebellum, and medulla oblongata. In addition, identify the following:

1. *Olfactory nerve tracts* cross the ventral surface of the hemispheres, ending in two enlarged areas—the *olfactory bulbs.*
2. *Optic nerves* cross over to form *optic tracts;* the crossover point is the *optic chiasma.*
3. The *pituitary gland,* an endocrine gland, is located

behind the optic chiasma. If it is present, it is about the size of a pea. If it has been removed, it leaves a small tissue mass, the *infundibulum,* which connects the pituitary to the brain.

4. *Mammillary bodies* are small U-shaped regions in the midline, posterior to the pituitary gland.
5. *Cerebral peduncles* are the masses surrounding the mammillary bodies and extending back about 1 inch. This area gives rise to the third pair of cranial nerves.
6. The *pons varolii* is a bridgelike area in the cerebellar region.
7. The *medulla oblongata* has already been identified on the dorsal surface and can also be seen clearly on the ventral surface. Posterior to it is the spinal cord.
8. The *trapezium* is the right or left side of the medulla of the sheep brain.

Also note the large vascular network on the ventral surface. Blood flows toward the brain in the *basilar artery,* which divides in the region of the cerebral peduncles to form circular vessels around the pituitary and the optic chiasma. This circuit, which has other contributing sources, is termed the *circle of Willis.* It is discussed in more detail in a later exercise.

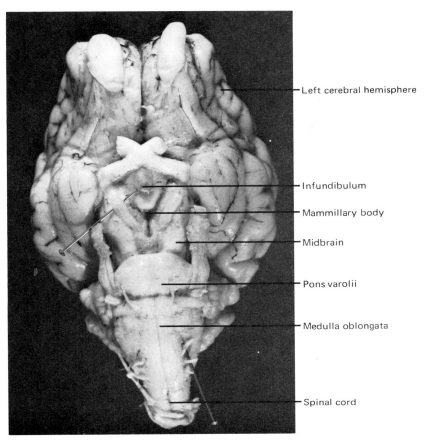

— Left cerebral hemisphere

— Infundibulum

— Mammillary body

— Midbrain

— Pons varolii

— Medulla oblongata

— Spinal cord

Figure 24-2 Sheep brain (ventral view) (with pituitary removed). The pins underline right cranial nerve III and left cranial nerves IX, X, and XI.

CRANIAL NERVES

The cranial nerves exit from the brain on the ventral surface. They can be identified with the aid of Figures 24-2 and 24-3. Compare the cranial nerves with their counterparts in the human brain (Figure 24-4).

I. *Olfactory nerves* contain nerve fibers from the nasal membranes and cannot be seen in this preparation. The nerve fibers synapse in the olfactory bulb. The second set of neurons continues in the *olfactory tract* and terminates in the olfactory areas of the cerebral hemispheres.

II. *Optic nerves* arise in the retina of the eye and cross over anterior to the pituitary gland to form the *optic chiasma*. They enter the brain as the *optic tracts*. Some fibers within the nerve cross over at the chiasma; others do not. The optic nerves are the only cranial nerves that cross over.

III. *Oculomotor nerves* are medium-sized nerves posterior to the mamillary bodies. They supply four of the six pairs of extrinsic eyeball muscles with nerve fibers.

IV. *Trochlear nerves* are small and arise from the midbrain. They project from the cleft between the posterior end of the cerebrum and the pons varolii. They supply nerve fibers to one pair of the extrinsic eyeball muscles.

V. *Trigeminal nerves* are large and arise laterally from the pons. Together with cranial nerve VII, they supply nerves to much of the head and face.

VI. *Abducens nerves* are small and arise medial to the trigeminal nerves from the anterior end of the medulla oblongata, at the border of the trapezium. They supply nerves to one pair of extrinsic eyeball muscles.

VII. *Facial nerves* are medium-sized and arise directly behind the trigeminal nerves from the trapezium area. They supply nerves to the head and face.

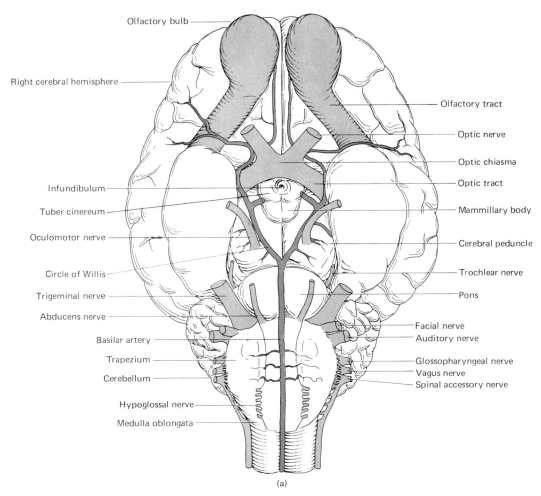

Olfactory bulb

Right cerebral hemisphere

Infundibulum

Tuber cinereum

Oculomotor nerve

Circle of Willis

Trigeminal nerve

Abducens nerve

Basilar artery

Trapezium

Cerebellum

Hypoglossal nerve

Medulla oblongata

Olfactory tract

Optic nerve

Optic chiasma

Optic tract

Mammillary body

Cerebral peduncle

Trochlear nerve

Pons

Facial nerve

Auditory nerve

Glossopharyngeal nerve

Vagus nerve

Spinal accessory nerve

(a)

Figure 24-3 Sheep brain (ventral view): (a) Diagram shows pituitary removed and cranial nerves identified by name.

VIII. *Acoustic (auditory, statoacoustic, or vestibulo-cochlear) nerves* are large, short nerves that arise from the lateral trapezium area, posterior to cranial nerve VII. They are associated with the cochlea (hearing) and the semicircular canals (equilibrium).

XI. *Glossopharyngeal nerves* are small nerves that arise laterally from the medulla oblongata. They are associated primarily with the tongue and pharynx regions.

X. *Vagus nerves* are small and arise from the lateral medulla posterior to the glossopharyngeal nerves.

They supply nerves to the pharynx and larynx and extend posteriorly into the thoracic and abdominal cavities to supply visceral organs with parasympathetic fibers.

XI. *Accessory nerves* have several roots of origin in the most posterior part of the lateral medulla. They innervate several neck and shoulder muscles.

XII. *Hypoglossal nerves* also arise from several roots in the medulla oblongata, medial to cranial nerves XI. They are associated with the tongue muscles.

(b)

1. Olfactory bulb
2. Olfactory tracts
3. Optic nerve (II)
4. Optic chiasma
5. Optic tract
6. Oculomotor nerve (III)
7. Pituitary gland
8. Trochlear nerve (IV)
9. Trigeminal nerve (V)
10. Abducens nerve (VI)
11. Facial nerve (VII)
12. Acoustic nerve (VIII)
13. Glossopharyngeal nerve (IX)
14. Vagus nerve (X)
15. Spinal accessory nerve (XI)
16. Hypoglossal nerve (XII)

Figure 24-3 *(Continued)* (b) Photograph with highlighting and labels shows pituitary in place and cranial nerves identified by number.

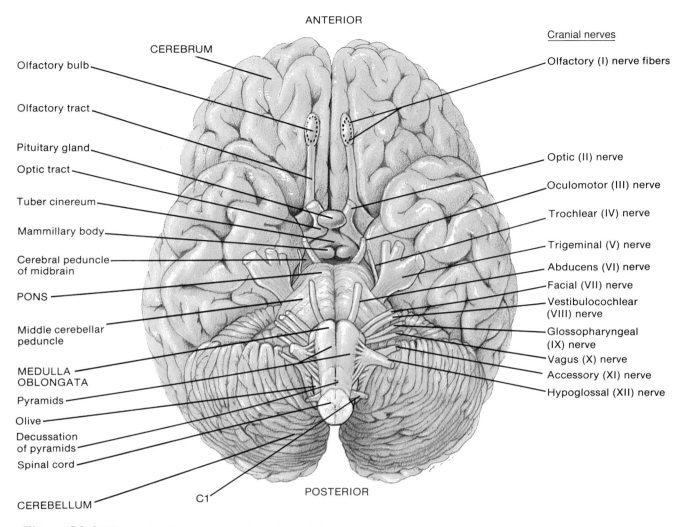

ANTERIOR

CEREBRUM

Cranial nerves

Olfactory bulb

Olfactory (I) nerve fibers

Olfactory tract

Pituitary gland

Optic (II) nerve

Optic tract

Oculomotor (III) nerve

Tuber cinereum

Trochlear (IV) nerve

Mammillary body

Trigeminal (V) nerve

Cerebral peduncle of midbrain

Abducens (VI) nerve

Facial (VII) nerve

PONS

Vestibulocochlear (VIII) nerve

Middle cerebellar peduncle

Glossopharyngeal (IX) nerve

MEDULLA OBLONGATA

Vagus (X) nerve

Pyramids

Accessory (XI) nerve

Olive

Hypoglossal (XII) nerve

Decussation of pyramids

Spinal cord

POSTERIOR

CEREBELLUM

C1

Figure 24-4 Human brainstem: ventral surface of the brain, showing the structure of the brainstem in relation to the cranial nerves and associated structures. (From G. J. Tortora and S. R. Grabowski, *Principles of Anatomy and Physiology*, 7th ed., New York: HarperCollins, 1993.)

SHEEP BRAIN IN MIDSAGITTAL SECTION

Brain Ventricles or Cavities

Cut the sheep brain in half, using a sharp, long-bladed knife. Examine the cut surface, and identify as many parts as possible, referring to Figures 24-5 and 24-6. Compare the midsagittal section of the sheep brain with similar views of the human brain (Figures 24-7 and 24-8).

Within the brain are cavities filled with cerebrospinal fluid. These are *brain ventricles* and are numbered or named as follows:

I and II (lateral ventricles)—located in the cerebral hemispheres

III (third ventricle)—located in the diencephalon region

IV (fourth ventricle)—located in the medulla oblongata

The lateral ventricles connect with the third ventricle by channels termed the *interventricular foramina* or *foramina of Munro*. Ventricle III is connected to ventricle IV by the *cerebral aqueduct* or *aqueduct of Sylvius*. Figures 24-8 and 24-9 show the location of each ventricle in the human brain. The locations and shapes of the ventricles in the sheep and pig brains are similar.

1. Cerebrum
2. Septum pellucidum
3. Fornix
4. Pineal gland
5. Midbrain
6. Cerebellum
7. Spinal cord
8. Medulla oblongata
9. Pons
10. Pituitary
11. Mammillary body
12. Optic chiasma
13. Massa intermedia
14. Corpus callosum

Figure 24-5 Sheep brain (midsagittal section).

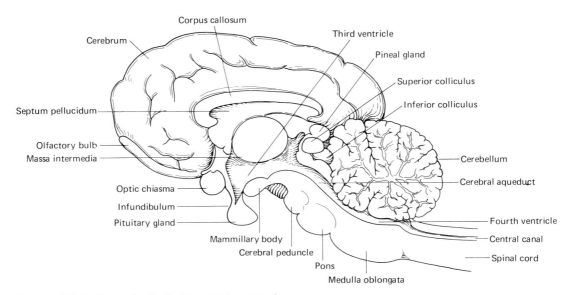

Figure 24-6 Sheep brain (midsagittal section).

Figure 24-7 Human brain: principal parts (seen in sagittal section). The infundibulum and pituitary gland are discussed, in conjunction with the endocrine system, in Exercise 27. (From G. J. Tortora and S. R. Grabowski, *Principles of Anatomy and Physiology,* 7th ed., New York: HarperCollins, 1993.)

Brain Parts

Telencephalon—Cerebrum

This region becomes highly developed in all mammals and particularly so in the human. It is composed of the right and left *cerebral hemispheres* and contains the lateral ventricles. (See "cerebrum" on Figures 24-7, 24-8, and 24-9.)

Note the extent of folding of the *cerebral hemispheres* back over the brainstem. The large mass of cerebral tissue growing back over other parts of the brain is an evolutionary phenomenon. It has reached its greatest extent in the human brain. Near the floor of the cerebrum is a tough white area termed the *corpus callosum,* the fibers of which connect the right and left hemispheres (see Figure 24-8).

Diencephalon

The *diencephalon* (Figure 24-7) is a less distinct region than the cerebrum. It consists of several structures:

1. *Pineal gland*—a small, dorsal projection of endocrine nature
2. *Infundibulum*—a ventral downgrowth attached to the hypophysis of the pituitary gland
3. *Anterior choroid plexus*—a network of blood vessels roofing the third brain ventricle (see Figure 24-5). (Together with other choroid plexuses, it contributes to the formation of cerebrospinal fluid.)
4. *Fornix*—part of the olfactory pathways of the brain
5. *Massa intermedia* or *intermediate mass*—gray matter that connects the right and left thalamus regions (centrally located)

Figure 24-8 Meninges and ventricles of the brain: brain, ventricles, spinal cord, and meninges (seen in sagittal section). Arrows indicate the direction of flow of cerebrospinal fluid. (From G. J. Tortora and S. R. Grabowski, *Principles of Anatomy and Physiology,* 7th ed., New York: HarperCollins, 1993.)

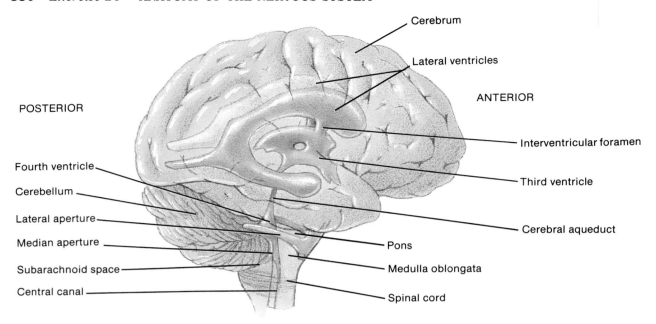

Figure 24-9 Human brain, ventricles. (From G. J. Tortora and S. R. Grabowski, *Principles of Anatomy and Physiology,* 7th ed., New York: HarperCollins, 1993.)

6. *Thalamus*—represents the lateral walls of the diencephalon, which are difficult to see in a midsagittal section
7. *Hypothalamus*—the floor of the diencephalon, which gives rise to the infundibulum
8. *Epithalamus*—rooflike portion made up of a vascular network termed the *anterior choroid plexus*

Mesencephalon

The *mesencephalon* or *midbrain* is part of the brainstem and consists of the following parts (featured in Figure 24-6, for sheep brain):

1. *Corpora quadrigemina*—located dorsally and composed of (a) the superior colliculi, which are visual reflex centers; and (b) the inferior colliculi, which are auditory reflex centers.
2. *Cerebral peduncles*—located ventrally

Metencephalon

The embryonic metencephalon develops into the following adult parts.

1. *Cerebellum*—The cerebellum is the second largest part of the brain. It is dorsally located and highly convoluted, and its gray cortex and inner white matter are distinct. The branching patterns of the white matter within the cerebellum are termed the *arbor vitae* because they resemble the branches of

a tree. The middle region of the cerebellum is the *vermis.*
2. *Pons*—This is an important relay center in the adult brain. It is located ventrally (Figure 24-4) and consists primarily of fiber tracts.
3. *Trapezium*—The trapezium is a fiber-tract area, which is considered a part of the medulla oblongata in the pig and sheep (Figure 24-3a) but is a part of the human pons.

Myelencephalon

The *medulla oblongata* is the most posterior part of the sheep brain (Figure 24-6). It merges with the spinal cord. Within it is the fourth ventricle. Roofing ventricle IV is another blood vessel plexus, the *posterior choroid plexus* (see Figure 24-8, showing human brain). Large motor tracts descending from the cerebrum cross over in the medulla so that skeletal muscle fibers on the right side of the body are controlled by the left side of brain. Thus, left-side muscle function is controlled by the right cerebral hemisphere.

Notice that the brain ventricles connect with the central canal of the spinal cord. The first and second ventricles can be seen by making a cross-section of the cerebral hemispheres. Refer to Figure 24-8.

The *pons varolii* is a bridgelike region of the same embryonic midbrain area. It consists primarily of fiber tracts.

THE HUMAN BRAIN: SURFACE ANATOMY OF THE CEREBRUM

Study the external anatomy of the cerebral hemispheres, using Figure 24-10. If a human brain is not available, use the sheep brain. The convolutions of the brain form grooves and prominences. The grooves may be either shallow (*sulci*-singular, *sulcus*) or deep (*fissures*). The *central sulcus* forms a key landmark on the brain, separating the lobes. The rounded prominences on either side of the grooves are termed *gyri*. Two prominent gyri are as follows:

Precentral gyrus—located anterior to the central sulcus; represents the location of an important motor area of the cortex

Postcentral gyrus—located posterior to the central sulcus; represents the location of important sensory areas of the cortex

(a)

(b)

Figure 24-10 Lobes and fissures of the human cerebrum: (a) Diagram of right lateral view; because the insula cannot be seen externally, it has been projected to the surface. (b) Photograph of right lateral view (courtesy of Martin Rotker, Taurus Photos).

(c)

(d)

Figure 24-10 *(Continued)* Lobes and fissures of the cerebrum: (c) Diagram of superior view; the insert to the left indicates the relative differences among a gyrus, a sulcus, and a fissure. (d) Photograph of superior view (courtesy of Martin Rotker, Taurus Photos).

Each hemisphere of the brain may be divided into four basic areas, termed *lobes,* as follows:

1. *Frontal lobes*—parts of the brain that constitute the area beneath the frontal bone, located anterior to the central sulcus
2. *Parietal lobes*—paired areas located posterior to the central sulcus and the frontal lobes
3. *Temporal lobes*—located laterally and separated from the frontal lobes by the *lateral cerebral sulcus*
4. *Occipital lobes*—occupying the area beneath the occipital bone. They are separated from the parietal lobe by the *parietooccipital sulcus*

Each hemisphere is separated from the other by the *longitudinal fissure.* The cerebrum is separated from the cerebellum by the *transverse fissure.*

HUMAN SPINAL CORD AND SPINAL NERVES

The human spinal cord and its 31 pairs of spinal nerves should be studied from models and charts (see Figure 24-11). The cord begins posterior to the medulla oblongata and continues down the middle of the back for about 16 to 18 inches. There are two enlargements, one in the neck region (*cervical enlargement*) and the other in the midback region (*lumbar enlargement*). During embryonic development, the spinal cord does not keep pace with body growth in length, and it terminates at the level of the second lumbar vertebra. The last seven or eight pairs of nerves run posteriad beyond the spinal cord until they reach their designated exits, creating what looks like a bundle of hairs, appropriately named the *cauda equina,* or horse's tail. The most inferior portion of the cord, in the region of the second lumbar vertebra, is the *conus medullaris.* Extending from the conus medullaris to the last sacral vertebra is a strand of connective tissue named the *filum terminale.*

The names and numbers of each spinal nerve correspond to the names and numbers of the vertebra at their point of exit (Figure 24-11). They are classified as follows:

1. *Cervical nerves* (8 pairs)—The first cervical nerves (C_1) exit between the skull and the first cervical vertebra. The eighth cervical nerves (C_8) exit from the intervertebral foramina between the seventh cervical vertebra and the first thoracic vertebra.
2. *Thoracic nerves* (12 pairs)—The first thoracic nerves (T_1) exit from the intervertebral foramina between the first and second thoracic vertebrae. The twelfth thoracic (T_{12}) nerves exit between the twelfth thoracic vertebra and the first lumbar vertebra.
3. *Lumbar nerves* (5 pairs)—The first lumbar nerves (L_1) exit from the intervertebral foramina between the first and second lumbar vertebrae. The fifth lumbar nerves (L_5) exit between the fifth lumbar vertebra and the sacrum.
4. *Sacral nerves* (5 pairs)—The first sacral nerves (S_1) exit between fused first and second sacral vertebrae. The fifth sacral nerves (S_5) exit inferior to the sacrum.
5. *Coccygeal nerves* (1 pair)—The coccygeal nerves (Co_1) exit at the coccyx.

NERVE PLEXUSES

A nerve *plexus* is a place where two or more nerves join and exchange nerve fibers. Figure 24-12 shows an example of the type of network formed by adjacent spinal nerves as they leave the spinal column. Such networks are formed on both sides of the body. Figure 24-12 shows the lumbar plexus in detail.

There are several important nerve plexuses. Using your text and charts (Figures 24-11 and 24-12), identify the following:

1. *Cervical plexus*—involves nerves C_1 through C_5
2. *Brachial plexus*—involves nerves C_4 through C_8 and T_1 and T_2
3. *Lumbar plexus*—involves nerves L_1 through L_4
4. *Sacral plexus*—involves nerves L_4, L_5, and S_1 through S_4

BRAIN AND SPINAL CORD OF THE PIG

The fetal pig nervous tissue is extremely soft and cannot be removed easily from its bony enclosure. If you want to observe the brain, remove the bone and muscle from the dorsal surface, according to the following instructions:

Pig Brain

Remove all skin and muscle from the dorsal surface of the pig's shoulder, neck, and head, as shown in Figure 24-13. Begin at the snout, and continue back to the shoulder region. There is a great deal of muscle tissue in the pig's neck region (see Figure 24-14). Remove this tissue, and expose the underlying bone.

When all of the soft tissue has been removed, begin the removal of the bone. Most of the head and neck area is covered with a thin layer of bone, but you will need sharp dissecting tools and bone forceps to chip it off. Remember that the brain below is very soft and is

CERVICAL PLEXUS (C1–C5)

Lesser occipital nerve

Ansa cervicalis

Transverse cervical nerve

Supraclavicular nerve

Phrenic nerve

BRACHIAL PLEXUS (C5–T1)

Musculocutaneous nerve

Median nerve

Axillary nerve

Radial nerve

Ulnar nerve

INTERCOSTAL
(THORACIC NERVES)

LUMBAR PLEXUS (L1–L4)

Iliohypogastric nerve

Ilioinguinal nerve

Genitofemoral nerve

Lateral femoral
cutaneous nerve

Femoral nerve

Obturator nerve

SACRAL PLEXUS (L4-S4)

Common peroneal nerve

Tibial nerve

Posterior femoral cutaneous nerve

Pudendal nerve

Atlas (first cervical vertebra)

C1
C2
C3
C4
C5
C6
C7
C8

Cervical nerves (8 pairs)

Cervical enlargement

First thoracic vertebra

T1
T2
T3
T4
T5
T6
T7
T8
T9
T10
T11
T12

Thoracic nerves (12 pairs)

Lumbar enlargement

First lumbar vertebra

Conus medullaris

L1
L2
L3
L4
L5

Lumbar nerves (5 pairs)

Cauda equina

Ilium

S1
S2
S3
S4
S5

Sacrum

Sacral nerves (5 pairs)

Coccygeal nerves (1 pair)

Filum terminale

Figure 24-11 Spinal cord and spinal nerves (seen in posterior view). (From G. J. Tortora and S. R. Grabowski, *Principles of Anatomy and Physiology,* 7th ed., New York: HarperCollins, 1993.)

Location

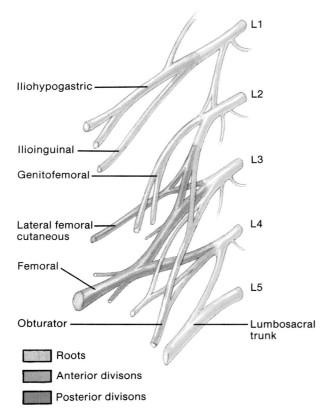

Iliohypogastric

Ilioinguinal

Genitofemoral

Lateral femoral
cutaneous

Femoral

Obturator

L1

L2

L3

L4

L5

Lumbosacral
trunk

	Roots
	Anterior divisons
	Posterior divisons

Figure 24-12 Human lumbar plexus (anterior view).
(From G. J. Tortora, *Principles of Human Anatomy*,
6th ed., New York: HarperCollins, 1992.)

easily damaged. Chip off thin pieces of skull with your
scalpel or forceps.

When all the bone over the brain has been removed,
observe the *dura mater,* which is the outer membrane
(meninx) surrounding the central nervous system. Review the *meninges* (membranes) discussed in the earlier
section on the sheep brain; these are the same in the
pig. Leave the dura mater in place until you are ready
to identify the parts of the brain. When the bone removal
is completed, use scissors to cut off the dura mater.
The *arachnoid membrane* and the *pia mater* are thin
and might not be visible. Use Figures 24-15 and 24-16
as guides to identify the following:

Olfactory Bulbs—These paired swellings forming
the most anterior region of the brain are involved
with the sense of smell. Lift the brain slightly to
see these lobes.

Cerebral Hemispheres—These two large lobes
(representing more than half of the visible brain)
are separated from one another by a *longitudinal
fissure.* Separate the two lobes completely by inserting the probe between them along their entire
length. Note that the *cerebrum* is made up of *gyri*
(convolutions) separated by *sulci* (fissures). The cerebral hemispheres are made up of an outer *gray
matter* and an inner *white matter.* Take a razor
blade, and cut out a slice of the cerebrum. Examine
it with the naked eye and then with a dissecting
microscope. Distinguish the gray matter from the
white matter.

Cerebellum—This relatively large area posterior to
the cerebrum is also convoluted, but its folds are
much smaller than those of the cerebrum.

Medulla oblongata—Posterior to the cerebellum is
the small triangular *medulla,* which merges with
the spinal cord posteriorly.

Other brain regions may not be easily seen on dissection
of the fetal pig. For example, the *midbrain or corpora
quadrigemina* are four bodies that can be seen by lifting
the posterior part of the cerebral hemispheres. The two
anterior lobes are the *superior colliculi,* and the two
posterior lobes are the *inferior colliculi.* The *diencephalon* is a small region anterior to the midbrain. The
pineal body is a derivative of its dorsal surface, and the
pituitary gland is a derivative of the ventral surface.

Finally, there are 12 pairs of *cranial nerves* (see
Figure 24-16), as in other mammals. These nerves
emerge from the ventral and ventrolateral surfaces of
the brain. Gently pull the brain to the side, and identify
as many as you can (Figure 24-16). Try to observe the
pituitary gland, which appears to grow downward from
the diencephalon in the midline. The pituitary is lodged
in the *sella turcica* of the *sphenoid bone.*

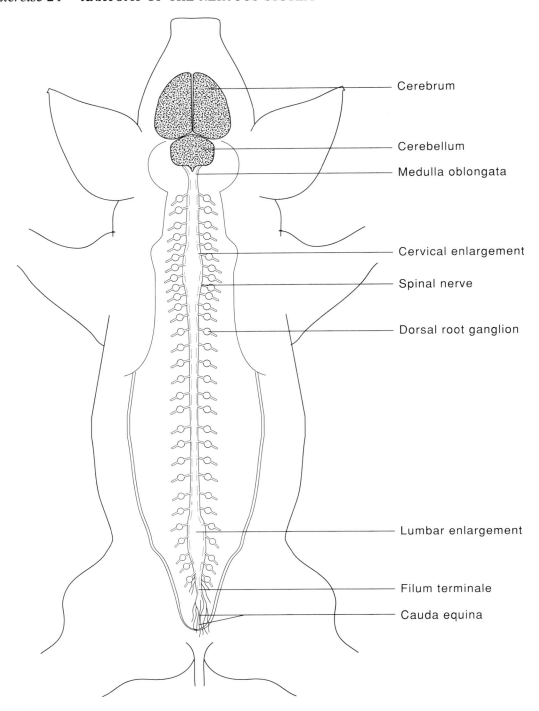

Figure 24-13 Brain and spinal cord of the fetal pig.

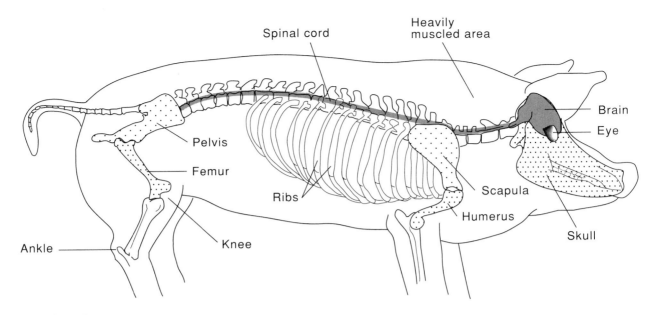

Figure 24-14 Side view of an adult pig, showing location of brain and spinal cord.

Figure 24-15 Pig brain (dorsal view).

Optic nerve (II)

Optic tract

Oculomotor nerve (III)

Trochlear nerve (IV)

Trigeminal nerve (V)

Facial nerve (VII)

Auditory nerve (VIII)

Glossopharyngeal nerve (IX)

Vagus nerve (X)

Spinal accessory nerve (XI)

Hypoglossal nerve (XII)

Olfactory bulb (I)

Optic chiasma
(the crossover point)

Mammillary body

Cerebral peduncle

Pons

Abducens nerve

Medulla

Cerebellum

Figure 24-16 Pig brain (ventral view), featuring the cranial nerves.

Spinal Cord

To expose the spinal cord, remove all bone and muscle from the surface between the medulla oblongata and the tail (see Figures 24-13 and 24-14). Snip off the vertebral spines with scissors, then cut back on either side of the spinal cord to the lumbar region. (Remember, this is very soft tissue.) Remove the strip of bone over the spinal cord. Also, remove the *dura mater,* being very careful not to tear the *spinal nerves,* which exit in pairs along the length of the spinal cord (Figure 24-13).

When you have exposed the entire length of the spinal cord, dissect an area at the level of one of the pairs of spinal nerves. With a sharp scalpel or a razor blade, cut out a section of the spinal cord, taking care to keep a pair of spinal nerves intact. Once more, observe the gray matter and the white matter. In the spinal cord, the gray matter is internal, and the white matter is external (Figure 24-17). Use a dissecting microscope, if necessary, to make the following observations.

Using Figure 24-17, find the dorsal and ventral roots of the spinal nerves. The canal in the spinal cord is the *central canal.* In the middorsal region, observe the *dorsal fissure,* and in the midventral region, note the *ventral fissure.* Demonstrate this dissection to your instructor.

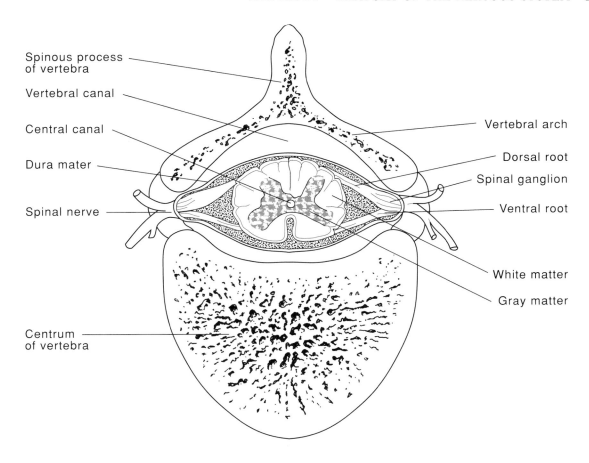

Figure 24-17 Cross-section of pig vertebra, showing spinal cord and nerves.

LABORATORY REVIEW 24

ANATOMY OF THE NERVOUS SYSTEM

COMPLETION, FILL-IN, AND DEFINITION

1. Fill in the names of the cranial nerves.

   I. _____

   II. _____

   III. _____

   IV. _____

   V. _____

   VI. _____

   VII. _____

   VIII. _____

   IX. _____

   X. _____

   XI. _____

   XII. _____

2. Define:

   a. Nerve plexus _____

   _____

   b. Ganglion _____

   _____

   c. Nerve _____

   _____

   d. Neuron _____

   e. Nerve fiber _____

   _____

   f. Brain ventricle _____

   _____

   g. Meninges _____

   _____

3. Name the lobes of the cerebral hemispheres:

    a. _____

    b. _____

    c. _____

    d. _____

4. How does a nerve differ from a nerve tract? _____

    _____

    _____

5. Name four nerve plexuses:

    a. _____

    b. _____

    c. _____

    d. _____

6. What is a choroid plexus? _____

    _____

    _____

7. What is the difference between a gyrus and a sulcus?

    _____

    _____

    _____

    _____

8. What is the difference between a sulcus and a fissure?

    _____

    _____

    _____

    _____

Demonstration of Selected Cranial Nerve and Cerebellar and Cerebral Functions in the Human

After completing this exercise, you should be able to

1. Explain the events taking place in a reflex arc
2. List the components of the reflex arc
3. List the functions of the cerebellum
4. Name and list the functions of each cranial nerve
5. Perform simple tests to demonstrate the functions of the cerebellum and selected cranial nerves
6. Briefly describe the essential processes of an overall memory system
7. Describe basic differences between short-term memory and long-term memory
8. Use selected techniques to demonstrate the influence and involvement of organizational processes with memory

1. *Receptor*—the distal end of a dendrite or sensory neuron that responds to a stimulus by initiating a nerve impulse in the sensory neuron
2. *Sensory neuron*—a nerve cell that passes the nerve impulse from the receptor to the central nervous system.
3. *Center*—a region within the central nervous system where an incoming sensory impulse generates an outgoing motor impulse. This event occurs when the sensory neuron synapses with an association or a motor neuron.
4. *Motor neuron*—a nerve cell that transmits the impulse generated by the sensory or association neuron in the center to the effector body part that will respond, such as a muscle or gland.
5. *Effector*—the effector body part that responds to the motor-nerve impulse. (An impulse traveling to the nerve fiber in an involved muscle causes it to contract.)

Reflex actions are rapid automatic responses to stimuli, over which an individual appears to have little or no control. Such responses attempt to restore the body to *homeostasis*. An example of a well-known reflex is the knee-tendon reflex, which occurs when the knee is hit by a small rubber mallet. The lower leg jerks in a characteristic way, which appears to be uncontrolled. For a reflex to occur, a nerve impulse must travel along a neural pathway termed a *reflex arc* (Figure 25-1), the functional unit of the body's nervous system. Figure 25-1 shows the direction or path of a nerve impulse, from its origin in one part of the body (*receptor*)to its termination elsewhere in the body (*effector*). The basic components of this *conduction pathway* in order of involvement are as follows:

Reflexes can be categorized according to the effector organ involved. Thus, reflexes that result in the contraction of skeletal muscles are termed *somatic reflexes*. Those that involve brain centers and cranial nerves are known as *cranial reflexes*. Finally, reflexes that either cause the contraction of cardiac or smooth muscle or cause gland secretions are *visceral* (autonomic) reflexes.

The cerebellum plays an important role in balance, posture, and coordination. It receives a steady flow of impulses from receptors embedded in joints and muscles, as well as from the eyes and portions of the ear. It registers the body's position, as well as its rate of movement and the coordination of muscular activities. This laboratory exercise includes a number of tests to demonstrate these functions of the cerebellum and of selected cranial nerves.

Figure 25-1 A generalized reflex arc. The components of the reflex arc and the direction of the conduction pathway (arrows) are shown. (From G. J. Tortora and S. R. Grabowski, *Principles of Anatomy and Physiology,* 7th ed., New York: HarperCollins, 1993.)

Table 25-1   A General Summary of the Cranial Nerves

Nerve Designation	Composition	Associated Functions
I. Olfactory	Sensory	Smell
II. Optic	Sensory	Sight
III. Oculomotor	Motor and sensory	Eye movement and focusing
IV. Trochlear	Motor and sensory	Nerve supply to eye muscle
V. Trigeminal	Motor and sensory	Chewing, sensation from the head region, between eyes and teeth
VI. Abducens	Motor and sensory	Eye movement
VII. Facial	Motor and sensory	Facial expression and taste
VIII. Statoacoustic	Sensory	Posture and hearing
IX. Glossopharyngeal	Motor and sensory	Swallowing, taste, and general sensation
X. Vagus	Motor and sensory	Visceral muscle movement and sensation
XI. Accessory	Motor and sensory	Swallowing and head movement
XII. Hypoglossal	Motor and sensory	Speech and swallowing

The brain gives rise to 12 pairs of cranial nerves that supply motor and sensory fibers to the head, neck, and some visceral organs. These nerves form part of the peripheral nervous system that connects the brain with receptors, muscles, and glands. Table 25-1 lists the cranial nerves and their composition and function.

This exercise also considers a fascinating integrative function of the cerebrum, namely, *memory*. The

Figure 25-2 A magnetic resonance image (MRI) of the human brain, showing portions of the cerebrum. The frontal lobe forms the anterior portion of each cerebral hemisphere.

frontal lobe (Figure 25-2), as well as other brain portions—such as the parietal, occipital, and temporal lobes, and the diencephalon—are associated with memory.

Memory is the ability of an individual to recall thoughts. For an experience to become part of memory, it must cause changes in the CNS that represent the experience.

The essential processes of an overall memory system are encoding, storage, and retrieval. Together, these three components form the foundation of the major information-processing sequence in the body. While these components have not been defined to everyone's satisfaction, the following definitions should be sufficient for purposes of this exercise. *Encoding* refers to processes by which a memorial representation of physical objects and events is developed. *Storage* may be defined as the persistence of information over time, while *retrieval* refers to the search for and utilization of stored information. It should be noted that even information that is stored may not be easily located and used. For example, an individual may know all of the states in the United States, but at any given moment, may not be able to retrieve all of them.

Two types of memory are generally recognized: short-term (*primary memory*) and long-term (*secondary memory*). With *short-term memory*, every item (stimulus) that is perceived is kept temporarily in the conscious mind. Such items may be lost or forgotten unless they are rehearsed. If an item (such as a phone number, name, etc.) is rehearsed, it remains in the short-term memory, and may enter the subconscious mind or long-term memory. Information stored in *long-term memory* is relatively permanent and is subject to a very slow decay process.

Both long-term and short-term memory are very flexible systems. Each is capable of encoding materials in a number of ways. Short-term memory is kept in mind by the processes of reverberation and synaptic facilitation. With *reverberation,* thoughts continue to cycle in closed chains of neurons until the threshold for excitation is no longer reached. In *synaptic facilitation,* neurotransmitting chemicals temporarily are produced in larger amounts, facilitating the transfer of impulses between particular neurons. The growth of synaptic contacts and the formation of memory molecules (specific protein and ribonucleic acid molecules, either individually or together) retain long-term memory.

From the results of various investigations, it appears that the overall memory system of the human is governed by principles of organization and the grouping of perceptions. Information is not passively stored in exactly the same form in which it is received. It is structured, interrelated with other items, and organized so that both the storage and the retrieval of the information will be efficient. Several techniques may be used to demonstrate and measure such activities and/or processes. These include category *clustering* (grouping related information) *for recall,* and *visual coding* (encoding information in terms of visual images). Several approaches are used in order to demonstrate the influence of selected organizational processes of memory.

SELECTED CRANIAL NERVE AND CEREBELLAR FUNCTIONS

Materials

The following materials should be provided for class use:

1. Containers of
   a. Garlic
   b. Ground coffee
2. Tuning forks
3. Sterile cotton swabs and tongue depressors
4. Timers or stopwatches that have an audible ticking sound
5. Flashlights or penlights

Procedure 1: Cranial Nerve Function Tests

This procedure should be performed by students in pairs.

1. Take turns in performing the tasks listed for each cranial nerve. Describe your responses and those of your laboratory partner in the results and observations section.

2. Read each of the tasks first before performing them.
   a. *Olfactory nerve.* Close your eyes while your partner passes containers of garlic and ground coffee by your nose. Can you distinguish the aromas?
   b. *Optic nerve.* Determine how many words on one page of a textbook or other publication you can read in 1 minute.
   c. *Oculomotor nerve.* Keep your head still while your partner slowly moves a pencil up and down in front of you. Can you follow the movement with your eyes?
   d. *Trochlear and abducens nerves.* Look straight ahead while your partner holds a pencil about 1 meter from the side of your head and moves it forward and backward. Indicate at what point you cannot see the pencil. Have your partner determine whether the pupil opens or closes more during this task.
   e. *Trigeminal nerve.* Close your mouth and clench your teeth. Have your partner hold one hand firmly under your chin. Try to open your mouth. (This is a test for a motor response.)
   f. *Facial nerve.* Carry out the following facial movements: smile and show your teeth, inflate your cheeks, wrinkle your forehead, and raise either one or both eyebrows. Can you raise each eyebrow separately?
   g. *Vestibulocochlear (acoustic) nerve.* Close your eyes. Have your partner hold a timer or stopwatch about 1 meter away from the left or right side of your head. Can you determine the direction from which the sound is coming? Is this a test of the cochlear or the vestibular portion of the nerve?
   h. *Glossopharyngeal and vagus nerves.*
      i. Open your mouth wide. Have your partner carefully touch the back of your throat. What is the response?
      ii. Have your partner hold down your tongue with a sterile tongue depressor and direct a penlight into your mouth, while you say "ah." What happens to the uvula (the fleshy mass near the back of the mouth)?
   i. *Spinal accessory nerve.*
      i. Sit down, and have your partner push firmly on your shoulders. Try to raise your shoulders. (Note that this does not mean to stand up.) What muscles are involved?
      ii. Have your partner place his or her hands firmly on the sides of your face in front of the ears. Try to turn your head. What muscles are involved?
   j. *Hypoglossal nerve.* Stick your tongue out and try to keep it as straight as you can.

Procedure 2: Demonstration of Cerebellar Function

This procedure should be performed by students in pairs.

1. Take turns in performing the following tasks. Describe your responses and those of your laboratory partner in the results and observations section.
2. Have your laboratory partner stand approximately 2 meters in front of you. You will be the examiner. Read each of the tasks out loud, and determine the ease with which they are performed.
   a. Stretch both arms to the side and move fingers rapidly.
   b. Stretch both arms to the side and snap the fingers of both hands.
   c. Turn to the side and, looking straight ahead, walk a straight line placing the heel of one foot directly in front of the toe of the other foot.
   d. Close your eyes and stand erect for 1 minute.
   e. Close your eyes and touch your nose with the index finger of your left hand.
   f. Close your eyes and touch your nose with the index finger of your right hand.
   g. Touch the index finger on my right hand. (Examiner should move the right hand accordingly.)
   h. Stand erect and move the heel of your right foot first up and then down the shin of your left leg.

RESULTS AND OBSERVATIONS

Cranial Nerve Function Tests

1. Enter your responses and those of your partner in Table 25-2.
2. How would you test for the sensory responses of the trigeminal nerve? _____

_____

3. How could you distinguish between the sensory responses of the vestibular and the cochlear portions of the vestibulocochlear nerve? _____

_____

_____

Demonstration of Cerebellar Function

1. Enter your responses and those of your partner in Table 25-3. Indicate whether the task was performed and whether any difficulties were observed.
2. Suggest two other possible tasks to test cerebellar function. _____

_____

Table 25-2

Cranial Nerve Tested	Your Response	Partner's Response
Olfactory		
Optic		
Oculomotor		
Trochlear and abducens		
Trigeminal		
Facial		
Vestibulocochlear (acoustic)		
Glossopharyngeal and vagus		
Spinal accessory		
Hypoglossal		

Table 25-3

Task	Your Response	Partner's Response
Move fingers quickly		
Snap fingers		
Walk a straight line in tandem without loss of balance		
Stand erect with feet together and eyes closed without loss of balance		
Touch index finger of each hand to the nose while eyes closed		
Touch examiner's index finger on command		
Move heel of one foot up and down shin of opposite leg		

CATEGORY CLUSTERING (GROUPING PHENOMENON)

Materials

List of words supplied in Table 25-4.

Procedure: This procedure should be performed by students individually.

1. Read the list of 30 words in Table 25-4 only once.
2. Turn immediately to the results and observations section, and write down as many words as you can recall. Use any order you wish.
3. After you cannot recall any additional terms, count the correct responses, and enter the number in the results and observations section.
4. Answer the questions concerning this experiment in the results and observations section.

VISUAL VERSUS VERBAL CODING

Materials

The following materials should be provided for general class use.

1. Two sets of 20 anatomical diagrams or photographs of body structures (each item should be numbered)
2. Two sets of 20 cards or sheets, each containing one term for an anatomical structure

Procedure

This procedure is to be performed by students in pairs.

1. Have your laboratory partner sit comfortably, with his or her laboratory manual opened to the results and observation section for this exercise.

Table 25-4   Grouping Phenomenon

Snake	Sycamore	Red	Pumpkin	blue	Elm
London	Brown	New York	Maple	Fox	Lettuce
Yellow	Tomato	Hemlock	Redwood	Broccoli	Black
Orange	Pine	Peas	Parrot	Moose	Paris
Los Angeles	Cow	Zebra	Moscow	Seattle	Carrot

2. Obtain one set of 20 anatomical photographs or diagrams. Show the illustrations to your partner one at a time and only once.

3. After you have shown your partner all the illustrations, ask your partner to write the names of all the structures he or she can recall in the results and observations section.

4. Obtain one set of 20 cards or sheets containing individual terms for anatomical structures. Show the cards to your partner one at a time and only once.

5. When you have shown your partner all the cards, ask your partner to write all the terms he or she can recall in the results and observations section.

6. Change places with your partner, and repeat Steps 1 through 5 with a second and different set of illustrations and word cards.

RESULTS AND OBSERVATIONS

Category Clustering

1. Write the words recalled from Table 25-4.

    1. _____   16. _____
    2. _____   17. _____
    3. _____   18. _____
    4. _____   19. _____
    5. _____   20. _____
    6. _____   21. _____
    7. _____   22. _____
    8. _____   23. _____
    9. _____   24. _____
    10. _____ 25. _____
    11. _____ 26. _____
    12. _____ 27. _____
    13. _____ 28. _____
    14. _____ 29. _____
    15. _____ 30. _____

    Correct number recalled: _____

2. a. Were the words toward the end of the list recalled better than earlier ones? _____

   b. What type of memory accounts for your response: short-term or long-term? _____

3. Were the items in the middle of the list difficult to recall? _____

4. What categories or groups were formed in the list? _____

5. Were you consciously aware of grouping during the listing process? _____

6. Compare your results with those of a laboratory partner. Were they comparable as to groupings? _____

Visual Versus Verbal Coding

1. Write the names of the anatomical structures recalled from the illustrations shown.

    a. _____   k. _____
    b. _____   l. _____
    c. _____   m. _____
    d. _____   n. _____
    e. _____   o. _____
    f. _____   p. _____
    g. _____   q. _____
    h. _____   r. _____
    i. _____   s. _____
    j. _____   t. _____

    Correct number recalled: _____

2. Write the anatomical terms recalled from the cards shown.

    a. _____   k. _____
    b. _____   l. _____
    c. _____   m. _____
    d. _____   n. _____
    e. _____   o. _____
    f. _____   p. _____
    g. _____   q. _____
    h. _____   r. _____
    i. _____   s. _____
    j. _____   t. _____

    Correct number recalled: _____

3. In which situation was the larger number of correct recall responses found? _____

4. Were the recall responses of your partner similar?

   _____

5. What conclusion can you draw from this experiment? _____

   _____

   _____

LABORATORY REVIEW *25*

NAME _____

LAB SECTION _____ DATE _____

DEMONSTRATION OF SELECTED CRANIAL NERVE AND CEREBELLAR AND CEREBRAL FUNCTIONS IN THE HUMAN

COMPLETION AND DEFINITION

1. Where is the cerebellum located in the human with respect to other parts of the brain? _____

   _____

2. List two functions of the cerebellum.
   a. _____

   b. _____

3. What are the cranial nerves? _____

   _____

4. How many cranial nerves are there in the human? _____

5. Give the functions of the cranial nerves listed.

   a. Trochlear _____

   b. Facial _____

   c. Glossopharyngeal _____

   d. Vagus _____

   e. Hypoglossal _____

6. For each item listed, indicate which specific cranial nerve in the human could be tested by using the item.

   a. Odor of fresh brewed coffee _____

   b. Reading letters on an eye chart _____

   c. Having a penlight flashed into an eye _____

   d. Moving a timer or stopwatch with an audible ticking sound on either side of an individual whose eyes are closed _____

   e. Touching the individual's uvula _____

7. List and briefly explain the components of the overall memory system.

   a. _____

   b. _____

   c. _____

   d. _____

8. Distinguish between short-term memory and long-term memory. _____

   _____

   _____

ANSWERS

1. _____
2. _____
3. _____
4. _____
5. _____
6. _____
7. _____
8. _____
9. _____
10. _____
11. _____
12. _____
13. _____
14. _____

9. What is category clustering? _____

_____

10. Is the memory for pictures better than the memory for names of the same pictures?

_____

MULTIPLE CHOICE

Insert the correct answers in the spaces provided.

1. Which of the following is the specific part of a reflex that responds to a motor impulse?
   a. effector
   b. center
   c. sensory neuron
   d. receptor
   e. dendrite

2. Which of the following is the part of a neuron that conducts nerve impulses away from the cell body?
   a. effector
   b. center
   c. sensory neuron
   d. receptor
   e. dendrite

3. What is the type of neuron that conducts nerve impulses toward the central nervous system?
   a. efferent
   b. internuncial
   c. association
   d. central
   e. afferent

4. What is the distal end of a sensory neuron that responds to a stimulus by starting a nerve impulse?
   a. efferent
   b. internuncial
   c. association
   d. central
   e. afferent

5. What is the region within the central nervous system where an incoming sensory impulse generates an outgoing motor impulse?
   a. motor neuron
   b. receptor
   c. sensory neuron
   d. center
   e. effector

6. What kind of reflex results in the contraction of skeletal muscles?
   a. somatic
   b. visceral
   c. cranial
   d. cardiac

7. What kind of reflex involves brain centers and cranial nerves?
   a. somatic
   b. visceral
   c. cranial
   d. cardiac
8. What are reflexes causing smooth muscle contraction called?
   a. somatic
   b. visceral
   c. cranial
   d. cardiac
9. What are reflexes that cause heart muscles to contract considered to be?
   a. somatic
   b. visceral
   c. cranial
   d. cardiac
10. Which of the following cerebral components is involved with responses related to memory, reasoning, and judgment?
    a. temporal lobe
    b. parietal lobe
    c. occipital lobe
    d. all of these
    e. only choices *a* and *b*.
11. In addition to *encoding* and *storage*, what is an essential process of an overall memory system?
    a. short-term memory
    b. primary memory
    c. retrieval
    d. organization
    e. long-term memory
12. What is the process by which a memory representation can be developed for physical objects and events?
    a. encoding
    b. storage
    c. retrieval
    d. short-term memory
    e. long-term memory
13. Which of the following is the search for and use of stored information?
    a. encoding
    b. storage
    c. retrieval
    d. short-term memory
    e. long-term memory
14. Short-term memory is kept in mind primarily by which of the following?
    a. reverberation
    b. memory molecule formation
    c. synaptic facilitation
    d. RNA molecules
    e. choices *a* and *c* only.

Demonstration of Reflexes in the Human

After completing this exercise, you should be able to

1. List at least five commonly encountered reflex actions in the human
2. Perform simple procedures to identify and to demonstrate common reflex actions

As indicated in Exercise 25, discrete nervous pathways to the spinal cord or brain are involved in the performance of a reflex action. The knee-jerk response is a well-known example. Another situation involving a reflex action familiar to most is the response to touching a hot object. A stimulus received by the heat- or pain-sensitive endings in the finger is transmitted as an impulse to the spinal cord, where it relays the signal to a motor neuron. The motor neuron completes the arc by stimulating muscle contraction in the finger, moving it away from the unpleasant stimulus. Reflexes carried out by the spinal cord are *spinal reflexes.*

Because normal reflexes depend on normal neuron functions, reflexes often are utilized to obtain information concerning the condition of the nervous system. An anesthesiologist, for example, may use a reflex in a patient being anesthetized to determine how a particular drug is affecting the nervous system. Some specific reflexes also are of importance in diagnosing nervous-system disorders and in locating injured tissue. Lesions of the central nervous system may occur at all levels and may result from various causes, including physical injuries, infections, and biochemical changes. In some situations, exaggerated spinal reflexes result, while in others, underactive reflexes occur. Examples of reflexes of diagnostic significance include the abdominal reflex, Archilles reflex (ankle jerk), Babinski sign, biceps reflex,

patellar reflex (knee jerk), and the triceps reflex. Table 26-1 briefly describes both the normal and the abnormal responses for these reflexes and some of the conditions responsible for abnormal reactions. The specific spinal segments involved also are indicated in this table.

This exercise considers several types of reflex actions. These include *deep* and *superficial* reflexes. *Deep reflexes,* also termed *somatic reflexes,* result in the contraction of skeletal muscles and are triggered by a sharp tap on an appropriate muscle or tendon. The associated receptors for these reflexes are situated in the muscle and not in the tendon. Tapping the tendon causes the muscle to be stretched; this stretching, in turn, activates the muscle spindle, thus triggering the response. Somatic reflexes are named according to the response they produce and include the *stretch reflex* and the *withdrawal* (or *flexor*) *reflex.*

A stretch reflex involves only two neurons and one synapse in the pathway, and is therefore termed a *monosynaptic reflex arc.* Because the receptor and effector organs of the stretch reflex involve structures on the same side of the spinal cord, the reflex arc is referred to as an *ipsilateral reflex.* The patellar (knee-jerk) reflex (Figure 26-1) and all other monosynaptic reflex arcs are examples of an ipsilateral reflex.

A *withdrawal* or *flexor reflex* consists of a *polysynaptic reflex arc.* These reflexes involve association neurons (termed *interneurons*), in addition to sensory and motor neurons. Visceral reflexes, which also are polysynaptic because at least two motor-neuron synapses are always involved, are included in this exercise.

Superficial reflexes, also referred to as *cutaneous reflexes,* are caused by irritating or tactile stimulation. Stroking or scratching the skin surface triggers these reflexes. Figure 26-2 shows the features of clinically important reflexes.

Abnormal responses to stimuli may be *reduced (hypoflexia), exaggerated (hyperreflexia),* or *pathological.*

Table 26-1   Normal and Abnormal Reflex Actions

Reflex	Spinal Segment[a]	Normal Response	Abnormal Response
Abdominal	T_8, T_9, T_{10} (above umbilicus) T_{10}, T_{11}, T_{12} (below umbilicus)	Abdominal muscles underlying the skin contract in response to stroking the side of the abdomen. Movement of the umbilicus toward the region stimulated also occurs.	Absence of reflex. Conditions causing abnormal reactions include lesions of peripheral nerves, lesions of reflex centers in the chest region of the spinal cord, and multiple sclerosis.
Achilles	S_1, S_2	Extension of the foot in response to tapping the Achilles tendon.	Absence of reflex or exaggerated reflex. Conditions causing first of abnormal reactions include damage to nerves supplying posterior leg muscles, damage to nerve cell in spinal cord's lumbosacral region, chronic diabetes mellitus, neurosyphilis, alcoholism, and subarachnoid bleeding. Exaggerated responses may be due to injuries of the cord's sacral region.
Babinski sign (positive)	L_4, L_5, S_1, S_2	Extension of the great (big) toe in response to a light stimulation of the outer margin of the sole of the foot. Other toes may or may not respond. Children under 1½ years of age exhibit the sign because of an incomplete development of the nervous system (corticospinal tract) at this time.	Presence of reflex after 1½ years of age. Conditions causing abnormal reaction include interruption of corticospinal tract development and lesion in the upper corticospinal tract.
Biceps reflex	C5, C6	Flexion of the forearm occurs in response to tapping the biceps tendon.	Absence of reflex or exaggerated response. Conditions causing abnormal reactions include injury to the cervical segments of the spinal cord.
Patellar	L_2, L_3, L_4	Extension of the leg in response to tapping the patellar ligament.	Absence of reflex or exaggerated response. Conditions causing abnormal reactions include injury to nerves in the lumbar segments of the spinal cord, chronic diabetes, and neurosyphilis. Exaggerated responses may result from injury or disease in the corticospinal tract.
Triceps reflex	C7, C8	Extension of the forearm occurs in response to tapping the triceps tendon.	Absence of reflex or exaggerated response. Conditions causing abnormal reactions include injury to the cervical segments of the spinal cord.

[a]C = cervical; L = lumbar; S = sacral; and T = thoracic.

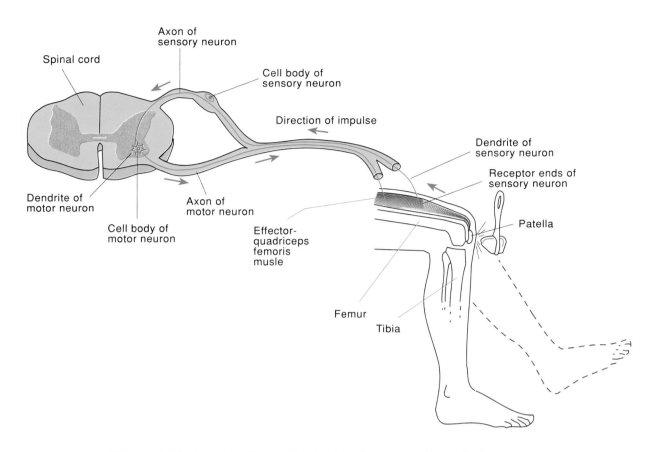

Figure 26-1 Simple reflex arc involved in the knee-jerk (patellar) response.

Aging, injury to neurons, and malnutrition are among the causes of hypoflexia. Hyperreflexia may result from the loss of inhibitory control by the motor cortex and from some types of poisonings. Pathological reflex responses involve one or more muscles other than the muscle where the stimulus originates.

Materials

The following materials should be provided for class use:

1. Rubber mallets (percussion hammers)
2. Sterile cotton swabs
3. Plastic disposable tubes (15 ml capacity)
4. Lemon juice or a safe substitute
5. 5 percent glucose solution
6. Small pieces of ice
7. Container of disinfectant
8. Marking pencils or pens

Procedure: Demonstrating Common Reflex Actions

The following tests should be performed by students in pairs (except Test 6, which requires three students). Enter brief descriptions of the reactions produced in Table 26-2, and refer to Figure 26-2, as needed.

SOMATIC REFLEXES OF THE LEG

1. **ACHILLES TENDON REFLEX (ANKLE JERK)**
   a. Have your laboratory partner kneel on a chair or table so that both feet hang over the edge.
   b. Locate the Achilles tendon of the right foot, and tap it with a rubber hammer or the lateral edge of your palm. Describe the response in the results and observations section of this exercise.
   c. Change places with your partner, and repeat Steps (a) and (b). (Refer to Figure 26-2a.)

2. PATELLAR TENDON REFLEX (KNEE JERK)

a. Have your laboratory partner sit on a table or desk so that both feet are off the floor and hang freely.

b. Locate the patellar tendon (the depression just below the kneecap; refer to Figure 26-1d) of the right leg, and tap the tendon with the rubber mallet or the edge of your hand. Note the response in the results and observations section.

c. Have your partner count backward from 99 to 32. Tap the patellar tendon of the right leg again. Was the response greater or less than the first time?

d. Repeat Steps (b) and (c) with the left leg. Note the extent of the response.

e. Change places with your partner, and repeat Steps (a)–(d).

SOMATIC REFLEXES OF THE ARM

3. BICEPS REFLEX

a. Have your laboratory partner rest his or her arm on a table or desk.

b. Place your thumb on the biceps tendon, as shown in Figure 26-2b.

c. Tap the first digit of your thumb with the reflex hammer. Note the extent of the response.

d. Change places with your partner, and repeat Steps (a)–(c).

4. TRICEPS REFLEX

a. Have your laboratory partner flex one arm at the elbow.

b. Hold the wrist of that arm, as shown in Figure 26-1e.

c. Tap the triceps tendon above the elbow, using the pointed end of the reflex mallet, and note the response.

d. Test each arm in a flexed, relaxed, state.

e. Change places with your partner, and repeat Steps (a)–(d).

SUPERFICIAL REFLEX OF THE FOOT

5. BABINSKI SIGN

a. Have your partner remove the shoe and sock from the right foot.

b. Take the tip of a rubber mallet or the blunt end of a pen, and run it on the sole of the foot firmly in one upward movement from the heel to the transverse arch. (Refer to Figure 26-1c.) Record your findings in the results and observations section.

c. Change places with your partner, and repeat Steps (a) and (b).

VISCERAL REFLEXES

6. CILIOSPINAL REFLEX

a. A total of three persons are needed for this activity. Have one laboratory partner sit comfortably in a chair. Observe this partner's pupils.

b. Have another laboratory member unexpectedly touch a small piece of ice on the back of the seated individual. Note the response of the pupils. Record your findings in the results and observations section.

c. Change places with your partners, and repeat Steps (a) and (b).

7. SALIVARY REFLEX

a. Have your partner rinse his or her mouth with water.

b. Collect all saliva secreted during a 3-minute period in a plastic tube. Mark the top level of fluid in the tube. Dispose of the tube, as indicated by your instructor.

c. Dip a cotton swab in the glucose solution and place the swab under your partner's tongue. Collect the saliva secreted during a 3-minute period and mark the fluid level.

d. Dip another swab in lemon juice, and repeat Step (c). In which of the three 3-minute intervals was more saliva produced? Enter your findings in the results and observations section.

e. Change places with your partner, and repeat Steps (a)–(d).

RESULTS AND OBSERVATIONS

1. Enter in Table 26-2 brief descriptions of your reactions and those of your partner in the reflex-testing procedures.

2. Were differences in patellar tendon responses observed? Explain. _____

_____

3. Were there any differences in the flexed versus the relaxed state of biceps tendon noted? Explain.

_____

_____

(a) The Achilles reflex.

(b) Biceps reflex

Pen

Stroke pattern

Toe spreading

Pen

Appearance of Babinski's sign

(c) Babinski's sign

Mallet

(d) The patellar reflex.

(e) Triceps reflex

Figure 26-2 Selected clinically important reflexes.

4. What is the medical significance of the Babinski response? _____

_____

5. What type of nerve is stimulated in the ciliospinal reflex? _____

_____

6. Can the salivary reflex be controlled? Explain.

_____

_____

_____

Table 26-2

Reflex Procedure	Your Response	Laboratory Partner's Response
Somatic reflexes of the leg Achilles tendon reflex		
Patellar tendon reflex Right leg (initial)		
Right leg (after counting)		
Left leg (initial)		
Left leg (after counting)		
Somatic reflexes of the arm		
Biceps reflex		
Triceps reflex		
Flexed arm		
Relaxed arm		
Superficial reflex of the foot: Babinski sign		
Visceral reflexes		
Ciliospinal reflex		
Salivary reflex Normal		
With glucose solution		
With lemon juice		

LABORATORY REVIEW *26*

DEMONSTRATION OF REFLEXES IN THE HUMAN

COMPLETION AND DEFINITION

1. What is a reflex? _____

_____

2. Diagram a reflex arc.

3. In the following tests for reflex action, what body structure is stimulated?

   a. Knee jerk _____

   b. Babinski sign _____

   c. Ankle jerk _____

4. With the aid of your text and other references, define or explain the following.

   a. Tonus _____

   _____

   b. Flaccid _____

   c. Ipsilateral _____

   d. Reciprocal innervation _____

   _____

5. What is a flexion or a withdrawal response? _____

_____

6. Distinguish between deep and superficial reflexes. Give one example of each type.

   a. _____

   b. _____

ANSWERS

1. _____
2. _____
3. _____
4. _____
5. _____
6. _____
7. _____
8. _____
9. _____
10. _____

MULTIPLE CHOICE

Insert the correct answers in the spaces provided.

1. The extension of the forearm by tapping the tendon of the triceps indicates the proper functioning of which of the following spinal-cord segments?
   a. sacral 1 and 2
   b. cervical 7 and 8
   c. cervical 5 and 6
   d. lumbar 2, 3, and 4
   e. lumbar 4 and 5, and sacral 1 and 2

2. A normal extension of the leg by tapping the patellar ligament shows the proper functioning of which of the following spinal-cord segments?
   a. sacral 1 and 2
   b. cervical 7 and 8
   c. cervical 5 and 6
   d. lumbar 2, 3, and 4
   e. lumbar 4 and 5, and sacral 1 and 2

3. Which of the following spinal-cord segments is responsible for the Achilles reflex?
   a. sacral 1 and 2
   b. cervical 7 and 8
   c. cervical 5 and 6
   d. lumbar 2, 3, and 4
   e. lumbar 4 and 5, and sacral 1 and 2

4. Which of the following spinal-cord segments is responsible for the biceps reflex?
   a. sacral 1 and 2
   b. cervical 7 and 8
   c. cervical 5 and 6
   d. lumbar 2, 3, and 4
   e. lumbar 4 and 5, and sacral 1 and 2

5. What is the term for reflexes that result from skeletal-muscle contraction, which are triggered by a sharp tap on an appropriate area?
   a. superficial
   b. hyperreflexia
   c. cutaneous
   d. hypoflexia
   e. somatic

6. What do abnormal reduced reflex responses exemplify?
   a. superficial reflexes
   b. hyperreflexia
   c. cutaneous responses
   d. hypoflexia
   e. withdrawal reflexes

7. What do abnormal exaggerated reflex responses exemplify?
   a. superficial reflexes
   b. hyperreflexia
   c. cutaneous responses
   d. hypoflexia
   e. withdrawal reflexes

8. Stretch reflexes involve (how many?) neurons and (how many?) synapses.
   a. 2, 1
   b. 1, 1
   c. 2, 2
   d. 3, 1
   e. 3, 2

9. Which of the following do somatic reflexes include?
    a. withdrawal reflexes
    b. flexor reflexes
    c. stretch reflexes
    d. choices *a* and *b* only
    e. choices *a, b,* and *c*
10. Of what is the patellar reflex an example?
    a. ipsilateral reflex
    b. polysynaptic reflex
    c. withdrawal reflex
    d. flexor reflex
    e. superficial reflex

The Endocrine System

*I*n addition to the nervous system, the endocrine glands provide a communication and regulatory system. While the two often supplement one another and frequently perform separately, the interplay between endocrine secretions and nerve impulses often functions to control total human behavior.

The endocrine secretions, or *hormones,* do not belong to any one category of chemical compounds. They have varying chemical structures. Some are proteins, while others are amines (amino acid derivatives) or steroids (types of lipids).

The concentration of a hormone in the blood is controlled by a negative-feedback mechanism that regulates production, storage, and elimination processes. Sensors read the status of the body's internal environment and then direct the appropriate responses. The presence of a greater than normal concentration of a hormone will initiate a response to inhibit the gland's production. Glands may be under direct neural control or under the control of other hormones or both.

The exercises in this section point out the location of the human endocrine glands and demonstrate selected properties of the human endocrine system.

GENERAL REFERENCES

Gorbman, A., W. W. Dickhoff, S. R. Vigna, N. B. Clark, and C. L. Ralph. *Comparative Endocrinology.* New York: Wiley, 1983.

Greenspan, F. S., and P. H. Forsham (Eds.). *Basic and Clinical Psychology.* Los Altos, CA: Lange Medical Publications, 1983.

Hardy, R. N. *Endocrine Physiology.* Baltimore: University Park Press, 1982.

Tortora, G. J., and S. R. Grabowski. *Principles of Anatomy and Physiology,* 7th ed. New York: HarperCollins, 1993.

SPECIFIC REFERENCES

Carmichael, S. W., and H. Winkler. "The Adrenal Chromaffin Cell." *Scientific American,* 253: 40–49, 1985.

DeVos, A. M., M. Ultsch, and A. A. Kossiakoff. "Human Growth Hormone and Extracellular Domain of Its Receptor: Crystal Structure of the Complex." *Science,* 255: 306–312, 1992.

Dunn, J. T. "Iodine Deficiency—The Next Target for Elimination." *The New England Journal of Medicine,* 326: 267–268, 1992.

Dupre, J. "Insulin Therapy: Progress and Prospects." *Hospital Practice,* 18: 171–179, 1983.

Golde, D. W., and J. C. Gasson. "Hormones That Stimulate the Growth of Blood Cells." *Scientific American,* 259: 62–70, 1988.

Reichlin, S. "Somatostatin." *The New England Journal of Medicine,* 309: 1495–1501, 1983.

Reiter, R. J. (Ed.). *The Pineal Gland.* New York: Raven Press, 1984.

Volpe, R. "Autoimmune Thyroid Disease." *Hospital Practice,* 19: 141–158, 1984.

Location and Histology of Human Endocrine Glands

After completing this exercise, you should be able to

1. Name the major endocrine glands and state their locations
2. Associate glands with the hormones that they produce
3. Recognize microscopic features of selected endocrine glands

Many endocrine glands are *organs* that secrete chemical regulators, or *hormones,* into the blood. The approximate locations of endocrine glands are shown in Figure 27-1. Hormones are carried by the vascular system to all the organs of the body. Some hormones, however, only affect the metabolism of specific organs; these organs are termed the *target organs* of the hormone (Figure 27-1). Other hormones influence every body cell. Hormones are not enzymes, but they affect the metabolism and activities of their target organs by regulating the activity of specific enzymes in these and related structures. In this way, the endocrine system plays an important role in the regulation of (1) growth and development, (2) metabolic homeostasis, and (3) reproduction. Because hormones are powerful chemicals, their circulating levels must be carefully controlled. When the level of a hormone is adequate to meet existing body needs, its further release is stopped. This form of control is termed *negative feedback.*

Both overactivity and underactivity of an endocrine gland are examples of malfunction. Conditions of this type cause endocrine diseases such as those listed in Figure 27-2. When a gland secretes an excessive amount of a specific hormone, it is *hyperactive.* This type of

activity may be caused by various conditions, including an enlarged gland or a tumor. A gland that fails to secrete its hormones or that simply secretes inadequate quantities is considered to be *hypoactive.* Conditions that may cause such limited activity include infection, tumors, or the unfavorable effects of surgical or radioactive treatments. Glands that are decreased in size and consequently secrete inadequate amounts of hormones are referred to as being *atrophied.*

This exercise presents the location and the microscopic features of normal endocrine glands, and some endocrine disorders and diseases commonly found in humans.

ENDOCRINE GLANDS—LOCATION AND BRIEF DISCUSSION

Materials

The following materials should be provided for class use:

1. Human torso models
2. Head models
3. Human skull models
4. Charts showing endocrine gland locations

Endocrine glands secrete into the blood and do not have ducts. They therefore are termed the *ductless glands.* All other large secretory masses produce substances that are carried to their specific destinations by one or more ducts. These are termed *ducted glands* or *exocrine glands.* The products of endocrine glands are termed *hormones,* while products of exocrine glands are given such specific names as saliva, bile, and pancreatic juice. The advantage of an endocrine gland is obvious. Once in the blood, the secretions may be carried to all cells of the body.

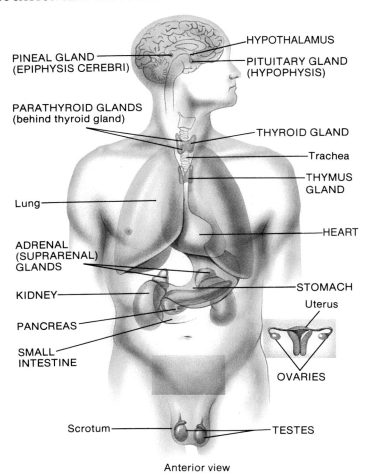

PINEAL GLAND (EPIPHYSIS CEREBRI)
HYPOTHALAMUS
PITUITARY GLAND (HYPOPHYSIS)
PARATHYROID GLANDS (behind thyroid gland)
THYROID GLAND
Trachea
THYMUS GLAND
Lung
HEART
ADRENAL (SUPRARENAL) GLANDS
STOMACH
Uterus
KIDNEY
PANCREAS
SMALL INTESTINE
OVARIES
Scrotum
TESTES

Anterior view

Figure 27-1 Location of human (male and female) endocrine glands. (From G. J. Tortora and S. R. Grabowski, *Principles of Anatomy and Physiology,* 7th ed., New York: HarperCollins, 1993.)

Pituitary Gland

The pituitary gland (Figure 27-3) is located at the base of the brain, within the hypophysial fossa of the sella turcica of the sphenoid bone. It is suspended by a stalk termed the *infundibulum* and is composed of two major parts: the anterior and posterior lobes.

The pituitary gland arises embryonically from two different sites. The *anterior lobe (adenohypophysis)* is a derivative of part of the embryonic mouth epithelium, termed *Rathke's pocket.* The *posterior lobe (neurohypophysis)* is a derivative of the diencephalon region of the embryonic brain. There is an *intermediate lobe* in the pituitary's early development, but it fuses with the anterior lobe. Its embryonic origin is the same as that of the anterior lobe.

Both the anterior and the posterior lobes are attached to the inferior surface of the brain by the *infundibulum.* The hormones found in the posterior lobe of the pituitary are synthesized in the *hypothalamus* of the brain. They are merely stored in the pituitary, but they are still thought of as posterior-lobe hormones. Some of the hormones produced by the pituitary gland and the actions of these hormones are listed in Table 27-1.

1. **ANTERIOR LOBE (ADENOHYPOPHYSIS) HORMONES**

   a. *Somatotropin or human growth hormone (HGH).* This hormone regulates growth of body tissue but specifically acts on the epiphysial cartilages of the long bones (refer to Exercise 13) to effect growth in height. Hypersecretion of this hormone in childhood may result in *gigantism,* and hyposecretion may result in a *midget* or *pituitary dwarf.* A hypersecretion in the adult may cause *acromegaly,* a disfiguring condition in which some growth areas in the skeleton still respond and grow absurdly out of

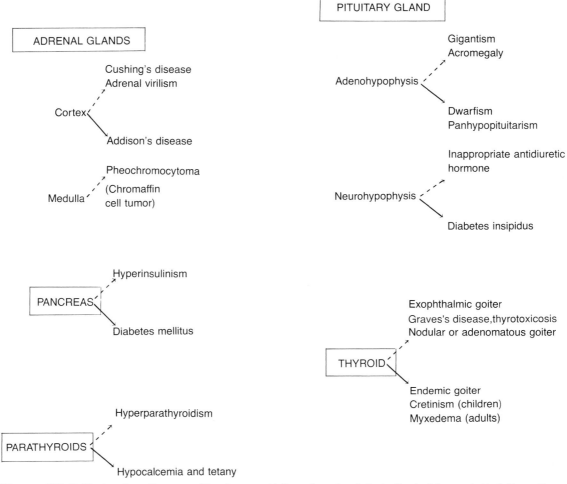

Figure 27-2 Endocrine diseases. The hyperactivity of a gland is indicated by a dotted line; the hypoactivity is indicated by a solid line.

shape, and others, such as those that affect height, do not continue to grow.

b. *Prolactin.* This hormone acts on the mammary glands, causing them to produce milk before and after childbirth.

The following tropic hormones are those that act as stimuli on other endocrine tissue:

c. *Thyrotropic hormone (TH).* This hormone regulates the growth, development, and activity of the *thyroid gland.*

d. *Adrenocorticotropic hormone (ACTH).* This hormone regulates the growth and development of the *adrenal cortex.*

e. *Follicle-stimulating hormone (FSH).* This is one of two *gonadotropic hormones.* It stimulates and regulates the early developmental stages of the ovarian (graafian) follicle in the female and

of the spermatozoa in the seminiferous tubules of the male testis. It has other functions as well.

f. *Luteinizing hormone (LH).* This hormone finishes what FSH has begun in the ovary or the testis. The follicle containing the ovum (or egg) completes its development and ovulates. The spermatozoa are anatomically completed and released from the seminiferous tubules. In the male, the luteinizing hormone often is termed the *interstitial-cell-stimulating hormone (ICSH).*

2. **POSTERIOR LOBE (NEUROHYPOPHYSIS) HORMONES**

a. *Oxytocin.* This hormone is found only in the female and stimulates uterine contractions. It is one of the major factors in initiating labor at the time of birth.

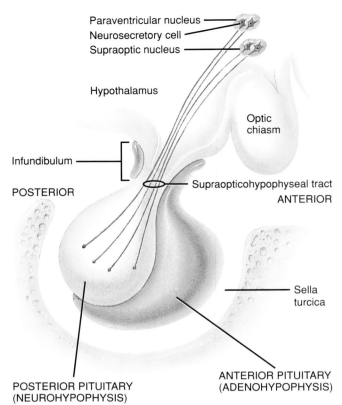

Paraventricular nucleus
Neurosecretory cell
Supraoptic nucleus

Hypothalamus

Optic chiasm

Infundibulum

POSTERIOR

Supraopticohypophyseal tract
ANTERIOR

Sella turcica

Figure 27-3 Hypothalamic-hypophysial tract. (From G. J. Tortora and S. R. Grabowski, *Principles of Anatomy and Physiology,* 7th ed., New York: HarperCollins, 1993.)

POSTERIOR PITUITARY (NEUROHYPOPHYSIS)

ANTERIOR PITUITARY (ADENOHYPOPHYSIS)

Table 27-1   Hormones of the Pituitary Gland

Gland	Hormones Produced	Action
Anterior lobe	Somatotropic hormone or human growth hormone (HGH)	Regulates normal body growth, particularly that of the skeleton
	Thyroid-stimulating hormone (TSH)	Regulates growth, development, and normal function of thyroid gland
	Adrenocorticotropic hormone (ACTH)	Regulates growth, development, and normal function of adrenal cortex
	Follicle-stimulating hormone (FSH)	Initiates development of ova in the female and of sperm in the male
	Luteinizing hormone, termed luteotropin (LH) in the female and interstitial cell stimulating hormone (ICSH) in the male	Stimulates ovulation and development of corpus luteum in the female and development of interstitial cells to produce testosterone in the male
	Prolactin (PR)	Stimulates milk production by mammary glands
	Melanocyte-stimulating hormone (MSH)	Increases skin pigmentation
Posterior lobe	Oxytocin	Stimulates contraction of uterine muscle
	Antidiuretic hormone (ADH) or vasopressin	Causes kidneys to remove water from urine from distal convoluted tubules Causes constriction of arterioles, to produce rise in blood pressure

b. *Antidiuretic hormone (ADH)* (also termed *vasopressin*). This hormone causes reabsorption of water from the distal convoluted tubules in the kidney, thereby preventing severe water loss. The hormone also causes a rise in blood pressure. Hyposecretion leads to *diabetes insipidus.*

Thyroid Gland

The thyroid gland (Figure 27-4) is a large mass of tissue on the anterior surface of the trachea just inferior to the larynx. It is composed of right and left lateral lobes, connected by a slim *isthmus* on the anterior surface of the trachea. The thyroid gland is an endodermal derivative of the embryonic gut.

The gland contains a large quantity of iodine, which it uses to produce three hormones:

1. Thyroxin (T4)
2. Triiodothyronine (T3)
3. Thyrocalcitonin

Thyroxin and triiodothyronine control the metabolic rate of all cells in the body, while thyrocalcitonin regulates blood calcium.

Hypothyroidism results in a slow-moving, sleepy, obese individual, whereas *hyperthyroidism* produces the opposite. A severe lack of the thyroid hormones in the fetus or in the child results in *cretinism.* In the adult, a similar physiological need produces an individual with *myxedema.* Lack of iodine in the diet also produces an abnormal gland that increases in size, due to *compensatory hypertrophy,* creating a *goiter,* which is an enlarged thyroid gland. A severe hyperthyroid condition causes another type of goiter, sometimes termed *exophthalmic goiter.* The two types of goiter are not the same because one is *endemic* and caused by iodine lack in certain geographic areas, whereas the other is a failure of the thyroid gland to function properly, even in the presence of iodine-rich soil and water.

Parathyroid Glands

The parathyroid glands (Figure 27-5) are usually four in number (two pairs embedded in the posterior medial surface of each thyroid lobe). The number is not constant, however. Some people have only one pair, and others have three pairs of the glandular masses. They arise from embryonic visceral arch tissue and produce the hormone *parathormone (PTH),* which controls calcium and phosphate metabolism in the body. *Hypoparathyroidism* results in brittle bones and low blood calcium and *hyperparathyroidism* results in soft bones, loss of teeth, and high blood calcium.

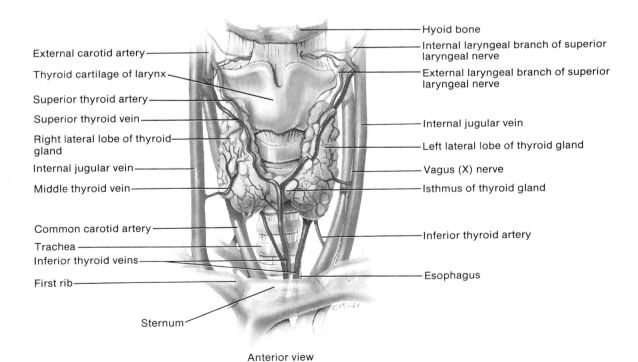

Anterior view

Figure 27-4 Location, blood supply, and innervation of the thyroid gland in anterior view. The pyramidal lobe of the thyroid, when present, may be attached to the hyoid bone by muscle. (From G. J. Tortora, *Principles of Human Anatomy,* 6th ed., New York: HarperCollins, 1992.)

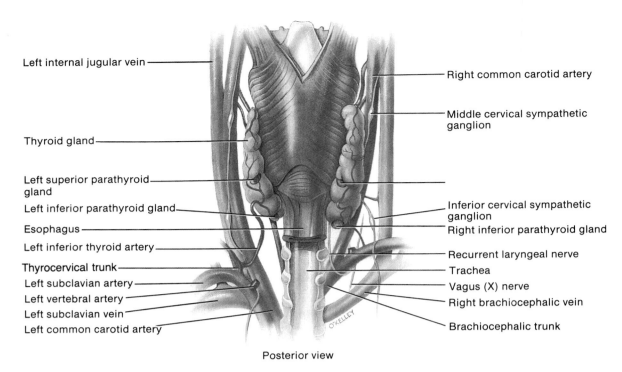

Left internal jugular vein

Right common carotid artery

Middle cervical sympathetic ganglion

Thyroid gland

Left superior parathyroid gland

Left inferior parathyroid gland

Inferior cervical sympathetic ganglion

Esophagus

Right inferior parathyroid gland

Left inferior thyroid artery

Recurrent laryngeal nerve

Thyrocervical trunk

Trachea

Left subclavian artery

Vagus (X) nerve

Left vertebral artery

Right brachiocephalic vein

Left subclavian vein

Brachiocephalic trunk

Left common carotid artery

Posterior view

Figure 27-5 Location, blood supply, and innervation of the parathyroid glands (posterior view). (From G. J. Tortora, *Principles of Human Anatomy,* 6th ed., New York: HarperCollins, 1992.)

Suprarenal or Adrenal Glands

The suprarenal glands (Figure 27-6) are located above the kidneys in the human. In most other animals, they are found lying beside the kidneys—hence the names *suprarenal* and *adrenal*. The gland has two distinct regions that have different embryonic origins:

1. The *suprarenal medulla* is the core of the gland, and it originates from embryonic neural crest cells. The hormones produced in this region are termed *epinephrine* (adrenaline) and *norepinephrine*. They have a function similar to that of the sympathetic division of the autonomic nervous system: They prepare the body for an emergency when secreted in large amounts.

2. The *suprarenal cortex* secretes many hormones, which are divided into categories:
   a. *Glucocorticoids—Cortisone* is an example. These hormones act as anti-inflammatory agents, function in protein digestion and storage, and provide resistance to stress.
   b. *Mineralocorticoids—Aldosterone* is an example. These hormones regulate sodium and potassium metabolism.
   c. *Adrenal sex hormones* or *gonadocorticoids*— These estrogens and androgens are secreted in minute quantities and have little effect in the normal body. However, they can cause sexual

anomalies when secreted in large amounts, as the result of a tumor. Hypersecretion leads to Cushing's syndrome. On the other hand, hyposecretion of the adrenocorticoids may result in a debilitating and progressive condition termed Addison's disease.

Islets of Langerhans of the Pancreas

The pancreas is a large exocrine gland (see Figure 27-7) that produces powerful digestive enzymes that are carried to the small intestine by ducts. Between the enzyme-producing tubules, there are islands of cells termed *alpha cells* and *beta cells* (the islets of Langerhans; see Color Plate 31). The beta cells produce *insulin,* which enables the body cells to obtain blood sugar needed for physiological oxidation. A hyposecretion of this hormone causes *diabetes mellitus*. The alpha cells secrete the hormone *glucagon,* which, by converting stored glycogen to glucose, increases the blood sugar level.

Gonads (Ovaries and Testes)

The paired ovaries (Figure 27-8) are the primary reproductive organs of the female and are located in the pelvic region of the body cavity (Figure 27-1). They have two types of hormone-producing tissue. The hormone *estrogen* is secreted by the cells of the *graafian follicle*

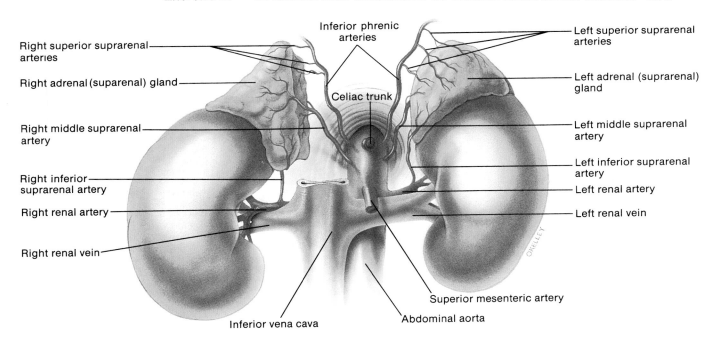

Figure 27-6 Location and blood supply of the adrenal (suprarenal) glands (anterior view). (From G. J. Tortora and S. R. Grabowski, *Principles of Anatomy and Physiology,* 7th ed., New York: HarperCollins, 1993.)

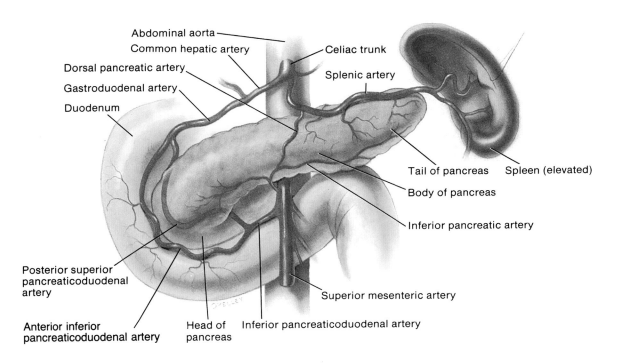

Figure 27-7 Location and blood supply of the pancreas (anterior view). (From G. J. Tortora and S. R. Grabowski, *Principles of Anatomy and Physiology,* 7th ed., New York: HarperCollins, 1993.)

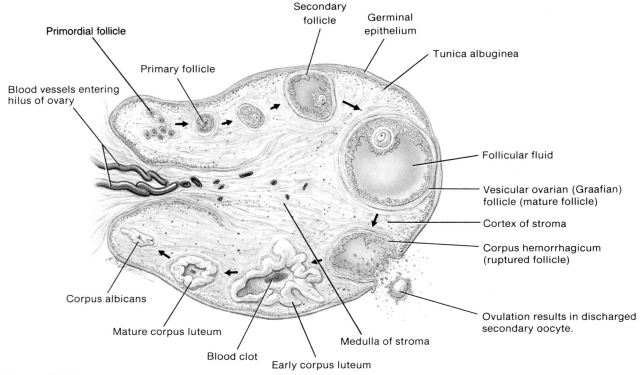

Figure 27-8 Human ovary (sectional view). (From G. J. Tortora and S. R. Grabowski, *Principles of Anatomy and Physiology,* 7th ed., New York: HarperCollins, 1993.)

and *progesterone* by the cells of the *corpus luteum.* There is a cyclic relationship between the ovarian hormones and the gonadotropic hormones of the pituitary gland. Estrogen is responsible for the development and maintenance of the female secondary sexual characteristics.

The paired testes (Figure 27-9) are the primary reproductive organs of the male. They lie outside the body cavity (see Figure 27-1), within a pouch of skin termed the *scrotum.* The testes develop in the embryo's pelvic region and descend into the scrotum before birth. The lower temperature (facilitated by external location) in the scrotum facilitates sperm development. In addition to producing and releasing spermatozoa, the testis contains interstitial cells that produce the male sex hormone *testosterone.* This hormone controls the growth and development of the male secondary sexual characteristics.

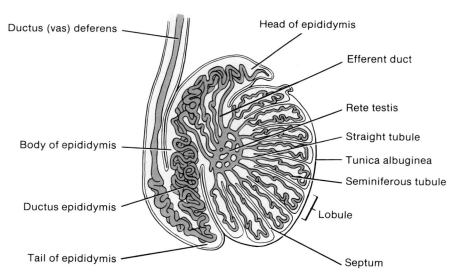

Figure 27-9 Human testis (sectional view). (From G. J. Tortora and S. R. Grabowski, *Principles of Anatomy and Physiology,* 7th ed., New York: HarperCollins, 1993.)

Pineal Gland

This small mass of endocrine tissue is a dorsal outgrowth of the diencephalon region of the brain (see Figures 24-5 and 24-6).

The products of the pineal gland are *melatonin,* which plays a part in ovarian function; *adrenoglomerulotropin,* which stimulates the adrenal cortex; and *serotonin,* which is found in normal brain tissue and acts as a neurotransmitter. The physiology of this gland has not been specifically defined.

Thymus Gland

The thymus gland is a dark, irregularly shaped, two-lobed structure. It is located behind the sternum and between the lungs (Figure 27-10). The two lobes are held together by connective tissue, and each lobe is enclosed by a connective tissue *capsule.* Microscopic examination of the gland shows that each lobe is divided further into smaller units termed *lobules.* The outer *cortex* of lobules is filled with lymphocytes (germinal centers), while the inner *medulla* regions hold fewer lymphocytes. (Color Plate 23 shows the structure of the thymus gland.)

The thymus gland is quite obvious in an infant. It reaches its maximum size at the time of puberty. After this period, much of the functional areas of thymic tissue are replaced by fat and connective tissue. Despite such replacement, the gland continues to be functional. The major role of the thymus gland is to help produce and distribute specialized cells, termed *T lymphocytes,* to other lymphoid organs (lymph nodes, spleen, etc.). Such cells are associated with the immune responses that result in the direct or indirect destruction of disease-causing microorganisms, cancers, and other foreign factors. Hormones produced by the thymus gland, such as *thymosin* and *thymopoietin,* promote the production and maturation of T lymphocytes.

HISTOLOGY OF SELECTED ENDOCRINE GLANDS

Materials

The following prepared slides should be provided for class use:

1. Adrenal gland
2. Pancreas
3. Thymus gland (prepuberty)
4. Thymus gland (adult)
5. Thyroid glands (normal and goiter specimens)
6. Parathyroid gland

Thyroid Gland

1. Examine the thyroid-gland slide with the low-power objective.
2. Identify the circular, single-celled layers termed *follicles* within the specimen. (Refer to Figure 27-11 and Color Plate 24.) Such follicles usually enclose solid-appearing materials known as *colloids,* which contain several proteins, including thyroxin. Some follicles will be empty.
3. Sketch and label a representative portion of the prepared slide in the results and observations section.

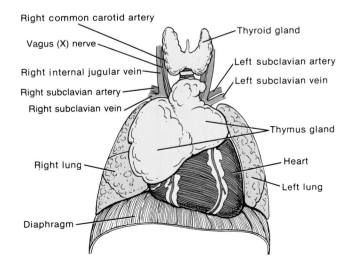

Right common carotid artery

Vagus (X) nerve

Right internal jugular vein

Right subclavian artery

Right subclavian vein

Thyroid gland

Left subclavian artery

Left subclavian vein

Thymus gland

Right lung

Heart

Left lung

Diaphragm

Figure 27-10 Location of the thymus gland in a young child. (From G. J. Tortora and S. R. Grabowski, *Principles of Anatomy and Physiology,* 7th ed., New York: HarperCollins, 1993.)

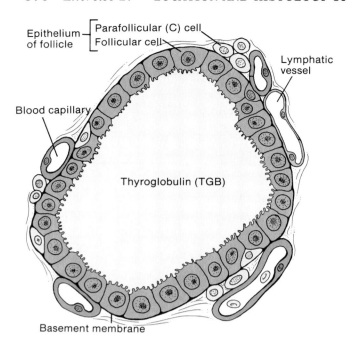

Figure 27-11 Histology of the thyroid gland, showing one follicle. (From G. J. Tortora and S. R. Grabowski, *Principles of Anatomy and Physiology,* 7th ed., New York: HarperCollins, 1993.)

Thyroid Gland and Goiter

1. Examine the thyroid gland preparation from a goiter case, with the low-power objective (Refer to Color Plate 25.)
2. Compare this specimen with the normal thyroid gland preparation. Sketch any obvious differences for a representative portion in the results and observations section.

Parathyroid

1. Examine the prepared slide of the parathyroid gland, with the low-power objective first and then with the high-power objective.
2. Identify the principal and oxyntic cells of the gland. (Refer to Color Plates 26 and 27.)
3. Sketch and label a representative portion of the gland in the results and observations section.

Adrenal Glands

1. Examine a prepared slide of an adrenal gland, with the low-power or scanning objective. Refer to your textbook, Figure 27-12 and Color Plates 28, 29, and 30 to identify the following parts.
   a. *Capsule*—thin, outer covering
   b. *Cortex*—outer section
   c. *Medulla*—central region

Figure 27-12 Histology of the adrenal (suprarenal) glands: photomicrograph of a section of the adrenal gland, showing its subdivisions and zones at a magnification of 45×. (Courtesy of Andrew Kuntzman.)

2. Examine the cortex more closely and locate its three zones.
   a. *Zona glomerulosa*—region next to the capsule containing thin layer of cells in clumps or even arcs
   b. *Zona fasciculata*—thicker central zone containing cells arranged in long cords
   c. *Zona reticularis*—zone next to the medulla containing cords of cells arranged in a netlike fashion.
3. Examine the medulla. Its cells are arranged in irregular groups, with several blood vessels separating them.
4. Sketch and label a general representation of the different parts of the adrenal gland in the results and observations section.

Pancreas

1. Examine the prepared slide of the pancreas, with the low-power objective. (Refer to Figure 27-13 and Color Plate 31.)

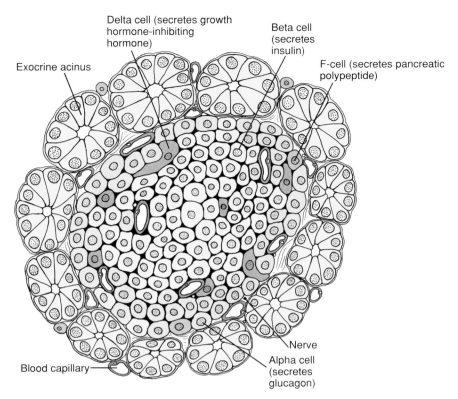

Delta cell (secretes growth hormone-inhibiting hormone)

Beta cell (secretes insulin)

F-cell (secretes pancreatic polypeptide)

Exocrine acinus

Nerve

Alpha cell (secretes glucagon)

Blood capillary

Figure 27-13 Islet of Langerhans. (From G. J. Tortora and S. R. Grabowski, *Principles of Anatomy and Physiology,* 7th ed., New York: HarperCollins, 1993.)

2. Identify the following portions of the gland.
   a. *Acini*—small tubules that consist of clusters of cells and a lumen (the exocrine part of the pancreas)
   b. *Islets of Langerhans*—clusters of alpha and beta cells, surrounded by capillaries and exocrine structures; small granules in alpha cells are darker than those found in beta cells
3. Sketch and label a general representation of the different parts of the pancreas in the results and observations section.

Thymus Gland

1. Examine the two forms of thymus gland preparation with the low-power and high-power objectives.
2. Note the cortex, the medulla, and the cell types in the preparation. (Refer to Color Plate 23.)
3. Answer the questions in the results and observations section.

RESULTS AND OBSERVATIONS

1. Thyroid gland

Normal

Goiter

a. Were there any differences between the normal thyroid and goiter specimens? _____

_____

b. What is a goiter? _____

_____

2. Parathyroid gland

3. Adrenal gland

Cortex

Medulla

4. Pancreas

5. Thymus

Adult

Puberty

a. Which cell types are more numerous in the adult thymus gland? _____

b. Which cell types are more numerous in the pre-puberty thymus gland? _____

c. Why are there such differences between the two preparations? _____

_____

_____

d. What are T lymphocytes? _____

_____

LABORATORY REVIEW 27

LOCATION AND HISTOLOGY OF HUMAN ENDOCRINE GLANDS

MATCHING AND FILL-IN

1. Match the hormones on the left with the endocrine glands that produce them.

   a. _____ Thyroxin

   b. _____ Parathormone

   c. _____ Aldosterone

   d. _____ Insulin

   e. _____ Epinephrine

   f. _____ Human growth hormone

   g. _____ Cortisone

   h. _____ Glucagon

   i. _____ Adrenocorticotropic hormone

   j. _____ Thyrocalcitonin

   k. _____ Estrogen

   l. _____ Prolactin

   m. _____ Testosterone

   n. _____ Melatonin

ANSWERS

1. Pituitary, anterior lobe
2. Pituitary, posterior lobe
3. Thyroid
4. Parathyroid
5. Adrenal cortex
6. Adrenal medulla
7. Pancreas
8. Ovary
9. Testis
10. Thymus
11. Pineal

2. Complete Table Q27-1 by indicating at least one specific abnormality that results from the hypo- or hypersecretion of the hormones indicated.

Table Q27-1

Hormone	Hormonal Level		Abnormality
Glucocorticoids	Hyposecretion	a.	
	Hypersecretion	b.	
Parathyroid hormone	Hyposecretion	c.	
	Hypersecretion	d.	
Thyroxin	Hyposecretion (child)	e.	
	Hypersecretion (adult)	f.	
Antidiuretic hormone	Hyposecretion	g.	
Human growth hormone	Hypersecretion (childhood)	h.	
	Hypersecretion (adult)	i.	

3. Complete Table Q27-2 from the one clue provided in each case.

Table Q27-2

Endocrine Gland	Hormone	Associated Activity
a.	b.	Regulates growth in height
c.	Follicle-stimulating hormone	d.
e.	f.	Stimulates contraction of the uterus
g.	h.	Stimulates water reabsorbtion by the kidneys
i.	Parathyroid hormone	j.
Thymus	k.	l.
m.	Prolactin	n.
o.	Insulin	p.
Alpha cells of the pancreas	q.	r.
s.	Serotonin	t.
u.	Thyroxin	v.

4. Distinguish between an *exocrine* and an *endocrine* gland. _____

_____

_____

5. Identify the parts (a)–(d) of the thyroid gland indicated in Figure Q27-1.

a. _____

b. _____

c. _____

d. _____

Figure Q27-1

MULTIPLE CHOICE

Enter the correct answers in the spaces provided.

1. Once hormones have reached levels that are adequate to meet existing body needs, further release is controlled by which of the following?
   a. hyperactivity of endocrine glands
   b. hypoactivity of endocrine glands
   c. hypoactivity of target organs
   d. hyperactivity of target organs
   e. negative feedback mechanism

2. What is the term for overactivity of an endocrine gland, in relation to hormone production?
   a. hypoactivity
   b. hyperactivity
   c. homeostasis
   d. negative feedback
   e. either choices *a* or *c*

3. Hyposecretion of HGH in childhood may result in which of the following conditions?
   a. gigantism
   b. acromegaly
   c. midget state
   d. exophthalmic goiter
   e. diabetes insipidus

4. Hypersecretion of HGH in childhood may result in which of the following states?
   a. gigantism
   b. acromegaly
   c. midget state
   d. exophthalmic goiter
   e. diabetes insipidus

5. Hypersecretion of HGH in an adult may produce which of the following conditions?
   a. gigantism
   b. acromegaly
   c. midget state
   d. exophthalmic goiter
   e. diabetes insipidus

6. Which of the following hormones is not produced by the posterior lobe of the pituitary?
   a. prolactin
   b. oxytocin
   c. FSH
   d. ACTH
   e. thyrotropic hormone

7. Diabetes insipidus results from the hyposecretion of which of the following hormones?
   a. oxytocin
   b. ADH
   c. FSH
   d. ACTH
   e. prolactin

1. _____
2. _____
3. _____
4. _____
5. _____
6. _____
7. _____
8. _____
9. _____
10. _____
11. _____
12. _____
13. _____
14. _____
15. _____

8. Which of the hormones is a major factor in initiating labor at the time of birth?
   a. oxytocin
   b. ADH
   c. FSH
   d. ACTH
   e. prolactin

9. Which of the following is known to produce estrogen?
   a. corpus luteum
   b. follicle cells associated with a graafian follicle
   c. islets of Langerhans
   d. thyroid glands
   e. thymus gland

10. Cretinism is associated with a lack of hormones produced by which of the following glands?
    a. corpus luteum
    b. follicle cells associated with a graafian follicle
    c. islets of Langerhans
    d. thyroid glands
    e. thymus gland

11. A goiter results from a deficiency or disorder in which of the following endocrine glands?
    a. thymus gland
    b. thyroid gland
    c. parathyroid glands
    d. adrenal gland
    e. pineal gland

12. Diabetes mellitus results from the hyposecretion of which of the following hormones?
    a. glucagon
    b. ADH
    c. cortisone
    d. testosterone
    e. insulin

13. What is a hormone known for its anti-inflammatory activity?
    a. glucagon
    b. ADH
    c. cortisone
    d. testosterone
    e. insulin

14. Which of the following hormones control calcium and phosphate metabolism in the body?
    a. insulin
    b. parathormone
    c. ADH
    d. aldosterone
    e. adrenocorticoids

15. Hypersecretion of which of the following hormones can lead to Cushing's syndrome?
    a. insulin
    b. parathormone
    c. ADH
    d. aldosterone
    e. adrenocorticoids

The Senses

*S*pecialized sensory receptors pick up environmental stimuli and transmit them to the central nervous system. Mechanical, chemical, and visual receptors each receive different types of stimuli. The nature of the sensation provoked by such stimuli is determined by the brain. Table VIII-1 describes several sensory receptors.

The exercises in this section consider the general structure of the eye and several procedures to test its normal functioning, the structure and functioning of the ear, the sense of smell and its relationship to taste, and various sensory properties of the skin. Attention also is given to several types of optical illusions, including those involving the interpretation of spatial relationships.

Table VIII-1   The Sensory Receptors

Sensory Receptor	Location	Type of Receptor and Type of Stimulus
Rods and cones	Retina of each eye	Visual—light
Hair cells		
For converting sound waves to nerve impulses	Organ of Corti in cochlear duct of inner ear	Mechanical—vibrations or movements of fluid in inner ear
For positioning of head and body	Utricle of inner ear	Mechanical—movements of $CaCO_3$ (calcium carbonate) particles in direction of gravity
For changes in movement of body	Ampulla of semicircular ducts of inner ear	Mechanical—rapid movement of $CaCO_3$ particles
Taste buds (Gustatory cells)	Papillae of tongue	Chemical—specific chemicals in solution
Olfactory cells	Olfactory epithelium in nasal cavity	Chemical—dissolved gaseous substances
Touch and pressure cell endings	Dermis of skin, smooth muscles of visceral organs	Mechanical—touch

Table VIII-1 *(Continued)*

Sensory Receptor	Location	Type of Receptor and Type of Stimulus
Vibration	Dermis of skin	Mechanical—touch
Temperature (heat and cold)	Dermis of skin, smooth muscles of visceral organs	Mechanical—hot and cold
Pain cell endings	Dermis of skin, smooth muscles of visceral organs	Mechanical—all intense stimuli
Proprioceptors	Muscles, tendons, joints	Mechanical—movement

GENERAL REFERENCES

Barlow, H. B., and J. D. Mollon (Eds.). *The Senses.* New York: Cambridge University Press, 1982.

Pickles, J. O. *An Introduction to the Physiology of Hearing.* New York: Academic Press, 1982.

Rodieck, R. W. *The Vertebrate Retina, Principles of Structure and Function.* San Francisco: Freeman, 1976.

Tortora, C. J., and S. R. Grabowski. *Principles of Anatomy and Physiology,* 7th ed. New York: HarperCollins, 1993.

SPECIFIC REFERENCES

Borg, E., and S. A. Counter. "The Middle Ear Muscles." *Scientific American.* 261: 74–81, 1989.

Cytowic, R. E. "Tasting Colors, Smelling Sounds." *Sciences,* September/October, 32–37, 1988.

Doty, R. L. "Olfactory Communication in Humans." *Chemistry of the Senses,* 6: 351–376, 1981.

Loeb, G. E. "The Functional Replacement of the Ear." *Scientific American,* 252: 104–111, 1985.

Koretz, J. F., and G. H. Handelman. "How the Human Eye Focuses." *Scientific American,* 259: 92–99, 1988.

Krachmer, J. H., and D. A. Palay. "Corneal Disease." *The New England Journal of Medicine,* 325: 1804–1806, 1991.

Mahowald, M. A., and C. Mead. "The Silicon Retina." *Scientific American,* 264: 76–82, 1991.

Nathans, J. "The Genes for Color Vision." *Scientific American,* 260: 42–49, 1989.

Wolfe, J. M. "Hidden Visual Processes." *Scientific American,* 248: 94–103, 1983.

Anatomy of the Eye and Ear

After completing this exercise, you should be able to

1. Identify the parts of the mammalian eye
2. Examine the muscles that move the eyeball
3. List and identify all parts of the ear
4. Trace the path of sound waves through the parts of the ear
5. Use an otoscope and become familiar with its applications

Materials

The following materials should be provided for class use:

1. Fresh eyes of sheep or cow
2. Preserved eyes of sheep or cow
3. Preserved fetal pigs
4. Model of human eye
5. Chart of human eye
6. Model of human ear
7. Demonstration—mammalian ear ossicles
8. Human skull

OUTER SURFACE OF EYEBALL AND EXTRINSIC EYEBALL MUSCLES

Instructions for Dissection of a Mammalian Eye

Because the fetal pig's eye is too small for easy study of the eye's anatomy, fresh cow or sheep eyes should be used for this dissection exercise.

Examine the surface of the eyeball. It is almost round, with a clear surface in front and with a muscular area surrounded by fat in the rear. Examine the posterior surface of the eyeball carefully before removing the fat. Locate the *optic nerve,* and isolate it, so that it will not be injured accidentally when removing the fat. Remove as much fat as possible, so that the *extrinsic eyeball muscles* can be studied before the eyeball is opened.

Mammalian eyes have six extrinsic muscles per eye, plus a set of *retractor muscles (retractor bulbi),* which are not present in the human. Using Figure 28-1, locate the following muscles:

1. Superior rectus
2. Inferior rectus
3. Lateral rectus
4. Medial rectus
5. Superior oblique
6. Inferior oblique
7. Retractor bulbi (not found in the human eye)

Opening the Eyeball

Hold the sheep or cow eye between thumb and forefinger, with the thumb on the posterior surface and the forefinger on the cornea (Figure 28-2). Make an incision about halfway between poles, dividing the eye into anterior and posterior halves. Continue this incision around the equator with your scissors. Separate the two halves on a dissecting tray or in a pan of water. Compare the parts of the eye of your dissected specimen with those shown in Figure 28-3, and answer the questions in the results and observation section.

The anterior portion of the eyeball contains the thick cornea and the circular ciliary body (muscles) and iris. The latter two are pigmented and appear black, in contrast to the translucency of the cornea. The aqueous fluid, or aqueous humor, in this part of the eye probably

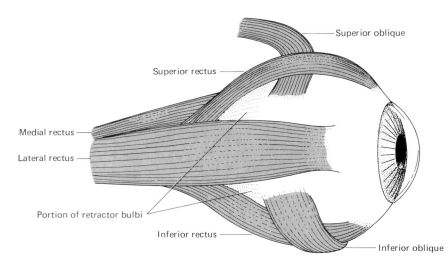

Figure 28-1 Extrinsic eyeball muscles of a sheep (lateral view).

will have drained away and will not be seen. The aqueous fluid in the living specimen is found in two chambers of the eye: the *anterior chamber* between the cornea and the iris, and the *posterior chamber* between the iris and the lens. This fluid is secreted by the cornea and drained into the canal of Schlemm, a venous sinus. Refer to your textbook for a more complete discussion.

The cavity within the eyeball, between the lens and the retina, contains the jellylike substance, the *vitreous body*. This viscous fluid creates pressure within the eyeball and keeps it from collapsing. Embedded in it is the large, transparent *lens*. In dissection specimens, the lens usually remains attached to the vitreous jelly by black strands, the *suspensory ligaments*.

Figure 28-2 Using scissors to open the eye around its circumference.

Remove the lens and vitreous body, and find the *retina,* a thin, transparent membrane forming the inner coat of the eyeball. It contains the sensory receptors (*rods* and *cones*). In a sheep or cow eye, the retina has a shiny, greenish-blue layer termed the *tapetum,* which improves night vision. The tapetum is not found in the human eye.

The region where the retinal fibers become the optic nerve is the *optic disc* or *blind spot*. In Exercise 29, you will demonstrate the blind spot in your own eyes. Close to the optic disc on the retina is the yellowish (in the living animal) *macula lutea*. In the center of the macula lutea is the *fovea centralis,* the region of sharpest vision, with a high concentration of the cone cells.

Observations on the Pig Eye

Locate the upper and lower eyelids of the fetal pig. Cut the skin back carefully at both lateral and medial angles (Figure 28-4). From the initial incision, cut completely around the eye, and remove both eyelids.

Observe the *nictitating membrane* at the median angle of the eye. This is the third eyelid of the pig and can be released to cover the entire corneal surface. Grasp the nictitating membrane with forceps, and free the eyeball as much as you can from the adjacent tissue.

When the eyeball has been somewhat loosened, cut the eyeball away from the eyeball muscles and the optic nerve. The names of these muscles are identical to those in the sheep eye, which you previously dissected. Be careful not to puncture the eyeball when removing it from the orbit.

After the eyeball has been removed, look into the bony orbit, and locate the cut ends of the six eyeball muscles. Like other skeletal muscles of the fetal pig, they are weak, and they tear easily. Distinguish the optic

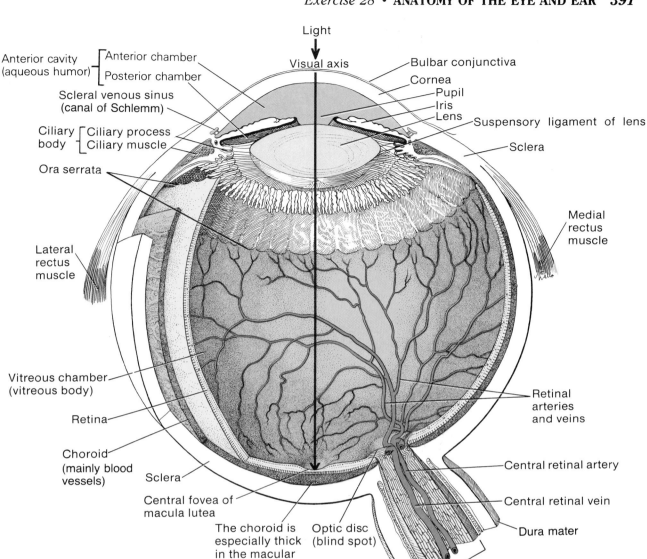

Light
Visual axis

Anterior cavity (aqueous humor) — Anterior chamber / Posterior chamber

Scleral venous sinus (canal of Schlemm)

Ciliary body — Ciliary process / Ciliary muscle

Ora serrata

Lateral rectus muscle

Vitreous chamber (vitreous body)

Retina

Choroid (mainly blood vessels)

Sclera

Central fovea of macula lutea

The choroid is especially thick in the macular region

Optic disc (blind spot)

Bulbar conjunctiva
Cornea
Pupil
Iris
Lens
Suspensory ligament of lens
Sclera

Medial rectus muscle

Retinal arteries and veins

Central retinal artery

Central retinal vein

Dura mater

Optic (II) nerve

Transverse section of left eyeball (superior view)

Figure 28-3 Structure of the human eyeball: diagram of gross structure in transverse section. (From G. J. Tortora and S. R. Grabowski, *Principles of Anatomy and Physiology,* 7th ed., New York: HarperCollins, 1993.)

nerve from the six extrinsic eyeball muscles. A *retractor bulbus* muscle surrounds the pig's optic nerve, as it does in the sheep, and it pulls the eyeball back against the wall of the orbit.

Clean the orbit of all extraneous material, such as latex from the injections, and study the patterns of the eyeball muscles. The eyeball of the pig is small, but if dissected carefully, it will show the same structures found in the sheep eye. Follow the dissection instructions described for the sheep eye, and make your observations. Use a dissecting microscope for detail.

Observations of the Human Eye

The structures of the human eye, shown in Figure 28-3, are not very different from the structures found in the pig or the sheep eyes. Using diagrams and models of the human eye, locate all the parts that have been seen in the sheep and pig eyes. In a frontal view of the human eye (Figure 28-5), look for the nictitating membrane, or the remnant of it, the *plica semilunaris*, in the median angle of the eye. It is seen lateral to a

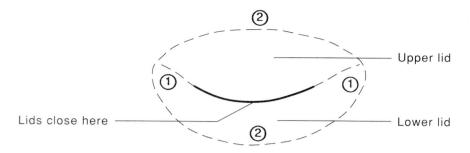

Figure 28-4 Frontal view of the fetal pig eye: a diagram showing incision lines.

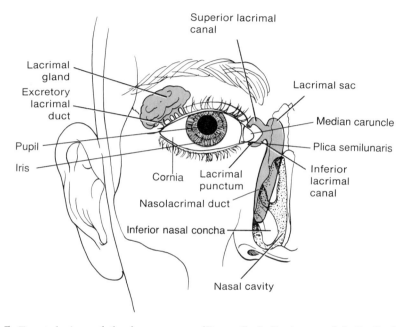

Figure 28-5 Frontal view of the human eye. (From G. J. Tortora and S. R. Grabowski, *Principles of Anatomy and Physiology,* 7th ed., New York: HarperCollins, 1993.)

rounded scarlike mass termed the *median caruncle.* Both of these parts are vestigial in the human.

The conjunctiva, which contains blood vessels, covers the cornea of the eyeball, which has no blood vessels of its own. What impact do you think this has on the physiology of the eye? Examine your partner's eyeball. Can you see these structures?

ANATOMY OF THE EAR

Materials

The following should be provided for class use:

1. Preserved fetal pigs
2. Model of human ear
3. Chart of human ear
4. Otoscope
5. Alcohol-treated pads and sterile cotton

Observations of the Fetal Pig Ear

Like the human, the pig has a tripartite hearing organ, which can be divided into the following:

1. *Outer ear*—This is composed of a large *pinna* and a short *external auditory canal.*
2. *Middle ear*—The ear ossicles of the middle ear are in contact with the tympanic membrane at the end of the auditory canal. In order, they are the *malleus,* the *incus,* and the *stapes,* with the malleus up against the tympanic membrane of the middle ear and the stapes positioned against the oval window of the inner ear.

3. *Inner ear*—This comprises the *cochlea* and has the same parts as other mammalian ears.

Cut off the fetal pig's pinna, to observe the external auditory canal. With a careful dissection, it is possible to see the *tympanic membrane.* Other parts are best studied in diagrams and models.

Anatomy of the Human Ear

Structurally, the human ear (like the pig ear) consists of three regions: the *external* or *outer ear,* the *middle ear,* and the *internal* or *inner ear* (Figure 28-6).

The *outer ear* is made up of the *auricle,* or *pinna,* and the *external auditory canal.* The auricle is the large, soft flap on the side of the head that is supported by cartilage. Within it is the opening to the external auditory canal, a tube about 2.5 cm (1 inch) long, which terminates at the tympanic membrane, or eardrum. The human auricle is vestigial, but the external auditory canal functions as a funnel that traps high-frequency vibrations and directs them to the middle ear. The skin lining this canal contains hair follicles and *cerumen* (or wax) *glands.*

The *middle ear* is an air-filled cavity in the temporal bone. It extends from the eardrum to the cochlea. The *tympanic membrane* arises from all three embryonic germ layers but is generally considered to be part of the middle ear. Attached to the tympanic membrane are three ear ossicles that have developed from the embryonic visceral arches and are therefore a part of the visceral skeleton. The ossicles, termed *malleus* (hammer), *incus* (anvil), and *stapes* (STĀP-eez) (stirrup), bridge the span of the middle ear and terminate at the *cochlea.* Also located in the middle ear is the opening to the *eustachian tube,* which leads to the pharynx.

The lower region of the *inner ear* is known as the *cochlea.* It contains the sensory apparatus for hearing and is shaped like a snail shell, coiled 2¾ times. The entire structure is enclosed in a bony labyrinth within the temporal bone.

The stapes of the middle ear lies up against the *oval window* of the cochlea. Sound waves passing over the ear ossicles cause the membrane to vibrate, setting into motion the fluid in the cochlear channels: the *scala vestibuli, scala tympani,* and *scala media* (see Figure 28-6). The membranes separating these fluid-filled canals are the *vestibular membrane* and the *basilar mem-*

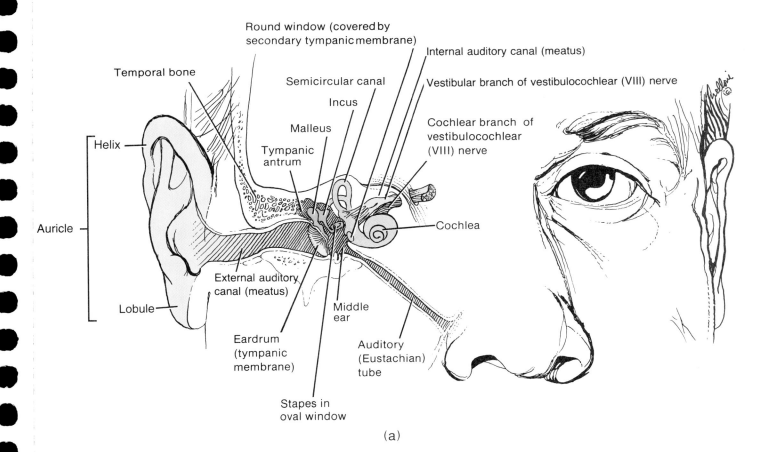

(a)

Figure 28-6 Human ear: (a) frontal section through right side of skull showing parts of ear.

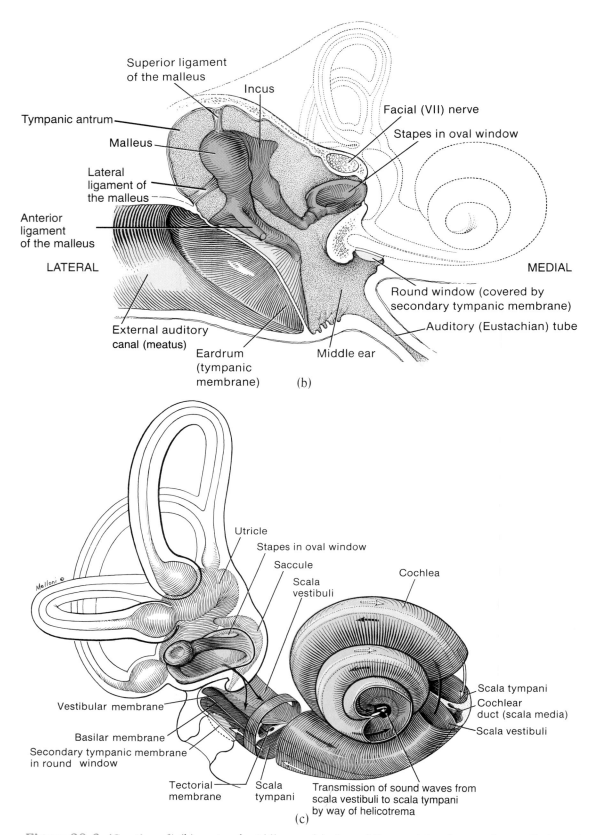

Figure 28-6 *(Continued)* (b) parts of middle ear; (c) view of the semicircular canals, vestibule and cochlea; (d) components of the vestibulocochlear nerve. (From G. J. Tortora and S. R. Grabowski, *Principles of Anatomy and Physiology,* 7th ed., New York: HarperCollins, 1993.)

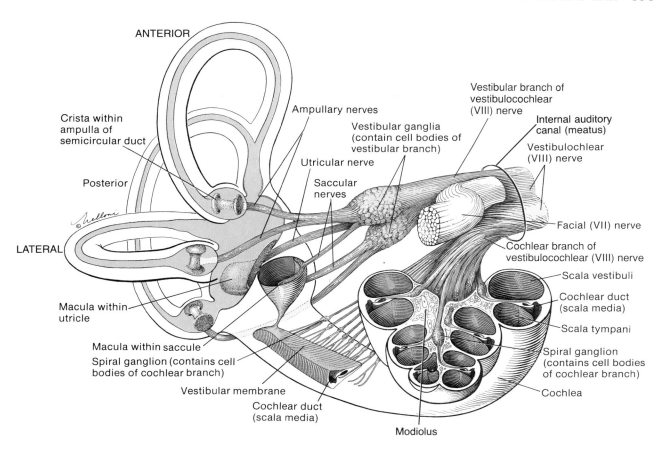

ANTERIOR

Crista within
ampulla of
semicircular duct

Posterior

LATERAL

Macula within
utricle

Macula within saccule
Spiral ganglion (contains cell
bodies of cochlear branch)

Vestibular membrane

Cochlear duct
(scala media)

Modiolus

Ampullary nerves

Vestibular ganglia
(contain cell bodies of
vestibular branch)

Utricular nerve

Saccular
nerves

Vestibular branch of
vestibulocochlear
(VIII) nerve

Internal auditory
canal (meatus)

Vestibulochlear
(VIII) nerve

Facial (VII) nerve

Cochlear branch of
vestibulocochlear (VIII) nerve

Scala vestibuli

Cochlear duct
(scala media)

Scala tympani

Spiral ganglion
(contains cell bodies
of cochlear branch)

Cochlea

(d)

brane. The basilar membrane supports the *organ of Corti* in the *scala media,* which is the sensory part of the ear in contact with cranial nerve VIII. At the end of the scala tympani is another membraneous window, the *round window,* which serves as an outlet for sound-wave vibrations in the cochlea. Just above the cochlea of the inner ear are three small semicircular canals. These function together as an organ of balance.

The parts of the ear are too small to make detailed dissection feasible and are almost inaccessible, lying deep within the petrous region of the temporal bone. The location of the parts of the ear may be studied using models and diagrams. Trace the path of a sound wave, beginning with the pinna and ending with the auditory nerve, using Figure 28-6.

EXAMINING THE EAR

If an otoscope (Figure 28-7) is available, use it to examine the external auditory canal and the tympanic membrane. Your instructor will explain the use and cleaning of the instrument. Carefully wipe the tip of the scope with a cotton square containing disinfectant. Let it air dry. Turn on the light, and look into the auditory canal of your lab partner. In the results and observations section, describe briefly what you see.

(a)

(b)

Figure 28-7 The otoscope: (a) the instrument; (b) use of the instrument.

LABORATORY REVIEW *28*

ANATOMY OF THE EYE AND EAR

IDENTIFICATION AND COMPLETION

1. Label the parts of the mammalian eye in Figure Q28-1. Use the following key:

   a. Sclerotic coat
   b. Choroid coat
   c. Retina
   d. Vitreous chamber
   e. Posterior chamber
   f. Anterior chamber
   g. Fovea
   h. Optic disc (blind spot)
   i. Cornea
   j. Lens
   k. Iris
   l. Ciliary body
   m. Suspensory ligaments
   n. Pupil
   o. Eyelid
   p. Plica semiluminaris
   q. Optic nerve

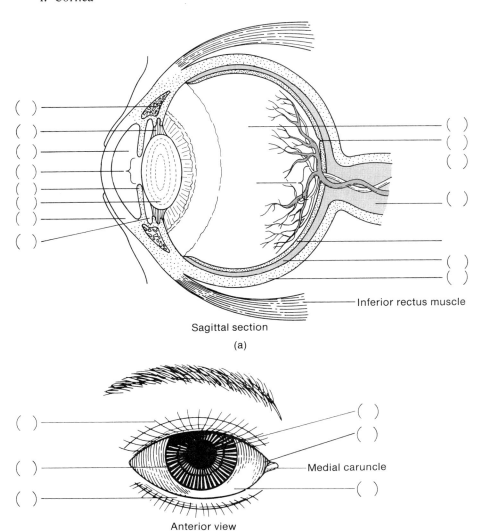

Inferior rectus muscle

Sagittal section
(a)

Medial caruncle

Anterior view
(b)

Figure Q28-1 The mammalian eye.

2. Trace the pathway of light from the outer surface of the eye to the brain. Some of the steps are provided. You fill in the rest.

   a. Conjunctiva _____

   b. _____

   c. _____

   d. _____

   e. _____

   f. Lens _____

   g. _____

   h. _____

   i. _____

   j. Brain _____

3. Label the parts of the ear in Figure Q28-2. Use the following key. (Refer to Exercise 30, as needed.)

   a. External auditory canal
   b. Tympanic membrane
   c. Cochlea
   d. Eustachian tube opening
   e. Oval window
   f. Round window
   g. Scala media
   h. Basilar membrane

   i. Organ of Corti
   j. Scala tympani
   k. Scala vestibuli
   l. Vestibular membrane
   m. Malleus
   n. Incus
   o. Stapes
   p. Tectorial membrane

Figure Q28-2 Parts of the ear. Diagramatic.

4. Trace the pathway of sound from a source outside the ear to the brain. Some of the steps are provided. You fill in the rest.

   a. <u>External auditory canal</u>

   b. _____

   c. _____

   d. <u>Incus</u>

   e. _____

   f. _____

   g. _____

   h. <u>Scala tympani</u>

   i. _____

   j. _____

   k. _____

   l. <u>Brain</u>

5. Describe the view through the otoscope.

   _____

   _____

   _____

   _____

The Physiology of Vision

After completing this exercise, you should be able to

1. Describe the focusing of light rays on the retina
2. Perform and interpret tests used to determine visual acuity
3. Perform two procedures to test pupillary reflexes
4. Perform a simple test to detect color blindness
5. Define or explain visual acuity, astigmatism, color blindness, and optical illusions

Light rays reflected from objects pass into the eye through the cornea and on through the anterior chamber and the lens. As light passes through these structures, it is focused on a light-sensitive region, the retina (Figure 29-1). The iris regulates the amount of light entering the eye, and the position and shape of the lens can be shifted slightly, to allow the eye to focus on objects at varying distances. This exercise includes a physical examination of the eye with the aid of an ophthalmoscope (Figure 29-3a).

Tiny inverted images are focused on the retina, and the varying intensities and wavelengths of the light energy stimulate combinations of the photoreceptors—the cones and rods—to initiate a series of nerve impulses. These impulses are coordinated by other neurons in the retina and are transmitted by the optic nerves to the optic centers of the brain, for final interpretation and integration. The normal human eye can sufficiently deflect light rays centering on an object 20 meters away to focus a clear object on the retina. An eye with a normal ability to refract light is referred to as being *emmetropic*. With a normal human lens, objects far in the distance, up to as close as 20 meters will form images on the light-sensitive area of the ret-

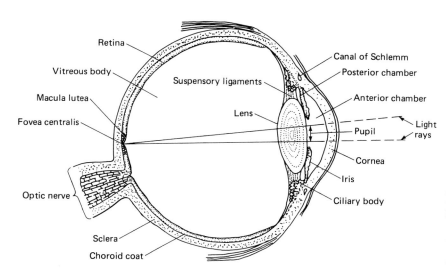

Figure 29-1 The components of the human eye (shown in sagittal—lengthwise—section): The focusing of light rays onto the fovea centralis, the area of sharpest vision is shown.

ina. With objects closer than 20 meters, the lens has to sharpen its focus, by means of ciliary muscles.

Various types of abnormalities related to improper *refraction* (deflecting light from a straight path) are known. These include *nearsightedness* (*myopia*), *farsightedness* (*hypermetropia*), and *astigmatism*. *Astigmatism* refers to irregularities in the surface of the cornea or lens, causing an inability of the eye to focus a ray of light sharply on the retina. Procedures commonly used to study the functioning of the eye related to the series of events described here are included in this exercise. They are a test of *visual acuity* (the degree of detail the eye can distinguish); a test of *astigmatism*, using the well-known astigmatism test chart (Figure 29-2), and the determination of the individual's location of the *blind spot*.

The opening in the center of the iris is the pupil, which controls light transmission. In the darkness, the diameter of the pupil is wide, but while in bright light, it decreases. The dimensions of the pupil are controlled by the muscles of the iris, which are under reflex control. Procedures to test pupillary reflexes are included in this exercise.

Sensation of any color can be obtained by the appropriate mixture of three colors of light: red, green, and blue. In color blindness, a sex-linked genetic disorder, the number of colors that may be distinguished by an individual is greatly reduced. With humans, the most common type is red–green color blindness. A less common type is an inability to distinguish between blues and yellows. In the most severe forms of the genetic disorder, all objects appear to be some shade of gray. Approximately 8.0 percent of males and 0.5 percent of females exhibit some degree of color blindness. A test for color blindness is included in this exercise.

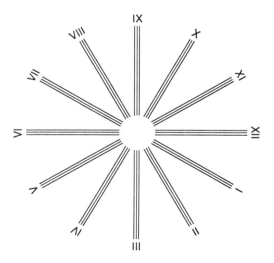

Figure 29-2 The astigmatism test chart.

Various drugs cause changes in the diameter of the pupil. For example, homatropine, which is a drug used in ophthalmic examinations, makes the pupil larger than normal. Drugs such as cocaine and epinephrine cause a similar reaction. In contrast, morphine brings about a constriction of the pupil size.

Individuals constantly make judgments related to their perception of the size and distance of the objects they see. Such estimations depend on several factors, including the relative positions of an individual's eyes, the size of the retinal images, the amount of detail seen, and the color and clarity of the outlines of objects. A false interpretation of a visual sensation results in an optical illusion. Examples of such illusions include objects appearing farther away or larger than they really are. A portion of this exercise considers depth perception measurement and demonstrates some optical illusions.

The retina is the location of the photoreceptor cells and is essentially an extension of the brain. Before light can reach the rods and cones, it must first penetrate the many nerve cell bodies and processes that form the remaining layers of the retina. These nerve fibers all come together in a thick bundle near the back of the eye, to form the *optic nerve*, which completely interrupts the retina. Because there are no receptors in this area, no sensation of vision is produced when light falls on it. This area is termed the *blind spot*. The existence of the blind spot is demonstrated in this exercise.

Most individuals are normally quite unaware of the large gap in their visual field because the movements of their eyes continuously place images on that part of the retina where the sharpest vision occurs, the central fovea in the macula lutea. The macula lutea lies to the side of the blind spot, and it forms a disk, the outer rim of which contains both rods and cones (see Figures 29-1 and 29-3c). Toward the center, the rods disappear, so that the central foveal area contains only the cones. This is the area transmitting the sharpest image to the brain.

The photoreceptive rods and cones contain light-sensitive pigments that decompose when they absorb light energy. Rods contain a purple pigment known as *rhodopsin*. In the presence of light, rhodopsin breaks down into the yellowish substance *retinene*, which is made from vitamin A, and a colorless protein known as *scotopsin*. This photochemical dissociation of rhodopsin is termed the *bleaching reaction*. In bright light, the sensitivity of retinal rods is greatly reduced.

Individual cones are composed of retinene, combined with one of three different types of visual proteins: red, blue, and green. The particular color experienced by an individual depends on the particular set or combination of cones stimulated by a light source. If all three types of cones are stimulated, a person will see white; if none of the cones is stimulated, a person will see black. Cones are less sensitive to light than rods are. Cones provide color vision and greater visual acuity.

In situations when an individual becomes adapted to a bright light in the form of a light bulb and then quickly views a blank surface such as a wall, the image of the light bulb will appear. This effect is termed a *positive afterimage* and results from the continued activity (firing) of the photoreceptors. Usually after a short period of time, the *negative afterimage,* a dark image of the bulb, will appear against a lighter background. This effect is caused by the bleaching of the visual pigment of the involved receptors. This exercise also considers a visual pigment activity.

EXAMINING THE EYE

Materials

The following materials should be provided for class use:

1. Model of the eye
2. Chart of the anatomy of the eye
3. Ophthalmoscope
4. A darkened room
5. Timers
6. Disinfectant and cotton

Procedure

This procedure is to be performed by students in pairs.

1. Review the locations of the components of the eye (Figure 29-1) and of the retina (Figures 29-3b and 29-3c, and Color Plate 32.).
2. Examine the parts of the ophthalmoscope. Locate all of the parts indicated in Figure 29-3a. Pay particular attention to the diopter window, lens-selection disk, rheostat control, and the viewing aperture.
3. Take the ophthalmoscope in one hand, and turn it on by depressing the rheostat lock button and then rotating the rheostat control. Hold the instrument about 5 cm (2 inches) above the desk, and rotate the rheostat control over its entire range of settings. Note how the intensity of the light changes from very bright to very dim with the movement of the control.
4. Set the diopter window on "0" and the light source on the green spot of the instrument. The ophthalmoscope is now ready for the eye-examination procedure.

5. Next, using Figures 29-4a through 29-4d, practice holding the ophthalmoscope and using the proper positions for examining both the right and left eyes. (Note that the ophthalmoscope is held in the right hand for the examination of the right eye and in the left hand for the left eye.)
6. In a darkened room, have your partner seated on your right and facing you.
7. Grasp the ophthalmoscope, resting your index finger on the lens-selection disk (see Figure 29-4c), and check to see that the "0" setting is in the diopter window and that the light is on.
8. Instruct your partner to set the timer for 1 minute and to look straight ahead at a fixed object or spot at his or her eye level. *(A safe time period for light exposure is 1 minute.)*
9. Begin viewing your partner's right eye at a distance of about 12 inches, as shown in Figure 29-4a. Shine the beam of light onto the pupil, and examine the lens and vitreous body of the eye. A red reflex (red-light reflection) will appear as you look through the pupil. If the image seen is not a sharp one, adjust the focus with the lens-selection disk of the ophthalmoscope.
10. Move to within 4 cm (2 inches) of your partner, while still keeping the pupil in focus (Figure 29-4b). Try to direct the ophthalmoscope's beam of light toward the pupil's edge rather than its center. (The red reflection should be more obvious.)
11. Look for the *optic disk* of the retina, and adjust the focus to obtain a sharp image. The blood vessels radiating from the optic disk should be clearly in view. Follow the path of at least two blood vessels to the edge of the retina.
12. Locate and identify the *macula* by having your partner look directly into the light beam for *1 second only.* Do not increase this exposure time. Can you see any blood vessels in the macula?
13. Examine the entire surface of the retina. Pay particular attention to surface depressions, protrusions (bumps) or any other irregularity. Bring any unusual condition to the attention of your instructor.
14. Keeping in mind the 1-minute beam-exposure time, examine the outer limits (periphery) of the retina. Ask your partner to look up, so that you can see the superior portion of the retina, and to look down so that the inferior portion of the retina can be observed.
15. Shut off the ophthalmoscope, and answer the questions in the results and observations section.
16. Change places with your partner, and repeat Steps 1 through 15.

Viewing
Aperture

Diopter
Window

Neck

Rheostat
Lock

Rheostat
Control

Handle

(a)

Blood
vessels

Macula
lutea

Optic
disc

Central
fovea

(b)

Blood vessels

Macula lutea

Central fovea

Optic disc

(c)

Temporal side

Nasal side

Figure 29-3 (a) The ophthalmoscope and its components; (b) the appearance of a retina, as seen through an ophthalmoscope; (c) diagram of the fundus of the right retina, as seen through an ophthalmoscope.

(a) The position for viewing the lens and vitreous body of the right eye.

(b) The position for examining the retinal surfaces of the right eye

(c) Examining the right eye. The ophthalmoscope is held with the right hand.

(d) Examining the left eye. The ophthalmoscope is held with the left hand.

Figure 29-4 Proper positions during eye examinations.

VISUAL ACUITY AND ASTIGMATISM TESTING

Materials

The following materials should be provided for class use:

1. Snellen chart
2. Astigmatism test chart
3. Blank 3- × 5-inch cards
4. Masking tape
5. Yardstick

Procedure 1: Snellen Visual Acuity Test

1. Place the Snellen chart (Figure 29-5) in a well-lighted area of the laboratory.
2. Use the yardstick to measure a distance of 20 feet on the floor from the Snellen chart. Put a short piece of masking tape on the 20-foot site.
3. If you are wearing corrective lenses (eyeglasses or contact lenses), remove them. Face the Snellen chart, and cover your left eye with a blank 3- × 5-inch card.
4. Read the letters of the fourth line from the top of the chart. Next, select and read the smallest line of letters that you can see clearly. As you call out the letters, have your laboratory partner write them in the space provided in the results and observations section.
5. Repeat Steps 3 and 4 with the right eye, reading from right to left.
6. If you normally wear corrective lenses, repeat Steps 3–5 with the lenses on.
7. Determine the visual acuity for each eye without and with glasses. The measure of visual acuity states the distance at which the chart is read (20 feet) and the distance indicated on the chart for the smallest line the individual can read. For example, an individual standing 20 feet from the chart, who can read line 2, but nothing smaller, is said to have a visual acuity of 20/100.
8. Compare the letters read with the actual letters on the Snellen chart. Enter the correct Snellen letter in the spaces provided in the results and observations section, in the correct left-to-right or right-to-left sequence.

Procedure 2: Astigmatism Testing

This procedure should be performed by students in pairs.

1. Place the astigmatism test chart (see Figure 29-2) in a well-lighted area of the laboratory.
2. Use the yardstick to measure a distance of 10 feet from the chart. Put a short piece of masking tape on the 10-foot site.

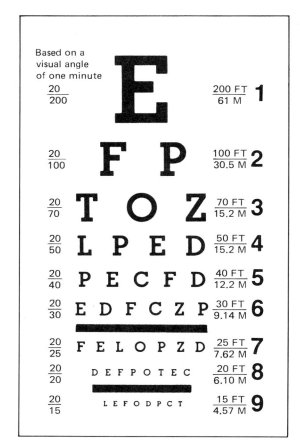

Figure 29-5 The Snellen chart: This chart is used to determine whether an individual has normal or abnormal vision; the standard of normal vision (20/20) is the ability to stand 20 feet from the chart and be able to read Line 8.

3. If you are wearing corrective lenses, remove them. Test each eye separately. Look at the chart, and determine which, if any, of the line groups appear exceptionally thicker and darker than others. Indicate such line groups by number, in the space provided in the results and observations section.
4. If you wear corrective lenses, repeat Step 3 with them on.

DEMONSTRATION OF BLIND SPOT

Materials

The following materials should be provided per each two students:

1. Five coins
2. Four blank 3- × 5-inch cards
3. A millimeter ruler

Procedure

This procedure should be performed by students in pairs.

1. Place the five coins in a straight line about 8 millimeters apart.
2. Look at the coins, and concentrate your view on the middle coin.
3. Close your right eye and hold a blank card over it. How many coins can you see? Have your partner record your findings in the results and observations section.
4. Repeat Step 3 with your left eye closed.
5. Assist your laboratory partner with this procedure, and record the findings in the results and observations section.

PUPILLARY REFLEXES

Materials

The following materials should be provided per each two students:

1. One small flashlight (or penlight)
2. A millimeter ruler

Procedure 1: Light Reflex

This procedure should be performed by students in pairs.

1. Hold the millimeter ruler under one eye of your laboratory partner, and determine the diameter of the pupil.
2. With the ruler in the same position, quickly and briefly flash a light into the eye. Quickly measure the diameter of the pupil.
3. Remove the light source, and record your findings in the results and observations section.
4. Change places with your partner, and repeat Steps 1–3.

Procedure 2: Accommodation Reflex

This procedure should be performed by students in pairs.

1. Hold the millimeter ruler under one eye of your laboratory partner, and determine the diameter of the pupil.
2. With the ruler in the same position, have your partner look first at an object located at the far end of the room and second at your shoulder. Measure the diameter of the pupil in each situation.
3. Record your findings in the results and observations section.
4. Change places with your partner, and repeat Steps 1–3.

DEPTH PERCEPTION TESTING

Materials

The following materials should be provided for class use:

1. Depth-perception testing system, with the following items in place.
    b. Calibrated base with adjustment string attached
    b. Two slider bases, equipped with arrow holders and arrows
2. Sunglasses
3. Commercially prepared disposable alcohol pads
4. Container for used alcohol pads

Procedure

This procedure should be performed by students in pairs.

1. Examine the depth-perception testing system provided. Make certain that all appropriate parts are operational. Note that one of the slider bases is attached to the adjustment string. Refer to Figure 29-6.
2. Place the testing system flat on a table or desk in a well-lighted area.
3. Have your laboratory partner sit comfortably on a chair situated 8 feet away from the depth-perception testing system. The zero end of the scale should be closest to the chair. If your partner wears glasses or contact lenses, have them removed.
4. Ask your partner to close his or her eyes, while the unattached slider base is moved to some position along the testing system's millimeter scale (Figure 29-6). Make certain that the arrows are facing one another and that they are at the eye level of your partner.
5. Give the adjustment string to your partner, to align the two arrows by moving the slider base attached to the string.
6. When your partner is satisfied with the alignment, note the position of the string-attached (movable) slider base relative to the unattached one. Record any difference shown on the depth-perception testing system scale in millimeters in Table 29-6 in the results and observation section.
7. Repeat Steps 4–6, using another position for the attached slider base on the testing system's scale.
8. Have your partner repeat Steps 4–7, first with the right eye closed. Record all findings in Table 29-6, in the results and observations section.
9. Repeat Steps 4–8, with your partner wearing a pair of sunglasses. Record all findings in Table 29-7.
10. Change places with your partner, and perform Steps 1–9.
11. Answer questions listed in the results and observations section.

Figure 29-6 A depth-perception tester. The components shown include arrow holders (AH), troughs (T), slider bases (SB), scale (S) and adjustment string (AS).

OPTICAL ILLUSIONS

Materials

The following materials should be provided for class use:

1. Millimeter rulers
2. Magnifying glasses (optional)

Procedure

This procedure should be performed by students individually.

1. Examine the three squares in Figure 29-7.
2. Determine visually which of the squares is smaller and which of the squares appears taller. Record your findings in the results and observations section.
3. Measure each square with a ruler. Record your findings in the results and observations section.
4. Examine the diagrams in Figure 29-8.
5. Determine visually whether the horizontal lines are of equal length. Record your findings in the results and observations section.
6. Measure the lines with a ruler. Record your findings in the results and observations section.
7. In the space provided in the results and observations section, construct an optical illusion diagram of your own making. Test it with your laboratory partner.

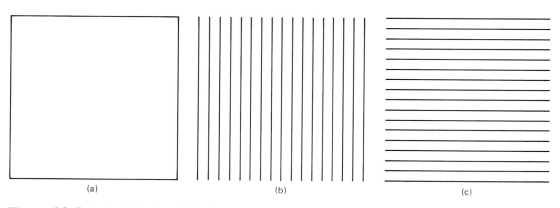

Figure 29-7 Optical illusion: Which square appears to be the smallest?

(a)   (b)

Figure 29-8 Optical illusion: Which horizontal line is longer?

TEST FOR COLOR-BLINDNESS

Materials

The following materials should be provided for students in groups of four:

1. One Holmgren Color Vision Test
2. One Ichikawa or Ishihara Color Blindness Test
3. Correct explanation keys for color-blindness tests

Procedure

This procedure should be performed by students in groups of four.

1. Each member of the group should take the Holmgren Color Vision Test.
   a. Select and match 12 different wool strands with those mounted in the binder. List your responses.
   b. Consult the key, and list the key's colors. Circle incorrect responses in the results and observations section.
2. Each member of the group should take the Ichikawa or Ishihara Color Blindness Test.
   a. Examine each of the plates in the test booklet.
   b. Enter your impressions as to the symbols on these plates and the key explanation for your responses in the results and observations section.
3. Compare the findings obtained with both color-blindness tests, and answer the questions in the results and observations section.
4. The instructor should compile test results for the whole class. Determine the number of individuals who are color-blind. Answer the questions in the results and observations section.

VISUAL PIGMENT ACTIVITY

Materials

The following materials should be provided for class use:

1. Laboratory lamps (40-watt bulbs) without covers or flashlights
2. Squares of bright, glossy red, yellow, green, and blue paper (4- × 4-inch) mounted in the center of the individual solid, black poster boards or papers (8½ × 11 inch)
3. Squares of bright, glossy, red, yellow, green, and blue paper (4- × 4-inch), with center-cut holes, mounted in the center of individual white poster boards or papers (8½ × 11 inch)
4. White poster board or paper sheets (8½ × 11 inch)

Procedure

This procedure should be performed by students in pairs.

1. Have your laboratory partner sit comfortably approximately 6 feet in front of a blank wall and a table with a lamp without a shade or cover.
2. Turn on the lamp, and direct your partner to stare at the bright light for 20 seconds and then immediately to look away toward the blank wall.
3. Ask what afterimages (positive and negative) appeared. Record the findings in the results and observations section.
4. Stare at the bright light once again for 20 seconds. Now have your partner immediately hold in front of you a white poster board with a red paper square. Record a description of the types of afterimages seen in the results and observations section.
5. Did you see the colored square?

6. Next, have your partner first hold a black poster board with a red paper square in front of you. Stare at the square for 60 seconds. Quickly have your partner hold a white poster board in front of you. Note the color of the afterimages. Did you see a colored square? Answer the questions in the results and observations section.
7. Repeat Step 6 with the yellow, green, and blue paper squares.
8. Change places with your partner, and repeat Steps 1–7.

RESULTS AND OBSERVATIONS

Examining the Eye

1. Description of the right eye, as seen through the ophthalmoscope. _____

_____

2. Description of the left eye, as seen through the ophthalmoscope. _____

_____

3. Were you able to find all the parts of the eye shown in Figures 29-3b and 29-3c? List the parts that you were able to observe. _____

_____

4. Were any unusual structures observed? _____

_____

Visual Acuity and Astigmatism Testing

1. Enter the results of the Snellen test in Table 29-1.
2. Enter the results of the astigmatism test in Table 29-2.

Demonstration of Blind Spot

1. Enter your findings and those of your laboratory partner in Table 29-3.
2. Would you expect to see more or fewer coins if the distance between coins were increased? _____

_____

3. What is the blind spot? _____

_____

Pupillary Reflexes

1. Enter your findings on the light reflex in Table 29-4.
2. Enter your findings on the accommodation reflex in Table 29-5.

Table 29-1

Eye Condition	Letters Read	Actual Sequence on Chart	Visual Acuity
Left eye (without glasses)			
Left eye (with glasses, if applicable)			
Right eye (without glasses)			
Right eye (with glasses, if applicable)			

Table 29-2

Eye Condition	Thicker and Darker Line Groups
Without glasses	
With glasses	

Table 29-3

Student	Number of Coins Seen	
	Right Eye Closed	Left Eye Closed

Table 29-4

Student	Pupil Diameter	
	Without Light	With Light

Table 29-5

Student	Pupil Diameter		
	At Start	Distant Object	Near Object

Depth Perception Testing

1. Enter your findings and those of your laboratory partner, obtained with conditions of normal lighting in Table 29-6.
2. Was alignment achieved in one or both attempts, with having both eyes open? _____

_____

3. Was alignment achieved with either the right eye or the left eye closed? _____

4. a. If alignment was not achieved with both eyes open, or with either eye closed, which situation produced a greater millimeter distance away from the fixed slider base? _____

_____

b. Offer an explanation for any differences that occurred. _____

5. Enter your findings and those of your partner, obtained with the use of sunglasses, in Table 29-7.

Table 29-6

| Student | Depth Perception Measurement Difference in Millimeters | | | | | |
| | Both Eyes Open | | Right Eye Closed | | Left Eye Closed | |
	1	*2*	*1*	*2*	*1*	*2*
(without glasses)						
(with glasses, if applicable)						
(without glasses)						
(with glasses, if applicable)						

Table 29-7

| Student | Depth Perception Measurement Differences in Millimeters | | | | | |
| | Both Eyes Open | | Right Eye Closed | | Left Eye Closed | |
	1	*2*	*1*	*2*	*1*	*2*
(without glasses)						
(with glasses, if applicable)						
(without glasses)						
(with glasses, if applicable)						

6. a. Was alignment more difficult with the sun-

glasses? _____

_____

b. Explain your findings. _____

_____

_____

Optical Illusions

1. Which of the squares appears to be smallest?

_____ Which of the squares appears to

be tallest? _____

2. Record the dimensions of the squares in Table 29-8.
3. Which of the horizontal lines appears to be longer?

_____

4. Record the length of the horizontal lines in Table 29-9.
5. Construct your own optical-illusion diagram in the space provided.

Table 29-8

Square	Height (in mm)	Length (in mm)
a.		
b.		
c.		

Table 29-9

Line	Length
a.	
b.	

2. Optical-illusion diagram.

Color-Blindness Tests

1. Enter the responses to the Holmgren Color Vision Test in Table 29-10.
2. Enter the responses to the Ichikawa or Ishihara Color Blindness Test in Table 29-11.

3. Were the findings obtained with the two color-blindness tests similar? If not, explain the results obtained. _____

_____

Table 29-10

Thread Color	Student Response				Key Color
	1	2	3	4	

Table 29-11

Plate	Student Response				Key Explanation
	1	2	3	4	
1					
2					
3					
4					
5					
6					
7					
8					
9					
10					
11					
12					
13					
14					
15					
16					
17					
18					
19					
20					
21					
22					
23					
24					

4. Summarize class results in Table 29-12.
5. How many color-blind individuals were unaware of their color-blindness? _____

Visual Pigment Activity

1. Did you see all color squares as afterimages in the situations of the experiment? If not, which colors were not seen? _____

_____

2. Were the center holes in any of the colored squares observed in the afterimages? If so, which ones?

_____

_____

3. What causes an afterimage? _____

_____

Table 29-12

Students	Number Color-Blind	Type of Color-Blindness	Percentage Color-Blind
Female			
Male			
All			

THE PHYSIOLOGY OF VISION

COMPLETION AND DEFINITION

1. Define or explain the following:
   a. Visual acuity _____

   _____

   b. Astigmatism _____

   _____

   c. Color-blindness _____

   _____

2. What is an optical illusion? _____

   _____

   _____

3. List five factors that can affect judgment of size and distance.

   a. _____

   b. _____

   c. _____

   d. _____

   e. _____

4. List two pupillary responses and the normal result for each.

   a. _____

   _____

   b. _____

   _____

5. Are the following Snellen chart test results considered normal, better than normal, or worse than normal?

   a. 20/20 _____

   b. 20/15 _____

   c. 20/100 _____

   d. 70/100 _____

ANSWERS

1. _____
2. _____
3. _____
4. _____
5. _____
6. _____
7. _____
8. _____
9. _____
10. _____
11. _____
12. _____

MULTIPLE CHOICE

Insert the correct answers in the spaces provided.

1. Which of the following parts of the eye holds the aqueous humor?
   a. macula lutea
   b. retina
   c. anterior chamber
   d. posterior chamber
   e. pupil

2. Which of the following findings of an optical examination would indicate the best visual acuity?
   a. 20/20
   b. 20/30
   c. 20/40
   d. 20/50
   e. 20/60

3. The condition of nearsightedness also can be referred to as which of the following?
   a. hyperopia
   b. astigmatism
   c. emmetropia
   d. myopia
   e. hypoopia

4. Which of the following is the small area of the retina where no image is formed?
   a. conjuctiva
   b. blind spot
   c. iris
   d. sclera
   e. cornea

5. Which portion of the eye regulates the amount of light entering the eye?
   a. conjuctiva
   b. blind spot
   c. iris
   d. sclera
   e. cornea

6. What is the term for the degree of detail the eye can distinguish?
   a. reflex control
   b. astigmatism
   c. macula lutea
   d. visual acuity
   e. myopia

7. In a dark environment, what does the diameter of the pupil do?
   a. increases
   b. decreases
   c. does not change

8. In bright light, what happens to the sensitivity of retinal rods?
   a. greatly increased
   b. greatly reduced
   c. is not changed

9. If all three types of retinal cones are stimulated, what will an individual see?
   a. black
   b. white
   c. red
   d. blue
   e. green

10. What does the central fovea contain?
    a. only rods
    b. only cones
    c. both rods and cones
11. In determining visual acuity with a Snellen chart, when reading the results, higher numbers following the "20/" (e.g., 20/50) indicate the relative size of the letter on the chart, for the individual being tested to see it clearly. Does a higher number indicate a larger or a smaller relative size?
    a. smaller
    b. larger
12. What is the term for an eye with a normal ability to refract light?
    a. myopic
    b. astigmatic
    c. hypermetropic
    d. emmetropic
    e. hyperopic

The Physiology of Hearing

After completing this exercise, you should be able to

1. Trace the path of sound waves through the parts of the ear
2. Perform experiments to demonstrate both normal and abnormal physiological aspects of hearing

The ear has two distinct functions: (1) The auditory component is concerned with the sensation of hearing, while (2) the vestibular portion provides information to the brain about the movement and position of the head. The sensory elements of both components are hair cells in the inner ear (see Figure 28-5). These are excited by the bending of hairs and are referred to as *mechanoreceptors*.

The ear responds to mechanical disturbances produced by a vibrating body. For example, if the arms of a tuning fork are struck, they will swing back and forth, and a mechanical disturbance of molecules will be created through the surrounding air in all directions. Such vibrations are perceived by the mechanoreceptors as sound. The human ear can distinguish among the properties of sound, which include intensity, pitch, and timbre. Intensity or loudness is measured in units termed *decibels*. *Pitch* is determined by the frequency of vibrations, with higher frequencies producing higher-pitched sounds. The range of frequencies to which the human ear is sensitive is from 20 to 20,000 hertz (vibration cycles per second). *Timbre* refers to the quality of sound and depends on the presence of several frequencies.

Sound is transmitted through the ear by first passing through the external auditory meatus (see Figure 30-1) and causing the tympanic membrane to vibrate.

These vibrations are transferred to the bones of the middle ear and are amplified as they reach the oval window. Vibrations of the oval window cause the fluid of the scala vestibuli to vibrate. This disturbance causes certain portions of the basilar membrane to bulge into the cochlear duct. Basilar membrane movement stimulates the hair cells of the organ of Corti, which results in the conduction of impulses along the cochlear nerve. Vibrations of the fluid of the scala vestibuli are also transferred to the scala tympani and are spread as the round window bulges toward the middle ear. Interference with or damage to this conduction system or to the nerve fibers or receptor cells results in hearing loss or deafness. *Deafness* is a severe form of hearing loss that makes everyday auditory communication difficult or impossible. Based on the part of the hearing mechanism that fails to function adequately, hearing loss or deafness can be classified into three main types: conductive, sensorineural, and central hearing loss.

Conductive hearing loss may be caused by simple situations such as the accumulation of wax in the ear. However, it also may be due to an inflammation and infection of the middle ear, known as *otitis media,* or to *otosclerosis,* a condition in which new bone growth may prevent the normal vibration of the stapes in the oval window.

Sensorineural hearing loss results from injury or disease involving the organ of Corti or the cochlear nerve. Causes of this form of hearing loss include intense noise, ear-damaging drugs, bacterial and viral infections of the inner ear, and aging.

Central hearing loss results from defects in the auditory tracts or centers in the brain. Several causes are associated with this condition and include traumatic injury, tumors, infections, blockage of blood vessels in the brain, and a progressive hearing loss that occurs with age, known as *presbyacusis*.

The vestibular apparatus, or *nonauditory labyrinth,* is the component of the inner ear that detects

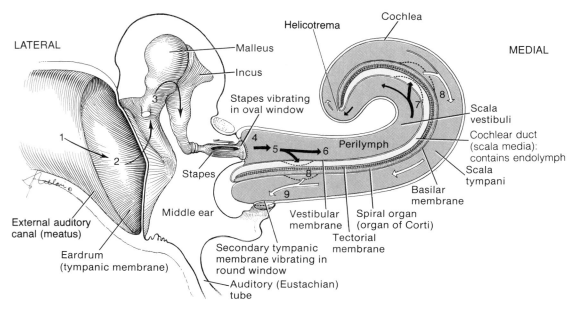

Figure 30-1 The mechanism of normal hearing: The arrows show the direction in which sound travels. (From G. J. Tortora and S. R. Grabowski, *Principles of Anatomy and Physiology,* 7th ed., New York: HarperCollins, 1993.)

specific movements and positions of the head. Compensatory responses of the muscles of the neck, trunk, and limbs are initiated by the vestibular apparatus and help to keep the body in balance and in a proper orientation with respect to gravity. Motion sickness (a temporary disturbance) and Meniere's disease are known to interfere with the functioning of the organs of Corti. Motion sickness results from repetitive changes in the rate and direction of movement and from conflicting visual and vestibular signals. Meniere's disease may be caused by allergies, infections, or trauma. It results in an irritation or injury to the inner ear, leading to the accumulation of an excess of endolymph (see cochlear duct on Figure 30-1).

Procedures in this exercise consider certain properties of sound, and an individual's sensitivity to sound, static equilibrium response, and spatial orientation to sound.

HEARING TESTS

Materials

The following materials should be provided for class use:

1. Watch, clock, or other small device with audible ticking sound
2. Tuning fork
3. Rubber reflex hammer
4. Individual cotton plugs
5. Yardstick
6. Masking tape

Procedure 1: Auditory Acuity

This procedure should be performed by students in pairs.

1. Have your laboratory partner sit down with eyes closed. Carefully insert a cotton plug into the left ear.
2. Hold a ticking watch or clock in line with your partner's right ear approximately 12 feet away.
3. As you move slowly toward your partner, have your partner indicate exactly when the ticking sound is heard.
4. Stop and measure the distance between the sound source and your partner's ear.
5. Repeat this procedure with the other ear, using a fresh cotton plug. Enter your findings in the results and observations section.
6. Change places with your partner, and repeat Steps 1–5.
7. Answer the questions in the results and observations section.

Procedure 2: Sound Localization

This procedure is to be performed by students in pairs.

1. Have your laboratory partner sit down with eyes closed.

2. Bring a ticking watch into the hearing range of your partner, from the following locations:
   a. Front of head
   b. Rear of head
   c. Left side of head
   d. Right side of head
   e. One arm's length above head on the right side
   f. One arm's length above head on the left side
3. Have your partner point to the position from which the sound is coming.
4. In the results and observations section, indicate whether the sound localization was correctly identified.
5. Change places and repeat Steps 1–4, but changing the order of the locations in Step 2.

Procedure 3: Rinne Test (Test for Middle-Ear Deafness)

This procedure is to be performed by students in pairs.

1. Have your partner sit down.
2. Locate the mastoid process of the temporal bone behind the left ear.
3. Strike a tuning fork and place the handle of the vibrating fork against the mastoid. (The sensation felt is bone conduction.)
4. Have your partner indicate when the vibrations are no longer heard, and immediately hold the tuning fork just outside the external ear. (In middle-ear deafness, no sound is heard by air conduction.)
5. Describe your partner's responses in the results and observations section.
6. Change places with your partner, and repeat Steps 1–5.

Procedure 4: Weber Test (Test for Middle-Ear Deafness and Nerve Defects)

This procedure is to be performed by students in pairs.

1. Have your partner sit down in a relatively quiet room.
2. Strike the tuning fork, and place the handle of the vibrating fork against the middle of your partner's forehead.
3. Have your partner indicate the direction from which the sound appears to be coming. (With normal hearing, the sound appears to be referred to the middle of the head. With middle-ear deafness, the sound is detected in the defective ear. With nerve defects, vibrations are received better with the normal ear.)
4. Describe your partner's responses in the results and observations section.
5. Change places with your partner, and repeat Steps 1–4.

RESULTS AND OBSERVATIONS

Hearing Tests

1. Enter your findings in the auditory acuity test in Table 30-1.
2. Was the auditory acuity of both ears the same for you? _____ For your partner? _____
3. Enter your findings in the sound-localization test in Table 30-2. Indicate correct sound localization by a " + " and an incorrect identification by a " − ."

Table 30-1

Student	Distance at Which Sound Is Detected (Right Ear)	Distance at Which Sound Is Detected (Left Ear)

Table 30-2

Student	Front	Rear	Left Side	Right Side	One Arm's Length to Left	One Arm's Length to Right

4. a. Describe your response to the Rinne test. ____

_____

b. Describe your partner's response. _____

_____

5. a. Describe your response to the Weber test.

_____

b. Describe your partner's response. _____

_____

STATIC EQUILIBRIUM AND SPATIAL ORIENTATION TESTS

Materials

The following materials should be provided for class use:

Rotating stools or swivel chairs without casters

Procedure 1: Static Equilibrium Response

The following tests should be performed by students in pairs (except Procedure 2, which will require three students):

1. Have one laboratory partner stand erect and still, with both feet as close together as possible. This position should be maintained for 2 minutes. Note the kinds and amount of movement necessary to keep the body in this position. Record your findings in the results and observations section.

2. Have your laboratory partner repeat Step 1, but with eyes closed. (*Note:* Two students should stand by to prevent a fall.) Enter your findings in the results and observations section.

3. Change places with your laboratory partner, and repeat Steps 1 and 2.

Procedure 2: Spatial Orientation

1. Have your laboratory partner stand erect, with both arms extended to the sides and then bring the tips of the index fingers together.

2. Repeat Step 1 with eyes closed until the index fingers are brought together successfully at least three times.

3. Have your partner sit in a swivel chair with eyes closed. Twirl your partner around rapidly 10 times. Repeat Step 2. Record your findings in the results and observations section.

4. Change places with your partner, and repeat Steps 1–3.

RESULTS AND OBSERVATIONS

1. In Table 30-3, enter brief descriptions of your reactions and those of your partner in the reflex-testing procedures.

2. What structures in the human body contribute to an individual's static equilibrium? _____

_____

Table 30-3

Reflex Procedure	Your Response	Laboratory Partner's Response
Static equilibrium Eyes open		
Eyes closed		
Spatial orientation Eyes open		
Eyes closed		
After twirling		

LABORATORY REVIEW *30*

NAME _____

LAB SECTION _____ DATE _____

THE PHYSIOLOGY OF HEARING

COMPLETION, DEFINITION, AND DESCRIPTION

1. Give the two distinct functions of the ear and the specific portions of the ear concerned with each.

   _____

   _____

   _____

2. What is sound? _____

   _____

3. List three specific properties of sound, which the human ear can distinguish.

   a. _____

   b. _____

   c. _____

4. List the three main types of hearing loss and the particular portion of the hearing mechanism involved.

   a. _____

   b. _____

   c. _____

5. Briefly describe two tests for the detection of middle-ear deafness.

   a. _____

   _____

   b. _____

   _____

6. What parts of the ear are associated with balance? _____

   _____

7. List and briefly describe two disturbances involved with the functions of the organs of balance.

   a. _____

   b. _____

8. Define or explain the following:

   a. Otitis media _____

   b. Otosclerosis _____

   c. Presbyacusis _____

   d. Decibel _____

ANSWERS

1. _____

2. _____

3. _____

4. _____

5. _____

6. _____

7. _____

8. _____

9. _____

10. _____

Insert the correct answers in the spaces provided.

1. What is the organ of hearing located within the inner ear?
   a. semicircular ducts
   b. oval window
   c. membranous labyrinths
   d. organ of Corti
   e. round window

2. Name the openings between the middle and inner ears:
   a. semicircular ducts
   b. oval window
   c. round window
   d. organ of Corti
   e. choices *b* and *c* only

3. What is the specific unit used to measure the loudness, or intensity, of sound?
   a. hertz
   b. decibel
   c. vibration
   d. pitch
   e. pitched sound unit

4. What are the sense organs of static equilibrium?
   a. semicircular ducts
   b. oval window
   c. membranous labyrinths
   d. organ of Corti
   e. round window

5. What is the unit used to express vibration cycles per second?
   a. hertz
   b. decibel
   c. vibration
   d. pitch
   e. pitched sound unit

6. Which of the following forms of hearing loss can be caused by the accumulation of cerumen?
   a. conductive
   b. sensorineural
   c. central hearing
   d. presbyacusis

7. Which of the following forms of hearing loss can be caused by otitis media?
   a. conductive
   b. sensorineural
   c. central hearing
   d. presbyacusis

8. What is the hearing loss that occurs with age called?
   a. otitis media
   b. myopia
   c. hypermetropia
   d. presbyacusis
   e. astigmatism

9. What is a medical term for a middle-ear infection?
   a. otitis media
   b. myopia
   c. hypermetropia
   d. presbyacusis
   e. astigmatism

10. Which of the following hearing losses would an intense noise over a period of time probably cause?
   a. conductive
   b. sensorineural
   c. central hearing
   d. presbyacusis
   e. otitis media

The Physiology of Smell and Taste

After completing this exercise, you should be able to

1. Indicate the location of the olfactory receptors
2. List the factors on which sensations of smell depend
3. Describe two roles of the sense of smell
4. Locate the areas of specific taste reception on the tongue

Chemoreceptors that respond to chemical changes in the external environment are referred to as *exteroceptors*. These include the sense of smell, or *olfactory receptors* that respond to gaseous molecules in the air, and the sense of taste, or *gustatory receptors* that respond to chemicals dissolved in food or drink.

The receptors for the sense of smell are located in the olfactory cells in the mucous membrane lining of the upper portion of the nasal cavity. Only volatile substances in solution can stimulate these receptors. The gaseous particles enter the nasal cavity by diffusion from the circulating air in the external environment. Glands that surround olfactory cells secrete fluid into which the particles dissolve, and this solution acts as a chemical stimulus to a nerve impulse (Figure 31-1).

In humans, the sense of smell is less important than in other animals. However, it does play a role in starting the flow of certain digestive juices at mealtime. What have you experienced when you are quite hungry and smelled the aroma of food? The sensation of smell also can play a role in reviving a person who has fainted. Smelling salts stimulate the endings of the trigeminal cranial nerve and produce pain. This effect initiates respiratory reflexes and eventually results in reviving individuals who have fainted.

The epithelial surface of the tongue contains taste receptors, which are specialized cells grouped into barrel-shaped structures termed taste buds. (See Figure 31-2a and Color Plate 33.) Each taste bud consists of a taste (*gustatory*) pore, which communicates with the external environment, supporting cells, and receptor cells, which have little gustatory hairlike processes that project through the taste pore (see Figure 31-2a). Molecules dissolved in saliva at the tongue's surface interact with the hairlike processes, causing receptor cells to release a neurotransmitting chemical. This response, in turn, triggers sensory nerve endings in the taste buds.

Despite the fact that individuals experience a variety and range of taste sensations, there are only four basic responses: sweet, sour, salty, and bitter. Figure 31-2b shows the general locations of the taste sensations.

Several experiments demonstrating the features of the sensation of smell and taste are considered in this exercise.

Materials

The following materials should be provided for class use:

1. Containers (with tops or plugs) of
   a. Freshly ground coffee
   b. Garlic powder
   c. Oil of cloves
   d. Oil of wintergreen
   e. Oregano
   f. Vanilla extract
   g. Vinegar
   h. Orange or lemon peels
   i. Perfume
   j. Distilled water
   k. 5 percent NaCl solution
   l. 5 percent sucrose solution
   m. 0.5 percent quinine-sulfate solution

Figure 31-1 The olfactory receptors: (a) Sagittal view of the skull shows location of receptor in the naval cavity (circled area); (b) an enlarged view of the receptors in action while a substance is being smelled—small arrows show the direction of the sensation. (From G. J. Tortora and S. R. Grabowski, *Principles of Anatomy and Physiology*, 7th ed., New York: HarperCollins, 1993.)

2. Equally cut slices of apple, onion, and potato in separate and covered containers
3. Cotton swabs
4. Tongue depressors
5. Container for the disposal of swabs and tongue depressors
6. Distilled water for drinking
7. Paper towels

Procedure 1: Demonstration of Olfactory Stimulus

This procedure is to be performed by students in groups of four.

1. Dip a cotton swab into the perfume, and gently squeeze out the excess fluid on the inner side of the container.
2. Apply the perfume to the forearm of one of your laboratory partners.
3. Note the time elapsed before each of your other laboratory partners first detects the aroma of the perfume after its application.
4. Note the time elapsed from application until the aroma is no longer detectable for each partner.
5. Record your findings and answer the questions in the results and observations section.

Procedure 2: Stimulus Identification

This procedure is to be performed by students in pairs.

1. Smell the contents of each of the containers.
2. Close all containers. Wait 5 minutes.
3. Close your eyes. Have a laboratory partner open each container, and pass it slowly about 3 inches from your nose. Your exposure should be no longer than 3 seconds for each container.
4. Identify each stimulus by name. Have your partner record whether each response is correct or incorrect in the results and observations section.
5. Immediately after the first round of testing, repeat Steps 3 and 4 two more times.
6. Repeat Steps 1–5 with each member of your group.
7. Answer the questions in the results and observations section.

Procedure 3: Taste Receptor Distribution

This procedure is to be performed by students in pairs.

1. Have your partner rinse his or her mouth with distilled water and then use a paper towel to dry the surface of the tongue. Dispose of the towel in an appropriate container.

(a)

(b)

Figure 31-2 Gustatory reception: (a) the general structure and organization of a taste bud; (b) general structure of the tongue—areas of specific taste reception on the tongue are indicated. (From G. J. Tortora and S. R. Grabowski, *Principles of Anatomy and Physiology,* 7th ed., New York: HarperCollins, 1993.)

2. Moisten a clean cotton swab with the 5 percent sucrose solution, and remove any excess fluid by pressing the swab against the inside surface of the container.

3. Touch several regions on your partner's tongue with the moistened swab, and ask your partner to indicate the type of sensation and where it is detected. Indicate the site(s) on the sketch of the tongue in the results and observations section.

4. Have your partner rinse her or his mouth with distilled water again, and repeat Steps 2 and 3 with the 5 percent NaCl, vinegar, and 0.5 percent quinine-sulfate solutions.

5. Change places with your partner, and repeat Steps 1–4.

6. Compare your responses.

7. Answer the questions in the results and observations section and those pertaining to this portion of the exercise in the laboratory review section.

Procedure 4: The Relationship Between Smell and Taste

This procedure is to be performed by students in pairs.

1. Have your laboratory partner sit comfortably with eyes closed and nostrils held shut.
2. Place a small piece of apple or potato on your partner's tongue. Instruct your partner not to swallow and to identify the material placed on the tongue. Note whether each identification is correct or incorrect in the results and observations section.
3. Have your partner rinse his or her mouth with water.
4. Repeat Step 2 with the other food, either apple or potato. Record your findings in the results and observations section.
5. Repeat Steps 2–4 with nostrils open.
6. Place a piece of onion on your partner's tongue. Remove it, and repeat Steps 1–5. Are the results obtained after the use of the onion similar to those obtained without the onion?
7. Change places with your partner, and repeat Steps 1–6.

RESULTS AND OBSERVATIONS

Demonstration of Olfactory Stimulus

1. Enter your findings on Table 31-1.
2. Did the laboratory partner on whom the perfume was applied detect the aroma earlier than the other

   laboratory partners? _____ Did the

   laboratory partner on whom the perfume was applied detect the aroma after the others could no

   longer smell it? _____

   Explain. _____

   _____

Stimulus Identification

1. Enter all findings in Table 31-2. Indicate correct responses by a "+" and incorrect responses by "−."
2. Did the number of correct responses increase or decrease with each repetition of the test?

   _____

Taste Receptor Distribution

1. Circle and label the tongue outline in Figure 31-3, to indicate the location where a particular taste sensation was detected.

Figure 31-3 Taste receptor distribution (to be labeled during Procedure 3).

Table 31-1

Student	Time Elapsed Until Odor Detected (in min)	Time Elapsed Until Odor No Longer Detected (in min)

Table 31-2

Stimulus	Student			Partner		
	Test 1	Test 2	Test 3	Test 1	Test 2	Test 3

2. a. Were all taste sensations present on your tongue?

   _____

   b. Were any taste sensations particularly more distinct? If so, which one(s)? _____

   _____

3. a. Did any of the taste receptor locations overlap? If so, which one(s)? _____

   _____

b. How can this situation be explained? _____

   _____

The Relationship Between Smell and Taste

Enter your findings in Table 31-3. Indicate correct identifications by a " + " and incorrect identifications by a " − ."

Table 31-3

Student	Before Onion				After Onion			
	Nostrils Closed		Nostrils Open		Nostrils Closed		Nostrils Open	
	Potato	*Apple*	*Potato*	*Apple*	*Potato*	*Apple*	*Potato*	*Apple*

NAME _____

LAB SECTION _____ DATE _____

THE PHYSIOLOGY OF SMELL AND TASTE

COMPLETION AND IDENTIFICATION

1. Where are the receptors for the sense of smell located?

   _____

2. What types of substances stimulate the receptors of smell?

   _____

3. On what physical process does the sensation of smell depend? _____

4. Identify the location of specific taste reception on the tongue in Figure Q31-1.

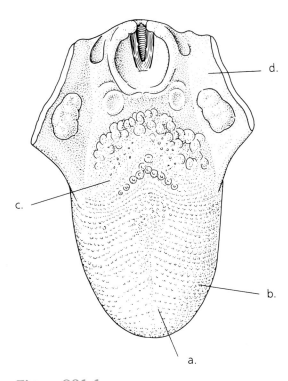

Figure Q31-1

a. _____   c. _____

b. _____   d. _____

5. Explain the relationship between the sensation of smell and the use of smelling salts
   (ammonium salts) to revive an individual who has fainted. _____

   _____

   _____

6. How would a nasal cold affect an individual's sense of taste?

   _____

   _____

ANSWERS

1. _____

2. _____

3. _____

4. _____

5. _____

MULTIPLE CHOICE

Insert the correct answers in the spaces provided.

1. Which part of the tongue reacts strongest to sweet tastes?
   a. tip
   b. center
   c. back
   d. sides
   e. choices *c* and *d* only
2. Where are the receptors for gustatory sensations?
   a. olfactory bulbs
   b. olfactory nerves
   c. taste pores
   d. taste buds
   e. supporting cells
3. Where are the taste buds most numerous?
   a. soft palate
   b. throat
   c. salivary glands
   d. tongue
   e. choices *a* and *c* only
4. Which part of the tongue reacts most strongly to sour flavors?
   a. tip
   c. center
   c. back
   d. sides
   e. choices a *and* c only
5. Which of the following is an example of a chemoreceptor that is an exteroceptor?
   a. taste bud
   b. Meissner's corpuscle
   c. pacinian corpuscle
   d. lamellated corpuscle
   e. touch corpuscle

Cutaneous Sensory Systems

After completing this exercise, you should be able to

1. List and give the locations of touch receptors
2. Identify a pacinian corpuscle in a prepared slide of human skin
3. Perform simple experiments to demonstrate the receptors of heat, cold, touch, pain, and pressure
4. Explain the importance of touch receptors

The sensations of touch include those of pressure, heat, cold, and pain. In humans and many other animals many of the receptors of these sensations are located in the skin (Figure 32-1). Touch receptors enable many animals to monitor their environments.

Structure and different functions serve as the basis on which sensory organs can be categorized. From a structural standpoint, the sense organ can be the dendrites of sensory neurons, which are either free, as in the case of those found in the skin mediating pain and temperature, or enclosed within nonneural structures, such as the pressure receptors in the skin. Other sense receptors, including the photoreceptors in the eye (Exercise 29) and taste buds (Exercise 31) synapse with sensory dendrites.

As indicated earlier, the structures of certain skin receptors for the different sensations are quite similar. The types of impulses they send to the brain are also very much alike. The impulse that registers heat, for example, is exactly the same as that for cold or for touch. Humans experience them as different sensations because they are transmitted to different centers in the brain. If the pathways were changed so that the cold receptors sent impulses to the heat center in the brain,

one would probably feel a sensation of heat every time a piece of ice was touched.

Pain receptors are nerve endings not covered by a myelin sheath. Upon injury of the nerve ending, an impulse is sent to the brain, which is interpreted as pain. Even if the nerve ending itself is not injured, it can pick up signals from those that are and can transmit the same message. In a condition termed *referred pain,* the sensation is not projected back to the point of stimulation. Pain receptors are more numerous than receptors for heat, cold, or touch, especially in humans.

Other cutaneous receptors are quite complex, with their dendritic endings encapsulated in other types of cells. *Meissner's corpuscles* (Figure 32-1 and Color Plate 34), commonly referred to as the *touch receptors* because they respond to light pressure, are located just beneath the epidermis in the dermal papillary layer of the skin. *Krause's end bulbs* and *Ruffini's corpuscles* are considered to be *thermoreceptors* that respond to cold and heat, respectively. All of these receptor types are quite similar, with the possible exception of the *pacinian corpuscles* (*deep pressure receptors*), which are anatomically more distinctive (Color Plate 35).

The application of cold at times has been used to reduce or eliminate pain in some clinical situations, such as times of tooth or mouth pain, bone marrow aspiration (removal), and postoperative recovery. The two forms of cold therapy currently in use are in ice massage (the direct application of ice with a circular or back and forth motion) and cold packs. The areas in which cold is applied include the skin site over or surrounding pain, a region between the pain and the brain, the opposite side of the body corresponding to the pain site, acupuncture points, and apparently unrelated areas. This exercise contains an experiment to test the effect of cold applications.

In this exercise, consideration is given to sensory receptors that are (1) *thermoreceptors,* which respond

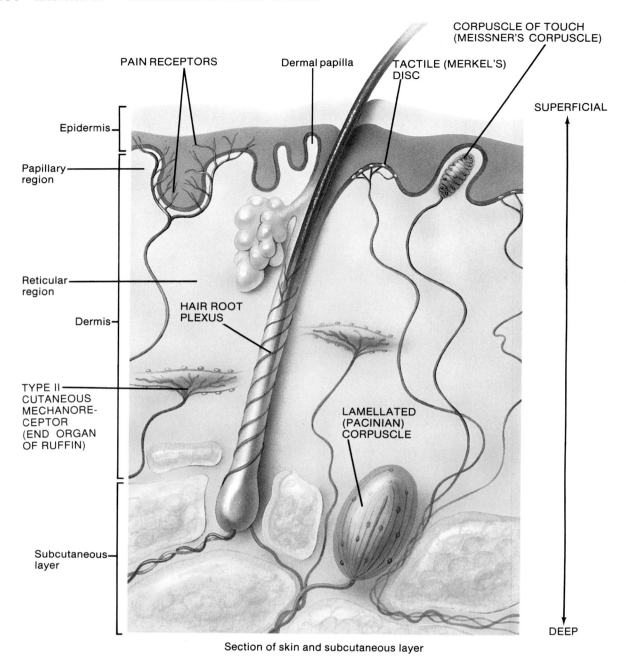

PAIN RECEPTORS

Dermal papilla

CORPUSCLE OF TOUCH
(MEISSNER'S CORPUSCLE)

TACTILE (MERKEL'S)
DISC

SUPERFICIAL

Epidermis

Papillary
region

Reticular
region

Dermis

HAIR ROOT
PLEXUS

TYPE II
CUTANEOUS
MECHANORE-
CEPTOR
(END ORGAN
OF RUFFIN)

LAMELLATED
(PACINIAN)
CORPUSCLE

Subcutaneous
layer

DEEP

Section of skin and subcutaneous layer

Figure 32-1 The structure and location of cutaneous receptors. (From G. J. Tortora, *Principles of Human Anatomy*, 6th ed., New York: HarperCollins, 1992.)

to heat and cold, (2) *mechanoreceptors,* such as those for touch and pressure, which are stimulated by deformation of the sensory dendrite cell membrane; and (3) *nociceptors,* or pain receptors, which are stimulated by chemicals released from damaged cells.

MICROSCOPIC IDENTIFICATION OF THE SKIN AND SKIN RECEPTORS

Materials

The following materials should be provided for class use:

1. Prepared slides of human skin, showing both epidermis and dermis
2. Prepared slides of human skin, showing skin receptors

Procedure

This procedure is to be performed by students individually.

1. Examine the prepared slides of human skin.
2. With the aid of Figure 32-1 and your textbook, identify the specific layers of the epidermis and the various glands typically found in the dermis.
3. Examine the preparation of human skin and identify the skin receptors present. Refer to Color Plates 34 and 35.
4. Answer the questions for review at the end of the exercise.

SKIN RECEPTORS

Materials

The following materials should be provided for class use:

1. Pens or markers with washable ink
2. Millimeter rulers
3. Metal compass or dividers with points blunted
4. Large beakers (500 ml) with water maintained at the following temperatures—
   a. 4°C (cold)
   b. 25°C (room temperature)
   c. 45°C (hot)
5. Metal probes or teasing needles
6. Ice with or without plastic cups
7. Paper or cloth towels
8. Timers

Procedure 1: Heat and Cold Receptors

This procedure is to be performed by students in pairs.

1. On the underside of your laboratory partner's forearm, use a fine-tip marking pen to draw a 4 × 14 mm rectangle, divided into 2-mm squares (total of fourteen 2-mm squares). Allow the grid to dry.
2. Place a metal probe in each of the temperature-controlled beakers of water. Allow probes to remain in the beakers for 5 minutes before use.
3. Touch one square with a probe from one of the beakers. Touch the same square with the probes from the other beakers, one at a time. Ask your partner to identify each touch as cold, room, or hot, and record in the results and observations section whether each response is correct or incorrect.
4. To maintain the temperature of the probes, return them to the appropriate beaker frequently during the experiment.
5. Repeat Step 3 on each of the remaining squares. Vary the order in which the probes for the three beakers are used.
6. Change places with your partner, and repeat Steps 1–5.

Procedure 2: Touch Discrimination (Two-Point Sensibility)

This procedure is to be performed by students in pairs.

1. Have your laboratory partner sit with eyes closed and one arm resting comfortably on a flat surface.
2. Adjust the compass so that the points are about 3 cm apart. Press it down gently on the resting arm. Can your partner feel *two* points?
3. Move the points of the compass closer together, and again press it gently on the arm. Continue to bring the points closer together until your partner reports feeling only one point. Then move the points just far enough apart to be detected as two when pressed on your partner's arm. This is the *two-point threshold.* Measure the distance between the two points of the compass in millimeters, and record it in the results and observations section.
4. Determine the two-point threshold for the tip of the index finger and thumb, the left and right palms, the back surface of both hands, and the back of the neck. Record your findings in the results and observations section.
5. Change places with your partner, and repeat Steps 1–4.

Procedure 3: Pressure Detection

This procedure is to be performed by students in pairs.

1. Place one hand palm down on a flat surface.
2. Observe the fine hairs on the back of the hand.
3. Have your partner move one hair in several directions with a metal probe. Can you feel this movement?
4. Repeat this step with a hair on the forearm.
5. Have your partner flick a hair rapidly several times. Can you feel a sensation?
6. Describe your sensations in the results and observations section.
7. Change places with your partner, and repeat Steps 1–6.

Procedure 4: The Effect of Cold or Pain

This procedure is to be performed by students in pairs.

1. Obtain a plastic cup with ice and a supply of paper towels.
2. Have your laboratory partner sit comfortably and place his or her right hand on a flat surface.
3. Pinch the web region between the thumb and index finger with your fingers until your partner indicates that a fairly intense level of pain is felt. *Do not inflict unnecessary pain.*
4. Immediately, start timing and ask your partner to indicate when the pain has disappeared completely. Record the time in the results and observations section. This reaction will serve as the baseline for comparisons with the steps that follow.
5. Repeat Step 3 with the left hand. As soon as the intense pain is felt, have your partner apply ice with a circular motion to the site. Use the paper towels to absorb the water runoff from the ice.
6. Immediately, start timing and ask your partner to indicate when the pain has disappeared completely. Use the paper towels to absorb the water runoff from the ice.
7. Record the time in the appropriate space in the results and observations section.
8. Repeat Steps 3 and 4, again using the opposite hand, and then apply the ice with a circular motion directly below the elbow of the same side.
9. Change places with your laboratory partner, and repeat Steps 1–8.

RESULTS AND OBSERVATIONS

Heat and Cold Receptors

1. Enter your findings in Table 32-1. Indicate correct temperature responses with a "+" and incorrect with a "−."
2. Were all temperatures identified correctly? _____
3. Were there any areas on which only one or two temperatures were not identified correctly? Explain. _____

   _____

4. Were your laboratory partner's responses significantly different from yours? Explain. _____

   _____

Touch Discrimination

1. Enter your findings in Table 32-2.
2. Which of the areas tested appeared to be the most sensitive? _____
3. Were the two-point threshold sensitivities of your partner similar to yours? Explain. _____

   _____

   _____

Pressure Detection

1. Describe the sensations felt with the movement of hairs. _____

   _____

   _____

2. Were there differences in sensations felt on the back of the hand and on the forearm? Explain. _____

   _____

   _____

Table 32-1

Square Number	Temperature Tested		
	Room	Cold	Hot
1			
2			
3			
4			
5			
6			
7			
8			
9			
10			
11			
12			
13			
14			

Table 32-2

Skin Area Tested	Laboratory Partner's Two-Point Threshold (in mm)	Your Two-Point Threshold (in mm)
Forearm		
Tip of index finger		
Tip of thumb		
Palm of left hand		
Palm of right hand		
Back surface of left hand		
Back surface of right hand		
Back of neck		

The Effect of Cold on Pain

1. Enter your findings in the first column and those of your laboratory partner in the second column of Table 32-3.
2. Did the cold application reduce the time the pain was felt, when compared to the situation without a cold application? _____

3. Were your responses similar to those of your laboratory partner? _____ If not, offer possible explanations of this finding. _____ _____ _____

Table 32-3

Site With or Without Cold Application	Reduction or Elimination of Pain Sensation (in seconds)	
	1	2
Pain site without cold		
Cold application over pain site		
Cold application between site and brain		

LABORATORY REVIEW *32*

CUTANEOUS SENSORY SYSTEMS

COMPLETION, DEFINITION, AND IDENTIFICATION

1. Identify the parts of the skin in Figure Q32-1.

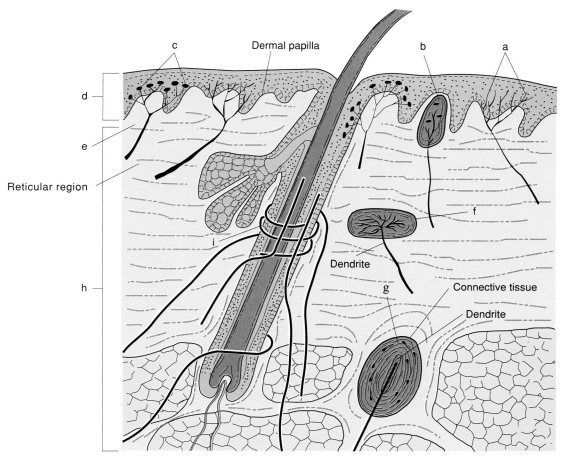

Figure Q32-1 Diagrammatic view of the skin.

a. _____

b. _____

c. _____

d. _____

e. _____

f. _____

g. _____

h. _____

i. _____

ANSWERS

2. What is a receptor? _____

_____

3. List the specific cutaneous sensations

    a. _____

    b. _____

    c. _____

    d. _____

    e. _____

4. What function do pain receptors serve? _____

_____

5. List two sense receptors in the skin that are free:

    a. _____

    b. _____

6. List a sense receptor in the skin that is enclosed within a non-neural structure.

_____

7. Briefly describe the functions and/or activities of the following sense receptors in skin.

    a. Thermoreceptor _____

    b. Mechanoreceptor _____

    c. Nociceptor _____

MULTIPLE CHOICE

Insert the correct answers in the spaces provided.

1. What may be considered tactile sensations, in addition to pain and pressure?
    a. sound
    b. heat
    c. cold
    d. choices *a* and *c*
    e. choices *b* and *c*

2. Which of the following sensations are detected by mechanoreceptors?
    a. smell
    b. taste
    c. gustation
    d. heat
    e. pressure

3. What would be an example of a touch receptor?
    a. vibration corpuscle
    b. pacinian corpuscle
    c. Meissner's corpuscle
    d. thermoreceptive corpuscle
    e. nociceptor

1. _____

2. _____

3. _____

4. _____

5. _____

4. What would be an example of a pain receptor?
   a. vibration corpuscle
   b. pacinian corpuscle
   c. Meissner's corpuscle
   d. thermoreceptive corpuscle
   e. nociceptor
5. What would be an example of a pressure receptor?
   a. vibration corpuscle
   b. pacinian corpuscle
   c. Meissner's corpuscle
   d. thermoreceptive corpuscle
   e. nociceptor

Fetal Pig Dissection: An Introduction to the Visceral Organs

*I*t is not always possible to dissect human cadavers. The fetal pig is a good substitute because its anatomy is similar to that of the human, and it is relatively inexpensive and easily available. The perfect situation is for each student to dissect his or her own specimen, but if finances do not permit this, it is possible to assign two or three students to each specimen. Throughout these exercises, references are made to human anatomy, especially where it differs in important ways from the anatomy of the pig. Frequent comparisons with photos and line drawings of the human make it possible to discuss human organ systems while working on the pig. Never forget that the ultimate goal in the class is to learn the human anatomy.

This exercise can be used to acquaint the student with the internal organs. Major organs can readily be identified, regions delineated, and organ systems pointed out. Gently move the exposed organs about, so that you can appreciate their depth. Do not cut anything out unless specifically told to do so by your instructor. Carefully separate one part from another without tearing membranes, and try to find as many of the organs mentioned in Exercise 33 as possible before going on to detailed discussion of the anatomy and physiology of each system in other exercises.

Opening the Body Cavity of the Fetal Pig

After completing this exercise, you should be able to

1. Open the abdominal and thoracic cavities
2. Examine the contents of the body cavities, and recognize many of the more prominent organs before making more comprehensive dissections

Materials

The following materials should be provided for class use:

1. Fetal-pig dissection specimens
2. Dissecting tools
3. Bone forceps

OPENING THE ABDOMINAL CAVITY

Place the pig on its back, and spread its arms and legs. Tie string above each foot and slip it under the dissecting tray (Figure 33-1). The string will not have to be removed at the end of each exercise. Simply pull it away from the pan, and tie the end to an identification card before putting it away at the end of the laboratory period (see Figures 33-1 and 33-2.)

From a point between the legs, make a ,U-shaped incision line that runs anteriorly around the umbilical cord (incision line 1). Then continue anteriorly with cut 2 just slightly to the left of the midline until you reach the diaphragm. Cut laterally (incision line 3) just below the diaphragm as far as possible. Posteriorly, make incision line 4, and fold the flaps laterally. In order to free the umbilical cord, cut the single umbilical vein—which is injected with blue latex—and pull the external flap back (incision line 1).

Continue incision line 2 anteriorly. You will meet with a little resistance from the costal (rib) cartilages. Continue forward to a point just anterior to the heart (see Figure 33-2). Make lateral incision line 5 above the diaphragm, and another from the point above the heart to the arms (incision line 6). Be very careful not to damage tissues in this region, especially blood vessels, which you will be dissecting in Exercise 35.

Incision line 2 may be extended forward to the chin. Make transverse cuts (incision line 7) if necessary.

Figure 33-1 Ventral view of the fetal pig ready for dissection. Note the way in which the animal is tied to the dissecting tray.

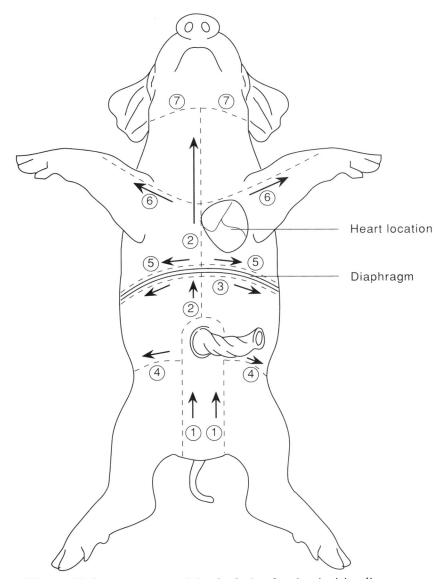

Figure 33-2 Ventral view of the fetal pig, showing incision lines.

Figure 33-3 (a) Ventral view of female fetal-pig organs (photograph).

OPENING THE THORACIC CAVITY

In order to complete the opening of the body cavity, either now or at a later time, continue incision line 2 to the neck, and cut laterally (incision line 7), to create another flap. Break the ribs back so that you have a good view of all the organs in the thoracic cavity, as well as in the abdominal cavity (Figures 33-3 and 33-4).

1. Larynx
2. Trachea
3. Thyroid gland
4. Thymus (posterior portion, reflected)
5. Right lung
6. Right atrium
7. Right ventricle
8. Pulmonary trunk (artery)
9. Left lung
10. Left atrium
11. Coronary artery
12. Left ventricle
13. Diaphragm
14. Liver
15. Gallbladder (pulled out of liver lobes)
16. Small intestine
17. Spleen
18. Colon
19. Right renal artery
20. Descending abdominal aorta
21. Left kidney
22. Left ureter
23. Rectum
24. Umbilical arteries
25. Urinary bladder
26. Ovaries
27. Uterine horns

Figure 33-3 (b) Ventral view of female fetal-pig organs (line drawing).

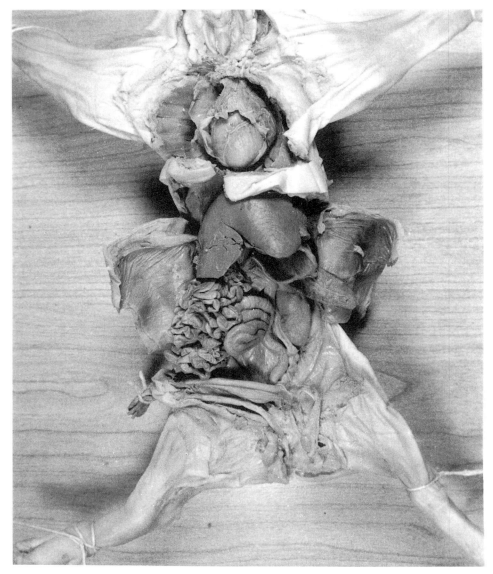

Figure 33-4 (a) Ventral view of male fetal-pig organs (photograph).

1. Larynx
2. Thymus gland
3. Trachea
4. Thyroid gland
5. Right lung
6. Pericardium (cut)
7. Right ventricle
8. Pulmonary trunk (artery)
9. Coronary vessels
10. Left ventricle
11. Left lung
12. Diaphragm
13. Liver
14. Stomach
15. Small intestine
16. Spleen
17. Left kidney
18. Colon
19. Ureter
20. Rectum
21. Umbilical arteries
22. Urinary bladder
23. Vas deferens
24. Right testis
25. Left epididymis

Figure 33-4 (b) Ventral view of male fetal-pig organs (line drawing).

Once you have located all the major organs, turn
to the next exercises for further detail.

LABORATORY REVIEW *33*

OPENING OF THE BODY CAVITY OF THE FETAL PIG

ANSWERS

Mouth
Anus

COMPLETION

1. In the right-hand column, arrange the organs listed on the left, in order, according to the passage of food through the digestive tract from mouth to anus. Locate all organs in the fetal pig.
   a. Ileum
   b. Ascending colon
   c. Pharynx
   d. Descending colon (rectum)
   e. Esophagus
   f. Jejunum
   g. Mouth
   h. Transverse colon
   i. Stomach
   j. Cecum
   k. Duodenum
   l. Anus

2. Name six organs visible in the thoracic cavity of the fetal pig.

   a. _____

   b. _____

   c. _____

   d. _____

   e. _____

   f. _____

The Cardiovascular System

*T*he human circulatory system performs a number of essential functions that include transporting various substances from one part of the body to another, defending the body against a wide variety of disease-causing microorganisms, assisting in the repair of damaged tissue, playing an important role in acid–base balance, and regulating body temperature. The cells of the body depend on the circulatory system for oxygen and essential nutrients and the removal of wastes. If the circulation to any region of the body is stopped, all the cells within that region soon die.

The cardiovascular system is the portion of the circulatory system that includes the heart and blood vessels. This system serves the body by moving blood between the cells of the body and the organs of the digestive, endocrine, integumentary, respiratory, and urinary systems, which communicate with the external environment. In performing this function, the heart works as a pump, forcing blood through blood vessels. These blood vessels, which form a closed system of channels, transport blood and allow for the exchange of gases, nutrients, and wastes between the blood and the cells of the body. Although slight variations in heart activity occur normally, marked changes may be indications of cardiac abnormalities or disease. Specific exercises demonstrate the structure and organization of the components of the cardiovascular system. Other exercises in this section include the procedures used to demonstrate the normal and abnormal activities and processes associated with the cardiovascular system.

GENERAL REFERENCES

Cohn, P. F. *Clinical Cardiovascular Physiology.* Philadelphia: Saunders, 1984.

Netter, F. H. *The CIBA Collection of Medical Illustrations, Vol. 5, The Heart.* Summit, NJ: CIBA Pharmaceutical Products, 1969.

Opie, L. H. *The Heart: Physiology and Metabolism.* New York: Raven Press, 1991.

Tortora, G. J., and S. R. Grabowski. *Principles of Anatomy and Physiology,* 7th ed. New York: HarperCollins, 1993.

SPECIFIC REFERENCES

Brown, A. M. "Receptors Under Pressure: An Update on Baroreceptors." *Circulation Research,* 46: 1–100, 1980.

Farley, D. "High Blood Pressure, Controlling the Silent Killer." *FDA Consumer* 25: 28–33, 1991.

Fuster, V., L. Badimon, J. J. Badimon, and J. H. Chesebro. "Mechanisms of Disease and the Acute Coronary Syndromes." *The New England Journal of Medicine,* 326: 242–250, 1992.

Laks, M. M., and R. H. S. Selvester. "Computerized Electrocardiography—An Adjunct to the Physician." *The New England Journal of Medicine.* 325: 1803–1804, 1991.

Murphy, K. R. *ECG Essentials: A Pocket Reference for Systematic Interpretation.* Chicago: Quintessence Books, 1991.

Oliver, M. F. "Diet and Coronary Heart Disease." *British Medical Bulletin,* 37: 49–58, 1981.

Pickering, J. G., et al. "Blood Pressure During Normal Daily Activities, Sleep and Exercise." *Journal of the American Medical Association,* 247: 992–996, 1982.

Ross, J., Jr. "The Failing Heart and the Circulation." *Hospital Practice,* 18: 151–169, 1983.

Mammalian Heart and Associated Blood Vessels

After completing this exercise, you should be able to

1. Identify all parts of the external surface of the mammalian heart
2. Relate all major blood vessels supplying and draining the heart to their specific heart chambers
3. Trace blood through the heart to the lungs and back again

Materials

The following materials should be provided for class use:

1. Fetal-pig dissection specimens
2. Fresh sheep or beef hearts
3. Dissecting instruments
4. Model of the human heart
5. Charts of the human heart

FETAL-PIG HEART

Locate the heart in the thoracic cavity. It is surrounded by a tough membrane—the *pericardium* or *pericardial sac.* Examine the sac carefully. Note that a pair of nerves attach to it laterally. These are the *phrenic nerves,* which supply the muscle of the diaphragm. Free the phrenic nerves (or at least the phrenic nerve on the left side of the pig) from the pericardial sac. Slit the pericardial sac, and fold it back out of the way—or remove it entirely.

Identify all parts of the heart that can be seen without removing it from the body. Locate the *left* and

right ventricles, the *left* and *right atria* (seen externally as the flaplike auricles), and the *coronary artery* and its branches on the surface of the heart.

By lifting the ventricles slightly and pulling the heart toward the pig's left side, you can see the large veins entering the right atrium. Identify the *superior vena cava,* which returns blood to the heart from the head and shoulders, and the *inferior vena cava,* which returns blood from the organs below the diaphragm.[1] The *pulmonary artery,* which carries blood from the right ventricle to the lung, can be seen prominently in ventral view. Remove enough connective tissue from the anterior region to expose the large *aorta,* which arches to the left just anterior to the heart. The *pulmonary veins,* which empty into the left atrium, can be seen if the ventricles are lifted and pulled to the pig's right side.

SHEEP HEART

For a more complete study of the anatomy of the mammalian heart (which is remarkably similar in all species), use the freshly thawed or preserved sheep heart. As in the pig, the terms *superior* and *inferior* are used. Orient the heart so that you are looking at its ventral side. Find the following parts, using Figures 34-1 and 34-2: *left ventricle, right ventricle, left atrium, right atrium, pulmonary artery, aorta, superior vena cava, inferior vena cava,* and the *pulmonary veins.* There are four pulmonary veins, two from each lung, but they

[1]The student should be aware that the blood vessels in the fetal pig were named to correspond with those in the human, so that it is common to find the adjectives *superior* and *inferior* replacing the more correct terms *anterior* and *posterior* for the four-legged animal. To be consistent with terminology used in most textbooks, the terms *superior* and *inferior* are used in these exercises.

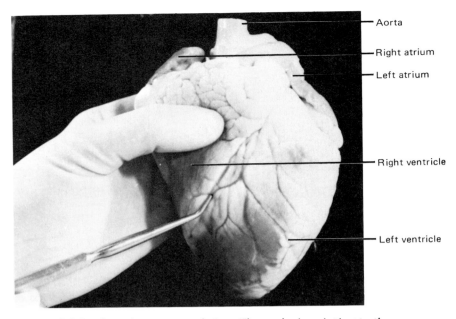

Figure 34-1 Sheep heart, ventral view: The probe is pointing to the coronary artery at the septum between the right ventricle and the left ventricle.

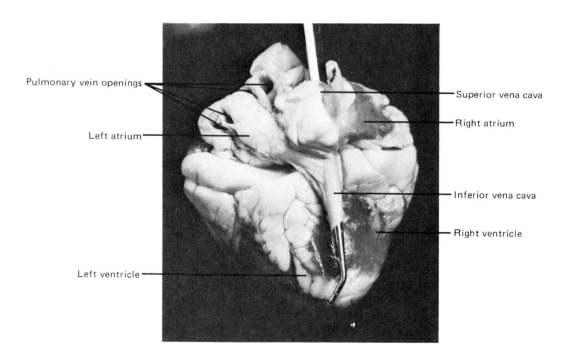

Figure 34-2 Sheep heart, dorsal view: The probe passes through the right atrium from the superior vena cava to the inferior vena cava.

may be cut off in your specimen. Insert a probe into all major blood vessels, to determine the chambers with which they connect.

To open the heart, begin at the end of the *pulmonary artery,* and cut back into the right ventricle (Figure 34-3a). The valve at the exit of the pulmonary artery from the right ventricle is the *pulmonary semilunar* valve.

Open the right ventricle carefully, and observe the delicate membranelike *tricuspid valve,* which separates the right ventricle from the right atrium (Figure 34-3b). The valve is held to the *papillary muscle* of the ventricle by connective tissue strands—the *chordae tendineae,* which prevent the valve from being forced into the atrium when the ventricle contracts.

Now cut through the tricuspid valve into the right atrium (Figure 34-3c). The muscle layer or *myocardium* of the atrium is thinner than that of the ventricle, and when not filled with blood, the atrial wall collapses. Open the atrium wide, and locate the openings from the three veins that empty into it: *superior vena cava, inferior vena cava,* and *coronary sinus.* The coronary sinus, located between the *venae cavae,* is the expanded end of the *coronary vein,* which drains the heart tissue. It may be difficult to locate if the specimen has been preserved. The walls between right and left sides of the chambers are *septa.* The *interatrial septum* separates right and left atria, and the *interventricular septum* separates the two ventricles. The muscle of the atrium is the *pectinate muscle,* which has a comblike appearance.

Turn to the left side of the heart. With your scalpel, cut through the *left atrium* and the *left ventricle,* through the *mitral* or *bicuspid valve,* to the apex of the heart. Like the tricuspid valve, the mitral valve is held to the ventricular wall by *chordae tendineae.* Opening into the left atrium are the four pulmonary veins, which return oxygenated blood to the heart from the lungs. (This is an example of a vein that carries oxygenated blood. Another example is the umbilical vein of the fetus, which returns blood from the *placenta.*)

Examine the left ventricle. Notice the thickness of the muscle. The left ventricle pumps blood through the aorta to distant parts of the body and must contract with greater force than the right ventricle. Locate the *aortic semilunar valve,* which leads to the aorta. Cut into the aorta and open it to see the two large openings near its base (Figure 34-3d). These are the openings into the *right* and *left coronary arteries,* which carry oxygenated blood to the heart muscle.

Examine the area where the *arch of the aorta* passes close to the *pulmonary artery.* Find a small ligament connecting the two (the *ligamentum arteriosum*). This is a remnant of the *ductus arteriosus,* an important fetal lung bypass, which closes and becomes ligamentous in the adult heart.

HUMAN HEART

A photograph of the human heart is shown in Figure 34-4. Its size and position in the thoracic cavity is much like that of the sheep. It is located in the *mediastinal cavity* and is surrounded by a tough pericardial sac, which is composed of two layers. The inner layer—the *epicardium*—corresponds to the visceral peritoneum of other visceral organs, and adheres tightly to the heart muscle. The outer pericardial layer corresponds to the parietal peritoneum and lies very close to the inner layer. A potential cavity, the *pericardial space,* exists between the two layers of the pericardium. This space is filled with fluid, to prevent friction. Figure 34-5 shows the relationship of the heart to the major blood vessels associated with it.

HEART STRUCTURE AND BLOOD PATHWAY IN THE HUMAN HEART

The human heart is a four-chambered muscle. The two superior chambers, the *atria,* are rather thin-walled, but the inferior chambers, the *ventricles,* are thick-walled. Separating the two atria is a muscular *interatrial septum.* Similarly, the ventricles are separated by a muscular *interventricular septum.* Valves between atria and ventricles (atrioventricular valves) are named for their anatomical configuration (Figure 34-6). The *tricuspid valve,* on the right side, has three flaps, and the *biscuspid valve,* on the left side, has two flaps. The bicuspid valve is often termed the *mitral valve.* Associated with the right atrium are three main veins: the *superior vena cava,* the *inferior vena cava,* and the *coronary sinus.* Blood returning to the heart through these veins passes through the tricuspid valve into the right ventricle. When the right ventricle contracts, and the tricuspid valve closes, the blood leaves the heart by way of the *pulmonary artery,* which terminates in pulmonary capillaries. Here, the gases oxygen and carbon dioxide are exchanged, and the newly oxygenated blood returns to the left atrium in the *pulmonary vein.* When the left atrium contracts, blood flows through the mitral valves into the left ventricle. The mitral valve closes when the left ventricle contracts, and blood leaves the heart through the aorta to be distributed to all parts of the body. This tracing of a single drop of blood through the pulmonary circuit can be noted in the diagram at the end of this exercise (Figure Q34-1). Of course, it must be remembered that in the living, beating heart, *both* atria contract simultaneously, and then *both* ventricles contract.

The tricuspid and mitral valves are made of fibrous tissue that is lined by the endothelial inner lining of

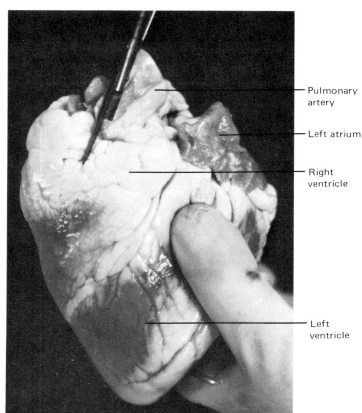

Pulmonary artery

Left atrium

Right ventricle

Left ventricle

Figure 34-3 (a) Use scissors to cut through the pulmonary artery into the right ventricle.

1. Right atrium
2. Tricuspid valve
3. Right ventricle
4. Chordae tendineae

Figure 34-3 (b) Right ventricle opened to show tricuspid valve.

1. Right atrium
2. Tricuspid valve
3. Right ventricle

Figure 34-3 (c) Right ventricle and right atrium opened to show the entire tricuspid valve and chordae tendineae. Probe runs beneath the tricuspid valve.

1. Openings to coronary artery
2. Aorta
3. Left ventricle
4. Left atrium
5. Chordae tendineae
6. Aortic semilunar valves

Figure 34-3 (d) Aorta cut lengthwise and held open to show the coronary artery openings.

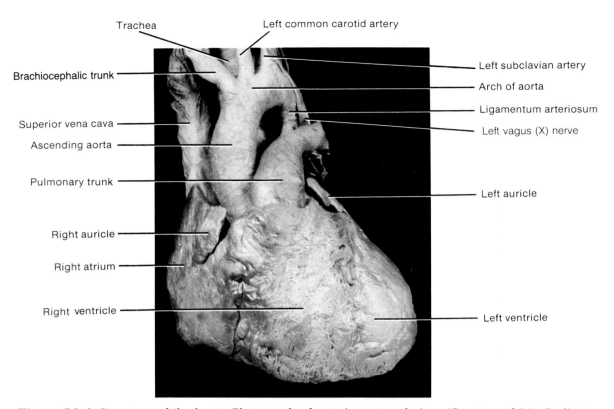

Trachea
Left common carotid artery
Brachiocephalic trunk
Left subclavian artery
Arch of aorta
Ligamentum arteriosum
Superior vena cava
Ascending aorta
Left vagus (X) nerve
Pulmonary trunk
Left auricle
Right auricle
Right atrium
Right ventricle
Left ventricle

Figure 34-4 Structure of the heart. Photograph of anterior external view. (Courtesy of J.A. Gosling, P.F. Harris, et al., *Atlas of Human Anatomy,* Gower Medical Publishing Ltd., 1985.)

Right common carotid artery
Right internal jugular vein
Right subclavian artery
Right subclavian vein
Brachiocephalic trunk
Right brachiocephalic vein
Superior vena cava
Right pulmonary artery
Right pulmonary veins
Right auricle
Right atrium
Right coronary artery
Coronary sulcus
Right ventricle
Inferior vena cava

Left common carotid artery
Left internal jugular vein
Left subclavian artery
Left subclavian vein
Left brachiocephalic vein
Arch of aorta
Ascending aorta
Ligamentum arteriosum
Left pulmonary artery
Pulmonary trunk
Left pulmonary veins
Left auricle
Left coronary artery
Anterior interventricular sulcus
Left ventricle
Descending aorta

(a)

Figure 34-5 Structure of the heart: (a) anterior external view; (b) posterior view; (c) anterior internal view.

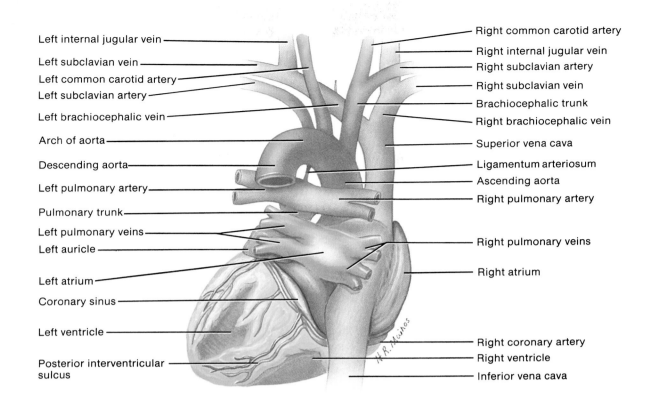

Left internal jugular vein

Left subclavian vein

Left common carotid artery

Left subclavian artery

Left brachiocephalic vein

Arch of aorta

Descending aorta

Left pulmonary artery

Pulmonary trunk

Left pulmonary veins

Left auricle

Left atrium

Coronary sinus

Left ventricle

Posterior interventricular sulcus

Right common carotid artery

Right internal jugular vein

Right subclavian artery

Right subclavian vein

Brachiocephalic trunk

Right brachiocephalic vein

Superior vena cava

Ligamentum arteriosum

Ascending aorta

Right pulmonary artery

Right pulmonary veins

Right atrium

Right coronary artery

Right ventricle

Inferior vena cava

(b)

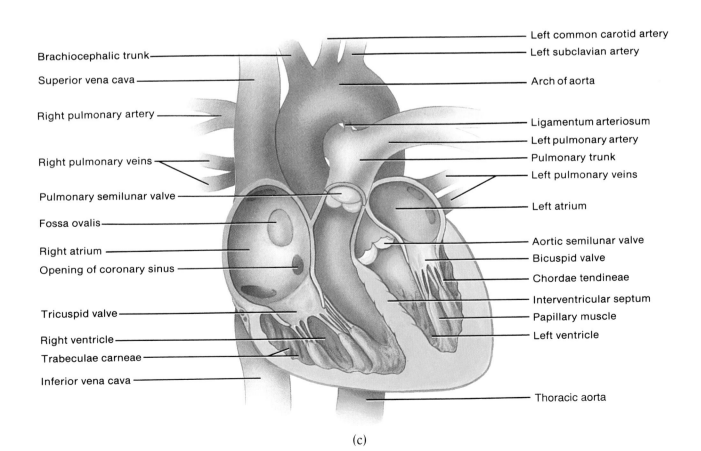

Brachiocephalic trunk

Superior vena cava

Right pulmonary artery

Right pulmonary veins

Pulmonary semilunar valve

Fossa ovalis

Right atrium

Opening of coronary sinus

Tricuspid valve

Right ventricle

Trabeculae carneae

Inferior vena cava

Left common carotid artery

Left subclavian artery

Arch of aorta

Ligamentum arteriosum

Left pulmonary artery

Pulmonary trunk

Left pulmonary veins

Left atrium

Aortic semilunar valve

Bicuspid valve

Chordae tendineae

Interventricular septum

Papillary muscle

Left ventricle

Thoracic aorta

(c)

467

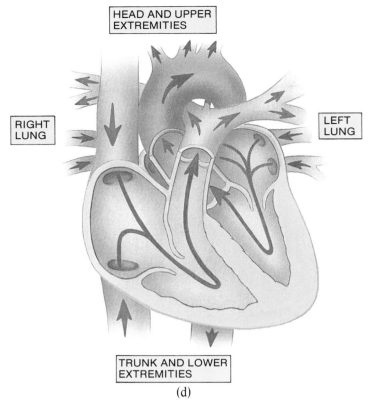

(d)

Figure 34-5 *(Continued)* (d) path of blood through the heart. (From G. J. Tortora and S. R. Grabowski, *Principles of Anatomy and Physiology,* 7th ed., New York: HarperCollins, 1993.)

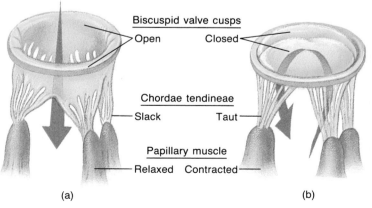

Figure 34-6 Atrioventricular (AV) valves: (a) bicuspid valve open; (b) bicuspid valve closed. (The tricuspid valve operates in a similar manner.) (From G. J. Tortora and S. R. Grabowski, *Principles of Anatomy and Physiology,* 7th ed., New York: HarperCollins, 1993.)

the heart. They are not muscular. Both heart valves are held down to the ventricular muscle by connective tissue cords, the *chordae tendineae.* The large vessels that leave the ventricle (the pulmonary artery and the aorta) also possess valves at their bases, to prevent backflow. These valves are the *semilunar valves.* Individually, they usually are referred to as the *pulmonary valve* and the *aortic valve.* All four valves are designed so that they permit blood flow in only one direction and do not allow for backflow. Damaged valves will, however, result in backflow.

The microscopic nature of the myocardium of the heart was studied in Exercise 6. It is striated, branched, and has centrally located nuclei and intercalated discs. The histology of cardiac muscle should be reviewed at this time. On opening the heart chambers, one finds a difference in the appearance of the muscle. In the atria, the *pectinate muscles* appear to be ridged, like the teeth of a comb. In the ventricles, the *papillary muscles* project as small cones. Irregular ridges and folds in the ventricular musculature are the *trabeculae carnae.*

CONDUCTION SYSTEM OF THE HEART

In addition to the extrinsic nerves from the autonomic nervous system, which regulate the beat of the heart, there is an *intrinsic* nervous system within the heart muscle, which originates from embryonic cardiac muscle cells, and which consists of the following parts (Figure 34-7):

Sinoatrial node (SA node)—the pacemaker in the right atrium, which initiates the beat

Atrioventricular node (AV node)—in the right atrium, near the ventricles

Atrioventricular bundle (bundle of His)

Right and left bundle branches

Purkinje cells—extending into the muscle of each ventricle

A chick embryo heart can be seen to beat before any nerve fibers have formed. If the beating heart of a pithed

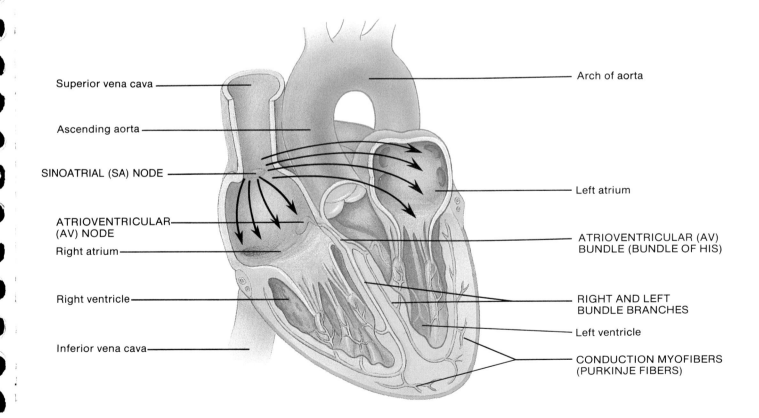

Figure 34-7 Conduction system of the heart: location of the nodes and bundles of the conduction system. The arrows indicate the flow of action potentials through the atria. Exercise 36 presents the normal cardiac cycle, using an electrocardiogram. (From G. J. Tortora and S. R. Grabowski, *Principles of Anatomy and Physiology,* 7th ed., New York: HarperCollins, 1993.)

frog is removed from the animal's body and placed in a dish of water, it will continue to beat. This indicates that nerve tissue is incorporated into the muscular tissue of the heart. This tissue cannot be seen with the naked eye. Use a stained slide of such tissue, if it is available.

THE FETAL HEART

Fetal-Pig Heart

The fetal pig uses the same two lung bypasses as the human (see Figures 34-8 and 34-9). Return to your dissection of the fetal-pig heart. Gently push away the tissue that binds the heart to other organs. Find the pulmonary artery that leaves the right ventricle, and trace it to the lungs. Be very careful of the other blood vessels in this region because they will be studied in Exercise 35.

After the pulmonary artery has been dissected, return to the aorta. As it leaves the heart, it is called the *ascending aorta*. It makes a sharp turn to the left as the aortic arch, and continues as the *descending aorta*. Expose this arch cleanly, along with 1 inch of the descending aorta. You will see the fetal shunt, the *ductus arteriosus,* which carries blood directly from the pulmonary artery to the aorta, thus bypassing the lung. The distal ends of the pulmonary arteries are small and nonfunctional at this stage.

At birth, they will fill with blood for the first time when the pig's lungs are inflated with the first inhalation of air, and the pulmonary circuit becomes functional. When a pig is born, just like any other mammal, the ductus arteriosus closes and becomes a strand of connective tissue, the *ligamentum arteriosum.* Study the changes before and after birth shown in Figure 34-9b.

Human Fetal Heart

The heart is fully developed before birth, with only two major differences, which are called the *lung bypasses.* These two lung bypasses enable the fetal heart to pump blood into the general circulation of the fetus without going first to the lung. The pulmonary circuit is ready to go but is not used by the fetus, which has no access to air. The fetus obtains its oxygen and nutrients from the maternal circulation and passes its wastes to the mother, whose organs will eliminate it. The arteries supplying the placenta are the right and left umbilical arteries, which carry unoxygenated blood, and the vein (single) from the placenta is the umbilical vein, which carries oxygenated blood. Figure 34-8 is a schematic illustration of blood flow before birth.

The two lung bypasses are

1. *Foramen ovale*—a hole in the interatrial septum, which allows blood to pass directly from right atrium to left atrium and thus bypass the lung. About 50 percent of the blood entering the right atrium passes through the foramen ovale.

2. *Ductus arteriosus*—a short artery, which connects the pulmonary artery with the aorta. The blood that enters the right ventricle first and then the pulmonary artery is diverted directly into the aorta, again bypassing the fetal lung.

Study Figure 34-8, and trace blood in the fetus from the right side of the heart to the umbilical arteries (branches of the internal iliac arteries), which take it to the placenta. There, exchange is made, such that wastes pass into the maternal circulation, and nutrients and oxygen enter the fetal circulation from the mother. Blood from the maternal circulation enters the single umbilical vein, which returns it to the fetal circulation, laden with the mother's oxygen and nutrients.

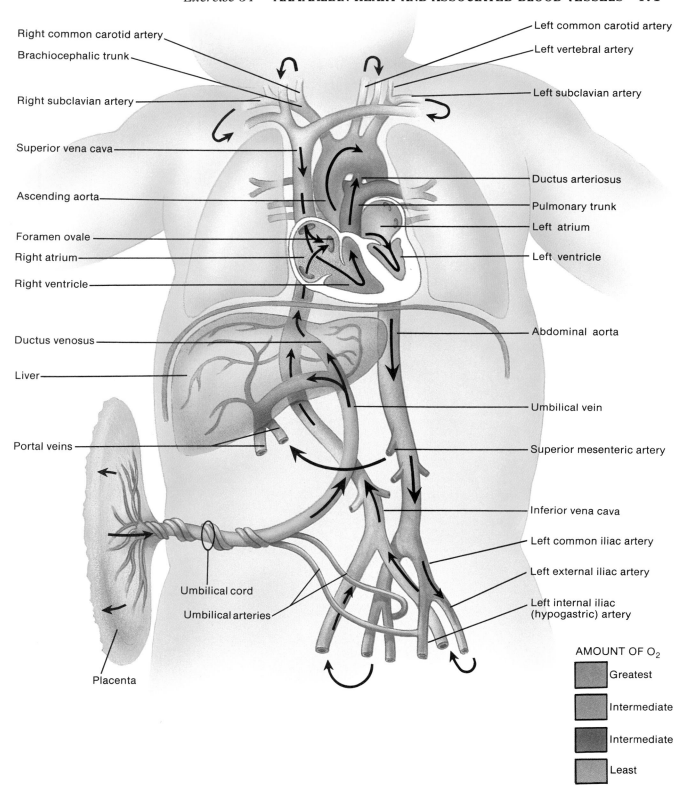

Figure 34-8 Fetal circulation: The various colors represent different degrees of oxygenation of blood ranging from greatest (red) to intermediate (two shades of purple) to least (blue). (From G. J. Tortora, *Principles of Human Anatomy,* 6th ed., New York: HarperCollins, 1992.)

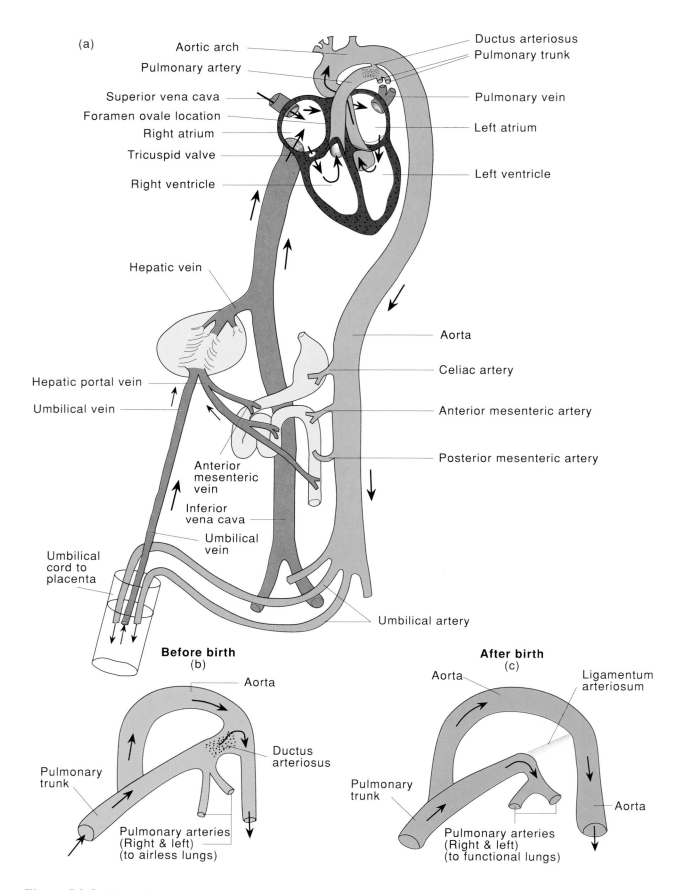

(a) Aortic arch
Pulmonary artery
Superior vena cava
Foramen ovale location
Right atrium
Tricuspid valve
Right ventricle

Ductus arteriosus
Pulmonary trunk
Pulmonary vein
Left atrium
Left ventricle

Hepatic vein

Aorta
Celiac artery
Anterior mesenteric artery
Posterior mesenteric artery

Hepatic portal vein
Umbilical vein

Anterior mesenteric vein
Inferior vena cava
Umbilical vein

Umbilical cord to placenta

Umbilical artery

Before birth
(b)
Aorta

Ductus arteriosus

Pulmonary trunk

Pulmonary arteries (Right & left) (to airless lungs)

After birth
(c)
Aorta
Ligamentum arteriosum

Pulmonary trunk

Aorta

Pulmonary arteries (Right & left) (to functional lungs)

Figure 34-9 (a) A schematic drawing of fetal-pig circulation; (b) ductus arteriosus before birth and (c) ligamentum arteriosum (connective tissue) after birth.

MAMMALIAN HEART AND ASSOCIATED BLOOD VESSELS

COMPLETION

1. Identify the parts of the mammalian heart in Figure Q34-1.

Anterior view

Figure Q34-1 The mammalian heart. (From G. J. Tortora and N. P. Anagnostakos, *Principles of Anatomy and Physiology*, 3rd, 4th, 5th, 6th eds., New York: HarperCollins, 1981, 1984, 1987, 1990.)

a. _____   i. _____

b. _____   j. _____

c. _____   k. _____

d. _____   l. _____

e. _____   m. _____

f. _____   n. _____

g. _____   o. _____

h. _____

2. Trace the pathway of a drop of blood through parts of the adult heart, cavities, and blood vessels from the superior vena cava to the aorta. Some steps are provided. You fill in the rest, being sure also to list the valves.

a. Superior vena cava _____

b. _____

c. _____

d. Right ventricle _____

e. _____

f. _____

g. _____

h. _____

i. Left atrium _____

j. _____

k. _____

l. _____

m. Aorta _____

3. Trace blood from the superior vena cava to the aorta in the fetal-pig heart. (Two pathways are possible.)

a. Superior vena cava _____   a. Superior vena cava _____

b. _____   b. _____

c. _____   c. _____

d. Right ventricle _____   d. Left atrium _____

e. _____   e. _____

f. _____   f. _____

g. _____   g. _____

h. Aorta _____   h. Aorta _____

Blood Vessels and Lymphatics

1. Locate and identify the major arteries and veins of the fetal pig, and compare with human blood vessels
2. Trace blood and other substances through the vascular system from one point to another
3. Indicate the general locations of lymph nodes and vessels in the fetal pig and the human
4. Understand the functions of the portal system.

Materials

The following materials should be provided for class use:

1. Fetal-pig dissection specimens
2. Dissecting tools
3. Charts of human blood vessels
4. Demonstration of the *portal system* in the fetal pig

MAJOR VEINS OF THE FETAL-PIG THORAX, NECK, AND SHOULDER

Veins return blood to the heart (Figure 35-3). Most of them transport unoxygenated blood, which will be pumped to the lung where the carbon dioxide will be exchanged for oxygen. In the fetal pig, however, the lung is not functional, and blood must be pumped to the *placenta,* where it picks up oxygen from the mother. Oxygenated blood is returned to the heart in the *umbilical vein.* This vein was cut in a previous dissection, in order to pull the umbilical cord posteriorly. After

birth, the pulmonary veins return oxygenated blood from the lungs directly to the left atrium of the heart. Blood flow in the umbilical vein then ceases.

In an injected pig specimen the veins of the thorax are so prominent that it is necessary to dissect them before studying the arteries. They lie over the main thoracic arteries and should be partially removed or moved aside on at least one side of the specimen to expose the arteries clearly. Ask your instructor for advice if you are not certain what to do or if the injection of the veins is poor, as it often is.

The logical place to begin the tracing of the veins is at the heart, but it must be kept in mind that as you move away from the heart, you are working back to tributaries of major veins and that the blood flow is always toward the heart. Locate the *superior vena cava* and follow it away from the heart (Figure 35-1). The small *internal mammary veins* from the chest wall empty into it on the ventral surface. These veins were cut when the thorax was opened. Lift the right side of the heart slightly and you will see a large vein emptying into the dorsal side of the *superior vena cava* from the abdominal cavity. This is the *azygos vein* from the muscles of the back and the intercostal muscles. Flowing into the superior vena cava about half an inch above the heart are the paired *brachiocephalic veins* from the head and arms.

Working back from the paired brachiocephalic veins, locate the

1. *Subclavian vein*—from the arm or foreleg
2. *External jugular vein*—from the shoulder, head, and neck
3. *Internal jugular vein*—from the arm and the head

The subclavian vein becomes the axillary vein in the underarm region and the brachial vein in the upper arm. The external jugular veins run alongside the vagus nerve. The external jugular vein receives the subscapular vein from the shoulder and can be seen lateral and

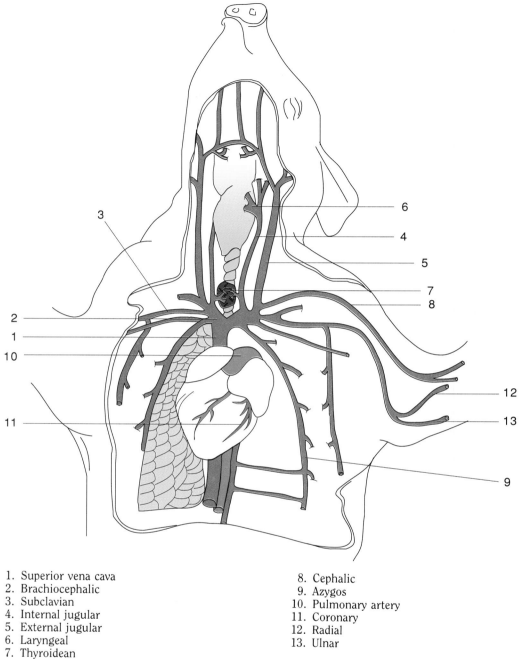

1. Superior vena cava
2. Brachiocephalic
3. Subclavian
4. Internal jugular
5. External jugular
6. Laryngeal
7. Thyroidean
8. Cephalic
9. Azygos
10. Pulmonary artery
11. Coronary
12. Radial
13. Ulnar

Figure 35-1 Major veins of the thorax, neck, and shoulder in the fetal pig.

superficial to the internal jugular. The internal jugular vein drains the smaller vessels that serve the brain, eye, and upper jaw. Sometimes, the internal jugular and the external jugular veins come together as one before entering the brachiocephalic (Figure 35-1).

MAJOR VEINS OF THE FETAL-PIG ABDOMINAL CAVITY

Locate the *inferior vena cava* again. Trace it through the diaphragm. In the abdominal cavity, the inferior vena cava receives the following veins (Figure 35-2):

1. *Hepatic*—from the superior part of the liver
2. *Adrenolumbar*—from the adrenal or suprarenal gland and muscles of the back
3. *Renal*—from the kidney
4. *Genital* or *gonadal*—from the reproductive organs
5. *Caudal*—from the tail
6. *Common iliacs*—from the legs and the organs in the pelvic region of the abdominal cavity (reproductive organs)

Most of the abdominal viscera are drained by the *portal vein,* which drains the stomach, small intestine, pancreas, liver, gallbladder, spleen, and colon (Figure 35-3). Veins from the kidney, adrenal gland, and reproductive ducts are all *retroperitoneal* (i.e., under the peritoneum) and flow into the inferior vena cava. The abdominal viscera in the peritoneal cavity are drained by capillaries, which ultimately flow into the *portal vein.*

The portal vein breaks up into capillaries within the liver, thus facilitating the passage of some specific products of the abdominal viscera into the cells of the liver, where they are further processed. This has enormous significance in the physiology of these products. A portal vein is one that begins and ends in capillaries. The pig and the human have only one such system, which is called the *hepatic portal system* (see Figure 35-3).

MAJOR ARTERIES OF THE FETAL-PIG THORAX

Return to the region of the heart. Locate the *pulmonary artery,* where it leaves the right ventricle (see Figure 35-1). Trace it until it branches into the following:

1. Right pulmonary artery to the right lung (nonfunctional)
2. Left pulmonary artery to the left lung (nonfunctional)

3. Ductus arteriosus—to the aorta, bypassing the lung

Because your dissection specimen is still an embryo and is living in a fluid environment (amniotic fluid), it must obtain its oxygen and rid itself of wastes by means of the *placenta.* Before birth, the lungs are not functional, so that blood from the heart is pumped directly to the aorta and then to the placenta (Figure 35-4). After birth, this shunt between the pulmonary artery and the aorta will shut down. A remnant, the *ligamentum arteriosum,* will remain.

The *ascending aorta* gives off the *coronary arteries.* It is difficult to see the place where they branch off, but if you have a well-injected pig you can see the coronary artery on the line between the right and left ventricles (interventricular sulcus).

At this point, move all thoracic veins aside to expose the *arch of the aorta,* which branches into all of the prominent arteries to the head, shoulder, and arms:

1. *Brachiocephalic*—This single vessel is located only on the right side. As the name implies, it branches into the *right subclavian* to the shoulder and arm, and the *right* and *left common carotids* to the neck and head (brachio = arm; cephalic = head). There are some variations here. Occasionally, both right and left common carotid arteries come off of the aortic arch.
2. Like the *right subclavian* artery, the *left subclavian* sends branches to the arm (axillary and brachial arteries), to the shoulder (*costocervical* and *internal thoracic*), and to the lower part of the neck and thyroid gland (the *thyrocervical*).

MAJOR ARTERIES OF THE FETAL-PIG ABDOMINAL CAVITY

Trace the *descending thoracic aorta* through the diaphragm (Figure 35-5). At this time, slit the diaphragm from its outer edge to the point where the great vessels penetrate it. Now trace the *descending abdominal aorta* and its branches (Figure 35-2):

1. *Celiac*—An unpaired artery, which comes off of the aorta near the diaphragm. It is a short vessel, which supplies branches to the liver (hepatic artery), stomach (gastric artery), spleen (splenic artery), and pancreas (*pancreatico duodenal* artery).
2. *Superior mesenteric*—unpaired artery that supplies the small intestine
3. *Adrenolumbar*—paired arteries supplying the adrenal (suprarenal) glands and some back muscles
4. *Renal*—paired vessels, which supply the kidneys
5. *Genital* or *gonadal*—paired vessels, which supply the gonads

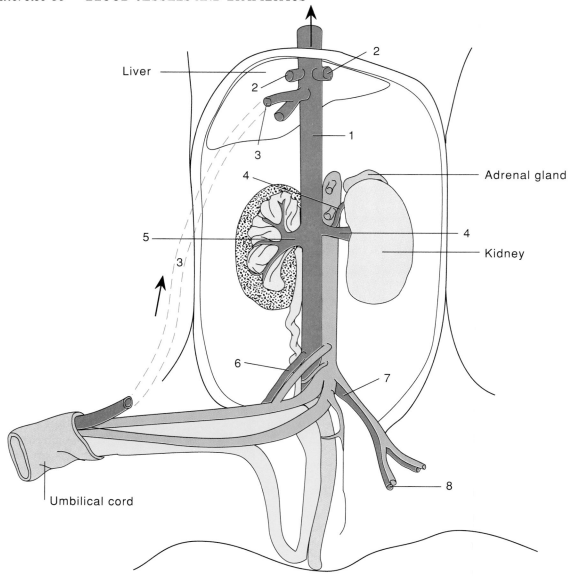

Liver

Adrenal gland

Kidney

Umbilical cord

1. Inferior vena cava
2. Hepatic
3. Umbilical
4. Adrenolumbar
5. Renal
6. Common iliac
7. External iliac
8. Femoral

Figure 35-2 Major veins of the abdomen in the fetal pig.

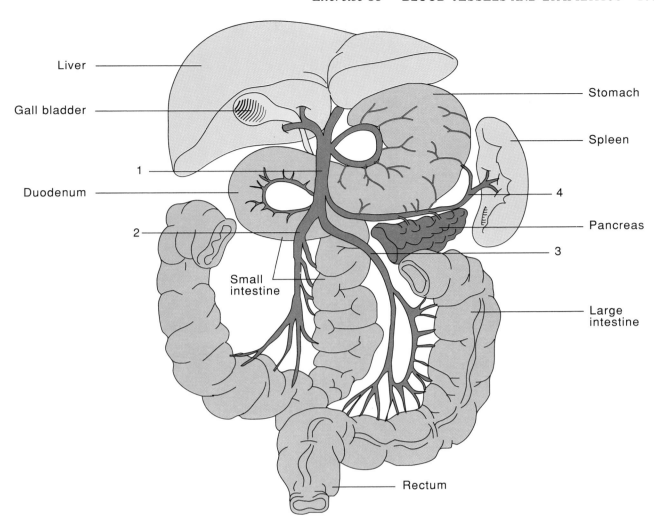

Liver

Gall bladder

Duodenum

1

2

Small
intestine

Stomach

Spleen

4

Pancreas

3

Large
intestine

Rectum

1. Portal vein
2. Superior mesenteric
3. Inferior mesenteric
4. Gastrosplenic

Figure 35-3 Major tributaries of the portal system in the fetal pig.

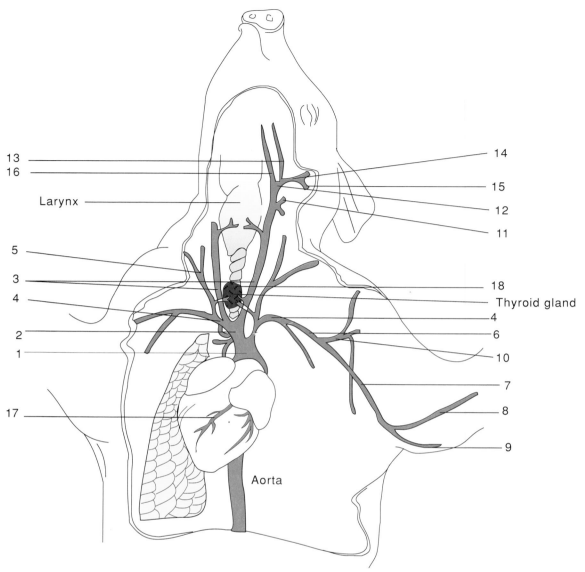

13
16
14
15
12
11
Larynx
5
3
4
2
1
18
Thyroid gland
4
6
10
7
8
9
17
Aorta

1. Aortic arch
2. Right Brachiocephalic
3. Common carotid
4. Subclavian
5. Thyrocervical
6. Axillary
7. Brachial
8. Radial
9. Ulnar
10. Subscapular
11. Internal carotid
12. External carotid
13. External maxillary
14. Internal maxillary
15. Temporal
16. Lingual
17. Coronary
18. Thyroidean

Figure 35-4 Major arteries of the fetal-pig thorax and neck.

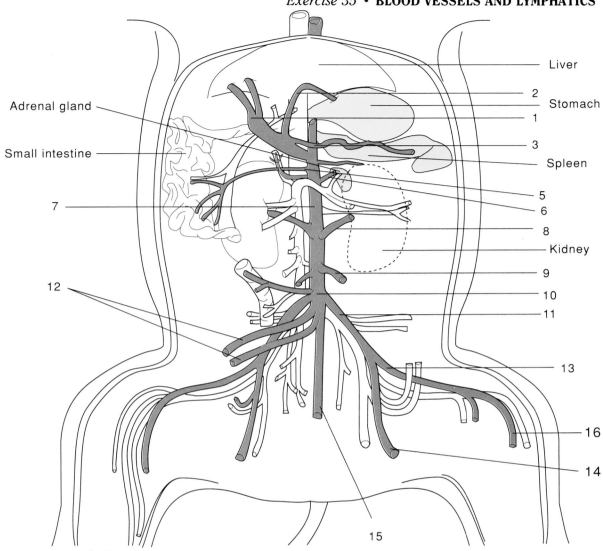

1. Hepatic
2. Gastric
3. Splenic
4. Celiac
5. Superior mesenteric
6. Adrenolumbar
7. Descending aorta
8. Renal
9. Genital
10. Inferior mesenteric
11. Common iliac
12. Umbilical
13. External iliac
14. Internal iliac
15. Median sacral
16. Femoral

Figure 35-5 Major arteries of the abdominal cavity in the fetal pig.

6. *Inferior mesenteric*—a single vessel, which supplies the colon and rectum
7. *External iliac*—paired arteries, which supply the legs
8. *Umbilical*—paired arteries, which carry blood to the placenta.

COMPARABLE HUMAN VENOUS SYSTEM

A diagram of the human venous system (Figure 35-6) is shown, for comparison with that of the fetal pig. Study the similarities and the differences. Compare as many veins as you can with the major veins seen in the fetal pig.

COMPARABLE HUMAN ARTERIES

Compare the arterial system of the fetal pig with that of the human shown in Figure 35-7. The relationship between the major arteries of the fetal pig and the human is strikingly similar.

OVERALL CIRCULATORY PATTERN IN THE HUMAN

Use the diagram in Figure 35-8 to trace blood from one part of the body to another. On the diagram, red indicates oxygenated blood, blue unoxygenated blood, and purple a capillary bed where oxygen and carbon dioxide are being exchanged at the tissue level. Arrows indicate the direction of flow. Figures 35-9 through 35-15 show in detail the vessels of the appendages, head, neck, and brain. Become familiar with the flow pattern in these areas. This will prepare you for the tracing questions in the laboratory review section at the end of this exercise.

THE LYMPHATIC SYSTEM

The lymphatic system is an accessory circulatory system composed of fluid (lymph), vessels (lymphatic capillaries and thin-walled conducting vessels), and lymph nodes. There is no pumping organ (e.g., the heart), and the fluid moves by means of skeletal-muscle contraction in the appendages and breathing movements in the thorax. All vessels in this system flow toward the heart and none away from the heart.

Numerous tiny valves along the length of the major lymph vessels close behind the moving fluid and prevent backflow, just as they do in some veins in the blood circulatory system. The walls of the lymphatic vessels are so thin that the presence of the valves causes a beadlike appearance in the major vessels.

Location of Lymph Vessels

Vessels in the lymphatic system are ubiquitous (see Figure 35-16), but most of them are too small to be seen with the naked eye. There are two prominent lymphatic ducts (Figure 35-16):

1. *Left thoracic duct*—This is the main duct of the system. It is located dorsolateral to the thoracic aorta. It presents a beadlike appearance and is usually yellowish in color. The left thoracic duct empties into the venous system at the left internal jugular vein near its junction with the left subclavian vein, or into the left subclavian vein itself.
2. *Right lymphatic duct*—This duct is similar in appearance to the thoracic duct, but it is shorter. It empties into the right external jugular vein.

Figure 35-16 shows the location and details of these ducts in the human.

Lymph Nodes

These large- to medium-sized nodules are located in all parts of the body (Figure 35-16a). They act as filters of the lymph that passes through them. Some of the more prominent lymph nodes are:

1. *Cervical nodes*—These large masses are found in the neck and jaw.
2. *Inguinal nodes*—These nodes vary in size and number. Most of them are small masses found in the groin region, near the inguinal canals.
3. *Axillary nodes*—These numerous small nodules in the underarm region are usually removed during surgery for breast cancer in the human, to prevent spread (metastasis) of the cancer cells from the mammary glands, via the lymphatics, to other parts of the body. Trace this route in Figure 35-16a.

Lymph Flow

An example of lymph flow is shown diagramatically in Figure 35-17. Lymph is taken up into lymph capillaries

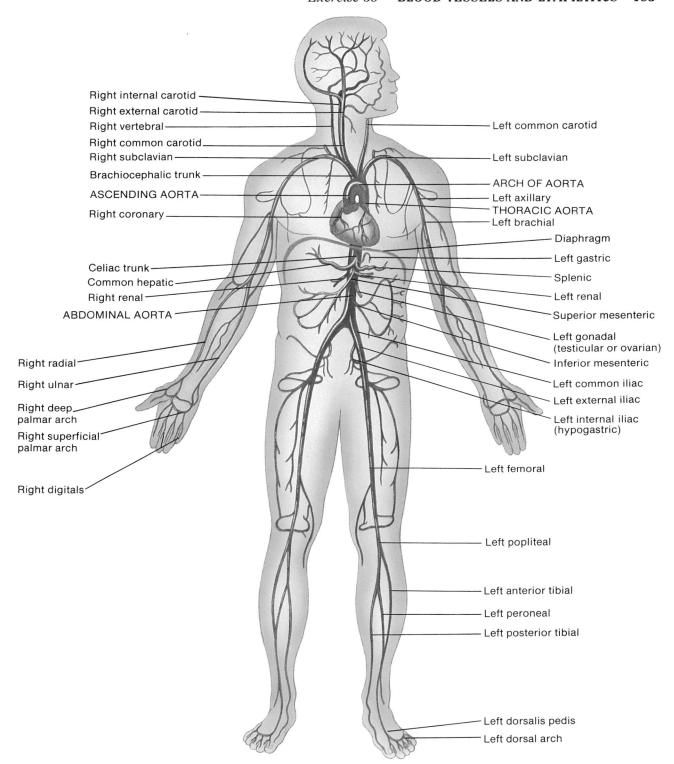

Right internal carotid

Right external carotid

Right vertebral

Right common carotid

Right subclavian

Brachiocephalic trunk

ASCENDING AORTA

Right coronary

Celiac trunk

Common hepatic

Right renal

ABDOMINAL AORTA

Right radial

Right ulnar

Right deep palmar arch

Right superficial palmar arch

Right digitals

Left common carotid

Left subclavian

ARCH OF AORTA

Left axillary

THORACIC AORTA

Left brachial

Diaphragm

Left gastric

Splenic

Left renal

Superior mesenteric

Left gonadal (testicular or ovarian)

Inferior mesenteric

Left common iliac

Left external iliac

Left internal iliac (hypogastric)

Left femoral

Left popliteal

Left anterior tibial

Left peroneal

Left posterior tibial

Left dorsalis pedis

Left dorsal arch

Figure 35-6 Principal human veins (anterior view). (From G. J. Tortora and S. R. Grabowski, *Principles of Anatomy and Physiology,* 7th ed., New York: HarperCollins, 1993.)

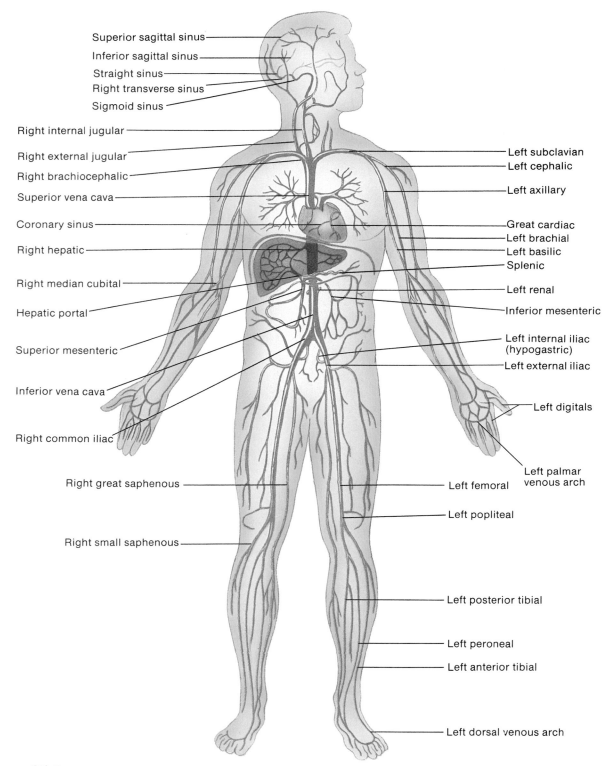

Figure 35-7 Human aorta and its principal branches (anterior view): major arteries in the human are shown. (From G. J. Tortora and S. R. Grabowski, *Principles of Anatomy and Physiology,* 7th ed., New York: HarperCollins, 1993.)

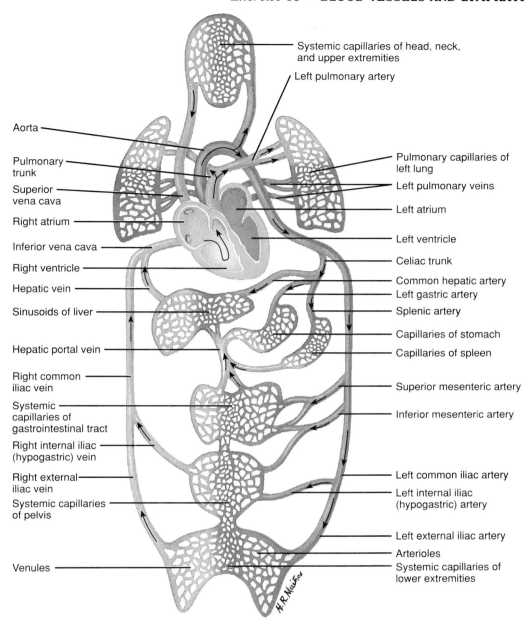

General plan of circulation

Figure 35-8 Human circulatory routes: Systemic circulation is indicated by heavy black arrows, pulmonary circulation by thin black arrows, and hepatic portal circulation by thin colored arrows. (From G. J. Tortora and S. R. Grabowski, *Principles of Anatomy and Physiology,* 7th ed., New York: HarperCollins, 1993.)

Figure 35-9 Human arteries of the right upper extremity (anterior view). (From G. J. Tortora and S. R. Grabowski, *Principles of Anatomy and Physiology*, 7th ed., New York: HarperCollins, 1993.)

Figure 35-10 Human veins of the right upper extremity (anterior view). (From G. J. Tortora and S. R. Grabowski, *Principles of Anatomy and Physiology,* 7th ed., New York: HarperCollins, 1993.)

L4

Right common iliac

Right external iliac

Abdominal aorta

Left common iliac

Right internal iliac (hypogastric)

Right femoral

Right popliteal

Right anterior tibial

Right posterior tibial

Right peroneal

Right dorsalis pedis

Right medial plantar

Right lateral plantar

Anterior view

Posterior view

Figure 35-11 Human arteries of the pelvis and right lower extremity. (From G. J. Tortora and S. R. Grabowski, *Principles of Anatomy and Physiology,* 7th ed., New York: HarperCollins, 1993.)

Inferior vena cava

L5

Right common iliac

Left common iliac

Right internal iliac (hypogastric)

Right external iliac

Right femoral

Right great saphenous

Right popliteal

Right small saphenous

Right small saphenous

Right peroneal

Right great saphenous

Right posterior tibial

Right dorsalis pedis

Right medial plantar

Right lateral plantar

Right dorsal venous arch

Right plantar arch

Anterior view

Posterior view

Figure 35-12 Human veins of the pelvis and right lower extremity. (From G. J. Tortora and S. R. Grabowski, *Principles of Anatomy and Physiology,* 7th ed., New York: HarperCollins, 1993.)

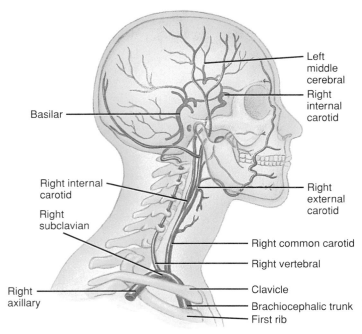

Figure 35-13 Human arteries of the head and neck (right lateral view). (From G. J. Tortora and S. R. Grabowski, *Principles of Anatomy and Physiology,* 7th ed., New York: HarperCollins, 1993.)

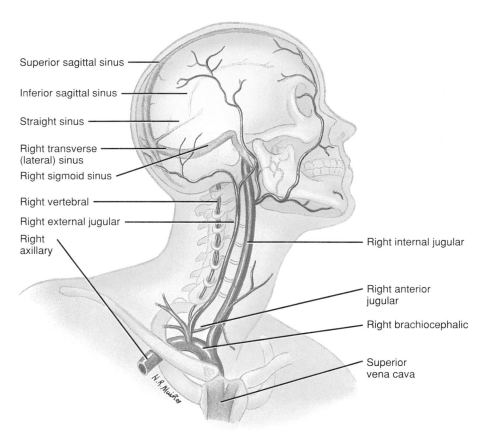

Figure 35-14 Human veins of the head and neck (right lateral view). (From G. J. Tortora and S. R. Grabowski, *Principles of Anatomy and Physiology,* 7th ed., New York: HarperCollins, 1993.)

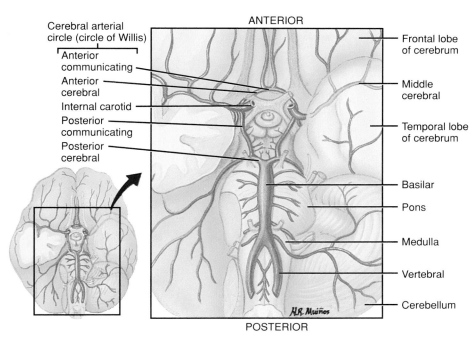

Figure 35-15 Human arteries of the base of the brain: note the arteries that compose the cerebral arterial circle (circle of Willis). (From G. J. Tortora and S. R. Grabowski, *Principles of Anatomy and Physiology,* 7th ed., New York: HarperCollins, 1993.)

from the surrounding tissue cells in all organs. It then passes through *lymph vessels* and *lymph nodes,* which filter the fluid until it reaches either the *left thoracic duct* or the *right lymphatic duct,* which flow into the venous system at or near the subclavian vein.

Path of Lymph in the Human

Lymph capillaries to

Lymph vessels to

Lymph nodes to one of the main ducts:
1. Left thoracic duct which collects lymph from
    1. Entire body posterior to the shoulder regions
    2. Left forelimb
    3. Left thorax, head, and neck
2. Right lymphatic duct, which collects lymph from
    1. right thorax, head, and neck
    2. right forelimb

Related Lymphatic Organs

These are large masses of lymphoid tissue that often have multiple functions.

Tonsils

1. *Palatine tonsils*—paired masses of lymphoid tissue covered with a mucous membrane which hang from the soft palate in the oropharynx
2. *Pharyngeal tonsil*—one large mass of lymphoid tissue located medially within the nasopharynx; prominent in children and also termed the adenoid
3. *Lingual tonsils*—aggregation of lymph follicles on the floor of the oropharynx at the base of the tongue

Spleen

In the human, the spleen is a large dark organ, located below the diaphragm on the left side of the body (Figure 35-19). The spleen has multiple functions:

1. It destroys red blood cells that are old (average length of life is 120 days) and worn out. It passes the breakdown products on to the liver, through the portal system, for further processing. Consult your textbook for more detail on the physiology of the spleen.

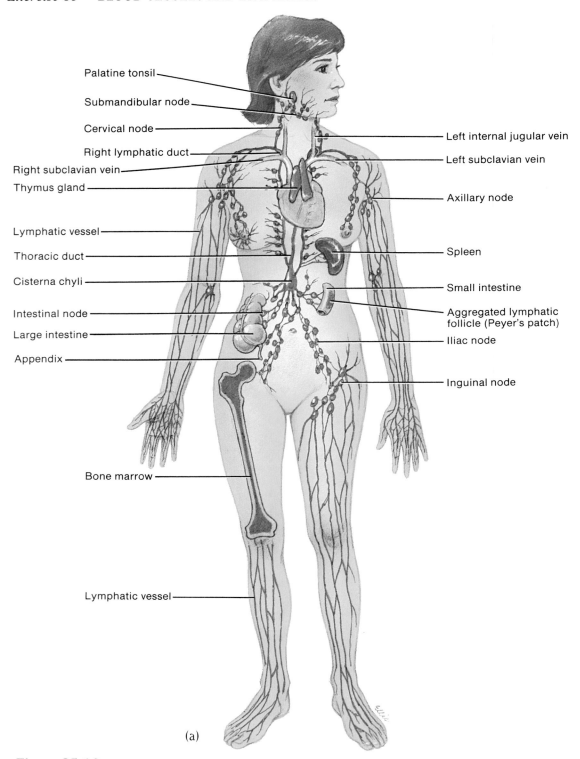

Palatine tonsil

Submandibular node

Cervical node

Right lymphatic duct

Right subclavian vein

Thymus gland

Lymphatic vessel

Thoracic duct

Cisterna chyli

Intestinal node

Large intestine

Appendix

Bone marrow

Lymphatic vessel

Left internal jugular vein

Left subclavian vein

Axillary node

Spleen

Small intestine

Aggregated lymphatic follicle (Peyer's patch)

Iliac node

Inguinal node

(a)

Figure 35-16 Human lymphatic system: (a) location of principal components of the lymphatic system; (b) details of the thoracic and right lymphatic ducts. (From G. J. Tortora and S. R. Grabowski, *Principles of Anatomy and Physiology*, 7th ed., New York: HarperCollins, 1993.)

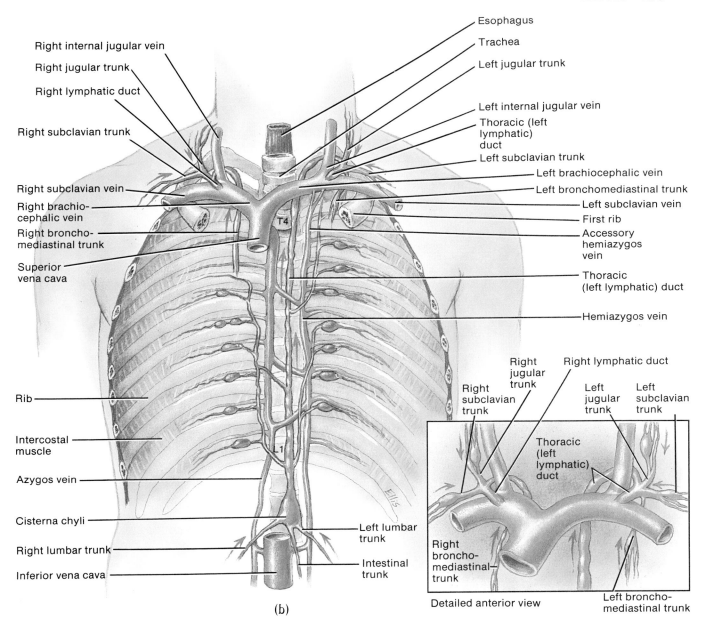

Esophagus

Trachea

Left jugular trunk

Left internal jugular vein

Thoracic (left lymphatic) duct

Left subclavian trunk

Left brachiocephalic vein

Left bronchomediastinal trunk

Left subclavian vein

First rib

Accessory hemiazygos vein

Thoracic (left lymphatic) duct

Hemiazygos vein

Right internal jugular vein

Right jugular trunk

Right lymphatic duct

Right subclavian trunk

Right subclavian vein

Right brachio-cephalic vein

Right broncho-mediastinal trunk

Superior vena cava

T4

Rib

Intercostal muscle

Azygos vein

Cisterna chyli

Right lumbar trunk

Inferior vena cava

L1

Left lumbar trunk

Intestinal trunk

(b)

Right jugular trunk

Right lymphatic duct

Right subclavian trunk

Left jugular trunk

Left subclavian trunk

Thoracic (left lymphatic) duct

Right broncho-mediastinal trunk

Left broncho-mediastinal trunk

Detailed anterior view

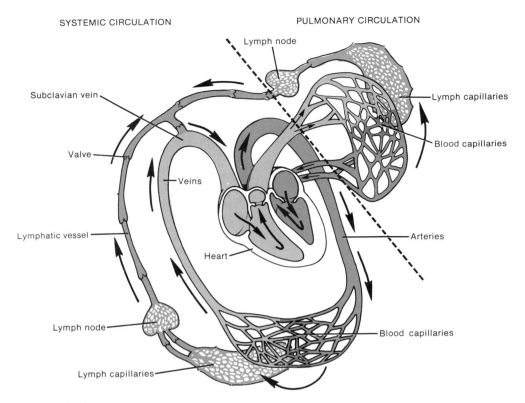

SYSTEMIC CIRCULATION   PULMONARY CIRCULATION

Lymph node

Subclavian vein

Lymph capillaries

Valve

Blood capillaries

Veins

Lymphatic vessel

Arteries

Heart

Lymph node

Blood capillaries

Lymph capillaries

Figure 35-17 Schematic representation of the relationship of the lymphatic system to the cardiovascular system, showing the flow pattern. (From G. J. Tortora, *Introduction to the Human Body*, 2nd ed., New York: HarperCollins, 1991.)

Esophagus (cut)

Spleen

Diaphragm

Splenic artery

Splenic vein

Tail of pancreas

Left renal vein

Left kidney

Inferior mesenteric vein

Colon

Figure 35-18 Gross structure of the spleen. The spleen in relation to abdominal viscera. (Courtesy of J. A. Gosling, P. F. Harris et al., *Atlas of Human Anatomy*, Gower Medical Publishing Ltd., 1985.)

2. It serves as a blood reservoir, which releases blood into the general circulation in times of stress.
3. It contains phagocytes, which cleanse the blood that passes through it.

Thymus Gland

This organ, described in Exercises 27 (human) and 33 (pig), is located anteroventral to the heart and to the major blood vessels in the mediastinal cavity. A very young animal has a large thymus gland. In humans, it reaches its maximum size in childhood, then begins to involute. It persists into old age, sometimes even increasing in size (hypertrophy). It produces lymphocytes (Exercise 27) and plays a role in one's immunologic responses throughout life.

BLOOD VESSELS AND LYMPHATICS

COMPLETION

1. Trace the pathway of a drop of blood from the head to the leg. Some of the steps are provided. You fill in the rest. Include the pulmonary circuit, cavities, and valves.

   a. <u>External jugular vein</u> _____

   b. _____

   c. _____

   d. _____

   e. _____

   f. <u>Right ventricle</u> _____

   g. _____

   h. _____

   i. <u>Lung capillaries</u> _____

   j. _____

   k. _____

   l. _____

   m. <u>Left ventricle</u> _____

   n. _____

   o. _____

   p. <u>Arch of aorta</u> _____

   q. _____

   r. _____

   s. _____

   t. _____

   u. <u>Femoral artery</u> _____

2. Trace the pathway of a drop of blood from the myocardium of the heart to the arm.

   a. <u>Coronary vein</u> _____

   b. _____

   c. _____

   d. _____

   e. <u>Right ventricle</u> _____

   f. _____

   g. _____

   h. _____

i. _____

j. Left atrium _____

k. _____

l. _____

m. _____

n. _____

o. _____

p. Brachiocephalic artery _____

q. _____

r. _____

s. _____

t. Radial artery _____

3. Fill in the following chart, specifying the relevant organ, artery, and vein for each row.

Supplied by	Organ	Drained by
Carotid Artery	Brain	Jugular vein (Example)
Hepatic artery	(1)	(2)
(3)	(4)	Great saphenous vein
(5)	Finger	(6)
(7)	Stomach	(8)
Inferior mesenteric artery	(9)	(10)
(11)	(12)	Renal vein
(13)	Spleen	(14)
Coronary artery	(15)	(16)
(17)	Adrenal gland	(18)
(19)	(20)	Axillary vein

4. Define the following terms:

a. Lymph _____

_____

b. Lymph vessel _____

_____

    c. Lymph node _____

    _____

    d. Right lymphatic duct _____

    _____

    e. Left thoracic duct _____

    _____

5. Make up your own tracing questions, one in the human, the other in the fetal-pig. Have your instructor check them.

    a. Trace a drop of blood in the human from the _____
    to the _____ .

    b. Trace a drop of blood in the fetal pig from the _____
    to the _____ .

6. Trace other products in the body:

    a. Food from the small intestine to the brain.

    b. Oxygen from the air to the lung to a cell in the toe.

    c. A metabolic waste product from a cell in the brain to the kidney.

Demonstration of the Normal Cardiac Cycle

1. List and describe the major components of an electrocardiogram (ECG)
2. Describe and perform a basic procedure for obtaining an ECG
3. Define *SA node, pacemaker, lead,* and *polygraph*

The heart is a specialized muscle that has a fundamental physiological property of rhythmicity. This feature accounts for the uniformly paced alternating periods of contraction and relaxation. In the human, the heartbeat is regulated by the *sinoatrial node,* known as the *SA node* or *pacemaker,* located in the wall of the right atrium.

In this exercise, we are concerned with the electrical activity associated with the contraction of heart muscle. Records of such electrical activity, referred to as *bioelectrical potentials,* may be made by suitable equipment and are of importance in the diagnosis and treatment of various disorders and disease states.

The *electrocardiograph* is the instrument that monitors electrical activity of the heart. It amplifies the impulses so that they are strong enough to activate a recording device. Electrodes attached to the surface of the body receive impulses from the heart. The amplified impulse causes a recording needle to move up and down on the recording paper. The needle may be a *pen,* which marks the paper with ink, or a *stylus,* which inscribes on special paper that does not require ink. The paper moves under the needle at a rate determined by the scale of the graph, and therefore the record appears as a series of waves (Figure 36-1). The record produced by the electrocardiograph is an *electrocardiogram,* abbre-

viated *ECG* or *EKG.* An ECG often is obtained with a *polygraph,* a recording device that operates on the principle just described but that is capable of monitoring impulses from several different physiological activities, in addition to that of the heart. Depending on the instrument used, the record is superimposed on either a standardized graph or other appropriate material. The vertical lines indicate time in seconds, and the horizontal lines amplitude in millivolts (mV). The space between two vertical lines represents 0.04 second; the space between two horizontal lines represents 0.1 mV.

The procedure used in taking an ECG involves the application of a jelly containing an electrolyte, which forms a low-resistance surface between the skin and the electrode. This material aids the conduction of electrical signals. Paper pads presaturated with an electrolyte solution also can be used in place of the jelly. The electrodes are held securely in place by means of straps. The particular arrangement of two electrodes is a *lead.* The position of the electrodes on the body, the manner in which they are wired to the recording system, and any and all extra movements of the individual undergoing testing will influence the final record.

The component waves of a normal ECG are shown in Figure 36-1. The major components include the *P wave,* which indicates the spread of excitation from the pacemaker over the atrial musculature; the *QRS wave (complex),* which represents the depolarization of the ventricular heart muscle that immediately precedes ventricular systole; and the *T wave,* which represents the repolarization of the ventricular muscle, which occurs just before ventricular diastole. A *U wave* is sometimes seen following the T wave and is thought to represent late repolarization of the ventricles. This exercise considers the equipment and procedure used in obtaining an ECG.

The direction of depolarization depends on the heart's orientation and on the particular instant the heart cycle is being recorded or examined. From a med-

Figure 36-1 The relationship of selected heart components to the conduction of electricity in the heart, as indicated by a normal electrocardiogram.

ical standpoint, it is sometimes desirable to record from more than two sites and even in two or three dimensions. With the particular technique of *vectorcardiography,* the continuous recording of the heart's electrical activity can be achieved. This type of continuous display of the depolarization–repolarization process as it moves over the heart may reveal important abnormalities. For example, an increase in the size (hypertroph) of one ventricle shifts the average or mean direction of depolarization during the cardiac cycle. This change occurs because it takes longer for the ventricle to depolarize and may be the result of a narrowing of the aortic semilunar valve.

In situations where there is a narrowing of the pulmonary semilunar valve or congenital defects such as the tetralogy of Fallot, a shift of the depolarization toward a hypertrophied right ventricle would occur.

Depolarization-pattern abnormalities also occur with a heart block, which is a delay of the electrical impulse or even a complete blockage of an impulse in its progress over the conduction system, or as a result of a heart attack.

This exercise also considers the determination of the mean electrical axis (direction) of depolarization of the cardiac cycle. This determination can be made by recording the QRS complex voltages from two different dimensions, using two different leads—namely, Leads I and III. Lead I provides a horizontal axis of observation (from left arm to right arm), while Lead III has an axis of about 120° (from left arm to left leg). The recordings from Leads I and III should show the normal mean

electrical axis of the ventricles to be about 59° (Figure 36-2a).

Materials

The following materials should be provided for class use or demonstration purposes:

1. Electrocardiograph systems and appropriate cables
2. ECG electrodes (metal or disposable paper pads)
3. Tubes of electrode jelly
4. Disposable abrasive pads or brushes
5. Commercially packaged sterile alcohol prep-pads
6. Sterile gauze pads (4 × 4 inch)
7. Paper towels
8. Examples of normal and abnormal electrocardiograms

Procedure 1: The Normal ECG

The instructor will demonstrate the principles, calibration, and correct use of the electrocardiograph system. This procedure should be performed on a group basis.

1. Before connecting the electrodes and taking an electrocardiogram, check the following items.
   a. Sufficient paper and recording ink (if needed) are available.
   b. The recording pen or device is properly centered.
   c. The recording speed has been set for 25 mm/second.
   d. The preamplifier has been set at 10 mm/mV.

(a)

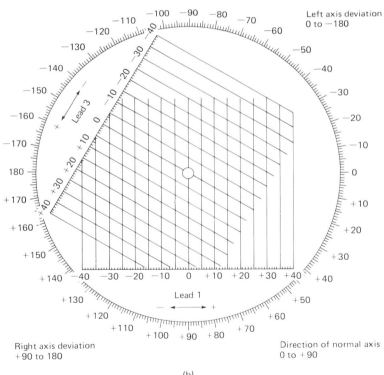

(b)

Figure 36-2 Grids used for the determination of the mean electrical axis of the ventricles: (a) grid showing an example of how to determine the mean electrical axis of the ventricles (results showing a value of 59°); (b) a grid to be used by each student, for Procedure 2.

2. Ask your laboratory partner to remove any materials that could interfere with taking a recording (remove wristwatch or bracelet, roll up sleeves, roll down stockings, and so forth). Have your partner lie down on a table or cot in a comfortable and relaxed position.

3. Disinfect both wrists and the area on the inside of the legs just above both ankles with sterile alcohol prep-pads.

4. Apply a small amount of electrode jelly to the skin where the electrodes will be placed and to the surface of the electrodes.

5. The following three leads will be used in this exercise.
   a. *Lead I*—electrodes attached to the left and right wrists
   b. *Lead II*—electrodes attached to the right wrist and left leg
   c. *Lead III*—electrodes attached to the left wrist and left leg

6. Obtain a record with each lead. Usually, six ECG complexes (six complete cycles of P wave, QRS wave, and T wave) for each lead will be sufficient. Be sure to mark the record with your partner's name.

7. Have your partner exercise vigorously for 3–5 minutes. Jogging, running in place, or going up and down stairs are suitable forms of exercise. Again, obtain a record with each lead.

8. After the recording is complete, remove the leads, and clean the electrodes. The subject should remove the electrode paste with gauze or paper towels.

9. Obtain representative segments of the ECG. Attach and label them in the results and observations section.

10. Change places with your partner, and repeat Steps 1–9.

Procedure 2: Mean Electrical Axis of the Ventricles

This procedure should be performed on a group basis, after the instructor has demonstrated the correct use of the electrocardiograph system.

1. Repeat Steps 1–4 of Procedure 1 in this exercise.
2. Leads I and III, with a sensitivity setting of 1, will be used in this procedure.
3. Obtain a record with each label. Be certain to mark the record with your partner's name.
4. With the Lead I record, obtain the following information items, and enter your findings and calculations in the appropriate spaces of the results and observations section.
   a. Find the QRS complex, and count the number

of small boxes (millimeters) that it projects above the baseline.
   b. Find the Q and S waves, and count the number of millimeters each projects below the top of the baseline. Add the sum of the two downward deflections.
   c. Algebraically add the two values obtained for the upward and downward deflections.

5. Repeat Step 4 with Lead III, and enter your findings and calculations in the appropriate spaces of the results and observations section.

6. Find and mark the number that corresponds to the total value obtained with Lead I on the horizontal axis in Figure 36-2b. With the aid of a ruler, make and extend a vertical line from your mark to the top of the figure.

7. Find and mark the number that corresponds to the total value obtained with Lead III on the slanted axis located on the left-hand side of Figure 36-2b. With the aid of a ruler, make and extend a straight line down to the right-hand side of this figure.

8. With the aid of a ruler, draw an arrow line from the centrally located circle in Figure 36-2b, which intersects the lines made earlier on the figure. Extend the arrow to the edge of the outer circle, and record the value that represents the mean electrical axis of the ventricles. Enter your finding in the results and observations section.

9. An example of this procedure is shown in Figure 36-2a. The values for Leads I and III were 10 and 10, respectively.

RESULTS AND OBSERVATIONS

The Normal ECG

1. Attach labeled segment of electrocardiograms made during relaxation period.
   a. Lead I

   b. Lead II

c. Lead III

2. Attach labeled segment of electrocardiograms made immediately after vigorous exercise.
   a. Lead I

   b. Lead II

   c. Lead III

3. Are there any noticeable differences in the records taken with the three leads? If so, how would you explain the differences? _____

_____

_____

The Mean Electrical Axis of the Ventricles

1. Enter your findings for Lead I in the spaces provided.
   a. Number of millimeters that the QRS complex projects *above* the baseline; upward deflections _____
   b. Sum in millimeters of the Q and S waves that project *below* the baseline; downward deflections _____
   c. The algebraic sum of the upward and downward deflections (if the resulting total is a negative number, enter it as such) _____

2. Enter your findings for Lead III in the spaces provided.
   a. Number of millimeters that the QRS complex projects *above* the baseline; upward deflections _____
   b. The sum in millimeters of the Q and S waves that project *below* the baseline; downward deflections _____
   c. The algebraic sum of the upward and downward deflections (if the result is a negative number, enter it as such)

   Total _____

3. Mean electrical axis of the ventricles was found to be _____

4. List four heart conditions that could produce depolarization-pattern abnormalities.

   a. _____

   b. _____

   c. _____

   d. _____

LABORATORY REVIEW *36*

DEMONSTRATION OF THE NORMAL CARDIAC CYCLE

COMPLETION, DEFINITION, IDENTIFICATION, AND INTERPRETATION

1. What is an electrocardiogram? _____
_____

2. What is a lead? _____
_____

3. Identify the specific components of the electrocardiogram shown in Figure Q36-1.

    a. _____

    b. _____

    c. _____

    d. _____

    e. _____

Figure Q36-1 ECG components.

4. With the aid of your textbook or other reference, indicate what each of the following components of an electrocardiogram represents.

    a. P wave _____

    b. QRS complex _____

    c. T wave _____

    d. Q–T interval _____

5. What is a normal cardiac cycle? _____
_____

6. With the aid of your textbook or other reference, determine the effect of the following chemical or physical factors on rate and strength of the heartbeat.

    a. Temperature _____

    b. Acetylcholine _____

    c. Epinephrine _____

    d. Digitalis _____

7. Define or explain the following:

   a. Fibrillation _____

   _____

   b. Heart block _____

   _____

8. Distinguish between diastole and systole. _____

   _____

9. a. What is a myocardial infarction? _____

   _____

   _____

   b. Can a myocardial infarction be detected in an ECG? If so, how is it indicated?

   _____

   _____

10. a. Would you consider the ECG in Figure Q36-2 to be normal? _____

   b. If not, what features are abnormal? _____

   _____

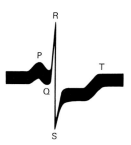

Figure Q36-2

11. Using the following recordings from Leads I and III of an electrocardiogram and the grid chart provided in Figure Q36-3, determine the electrical axis of the ventricles. Show your calculations.

Figure Q36-3

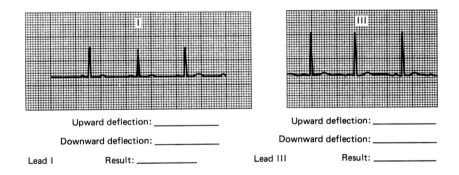

Upward deflection: _____

Downward deflection: _____

Lead I   Result: _____

Upward deflection: _____

Downward deflection: _____

Lead III   Result: _____

12. Considering that the normal mean electrical axis of the ventricles is 59°, was the value obtained in Question 11 normal? _____

ANSWERS

1. —————
2. —————
3. —————
4. —————
5. —————
6. —————
7. —————
8. —————
9. —————
10. —————

MULTIPLE CHOICE

Indicate the correct answers in the spaces provided.

1. In the human, which of the following regulates the heartbeat?
   a. T wave
   b. right atrium
   c. left ventricle
   d. QRS complex
   e. sinoatrial node

2. Where is the pacemaker of the heart located?
   a. T wave
   b. right atrium
   c. left ventricle
   d. QRS complex
   e. sinoatrial node

3. What is the record produced by an electrocardiograph?
   a. pulse
   b. lead
   c. bioelectrical potential
   d. electrocardiogram
   e. QRS complex

4. Which of the following is the particular arrangement of two electrodes in an EKG procedure?
   a. pulse
   b. lead
   c. bioelectrical potential
   d. electrocardiogram
   e. QRS complex

5. Which of the following will influence the EKG?
   a. extra movements by the individual being tested
   b. the position of electrodes on the body
   c. the way in which leads are wired to the recording system
   d. all of these
   e. choices *a* and *c* only

6. Which of the following component waves of a normal EKG indicates the spread of excitation from the pacemaker over the atrial musculature?
   a. U wave
   b. QRS wave (complex)
   c. T wave
   d. P wave
   e. R wave

7. Which of the following component waves of a normal EKG represents the depolarization of the ventricular heart that immediately precedes ventricular systole?
   a. U wave
   b. QRS wave (complex)
   c. T wave
   d. P wave
   e. R wave

8. Which of the following in a normal EKG represents the repolarization of the ventricular muscle that occurs just before ventricular diastole?
   a. U wave
   b. QRS wave (complex)
   c. T wave
   d. P wave
   e. R wave

9. When is a U wave seen?
   a. at same time as a T wave
   b. following a T wave
   c. at same time as a P wave
   d. following a P wave
   e. at same time as the QRS complex

10. With which of the following conditions can depolarization-pattern abnormalities occur?
    a. heart blocks
    b. tetralogy of Fallot
    c. narrowing of the pulmonary semilunar valve
    d. heart attack
    e. all of these

Blood Pressure and Pulse Determination

After completing this exercise, you should be able to

1. Identify the parts of instruments used to determine blood pressure and pulse
2. Perform procedures to determine blood pressure and pulse
3. List three factors that affect heart rate
4. Define Korotkoff sounds, pulse rate, bradycardia, and tachycardia

When the ventricles of the heart contract, blood is pumped under pressure into the pulmonary artery and aorta. This pressure exerts a force against the walls of the vessels through which blood flows. Within any blood vessel, artery, vein, or capillary, the pressure depends on the volume of blood present, the size of the vessel, the rate of blood flow, and any resistance to blood flow. Blood pressure is the basis for maintaining a functional and stable movement of blood from the heart, to the various regions of the body, and back into the heart. Because of its importance, a blood pressure determination is a major part of any physical examination. In a medical context, *blood pressure* refers to the pressure of the arterial system.

The routine method for measuring arterial blood pressure is indirect and usually employs a *sphygmomanometer*. This device basically consists of an inflatable cuff, an inflation pumping device, a pressure gauge, and an exhaust control valve.

A *stethoscope* is used to amplify the sounds within the artery used for the determination. Usually, the brachial artery is chosen, for convenience. In the procedure, the pressure cuff is first inflated around the upper arm, to stop the flow of blood temporarily (Figure

37-1). This collapses the artery. When a stethoscope is placed above the brachial artery, no sound can be heard (Figure 37-2a). The pressure in the cuff is gradually reduced until the blood flows through the vessel with each systole of the heart. When the cuff pressure is equal to the systolic pressure, the first *Korotkoff sound* is heard (Figure 37-2b). Korotkoff sounds are produced by the turbulent flow of blood through the partially constricted brachial artery. The reading on the pressure gauge when the first Korotkoff sound is detected corresponds to the individual's *systolic blood pressure*. The artery will now remain open during the systole of the heart. The sounds continue with each heart cycle (Figure 37-2c). Reducing the cuff pressure further enables the artery to remain open even during diastole. Blood flows freely now, and the thumping sound disappears. The reading at which the sound disappears corresponds to the *diastolic blood pressure* (Figure 37-2d).

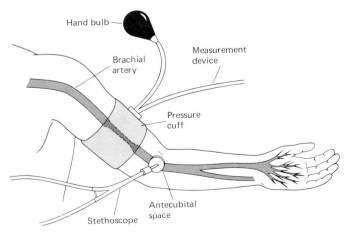

Figure 37-1 Measuring blood pressure: position of a sphygmomanometer cuff and stethoscope during measurements using the brachial artery.

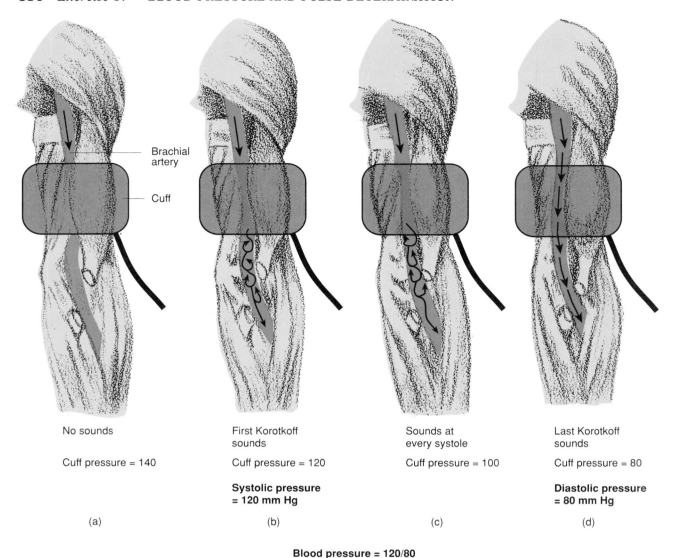

No sounds	First Korotkoff sounds
Cuff pressure = 140	Cuff pressure = 120
	Systolic pressure = 120 mm Hg
(a)	(b)

Sounds at every systole

Cuff pressure = 100

(c)

Last Korotkoff sounds

Cuff pressure = 80

Diastolic pressure = 80 mm Hg

(d)

Blood pressure = 120/80

Figure 37-2 Diagrammatic representation of the appearance of the brachial artery at different pressures in the cuff. Korotkoff sounds are produced by the turbulent flow of blood through the partially constricted brachial artery. These sounds will continue to be heard at every systole as long as the cuff pressure remains greater than the diastolic pressure.

Blood pressure usually is measured in millimeters (mm) of mercury (Hg). The arterial blood pressure is expressed as the systolic blood pressure value over the diastolic pressure value. Table 37-1 lists values for males and females. The difference between the systolic and diastolic pressures is known as the *pulse* pressure. Various factors influence arterial blood pressure, including age, emotional state, sex, weight, and strenuous exercise.

Today, additional methods have been developed for arterial pressure determinations. Electronic devices incorporating both sound and light indicators are available for rapid determinations (see Figure 37-3).

The rate and characteristics of the arterial pulse are extremely important because it is indicative of heart action. This pulsation of the blood through the arteries is in response to the contraction of the heart. The number of pulsations per minute is the *pulse rate* or *heart rate*. It can be felt in any artery that lies near the surface of the body or over a bone or other firm tissue. One of the sites commonly used to measure pulse rate is the undersurface of the wrist behind the thumb, where the radial artery pulsates against the radius bone. Individuals with a rapid heart rate are said to be exhibiting *tachycardia,* while those persons with a very slow rate exhibit *bradycardia.*

This exercise considers several approaches to determination of blood pressure and pulse rate.

Table 37-1 Selected Blood Pressure Values, Normal Range

Sex	Age	Systolic	Diastolic
Females	16–24	100–130	60–85
	25–29	102–130	60–86
	30–34	102–135	60–88
	35–39	105–140	65–90
	40–44	105–150	65–92
	45–49	105–151	65–96
	50–54	110–165	70–100
	55–59	110–170	70–100
	60–64	115–175	70–100
Males	16–18	105–135	60–86
	19	105–140	60–88
	20–24	105–140	62–88
	25–29	108–140	65–90
	30–39	110–145	68–92
	40–44	110–150	70–94
	45–49	110–155	70–96
	50–54	115–160	70–98
	55–59	115–165	70–98
	60–64	115–170	70–100

BLOOD PRESSURE DETERMINATIONS

Materials

The following materials should be provided for class use:

1. Stethoscope
2. Standard sphygmomanometer
3. Electronic sphygmomanometer
4. Astropulse 10 blood pressure monitor
5. Digimax 5000 electronic blood pressure monitor
6. Disinfectant
7. Sterile gauze pads or cotton

Procedure 1: Examination of Blood Pressure Determination Devices

This exercise is to be performed by students individually. Knowing the parts of equipment and their normal positions is extremely important if equipment is to be used effectively.

1. Examine the various devices provided for the measurement of blood pressure and pulse determinations.
2. With the aid of Figure 37-3, identify the parts and note the positions of these devices.

Procedure 2: Auscultation with Standard Sphygmomanometer

This procedure is to be performed by students in pairs.

1. Ask your laboratory partner to sit comfortably, and place the left forearm, palm up, on a table. The upper arm should be at heart level.
2. With Figure 37-3, identify the parts and note the positions of these devices.
3. Examine the *sphygmomanometer cuff* to see whether the area to be placed against the brachial artery is marked. If the cuff is not marked, hold the inflatable portion of the cuff on the inside of the upper arm, while you wrap the long end of the cuff around the arm. The cuff should fit snugly. Inflate the cuff until it does not fall. Ask the instructor for assistance if you have any problems with the placement of the cuff. Place the measurement gauge or mercury column in a position close enough to be read easily.
4. Insert the earpieces of the stethoscope into your ears, and place the stethoscope bell or diaphragm above the brachial artery at the antecubital space (see Figure 37-1). What do you hear? (Have a pencil and paper handy.)
5. Open the valve on the rubber bulb, and inflate the cuff to 150 mm Hg. (Do not go higher without the recommendation of your instructor.) Close the valve. (*Note:* You should work quickly from this point.)
6. Open the valve on the rubber bulb slightly and listen carefully for pressure sounds. As the cuff pressure is decreased, vibrations of the arterial wall are caused by the spurting of blood through the narrowed artery under the cuff into the fully opened artery below the cuff. These vibrations are *Korotkoff* sounds (Figure 37-2). The pressure at which these sounds are detected is the systolic pressure. Record this pressure in the results and observations section (or make note, for later reference).
7. Slowly open the valve farther while you listen for the total disappearance of all sounds. The point at which no sound can be detected is the *diastolic pressure*. Record this value in the results and observations section (or make yourself a note).
8. Determine the systolic and diastolic pressures with your partner standing and after running up and down several flights of stairs. Record all of your findings in the results and observations section.
9. Calculate the pulse pressure for each situation, using the following formula:

$$\text{Pulse pressure} = \text{systolic pressure} - \text{diastolic pressure}$$

Bowles chestpiece

(a)

Manometer gauge

BLADDER CENTER

PRE-GAGED made in
ADULT CUFF U.S.A.
the hoder corp. Princeton, N.J.
LIC. under PyMcN Corp Patent
ABM CATERIZED

(b)

Figure 37-3 Instruments used for blood-pressure measurement: (a) combination stethoscope with a *Bowles chestpiece;* (b) *Hare Deluxe Sphygmomanometer,* which offers several convenient features, including a cuff that indicates the right size for different individuals and a clearly marked bladder center, which shows where to position the cuff over the artery; (c) *electronic sphygmomanometer*—solid state electronic device that can be used to read systolic and diastolic blood pressure using both a flashing light and a sound, and which includes a highly sensitive and functional microphone in the cuff, replacing the use of the stethoscope (courtesy of Dyna-Med, Inc., Carlsbad, CA); (d) *Digimax 500 electronic blood pressure monitor,* which can provide a simultaneous digital readout of diastolic and systolic pressure and the pulse rate; (e) gauge of the *Astropulse 10 blood pressure monitor,* which uses electronic sound and light indicators, a standard needle indicator, a microphone in the cuff, eliminating the need for a stethoscope (courtesy of Dyna-Med, Inc., Carlsbad, CA).

Enter your findings, and answer the questions in the results and observations section.

10. Repeat the entire procedure using the right arm. Record your findings in the results and observations section.

11. Change places with your laboratory partner, wipe the stethoscope with disinfectant, and repeat Steps 1–10.

Procedure 3: Astropulse 10 Blood Pressure Monitor

This procedure is to be performed by students in pairs.

1. Check to see whether the unit is ready to use. The battery should be in place and the microphone jack plugged into the receptacle at the bottom of the unit.

Sound

Light

(c)

(d)

(e)

2. Ask your laboratory partner to sit comfortably with the left palm facing up and the upper arm in line with the heart.
3. Be certain that the end of the pressure cuff is inserted through the D ring, and slip the cuff over the left arm (Figure 37-4).
4. Turn the cuff so that its blue dot is on the inside of the arm just above the elbow. Pull the end of

the cuff so that it fits firmly on the arm, and wrap any extra cuff under and around the arm. Press the Velcro tape together to secure the cuff.
5. Place or hold the pressure gauge so that it can be easily read. Now you are ready to determine the systolic and diastolic pressures.
6. With the power off, close the air-flow valve on the bulb by turning it in a clockwise direction. Inflate

Figure 37-4 The pressure cuff and its application: Various parts of the pressure cuff, the air flow system, and the positioning of the arm are shown; the entire procedure can be performed by one individual.

the cuff quickly by repeatedly squeezing the bulb until the needle reaches the point marked 150 mm on the gauge. (In some cases, it may be necessary to inflate the cuff beyond 150 mm. Consult with your instructor.)

7. Turn on the power switch. You will hear a beep and see a light flash on the gauge as the power enters the unit. (The beep and light will not occur again until the pressure is reduced.)

8. Open the valve slightly in a counterclockwise direction so that the pressure falls at the rate of 2–4 mm/second (one mark on the gauge equals 2 mm).

9. Continue to deflate the cuff at the rate of 2–4 mm/second. When the pressure in the cuff is equal to the *systolic pressure* in the artery, the beep will sound, and the light will flash. Note the reading on the gauge.

10. Allow the pressure to continue to drop at the same rate by opening the valve. The beeping and flashing will continue. Watch the gauge. At the moment the beeping and flashing stop, take a reading. This is the *diastolic pressure value.*

11. Repeat Steps 1–10 until you can carry out the procedure easily.

12. Obtain readings for systolic and diastolic pressures in sitting and standing positions and after exercise. Record the values, and answer the questions in the results and observations section.

13. Change places with your partner, and repeat Steps 1–12.

14. Experiment and carry out the procedure on yourself.

PULSE-RATE DETERMINATIONS

Materials

The following materials should be provided for class use:

1. Electronic pulse monitor
2. Paper towels or other material to wipe fingers
3. Stopwatch

Procedure 1: Examination of Pulse-Rate Determining Device

This procedure is to be performed by students individually.

Figure 37-5 *Electronic pulse monitor:* The pulse rate of an individual can be determined simply by inserting the index finger into the sensitive photoelectric sensor(s); the pulse rate registers on the scale dial. (Courtesy of Dyna-Med, Inc., Carlsbad, CA.)

1. Examine the electronic pulse monitor provided.
2. With the aid of Figure 37-5, familiarize yourself with the parts of this device.

Procedure 2: Radial Pulse Determinations

This procedure is to be performed by students in pairs.

1. Ask your laboratory partner to sit comfortably, and hold the forearm of one hand with palm up toward you.
2. Take the wrist of this hand, and lightly place your index and middle fingers on the surface facing you in the area behind your thumb (Figure 37-6). Your thumb should be holding the underside of the wrist.
3. Feel for the beating sensation in this area. Do not use your thumb because it has a pulse of its own.
4. Count the number of beats that occur per 15 seconds. Multiply the beats by 4, to obtain the pulse rate for a minute.
5. Repeat this procedure three times, and obtain an average value. Record your findings in the results and observations section.
6. Take the pulse rate of your partner in the standing position and after exercise. Running up and down a flight of stairs three times or running rapidly in place for 2 minutes would be sufficient exercise.
7. Enter your findings, and answer the questions in the results and observations section.
8. Change places with your laboratory partner, and repeat Steps 1–7.

Procedure 3: Electronic Pulse Monitoring

This procedure is to be performed by students in pairs.

1. Turn on the electronic pulse device.
2. Ask your laboratory partner to sit comfortably next to the measuring device.
3. Insert your partner's index finger into the photoelectric sensor, and obtain a reading.
4. Repeat this step. Was the reading the same? Enter your findings in the results and observations section.
5. Obtain pulse rates with your partner standing and after exercise. Enter your findings in the results and observations section.
6. Change places with your partner, and repeat Steps 1–5.
7. Turn off the electronic device after you have completed this procedure.

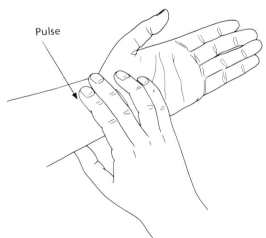

Figure 37-6 Position of the fingers for radial pulse determination.

Procedure 4: Carotid Pulse Determinations

This procedure is to be performed by students in pairs.

1. Ask your partner to run in place for 2 minutes and then sit down.
2. Place your fingers (no thumb) on either side of the partner's larynx. Gently press inward and toward the back of the neck. Continue to apply gentle pressure until a pulse is felt.
3. Once a pulse region is located, practice the procedure until only two fingers are necessary to make a pulse determination. Record your findings, and answer the questions in the results and observations section.
4. Repeat Step 1, and determine the radial pulse rate (refer to Procedure 2). Enter your findings in the results and observations section.
5. Change places with your partner, and repeat Steps 1–4.

RESULTS AND OBSERVATIONS

Blood Pressure Determinations

1. Enter the blood pressure readings determined by auscultation in Table 37-2. Record your own and your partner's systolic (S) and diastolic (D) pressure. Space is provided for measurements of other students.
2. Enter pulse-pressure results in Table 37-3.
3. Compare your pressures with those listed for your age group in Table 37-1. Are your values similar to those listed? _____
4. Are the values for your partner similar to those listed? _____
5. Did this procedure cause you any problems? Explain. _____
_____
6. Enter the blood pressure readings determined by the Astropulse 10 blood pressure monitor in Table 37-4.
7. Was the electronic device easier to use than the standard sphygmomanometer? Explain. _____
_____
_____
8. Were the determinations comparable to those obtained by other methods used? _____

Table 37-2

	Seated				Standing				After Exercise			
	Left Arm		Right Arm		Left Arm		Right Arm		Left Arm		Right Arm	
Student	*S*	*D*	*S*	*D*	*S*	*D*	*S*	*D*	*S*	*D*	*S*	*D*

Table 37-3

	Seated		Standing		After Exercise	
Student	Left Arm	Right Arm	Left Arm	Right Arm	Left Arm	Right Arm

9. Was your blood-pressure value within the normal range? _____

10. Was your laboratory partner's value within the normal range? _____

Pulse-Rate Determinations

1. Enter your findings obtained with the radial pulse determinations (R) and with the electronic monitor (E) in Table 37-5.

2. If an electronic monitoring device was provided, were the determinations comparable to those obtained by other methods used? _____

3. Enter your findings comparing radial and carotid pulse determinations in Table 37-6.

Table 37-4

Student	Seated		Standing		After Exercise	
	Systolic	Diastolic	Systolic	Diastolic	Systolic	Diastolic

Table 37-5

Student	Seated		Standing		After Exercise	
	R	E	R	E	R	E

Table 37-6

Student	Radial Pulse Determination	Carotid Pulse Determination

LABORATORY REVIEW *37*

BLOOD PRESSURE AND PULSE DETERMINATION

COMPLETION,
IDENTIFICATION, DEFINITION,
AND INTERPRETATION

1. List three factors that affect heart rate.

   a. _____

   b. _____

   c. _____

2. a. What is pulse rate? _____

   _____

   b. What is the average value for an adult? _____

   _____

3. What sound is heard with a stethoscope when the brachial artery is collapsed?

   _____

4. List the major conditions that should be met while taking an individual's blood pressure. _____

   _____

   _____

5. Identify the device shown in Figure Q37-1 and identify the specific labeled parts.

   a. Device _____

   b. _____

   c. _____

Figure Q37-1 (Courtesy of Dyna-Med, Inc., Carlsbad, CA.)

ANSWERS

6. Which artery is usually used to take an individual's pulse rate? _____

7. Complete the following formula.

   Pulse rate (beats/minute) × stroke volume (ml) = _____ (liters/minute)

8. What are Korotkoff sounds? _____

   _____

9. Define the following:

   a. Bradycardia _____

   b. Tachycardia _____

10. With the aid of your textbook and other references, indicate the effects of the following conditions on arterial blood pressure:

   a. Vasoconstriction _____

   b. Increased blood volume _____

   _____

   c. Arteriosclerosis _____

   _____

   d. Increased aldosterone concentration _____

   _____

11. What is hypertension? _____

   _____

MULTIPLE CHOICE

Insert the correct answers in the spaces provided.

1. In a medical context, to which system does *blood pressure* refer?
   a. venous
   b. arterial
   c. capillary
   d. diastolic
   e. systolic

2. Through which of the following does the turbulent flow of blood produce Korotkoff sounds?
   a. aorta
   b. subclavian artery
   c. pulmonary arteries
   d. brachial artery
   e. pulmonary veins

3. Which of the two blood-pressure measurements is the reading on a blood pressure gauge measuring when the first Korotkoff sound is detected?
   a. systolic
   b. diastolic
   c. brachial
   d. venous
   e. arterial

1. _____

2. _____

3. _____

4. _____

5. _____

6. _____

7. _____

8. _____

4. Which of the two blood-pressure measurements is the reading on a blood pressure gauge measuring when the Korotkoff sound disappears?
   a. systolic
   b. diastolic
   c. brachial
   d. venous
   e. arterial

5. Which is correct in an individual's blood pressure result?
   a. the systolic pressure is listed above or before the diastolic
   b. the diastolic pressure is listed above or before a systolic

6. Is the systolic pressure greater or less than the diastolic pressure when the Korotkoff sounds continue to be heard at every systole?
   a. less
   b. greater

7. What is the difference between the systolic and the diastolic pressure called?
   a. Korotkoff pressure
   b. pulse pressure
   c. heart pressure
   d. arterial pressure
   e. choices *a* and *b* only

8. What is the term used for individuals exhibiting a rapid heart rate?
   a. tachyardia
   b. bradycardia
   c. neutrocardia
   d. diastole
   e. systole

Blood and the Immune System

*T*he circulatory system, which consists of the blood, heart, and blood vessels, provides a link between the body's internal components and its external environment. The exercises in this section deal specifically with the blood and its components and demonstrate their involvement in providing vital support for cellular activities and the maintenance and protection of a favorable cellular environment. In addition, attention is given to conditions that interfere with the production of red blood cells or that lead to an increased level of red-blood-cell destruction.

The two major components of blood are *formed elements*—including erythrocytes (red blood cells), leukocytes (white blood cells), and platelets—and *plasma,* a pale yellow fluid. Plasma contains about 90 percent water, with the remaining 10 percent consisting of carbohydrates, hormones, lipids, proteins (albumin, enzymes, fibrinogen, immunoglobulins or antibodies, and other globulins), salts, and vitamins. Several laboratory procedures are used to determine the concentrations of blood components (Figure XI-1). Because the results of such tests are

(a)

(b)

Figure XI-1 Normal and abnormal blood cells: (a) This electron micrograph shows an activated plasma cell; this cell type is the immunoglobulin producer of the body; note the presence of the large quantity of endoplasmic reticulum (ER) within the cell (from D. Douer, A. Weinberger, M. Djaldetti, J. Asherov, I. Pick, and J. Pinkhas. *Acta Hematologica,* 62: 81–85, 1979). (b) This scanning micrograph shows a leukemia cell originating from lymphoid tissue.

of diagnostic importance, instrumentation improvements have been made so that procedures can be performed more rapidly and accurately.

Several exercises consider procedures that are used for demonstrating the presence and functions of the components of blood. In addition, attention is given to conditions that interfere with the production of red blood cells or that lead to an increased level of red-blood-cell destruction.

GENERAL REFERENCES

Hillman, R. S., and C. A. Finch. *Red Cell Manual,* 6th ed. Philadelphia: F. A. Davis Company, 1992.
Harmening, D. M. (Ed.). *Clinical Hematology and Fundamentals of Hemostasis,* 2nd ed. Philadelphia: F. A. Davis Company, 1991.
Koepke, J. A. (Ed.). *Practical Laboratory Hematology.* New York: Churchill Livingston, 1991.
Tortora, G. J., and S. R. Grabowski. *Principles of Anatomy and Physiology,* 7th ed. New York: HarperCollins, 1993.

SPECIFIC REFERENCES

Dvorak, A. M., and H. F. Dvorak. "The Basophil." *Archives of Pathology and Laboratory Medicine,* 103: 551–557, 1979.
Evans, V. J. "Platelet Morphology and the Blood Smear." *Journal of Medical Technology,* 1: 689–695, 1984.
Golde, D. W., and J. C. Grasson. "Hormones That Stimulate the Growth of Blood Cells." *Scientific American,* 259: 62–70, 1988.
Metcalf, D. "Control of Granulocytes and Macrophages: Molecular, Cellular, and Clinical Aspects." *Science,* 254: 529–533, 1991.
Morse, E. E., and B. Jacobs. "Blood Donation and Its Aftereffects." *American Scientist,* 73: 68–69, 1985.
Ranney, H. M. "The Spectrum of Sickle Cell Disease." *Hospital Practice,* 27: 133–163, 1992.
Roberts, H. R., and J. N. Lozier. "New Perspectives on the Coagulation Cascade." *Hospital Practice,* 27: 97–112, 1992.
Von Boehmer, H., and P. Kisielow. "How the Immune System Learns About Self." *Scientific American,* 265: 74–81, 1991.

Blood Smear
Preparation and
Differential Staining

After completing this exercise, you should be able to

1. Prepare and stain blood smears
2. Distinguish among blood cells
3. Perform a differential blood count
4. Recognize and identify certain abnormalities and disease agents in blood smears

Blood cells must go through a development cycle in the same way that a human matures during development. In healthy individuals, only mature cells are seen, upon examination of blood specimens. Immature or abnormal cells can be observed in various disease states. Thus, the examination of a patient's blood may be extremely important in either the detection of abnormalities or the diagnosis of disease.

An essential part of a hematological examination is a properly prepared *blood smear* (or *spread*). Such a smear is made by obtaining a drop of blood from the patient, placing it on a slide, and then spreading it, with the aid of a second slide, into a thin layer. After the preparation dries, any one of several staining procedures, such as Wright's or Giemsa, can be applied.

The examination of blood for abnormalities and the determination of the percentage of various types of cells may be diagnostic in themselves (Figure 38-1), or they may serve as a basis for additional tests. For example, the presence of sickled cells in the blood is diagnostic for the condition of sickle-cell anemia (Figure 38-1), whereas an elevated white-cell count may be a sign that a disease exists, but not an identification of the disease.

This exercise concentrates on formed elements of

Figure 38-1 The appearance of sickle cells: (a) a light microscopic view of a red-blood-cell preparation, showing a large number of sickled cells (arrows) (from N. M. Rumen, *Blood,* 45: 45–48, 1975); (b) a scanning micrograph of sickled cells.

blood. The formed elements of blood include red blood cells *(erythrocytes)*, white blood cells *(leukocytes)*, and platelets *(thrombocytes)*. *Erythrocytes* carry oxygen to tissues and remove carbon dioxide from them. *Leukocytes* function in inflammation and immune responses of the body. There are five types of leukocytes, divided into two general categories: granulocytes and agranulocytes. The *granulocytes*, cells containing cytoplasmic granules, consist of neutrophils, eosinophils, and basophils (so named for the staining characteristics of their granules; see Color Plate 36). *Neutrophils* serve as essential agents in an individual's defense against various bacteria and some fungi. *Eosinophils* have a role in defending against disease agents *(pathogens)* such as protozoa and worms. The exact function of *basophils* is not well understood. These cells are believed to be associated with allergic reactions and inflammation. The *agranulocytes*, cells without granules, consist of lymphocytes and monocytes (Color Plate 37). *Monocytes* are critical in the body's defense against intracellular disease-causing agents such as viruses and some bacteria. *Lymphocytes* also help in recognizing and destroying several types of pathogens. The platelets, together with coagulation factors in plasma, control the clotting of blood. The diameters of the formed elements are as follows: erythrocytes, approximately 7 μm; lymphocytes, 9–18 μm; monocytes, 24 μm; granulocytes, 10–14 μm; and platelets, 2–4 μm.

Determination of the proportion of each type of leukocyte in the blood is made by means of the *differential count* procedure. Out of a sample of 100 white blood cells observed on a stained blood smear, the normal number for each type of cell is

Basophils	0–1
Eosinophils	1–4
Neutrophils	60–70
Large lymphocytes	0–3
Small lymphocytes	25–30
Monocytes	4–8
	90–116

Detection of abnormalities associated with erythrocytes, such as anemias and malignancies, can be made from examination of blood smears. *Anemia* is a lower-than-normal number of erythrocytes or an abnormally low hemoglobin content in the cells. In general, anemia is caused by blood loss, by an interference with erythrocyte production in the bone marrow, by accelerated erythrocyte destruction, or by combinations of these factors. The appearance of the erythrocytes is one method of classifying anemias. The cells may be described in terms of size and hemoglobin content. Combinations of the following properties are encountered in examination of blood smears.

ERYTHROCYTE SIZE

Macrocytic—cells are larger than normal

Normocytic—cells are normal size

Microcytic—cells are smaller than normal

HEMOGLOBIN CONTENT

Hyperchromic—cells contain more than normal amount

Normochromic—cells contain normal amount

Hypochromic—cells contain less than normal amount

Anemias resulting from the decreased production of erythrocytes can be caused by various deficiencies, including iron or vitamin B_{12}; by destruction of bone marrow by chemical and physical agents; and by malignancies of different types. Anemias caused by the increased destruction of red blood cells occur in cases of severe infections, sickle-cell anemia, incorrect transfusions, and hemolytic disease of the newborn.

Various conditions also are associated with leukocytes. *Leukopenia*, or low white-blood-cell count, often is induced by viral infections and by treatments of malignancies by radiation or chemicals. An above-normal white-blood-cell count may be provoked by several conditions (see Table 38-1).

The leukemias are a group of diseases primarily affecting the lymphatic, mononuclear, and phagocytic systems and the bone marrow. Overproduction of leukoyctes and the presence of great numbers of immature and abnormal forms of white blood cells in the bone marrow (Color Plate 38) and blood-forming organs are characteristic features of these diseases. Many of the immature and abnormal leukemic cells gain access to the circulating bloodstream. These rapidly reproducing leukemic cells fill the bone marrow spaces, prevent immature red blood cells from maturing, and interfere with platelet formation. Leukemic cells also spill into regions of the body and accumulate in organs with no blood-forming capability.

Three types of leukemia are recognized: granulocytic (myelogenous), lymphatic, and monocytic (Color Plates 38–41). This classification is based on the particular leukocyte involved. Leukemias also are grouped according to other factors, such as the duration of the disease—acute (rapid onset) or chronic (of long duration)—and the prominence of immature and abnormal cell types in the bone marrow and blood. In granulocytic (myelogenous) leukemia, an excessive number of granulocytes and the bone marrow cells that give rise to them (precursors) appear in the blood. The precursors for granulocytes are called myeloblasts and myelocytes. With lymphatic leukemia, an abnormal number of lymphocytes and their precursors (lymphoblasts and lymphoid cells) appear in the blood and bone marrow.

Table 38-1   Conditions Associated with White-Blood-Cell Number

Cell Type	Condition Causing Increase in Cell Number
Basophil	Granulocytic leukemia
Eosinophil	Allergies, worm infections, Hodgkin's disease, some leukemias
Lymphocyte	Hepatitis, infectious mononucleosis, measles, mumps, tuberculosis, lymphocytic leukemia
Monocyte	Hodgkin's disease, monocytic leukemia, tuberculosis
Neutrophil	Granulocytic leukemia, pus-producing bacterial infections, poisons

In monocytic leukemia, an abnormal number of immature monocytic leukocytes are found in the blood (Figure 38-2a). One other type of leukemia, known as plasma-cell (multiple) myeloma, is included here, even though the cells involved normally do not circulate in the bloodstream. Plasma cells (described in the section introduction) are immunoglobulin-producing cells. When abnormal plasma cells spill over from locations such as the marrow of the ribs and skull into the peripheral blood, a condition of plasma-cell leukemia exists (Figure 38-2b). In this exercise, slides from several types of leukemia states are placed on demonstration, to emphasize the appearance of abnormal and immature leukocytes.

This exercise considers preparation of a blood smear, a differential-count procedure, and examination of normal and abnormal blood specimens. Exercise 39 contains procedures using blood-cell-counting equipment.

Materials

1. The following materials should be provided per each four students:
   a. Commercially prepared sterile alcohol prep-pads
   b. Sterile, disposable blood lancets
   c. Sterile gauze pads (2 × 2 inch)
   d. Wright's stain and buffer solution for Wright's stain (in dropper bottles or other suitable containers)
   e. Staining rack

(a) (b)

Figure 38-2 Examples of the leukemias: (a) leukemic cells with distinct monocytic morphology (from I. Mirchandani, P. Tabaczka, and M. Palutka, *American Journal of Hematology,* 12: 139–147, 1982); (b) the appearance of plasma-cell leukemia, in which the majority of the cells shown are plasma cells (from D. Douer, A. Weinberger, M. Djaldetti, J. Asherov, I. Pick, and J. Pinkhaus, *Acta Hematologica,* 62: 81–85, 1979).

f. Containers for the disposal of lancets, pads, and used slides
g. Disposable surgical gloves
h. Protective eyewear (optional)
2. The following materials should be provided for class use:
a. Clean, lint-free glass slides
b. Immersion oil
c. Xylene and lens paper
d. Demonstration slides of blood smears representing normal blood and noninfectious anemic states
   i. Normal blood smears
   ii. Sickle-cell anemia
   iii. Macrocytic, hyperchromic anemia
   iv. Microcytic, hypochromic anemia
   v. Hemolytic anemia
e. Demonstration slides of blood smears from selected microbial infections
   i. Representative blood stages of malaria parasites, *Plasmodium falciparum* or *P. vivax* (see Color Plate 42)
   ii. African sleeping sickness *(Trypanosoma brucei* variety *gambiense)* infection (see Color Plate 43)
   iii. Infectious mononucleosis
f. Demonstration slides of blood smears from selected leukemic conditions
   i. Acute myelomonocytic leukemia
   ii. Chronic granulocytic leukemia
   iii. Acute lymphoblastic leukemia
g. Twelve demonstration microscopes

Procedure 1: Blood Smear Preparation

This procedure is to be performed by students individually, but it will require the assistance of another student.

CAUTION

Working with blood or other body fluids will not pose any danger if the following precautions are observed:

1. Place any and all blood, or materials containing blood, into the disinfectant containers provided.
2. Dispose of any blood containing materials as indicated by the instructor.
3. In cases of blood spillage, wipe the area with a disinfectant-soaked paper towel, and dispose of all materials as indicated by the instructor.
4. Always wash your hands after handling blood and associated materials.
5. The use of surgical gloves and protective eye wear.

---

1. Wipe the finger from which a sample is to be taken with an alcohol-moistened (but not saturated) gauze pad. Do not use a finger that will be involved later in holding a pencil or other such activity. (See Exercise 5 for this procedure.)
2. Unwrap the blood lancet.
3. Hold the finger in a downward position, and squeeze it gently.
4. Have your laboratory partner puncture the ball of the finger with the lancet.
5. Discard the first two drops of blood by wiping them off with a fresh pad.
6. Keeping the finger in a downward position, touch one end of a clean glass slide to a newly formed blood drop. Put the slide on a flat surface, wet side up.
7. Place a fresh, sterile pad on the punctured area, and hold it tightly in place until the bleeding stops.
8. Pull one edge of a second slide into the blood drop (Figure 38-3a). Allow the blood to run across its width (Figure 38-3b).
9. With the second slide at about a 30° angle, move

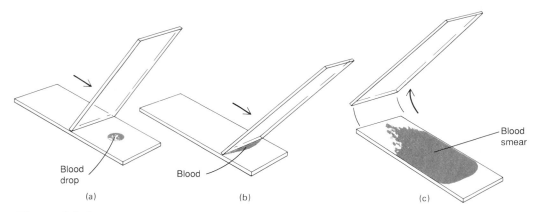

Figure 38-3 Blood smear preparation: (a) Pull the second slide into the drop of blood; (b) allow blood to spread across its width; (c) push the slide toward the end of the first slide.

it toward the opposite end of the first slide (Figure 38-3c). A good smear should have a feathered edge.

10. Scratch your initials in the center of the preparation, and allow it to air dry.

Procedure 2: Staining and Differential Count

1. Place your blood smear on the staining rack, or as otherwise directed by your instructor.
2. Apply enough Wright's stain to cover the entire preparation.
3. Allow the stain to remain in contact with the smear for approximately 1 minute.
4. Add the buffer solution (in drops) until the buffer and stain are in equal proportion. Blow gently on the slide to mix the two solutions.
5. Allow the resulting mixture to stay in contact with the smear for 4 minutes.
6. Pour off the solution, and flood the slide with buffer solution, in order to wash the preparation.
7. Air dry or blot dry the smear. No cover slip is needed.
8. Examine the demonstration slides of normal blood smears.
9. Perform a differential count—that is, count the first 100 white blood cells observed, and note the type of each cell. Enter your findings in the results and observations section.
10. Examine the red blood cells in your smear and in your laboratory partner's. Answer the relevant questions in the results and observations section. (Refer to Color Plates 3, 36, and 37.)

Procedure 3: Demonstration of Normal Blood and Noninfectious Anemic States

1. Your instructor will emphasize the distinguishing features of a normal blood smear, as compared to those of anemic states.
2. Examine all the demonstration slides provided for this portion of the exercise. Refer also to Color Plates 3, 36, and 37.

3. Make sketches of the distinguishing features noted in the results and observations section, and answer all questions pertaining to this exercise in the laboratory review section.

Procedure 4: Demonstration Slides of Blood Smears from Selected Microbial Infections

1. Your instructor will emphasize the major distinguishing features of the preparations.
2. Examine all the demonstration slides. Refer to Color Plates 42–44.
3. Note all distinguishing features. Make sketches of these in the results and observations section.
4. Note whether the infectious disease agents are found inside or outside the erythrocytes.

Procedure 5: Demonstration Slides from Selected Leukemic Conditions

1. Your instructor will emphasize the major features of the preparations. Also refer to Color Plates 38–41.
2. Pay particular attention to the presence of immature blood cells, as well as the smaller-than-normal size of others.
3. Make sketches of typical leukemic cells in the results and observations section, and answer questions associated with this portion of the exercise.
4. Answer any items pertaining to this and previous portions of this exercise in the laboratory review section.

RESULTS AND OBSERVATIONS

1. Enter your findings for the differential count in Table 38-2.

Table 38-2

Type of White Cell	Number per 100	Average Number per 100*
Lymphocyte		
Monocyte		
Eosinophil		
Basophil		
Neutrophil		

*Enter normal values here, as indicated in your textbook or other reference.

2. Did you observe any unusual cells in the demonstration slide? If so, describe them. _____

_____

_____

3. Were your red blood cells and those of your partner similar to those in the demonstration slides of a normal blood smear? Describe any differences.

_____

_____

4. Sketch distinguishing features of the blood preparations representing normal blood and noninfectious anemic states.

Normal smear

Macrocytic, hyperchromic anemia

Microcytic, hypochromic anemia

Sickle cell anemia

5. Sketch distinguishing features of blood smear preparations from selected microbial infections.

African sleeping sickness infection

Infectious mononucleosis

Malaria infection with *Plasmodium falciparum*

6. In which of the demonstration-slide preparations did you find disease agents actually inside erythrocytes? _____

7. Sketch distinguishing features of leukemic blood cells.

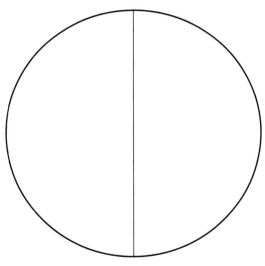

Acute myelomonocytic leukemia | Chronic granulocytic leukemia

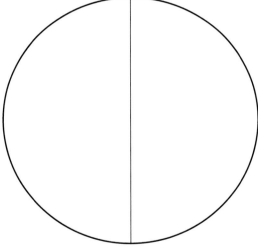

Acute lymphoblastic leukemia | Chronic lymphoblastic leukemia

LABORATORY REVIEW *38*

BLOOD SMEAR PREPARATION AND DIFFERENTIAL STAINING

COMPLETION, IDENTIFICATION, AND COMPARISON

1. Identify the blood cells shown in the blood smear in Figure Q38-1.

   a. _____

   b. _____

   c. _____

   d. _____

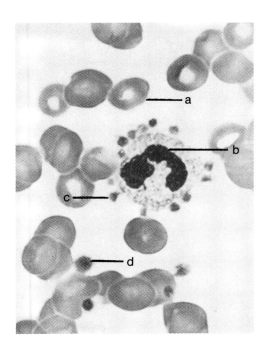

Figure Q38-1 A blood smear.

2. Give the function or explain the activity of the following:

   a. Differential stain _____

   _____

   b. Lymphocyte _____

   c. Monocyte _____

   d. Eosinophil _____

   e. Neutrophil _____

   f. Basophil _____

   g. Platelet _____

ANSWERS

3. Would you consider the micrograph in Figure Q38-2 typical of a normal blood smear? If not, explain your answer. _____

_____

Figure Q38-2

4. List three infectious diseases of the circulatory system.

    a. _____;

    b. _____;

    c. _____

5. Name two hereditary diseases of the blood.

    a. _____;

    b. _____;

6. Which of the micrographs in Figure Q38-3 contain macrocytic, normocytic, or microcytic erythrocytes? Use a micrograph of a normal smear as a guide.

    a. _____;

    b. _____;

    c. _____

MULTIPLE CHOICE

Insert the correct answers in the spaces provided.

1. Which of the following cell types is known to carry oxygen to tissues and to remove carbon dioxide from them?
   a. basophils
   b. monocytes
   c. neutrophils
   d. platelets
   e. erythrocytes

1. _____

2. _____

3. _____

4. _____

5. _____

6. _____

7. _____

8. _____

9. _____

10. _____

Figure Q38-3

2. Which of the following cell types is not an example of a leukocyte?
   a. basophil
   b. monocyte
   c. neutrophil
   d. lymphocyte
   e. platelet
3. Which of the following would be an example of a granulocyte?
   a. lymphocyte
   b. erythrocyte
   c. platelet
   d. basophil
   e. megakaryocyte
4. Which of the following would be an example of an agranulocyte?
   a. lymphocyte
   b. erythrocyte
   c. platelet
   d. basophil
   e. megakaryocyte
5. In a normal differential white-blood-cell count, which of the following cell types
   would be the most numerous?
   a. basophils
   b. eosinophils
   c. neutrophils
   d. lymphocytes
   e. monocytes

6. In a normal differential white-blood-cell count, which of the following cell types would be the least numerous?
   a. basophils
   b. eosinophils
   c. neutrophils
   d. lymphocytes
   e. monocytes

7. Which of the following terms describes erythrocytes that are larger than normal?
   a. macrocytic
   b. normocytic
   c. microcytic
   d. hypochromic
   e. hyperchromic

8. Which of the following terms describes erythrocytes that contain more than the normal hemoglobin content?
   a. macrocytic
   b. normocytic
   c. microcytic
   d. hypochromic
   e. hyperchromic

9. Which of the following terms describes red blood cells that contain less than the normal level of hemoglobin?
   a. macrocytic
   b. normocytic
   c. microcytic
   d. hypochromic
   e. hyperchromic

10. What are antibody-producing cells known as?
    a. erythrocytes
    b. lymphocytes
    c. plasma cells
    d. platelets
    e. monocytes

Blood Cell Counts, the Hematocrit (Packed Cell Volume), Coagulation Time, and Hemoglobin Determinations

After completing this exercise, you should be able to

1. Describe and use blood-cell-counting equipment
2. Perform standard total red-blood-cell and white-blood-cell counts
3. Carry out the calculations necessary to determine the numbers of red and of white blood cells in blood samples
4. Perform and interpret a standard hematocrit procedure
5. Describe the general stages in coagulation
6. Distinguish between serum and plasma
7. Perform and interpret standard clotting tests and hemoglobin determinations
8. Detect selected clotting and hemoglobin abnormalities

The total blood volume in an adult human is about 6.1 liters, or approximately 7.5 percent of body weight. About 45 percent of blood consists of formed elements: erythrocytes, leukocytes, and platelets. The remaining 55 percent, the fluid portion, is plasma.

The number of the formed elements bears a direct relationship to an individual's state of health. This exercise emphasizes both red and white blood cells. The total number of red blood cells in an adult human is approximately 25 trillion. These cells survive for 100–120 days within the circulatory system. In laboratory tests, the normal average red-blood-cell count is usually in the range of 4.5–5.5 million cells per cubic millimeter (mm^3) of blood. Males tend to have a red-blood-cell count about 20 percent higher than females at the same age and level of physical activity. The red-cell count increases proportionately as the level of physical activity increases over a period of time. Thus, a well-trained athlete can be expected to have a higher count than a bedridden hospital patient.

In the healthy adult, the normal number of circulating white blood cells is about 7000/mm^3 of blood. This value is somewhat higher in children. Newborns have counts of about 16,000/mm^3. If the white-blood-cell count is quite high (i.e., over 10,000), the condition of *leukocytosis* exists. If, on the other hand, the final count is 5000 or below, a condition of *leukopenia* exists. Leukocytosis occurs in situations of acute infections. It is also evident in leukemia, but in this case, white blood cells are abnormal. Leukopenia generally is associated with viral infections such as chickenpox, measles, mumps, and poliomyelitis. The survival or life span of white blood cells in the blood is difficult to determine exactly.

One laboratory method used to determine the number of blood cells incorporates the use of an instrument termed a *hemocytometer,* a counting chamber, and diluting pipettes (Figure 39-1). The counting chamber is designed to hold exactly 1 mm^3 of a diluted blood sample. On the bottom of the chamber is a glass surface

Figure 39-1 Equipment for blood cell counting: (a) hemocytometer counting chamber; (b) red-blood-cell pipette and tubing—note the presence of a mixing bead in the bulb portion; (c) enlargement of grid on the glass surface of the counting chamber. The areas used for red-cell counts are labeled R_1 through R_5 and those for white-cell counts are labeled W_1 through W_4.

etched in a precise grid. The blood cell determination involves counting cells in several squares of the grid and obtaining an average number. That number is multiplied by a factor that takes into account the amount of dilution.

More advanced approaches to making blood cell counts incorporate an electronic counting device. As a suspension containing either red or white blood cells passes through a narrow opening at high speed, each cell is counted automatically. Excellent reproducibility and accuracy generally are obtained with electronic counting devices.

The laboratory offers a number of procedures that serve as valuable aids in the diagnosis and management of various disease states. The *hematocrit,* or *packed cell*

volume, is one such procedure. It is an important measurement of red-blood-cell concentration and is used as a test for the detection of an anemic condition. The hematocrit is the ratio of red blood cells to whole blood by volume. When subjected to centrifugal force, red blood cells separate from plasma and the other constituents in blood; they become "packed" together. Once separated, the volumes can be measured to obtain the ratio (Figure 39-2).

In a normal condition, such a preparation consists of approximately 45 percent packed red blood cells, 1 percent white blood cells and platelets, and 54 percent plasma. The normal average hematocrit values for adult males is 47 percent (range 40–54) and for adult females (range 37–47). Hematocrits of 15 percent or lower are

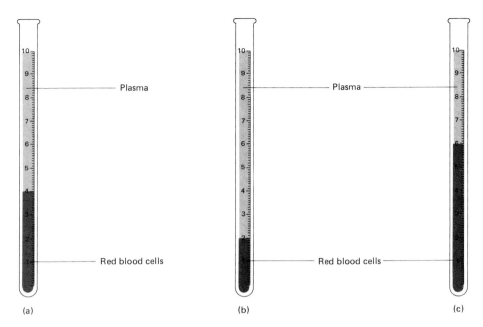

(a)   (b)   (c)

Figure 39-2 Hematocrit determination: A hematocrit is the ratio of red blood cells to whole blood. The appearance of three specimens after centrifuging in Wintrobe tubes is shown: (a) normal blood; (b) low hematocrit or serious anemia; (c) elevated hematocrit, as in a condition of dehydration resulting from a severe burn.

characteristic of anemia and significant red-blood-cell loss. Conditions such as polycythemia and malignancies involving the red bone marrow, in which there is an overproduction of red blood cells, produce hematocrits of 65 percent and higher. The hematocrit is also performed in this exercise.

Blood normally is maintained in liquid form as long as it remains in blood vessels. However, when blood comes out of a vessel, it is changed within a short time to a soft, jellylike mass, a *blood clot*. The liquid released by the clot is termed *serum*. The physiological process whereby blood is so changed is termed *coagulation*. Clot formation is a vital mechanism that prevents excessive blood loss from the body.

Examination of a magnified blood clot shows that it is composed of a network of delicate fibers, in which both red and white blood cells are enmeshed. The fibers consist of *fibrin*, a substance derived from one of the plasma proteins, *fibrinogen*. The conversion of fibrinogen to the solid fibrin is a complicated chemical process involving enzyme systems.

In recent years, the enzymatic reactions associated with clotting have been formalized into a "cascade" scheme involving at least 13 factors (Table 39-1). It is widely held that the different coagulation factors exist in plasma as inactive proenzymes that are converted to active enzymes during the clotting process. The reactants appear in sequence, with the product of the first reaction operating as an enzyme to catalyze the second reaction. The second reaction yields a product that is an enzyme to catalyze the third reaction, and so on. Once activated, the sequence must continue to completion.

The complex process of coagulation is often greatly simplified into five stages, with the first three related to the formation of the visible blood clot (Figure 39-3), the fourth to clot retraction, and the fifth to dissolution of the clot.

Stage 1. Formation or release of tissue factor (TF), the initiator of clotting

Stage 2. Prothrombin converted to thrombin (by thromboplastin and other coagulation factors)

Stage 3. Fibrinogen converted to fibrin (by thrombin and other coagulation factors)

Stage 4. Clot retraction or *syneresis* (consolidation and contraction of fibrin threads)

Stage 5. Fibrinolysis (clot dissolution)

All five stages may be occurring continuously in minute amounts in the normal coagulation process. Table 39-1 lists the plasma clotting factors.

Depending on the manner in which thromboplastin is introduced, the clotting process may proceed along one of two routes, the *extrinsic* or the *intrinsic* pathway (as shown in Figure 39-3). The extrinsic pathway begins

Table 39-1

Number*	Name
Plasma clotting factors	
I	Fibrinogen
II	Prothrombin
III	Tissue thromboplastin (extrinsic or intrinsic), the tissue factor (TF)
IV	Calcium ions (Ca^{2+})
V	Proaccelerin, or labile factor
VII	Proconvertin, serum prothrombin conversion accelerator (SPCA), or stable factor
VIII	Antihemophilic factor (AHF)
IX	Plasma thromboplastin component (PTC), or Christmas factor
X	Stuart–Prower factor, or Stuart factor
XI	Plasma thromboplastin antecedent (PTA)
XII	Hageman factor
XIII	Fibrin stabilizing factor (FSF)
Platelet factors (Pf)	
Pf_1	Platelet accelerator
Pf_2	Thrombin accelerator
Pf_3	Platelet thromboplastin factor
Pf_4	None assigned formally

*The substance that had been assigned number VI is no longer used in coagulation theory.

when a blood vessel is ruptured, and damaged tissues surrounding it or the ruptured area of the blood vessel itself release a lipoprotein termed *tissue thromboplastin*. This substance reacts with coagulation factors in plasma to form *extrinsic thromboplastin (extrinsic prothrombin activator)*.

The intrinsic pathway involves the clumping of platelets, their disintegration, and the release of platelet coagulation factors into the plasma. These factors in turn react with certain other coagulation factors to form *intrinsic thromboplastin* or tissue factor *(intrinsic prothrombin activator)*.

Coagulation or bleeding disorders may be the result of a variety of factors, such as platelet abnormalities, blood vessel (vascular) abnormalities, or coagulation-factor deficiencies. The laboratory tests available to detect or screen for these disorders are described in Table 39-2.

Three tests associated with clotting and related disorders, clotting time, prothrombin time (PT), and activated partial thromboplastin time (APTT), are considered in this exercise. The PT and APTT tests complement one another and can be used together with other laboratory tests to determine the basis of bleeding

Table 39-2   Coagulation-Disorder Screening Tests

Test	Brief Description
Clotting or bleeding time	A check for platelet aggregation and primary clot formation. This test is a helpful indication of platelet abnormality, whether in number or in function. Individuals with vascular problems will have an abnormal bleeding time.
Platelet count	Determination of the quantity of platelets.
Prothrombin time (PT)	A test to detect deficiencies in the extrinsic clotting system (Figure 39-3) and defects with Factors I, II, VII, or X in the presence of tissue thromboplastin. Individuals on anticoagulant therapy will have a prolonged PT.
Activated partial thromboplastin time (APTT)	A test to detect deficiencies in the intrinsic clotting system (Figure 39-3) and defects with Factors I, V, VIII, IX, X, XI, and XII. Individuals on anticoagulant therapy will have a prolonged APTT test result.

Stage 1

Figure 39-3 Blood-clotting mechanism: Both extrinsic (a) and intrinsic (b) pathways are shown. Roman numerals are plasma coagulation factors. (From G. J. Tortora and S. K. Grabowski, *Principles of Anatomy and Physiology,* 7th ed., New York: HarperCollins, 1993.)

Table 39-3 Clotting Factor Defects: Results with PT and APTT

Clotting Factor Defect	PT Result	APTT Result
V	Abnormal	Abnormal
VII	Abnormal	Normal
VIII	Normal	Abnormal
IX	Normal	Abnormal
X	Abnormal	Abnormal
XI	Normal	Abnormal
XII	Normal	Abnormal

The major fibrinolytic system, known as the *plasminogen–plasmin system,* consists of five components—plasminogen, plasmin, plasminogen activators, plasminogen activation inhibitors, and plasmin inhibitors. Normal serum contains less than 10 micrograms per milliliter (μg/ml) of FDP. Tests used to measure FDP levels are of value in cases involving recovery from abdominal operations, kidney transplant, or childbirth. Such tests also serve as an aid in the diagnosis of pulmonary embolism and renal disease. The Thrombo-Wellco test is a rapid and reliable procedure for the semiquantitative determination of FDP (Figure 39-4).

Hemoglobin is responsible for gaseous transport in the body. The hemoglobin content of the blood usually is expressed in grams per 100 milliliters (g/100 ml) or sometimes as a percentage of normal. The normal hemoglobin content of the blood is about 14.5–15 g/100 ml. Tests for the determination of hemoglobin content are included in this exercise.

Normal hemoglobin (HbA) is a conjugated protein consisting of *heme,* an iron-containing nonprotein, and a globin (protein) portion composed of two pairs of polypeptide chains. The amino acids and the sequence of their arrangement in each polypeptide are under genetic control. If a gene causes a substitution for an amino acid or a change in the amino-acid sequence in one of the polypeptides, an abnormal molecule of hemoglobin forms. Such genetic changes (mutations) may result in an anemia.

One of the better-known gene-controlled defects is sickle-cell anemia. In sickle-cell anemia, hemoglobin S (HbS) is produced; under conditions of low oxygen tension, HbS tends to form crystals in red blood cells that result in the characteristic sickle-shaped distortion. In *homozygous individuals* (individuals in whom both genes for hemoglobin code for HbS), severe hemolytic sickle c-cell anemia occurs. In *heterozygous individuals* (individuals who have one gene for HbS and one for HbA), the hemolytic anemia is comparatively mild.

disorders associated with prolonged clotting time. Table 39-3 shows test results that can occur in case of clotting factor defects.

This exercise also considers *fibrinolysis,* or clot dissolution. *Fibrinolysis* is the process responsible for the enzymatic degradation of fibrin thrombi, which results in the formation of *fibrin degradation products* (FDP).

Figure 39-4 Result patterns with the Thrombo-Wellco test/ Latex test. A negative result appears on the left. On the right, a positive clumping or agglutination reaction can be seen. In a positive test, latex particles coated with antibodies against *fibrin degradation products* (FDP) react with FDPs in serum from a blood sample. (Courtesy of Wellcome Reagents Division, Burroughs Wellcome Company, Research Triangle Park, NC.)

Screening tests for HbS are available. This exercise demonstrates this type of procedure.

BLOOD CELL COUNTS

Materials

1. The following materials should be provided per each two students:
    a. One *hemocytometer* and cover glass (a Neubauer hemocytometer is recommended)
    b. Two Unopette Reservoirs (Becton-Dickinson) for red-blood-cell determinations
    c. Two Unopette Reservoirs (Becton-Dickinson) for white-blood-cell determinations
    d. Four capillary pipette assembly systems (Becton-Dickinson)
    e. One microscope
    f. One hand counter
2. The following materials should be provided for class use:
    a. Individual, sterile alcohol prep-pads
    b. Sterile blood lancets
    c. Kim wipes or other laboratory tissue material
    d. Disposal container for blood contaminated items
    e. Surgical gloves
    f. Protective eyewear (optional)

CAUTION

Working with blood or other body fluids will not pose any danger if the following precautions are observed:

1. Place any and all blood or materials containing blood into the disinfectant containers provided.
2. Dispose of any blood-containing materials as indicated by the instructor.
3. In cases of blood spillage, wipe the area with a disinfectant-soaked paper towel, and dispose of all materials as indicated by the instructor.
4. Always wash your hands after handling blood and associated materials.
5. The use of surgical gloves and protective eyewear.

Procedure 1: Red-Blood-Cell Count

This procedure is to be performed by students in pairs, as in other procedures where partners change places, so that each does the experiment (still working in pairs, with materials distributed to pairs). Blood samples for both red-blood-cell and white-blood-cell counts can be obtained at the same time. (*Note:* Different-colored reservoirs are needed.)

1. Examine the Unopette Reservoir for red-blood-cell determinations and the capillary pipette assembly (Figure 39-5). Identify the reservoir chamber, di-

Figure 39-5 Components of a Unopette Reservoir System: The reservoirs used for red-blood-cell counts have a red undersurface; those used for white-blood-cell counts have a white undersurface.

luent fluid, protective shield on the capillary pipette, and the pipette itself. Note the color of the reservoir's bottom surface. It should be red.

2. Hold the reservoir on a flat surface in one hand, and hold the pipette assembly in the other hand. Push the tip of the pipette shield firmly through the diaphragm in the reservoir's neck (Figure 39-6a). Pull out the assembly unit and remove the protective shield from the pipette assembly with a twist.

3. Wipe the finger of your partner from which the blood sample is to be taken with an alcohol-moistened gauze square.

4. Hold the finger in a downward position, and prick it with a sterile lancet. Wipe away the first drop and gently squeeze the finger to provide a free flow of blood.

5. Holding the pipette almost horizontally, touch the tip of it to the forming blood drop (Figure 39-6b). The pipette will fill by capillary action. When the blood reaches the end of the capillary bore in the neck of the pipette, the filling action will stop. Carefully wipe any excess blood from the pipette's surface with a Kim wipe or similar wiping tissue.

6. Squeeze the reservoir slightly, to force out a small amount of air. While still maintaining pressure on the reservoir, cover the opening of the overflow chamber of the pipette with your index finger, and push the pipette securely into the reservoir neck (Figure 39-6c).

7. Release the pressure on the reservoir, and remove your finger from the pipette opening. Negative pressure will draw the blood into the diluent fluids.

8. Mix the contents of the reservoir chamber by squeezing the reservoir gently two or three times. (Squeeze gently so that the diluent is not forced out of the chamber.) Also invert the reservoir gently a few times (Figure 39-6d).

9. Remove the pipette from the chamber, reverse its position, and replace it on the reservoir (Figure 39-6e). This converts the system into a dropper assembly (Figure 39-6f).

10. Squeeze a few drops out of the system into a container, or wipe with tissue to clean the capillary bore. Now you are ready to fill the blood-counting chamber.

11. Wipe the glass cover of the hemocytometer, and position it as shown in Figure 39-7a.

12. Hold the end of the pipette, and squeeze the sides of the reservoir so that you can deposit a small drop of the diluted specimen onto the polished surface of the counting chamber next to the edge of the cover glass. *Be careful not to allow the chamber to overfill.* If the chamber overfills, charge (fill) the other side of the chamber.

13. Carefully place the charged hemocytometer on the microscope stage, and focus with the low-power objective, to bring the small (R) grid areas into clear view (Figure 39-1).

14. Then move the high-power objective into place,

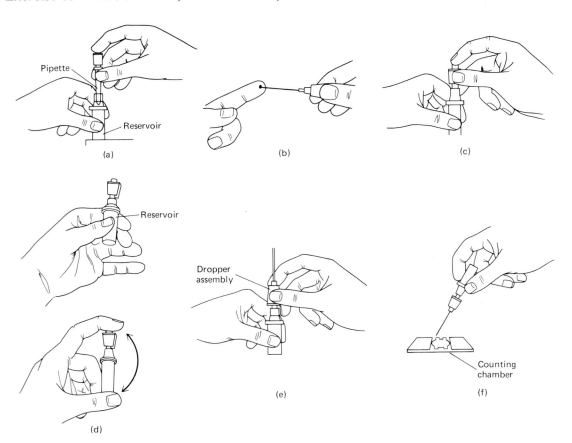

Figure 39-6 Preparation of the Unopette Reservoir System: (a) puncturing the diaphragm; (b) obtaining a blood sample; (c) introducing the sample; (d) mixing the contents of the system by squeezing and inverting; (e) converting to a dropper assembly; (f) filling (charging) the blood-counting chamber. (Modified from *RBC Determination for Manual Methods and WBC Determination for Manual Methods,* Becton-Dickinson, Division of Becton, Dickinson and Company.)

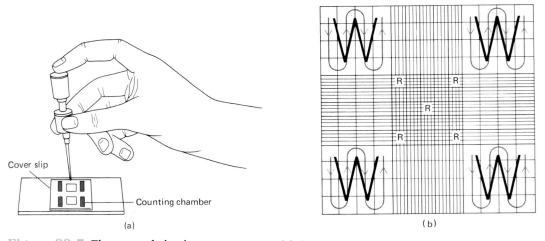

Figure 39-7 The use of the hemocytometer: (a) Positioning the Unopette and filling the chamber—it is important not to overfill the chamber; a properly filled chamber has a blood specimen only within the space between the cover glass and the counting area. (b) This shows the direction to follow for white-cell counts.

and, with a hand counter or by other means of counting, determine the number of cells in each of the five specified areas. On the edges of the squares, count only the cells that touch the lines on the *left* and *top* sides.

15. Multiply the total number of cells counted in the five squares by 10,000, to obtain the red-blood-cell count for 1 mm.³ Record your findings in the results and observations section.

16. Rinse the hemocytometer, wipe it dry, and repeat Steps 11–15. Was your second count similar to the first?

17. Repeat Step 16.

18. Have your partner repeat the entire procedure.

19. Dispose of all materials as directed by your instructor.

20. Answer the questions in the results and observations section.

Procedure 2: White-Blood-Cell Count

This procedure is performed by students in pairs.

1. Follow steps 1–12 of Procedure 1, but use the Unopette chamber for white-blood-cell determinations. It has a white bottom surface.

2. Place the hemocytometer on the microscope stage. With the low-power objective, focus on the chamber area to bring the four large (W) corner regions into view (see Figure 39-1).

3. With a hand counter or by other means, determine the number of cells in each of the four specified areas. On the edges of the squares, count only the cells that touch the lines on the *left* and *top* sides.

4. Multiply the total number of cells by 50 to obtain the white-blood-cell count for 1 mm³. Record your findings in the results and observations section.

5. Rinse the counting chamber, wipe it dry, and repeat Steps 1–4. Was your second count similar to the first?

6. Repeat Step 5.

7. Have your partner repeat the entire procedure.

8. Dispose of all materials as directed by your instructor.

9. Answer the questions in the results and observations section.

THE HEMATOCRIT (PACKED CELL VOLUME)

Materials

1. The following materials should be provided per four students:

a. Commercially prepared individual sterile alcohol prep-pads

b. Sterile, disposable blood lancets

c. Sterile gauze pads (2 × 2 inch)

d. Heparinized capillary tubes

e. Clay-Adams Seal-Ease

2. The following materials should be provided for class use:

a. Microhematocrit centrifuge

b. Microhematocrit tube reader (Clay-Adams) or a substitute

c. Containers of disinfectant

d. Disposable surgical gloves

Procedure

This procedure is to be performed by students in groups of four.

1. Laboratory partners should obtain blood specimens from one another. Wipe the finger from which the sample is to be taken with an alcohol-moistened sterile pad. Hold the finger in a downward position, and gently squeeze it. Puncture the ball of the finger with the lancet. Discard the first two drops of blood. (Refer to Exercise 5 for a description of the procedure.)

2. Place the marked (red) end of a heparinized capillary tube in the forming blood drop, and allow the blood to flow downward to fill about two-thirds of the tube (Figure 39-8a).

3. Put a finger over the unmarked end of the tube, and gently stick the red end into the Seal-Ease (Figure 39-8b). This seals the blood-containing end of the tube.

4. Place the prepared blood specimen into a slot of the microhematocrit centrifuge, with the sealed end of the capillary tube next to the rubber ring guard at the outer rim of the instrument (Figure 39-8c). Do not centrifuge the specimen until there are additional specimen tubes. There must be an even number of tubes for proper balance. Use the specimens of your laboratory partners.

5. Check to be sure that the centrifuge lid is properly fastened. A wrench is usually provided for this purpose.

6. Set the centrifuge time for 4 minutes, and turn on the instrument.

7. After centrifugation, place your capillary tube on or next to the tube reader, and determine the hematocrit reading. There are usually directions on the instrument to help you make your determination.

8. Record your findings, and answer the questions in the results and observations section.

Figure 39-8 General steps in the hematocrit procedure: (a) specimen collection; (b) sealing the capillary tube; (c) inserting the specimen into the hematocrit for centrifugation.

RESULTS AND OBSERVATIONS

1. Enter the results of both red-blood-cell (RBC) and white-blood-cell (WBC) counts in Table 39-4.
2. Indicate the calculation formula for the following:
   a. RBC determinations

   b. WBC determinations

3. Enter the hematocrit determinations for members of your laboratory group in Table 39-5.

COAGULATION (CLOTTING) TIME

Materials

The following materials should be provided for class use:

1. Sterile alcohol prep-pads
2. Sterile, disposable blood lancets
3. Sterile gauze pads
4. Nonheparinized capillary tubes (0.5-mm diameter)
5. Stopwatch or other timing device
6. Glass files
7. Container with disinfectant for disposal of glass and other blood-containing items
8. Disposable surgical gloves

Procedure

This procedure is to be performed by students in pairs.

1. Disinfect the surface of a finger with an alcohol prep-pad, and puncture it with the lancet, to produce a free flow of blood (as described in Exercise 5). Wipe away the first blood drop.
2. Hold the finger in a downward position, and place a nonheparinized capillary tube into and below the newly forming blood drop (Figure 39-9a). Allow the

Table 39-4

Student	Sex	Number of RBCs Counted	RBC Count per mm³	Number of WBCs Counted	WBC Count per mm³

tube to fill as completely as possible by capillary action. (Prevent air bubbles from forming by keeping the open end of the tube completely submerged in the blood drop.)

3. Have your partner start timing as soon as the tube is filled to capacity.

4. After 1 minute, use the file to make a small surface scratch near one end of the tube. Break the tube on the score line, and separate the small piece from the rest of the tube. Look for the presence of coagulation (Figure 39-9b). Continue to break off pieces of the tube at 1-minute intervals until coagulation is observed. Record your findings in the results and observations section.

Table 39-5

Student	Sex	Hematocrit	Normal or Abnormal

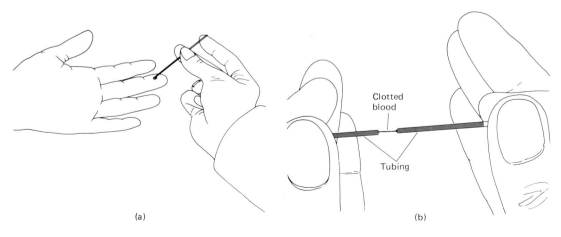

Figure 39-9 Steps in determining coagulation time: (a) obtaining a specimen; (b) breaking portions of capillary tube to find coagulation time.

5. Change places with your laboratory partner, and repeat Steps 1–4. Answer the questions in the results and observations section.
6. Dispose of the capillary tube as directed by your instructor.

PROTHROMBIN TIME (PT)

Materials

The following materials should be provided for class use:

1. Simplastin or other commercial source of tissue thromboplastin
2. Plasma samples kept at 37°C
3. Plastic test tubes
4. 1-ml pipettes, graduated in 0.1 ml, with rubber bulbs
5. 37°C water bath with a rack for tubes
6. Stopwatch or other timing device
7. Container with disinfectant for disposal of pipettes and other blood-containing items
8. Disposable surgical gloves

Procedure

This procedure is to be performed by students in pairs or in groups of four.

CAUTION

Working with blood or other body fluids will not pose any danger if the following precautions are observed:

1. Place any and all blood or materials containing blood into the disinfectant containers provided.
2. Dispose of any blood-containing materials as indicated by the instructor.
3. In cases of blood spillage, wipe the area with a disinfectant-soaked paper towel, and dispose of all materials as indicated by the instructor.
4. Always wash your hands after handling blood and associated materials.
5. The use of surgical gloves and protective eyewear.

---

1. Your instructor will review the use of a pipette.
2. Pipette 0.2 ml of the thromboplastin into a small test tube. Place the tube in the 37°C water bath.
3. Remove the tube containing the thromboplastin from the water bath, and pipette 0.1 ml of warmed plasma sample into it. Shake gently to mix.

4. Have your partner start timing as soon as the two materials are mixed.
5. Place the tube in the water bath again.
6. At least every 4 seconds, remove the tube, tilt it, and look for the formation of a fibrin clot.
7. Stop the procedure as soon as a clot appears. Record the time in the results and observations section.
8. Repeat Steps 2–7, and try to obtain a more accurate time for formation of the fibrin clot.
9. Answer the questions in the results and observations section.

ACTIVATED PARTIAL THROMBOPLASTIN TIME (APTT)

Materials

The following materials should be provided for class use:

1. Activated cephaloplastin
2. Calcium chloride ($CaCl_2$) 0.02 molar solution
3. Two plasma samples
4. Plastic test tubes
5. Test-tube racks
6. 1-ml pipettes, graduated in 0.1 ml, with rubber bulbs
7. 37°C water bath with a rack for tubes
8. Stopwatch or other timing device
9. Marking pens
10. Container for disposal of pipettes and other blood-containing items
11. Disposable surgical gloves

Procedure

This procedure is to be performed by students in pairs or in groups of four.

1. Your instructor will review the use of pipettes.
2. Obtain three test tubes. Label two of the tubes "1" and "2," to correspond to the labels on the plasma samples; label the third tube "control."
3. With a 1-ml pipette, introduce 0.1 ml of activated cephaloplastin into each tube. Dispose of the pipette as indicated by the instructor.
4. With separate pipettes, introduce 0.1 ml of the plasma samples into the correspondingly labeled tubes. Dispose of the pipettes as indicated.
5. Incubate all tubes at 37°C for 3 minutes.
6. With another 1-ml pipette, introduce 0.1 ml of warmed $CaCl_2$ into each tube, and immediately start the timer.
7. Observe the tubes for the appearance of fibrin-thread formation.

8. Place the tubes in the 37°C water bath, and gently shake them continuously. Examine the tubes at 20-second intervals. Holding and tilting the tubes against a bright light helps to detect the fibrin threads.
9. Note the time required for the reaction to occur. Normal reaction time is approximately 40 seconds or less.
10. Record your findings in the results and observations section, and answer the questions pertaining to this exercise in the laboratory review section.

RESULTS AND OBSERVATIONS

Coagulation (Clotting) Time

1. Enter your findings and those of your laboratory partner in Table 39-6. Indicate the presence of coagulation with a "+" and its absence by a "−."

2. What is a normal clotting time for blood? _____

_____

3. Were the coagulation times of most students within normal limits? (Check the clotting times of at least 12 students before answering this question.) What

variations did you observe? _____

_____

4. List three factors that increase clotting time.

    a. _____

    b. _____

    c. _____

PT

1. How long did it take for the plasma to clot? _____

_____

2. How would this test be performed with a blood sample? _____

_____

_____

3. List two factors that increase prothrombin time.

    a. _____

    b. _____

APTT

1. Enter your findings in Table 39-7.
2. Were the test results easy to read and interpret?

_____

3. Were any abnormal reactions observed? If so, which

ones? _____

_____

4. If an individual had a defect in Factor X, would you

expect to see an abnormal APTT reaction? _____

Table 39-7

Reaction Time	Sample 1	Sample 2	Control

DETECTION OF FIBRIN DEGRADATION PRODUCTS (FDP)

Materials

The following materials should be provided for the instructor's use:

1. Latex test suspension (contains polystyrene latex particles, coated with sheep antibodies against FDP)
2. Positive control serum
3. Negative control serum
4. Two unknown serum preparations, diluted 1:5 in buffer
5. Glycine saline buffer
6. Glass slides
7. Wooden applicator sticks
8. Clean Pasteur pipettes with rubber bulbs

Table 39-6

Student	Coagulation Time (in minutes)					
	1	2	3	4	5	6

9. Wax pencil or other marking device
10. Container with disinfectant for blood-containing items

Procedure

This procedure is to be demonstrated by the instructor.

1. Use the marking device to make and number four separate circles on a glass slide.
2. With separate Pasteur pipettes, place 1 drop of each specimen in the circles, as follows:
   a. Circle 1, unknown specimen 1
   b. Circle 2, unknown specimen 2
   c. Circle 3, known positive serum
   d. Circle 4, known negative serum
3. Gently shake the container of latex test suspension. Add 1 drop of the suspension to each circle.
4. Stir each combination with separate applicator sticks.
5. Rock the slide back and forth for 2 minutes, and observe for clumping (agglutination). Agglutination indicates the presence of more than 10 μg/ml of FDP.
6. Record your findings, and answer the questions in the results and observations section.
7. Dispose of slides and other items, as directed by your instructor.

RESULTS AND OBSERVATIONS

1. Enter your findings in Table 39-8. Indicate the presence of clumping by a " + " and its absence by a " − ."
2. Which of the unknown samples produced a positive result? _____
3. What does FDP represent? _____

_____

Table 39-8

Specimen	Presence of Agglutination
Unknown 1	
Unknown 2	
Known positive serum	
Known negative serum	

4. List three situations in which FDP determinations are useful.

   a. _____
   b. _____
   c. _____

HEMOGLOBIN DETERMINATIONS

Materials

The following materials should be provided for class use:

1. Commercial sterile alcohol prep-pads
2. Sterile, disposable blood lancets
3. Sterile gauze pads
4. Tallquist test paper
5. Tallquist reference scale
6. Hemo Cue hemoglobinometer
7. Hemo Cue microcuvettes
8. Kim wipes or other tissue material or lens paper
9. Isopropyl alcohol
10. Rubber globes (disposable surgical gloves)
11. Container of disinfectant

Procedure 1: Tallquist Method

This procedure is to be performed by students in pairs.

1. Remove a piece of test paper from a Tallquist booklet and place it on a flat surface before you.
2. Obtain a blood sample (as described in Procedure 1, Exercise 5). Wipe away the first drop of blood.
3. Touch the second forming blood drop to the center of the Tallquist paper.
4. Wipe the finger, and apply a sterile gauze pad to stop any additional bleeding.
5. After 15 seconds, compare your blood sample with the Tallquist reference scale. Record your hemoglobin estimate in grams of hemoglobin per 100 ml of blood in the results and observations section. (Refer to Color Plate 45.)
6. Repeat Steps 1–5 with your laboratory partner.
7. Answer the questions in the results and observations section.

Procedure 2: Hemoglobinometer Method

This procedure is to be performed by students in pairs.

1. Examine the hemoglobinometer and its parts (Figure 39-10a). Note the display window for hemoglobin measurements.

(a)

(b)

Figure 39-10 The *Hemo Cue*™ hemoglobinometer and specimen preparation: (a) the Hemo Cue and its parts (courtesy of Medical Equipment Designs, Inc., Laguna Hills, California); (b) filling the Microcuvette—note the positions of the fingers (courtesy of Medical Equipment Designs, Inc., Laguna Hills, California).

2. Have your partner put on a pair of rubber gloves before performing any blood work.

3. Obtain a blood sample (as described in Procedure 1, Exercise 5). Wipe away the first two drops.

4. Position the finger so that one large drop of blood forms. It should be big enough to fill the microcuvette completely.

5. Introduce the microcuvette's tip into the center of the blood drop, as shown in Figure 39-10b.

6. Fill the cuvette in one continuous process. Wipe off any excess blood on the outside surface of the cuvette. (*Note:* If a second sample is to be taken from the same finger, it should be done immediately after the first has been taken. Wipe away the remains of the first drop of blood, and take a second sample from a second drop.

7. Place the microcuvette in the cuvette holder, and push it into the Hemo Cue, to the stop point.

8. Wait 45 seconds, and read the hemoglobin value shown on the display window.

9. Enter your findings in the results and observations section. Note that the value is given in grams/liter.

10. Change places with your partner, and repeat Steps 2 through 9.

RESULTS AND OBSERVATIONS

1. Enter your findings and those of your laboratory partner in Table 39-9.

2. Did the hemoglobin values obtained with both methods agree? _____

3. If not, what factors could account for the differences? _____

SICKLE HEMOGLOBIN (HbS) ASSAY DEMONSTRATION

Materials

The following items should be provided for the instructor's use:

1. One HemoCard for an Hbs assay (see Figure 39-11)

2. Three whole blood samples, to include
   a. One containing hemoglobin AS (heterozygous for sickle-cell trait)
   b. One containing hemoglobin SS (homozygous for sickle-cell trait)
   c. One normal specimen
   d. Two unknowns

3. Controls
   a. Hb AA Control (HbS negative)
   b. Hb AS Control (HbS positive)

4. Containers of the following preparations
   a. Sample conditioner
   b. HbS reagent
   c. Wash solution

5. Seven test tubes

6. Seven sample cups

7. One test tube rack or HemoCard HbS Assay Acrylic Workstation

8. One 1-ml pipette

9. Eleven pipettes (5 μl capacity) with delivery device

10. Eight disposable transfer pipettes (1 ml capacity)

11. One pair of disposable surgical gloves

12. Container for pipette disposal

13. Container of disinfectant

14. Marking pen

Procedure

This procedure is to be demonstrated by the instructor (Figure 39-12).

1. Label one sample cup and one test tube with the following *codes*, to represent each of the different preparations to be used in the assay.
   a. *AS* (for heterozygous sickle-cell-trait blood sample)
   b. *SS* (for homozygous sickle-cell-trait blood sample)
   c. *N* (for normal sample)
   d. *1* (for unknown blood, Sample 1)
   e. *2* (for unknown blood, Sample 2)
   f. *CAA* (for HbAA control)
   g. *CAS* (for HbAS control)

Table 39-9

Student	Sex	Hemoglobin Value, Tallquist (in g/100 ml)	Hemoglobin Value, Hemoglobinometer (g/l)

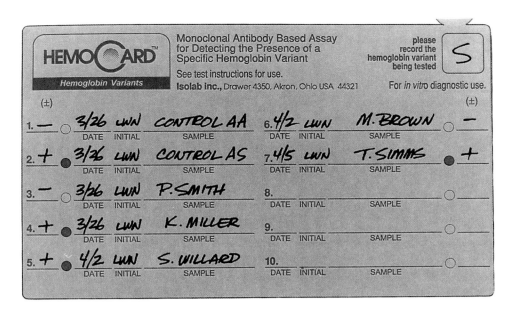

Figure 39-11 The HemoCard. (After IsoLab Inc. Akron, Ohio, 44321.)

Figure 39-12 Steps in the sickle hemoglobin S(HbS) detection procedure.

2. With a 5-ml pipette, add 2 ml of the sample conditioner to each test tube (Figure 39-12a).
3. With separate μl pipettes, add 2.0 μl of each blood sample and each control to their respectively labeled test tubes (Figure 39-12a).
4. With separate μl pipettes, transfer 5 μl of each sample and control to their corresponding sample cups (Figure 39-12b).
5. Next, squeeze 1 drop of the HbS reagent into each sample cup (Figure 39-12c).
6. Allow the samples to stand for 2 minutes.
7. During this time, fill out the HemoCard by entering the date and the code for each sample onto their respective places.
8. With separate disposable plastic pipettes, transfer the contents of the individual sample cups to the appropriate sample well on the HemoCard (Figure 39-12d).
9. Place the pipettes into the container of disinfectant.
10. Add 1 drop of wash solution to each sample well, and observe the results (Figure 39-12e). A pink to red color is a positive result for HbS.
11. Record your findings, and answer the questions in the results and observations section.

RESULTS AND OBSERVATIONS

1. Enter your findings in Table 39-10. Indicate samples positive for HbS by a " +," and with " −" for a negative one.
2. Which, if any, of the unknown blood samples showed a positive result for HbS? _____

Table 39-10

Sample	Color Reaction	HbS (+/−)
AS		
SS		
1		
2		
N		
CAA		
CAS		

LABORATORY REVIEW *39*

NAME _____

LAB SECTION _____ DATE _____

BLOOD CELL COUNTS, THE HEMATOCRIT (PACKED CELL VOLUME), COAGULATION TIME, AND HEMOGLOBIN DETERMINATIONS

COMPLETION, DEFINITION, AND INTERPRETATION

1. a. Are human red blood cells nucleated? _____

   b. Are all mammalian red blood cells nucleated? _____

2. Describe human erythrocytes by filling in the following information.

   a. Function _____

   b. Shape _____

   c. Diameter _____

   d. Average life span _____

   e. Normal number per mm^3 in the adult female _____

   f. Normal number per mm^3 in the adult male _____

3. a. What is hemoglobin? _____

   _____

   b. What heavy metal has as its chief function the synthesis of hemoglobin? ____

   _____

4. What is anemia? _____

   _____

5. List three causes of anemias.

   a. _____

   b. _____

   c. _____

6. Where are red blood cells formed in the human adult? _____

   _____

7. What is the study of blood called? _____

8. What is a hemocytometer? _____

   _____

9. In Table Q39-1, determine the final RBC counts, and indicate whether the results are normal or abnormal.

Table Q39-1

Microscopic Count	Sex	RBC Count (per mm³)	Normal or Abnormal
1000	Male	a.	b.
300	Female	c.	d.
150	Female	e.	f.
195	Male	g.	h.

10. In Table Q39-2, determine the final WBC counts, and indicate whether the results are normal or abnormal.

Table Q39-2

Microscopic Count	Sex	WBC Count	Normal or Abnormal
900	Male	a.	b.
100	Female	c.	d.
500	Female	e.	f.
62	Male	g.	h.

11. Are the dilution pipettes for RBC and WBC counts different? Explain. _____

_____

12. How many and which squares are counted in an RBC count? _____

_____

13. How many and which squares are counted in a WBC count? _____

_____

14. What is a hematocrit? _____

_____

15. What are the percentages of the components of a blood sample in a normal hematocrit? _____

_____

16. The results of 12 hematocrit determinations are listed in Table Q39-3. Indicate whether they are normal or abnormal.

Table Q39-3

Adult Male Patient	Hematocrit Value	Normal or Abnormal	Adult Female Patient	Hematocrit Value	Normal or Abnormal
1	36	a.	1	32	g.
2	45	b.	2	40	h.
3	54	c.	3	61	i.
4	63	d.	4	47	j.
5	42	e.	5	22	k.
6	40	f.	6	37	l.

17. Distinguish between plasma and serum. _____

_____

_____

18. What is the normal hematocrit value for the following?

    a. Adult female _____

    b. Adult male _____

19. Describe four laboratory tests associated with the clotting activity of blood.

    a. _____

    b. _____

    c. _____

    d. _____

20. Distinguish between the intrinsic and the extrinsic pathways of clotting. _____

_____

_____

_____

21. How many factors are known to be involved with the blood-coagulation process?

_____

22. List the stages of blood coagulation, and indicate which factors are associated with each.

    a. _____

    b. _____

    c. _____

23. Which of the blood-clotting factors require vitamin K for their formation? _____

_____

24. What are platelets, and where are they formed? _____

_____

ANSWERS

25. Indicate the PT and APTT reactions that can be expected with defects of the clotting factors listed in Table Q39-4.

Table Q39-4

Defective Clotting Factor	PT Reaction	APTT Reaction
VIII		
V		
X		
XI		
XII		
IX		
VII		

26. What is the physiological function of hemoglobin? _____

_____

27. How is the hemoglobin content of blood usually expressed? _____

28. Give the normal hemoglobin content ranges for the following:

    a. Adult female _____

    b. Adult male _____

    c. 4-year-old child _____

    d. 1-year-old child _____

29. What is sickle-cell anemia? _____

_____

MULTIPLE CHOICE

Insert the correct answer in the spaces provided:

1. What is the total blood volume of an adult human (in liters)?
   a. 5
   b. 6
   c. 8
   d. 2
   e. 2

2. What is the percentage of normal blood that consists of formed elements?
   a. 55
   b. 75
   c. 45
   d. 61
   e. 16

1. _____
2. _____
3. _____
4. _____
5. _____
6. _____
7. _____
8. _____
9. _____
10. _____
11. _____
12. _____

3. What is the percentage of plasma found in blood?
   a. 55
   b. 75
   c. 45
   d. 61
   e. 16

4. What is the usual range for the normal average RBC count, per cubic millimeter?
   a. 4.5–5.5 million
   b. 25–30 trillion
   c. 7000–8000
   d. 10,000–11,000
   e. 5000–6000

5. In a healthy adult human, what is the normal number of circulating WBCs per mm^3?
   a. 4.5–5.5 million
   b. 25 trillion
   c. 7000
   d. 10,000
   e. 5000

6. Which of the following procedures is used to determine the ratio of RBCs to the whole body by volume?
   a. hemocytometer
   b. coagulation time
   c. activated partial thromboplastin time
   d. hematocrit
   e. fibrin degradation products determination

7. What would a packed-cell-volume value of 49 for an adult male indicate?
   a. a significant blood loss
   b. an anemia
   c. polycythemia
   d. a malignancy
   e. a normal result

8. In the coagulation process, which of the following substances is converted to fibrin?
   a. thromboplastin
   b. fibrinogen
   c. thrombin
   d. prothrombin
   e. choices *a* and *b* only

9. In the coagulation process, which of the following substances is converted to thrombin?
   a. thromboplastin
   b. fibrinogen
   c. prothrombin
   d. fibrin
   e. proconvertin

10. Which of the following tests is used to detect deficiencies in the extrinsic clotting system?
    a. bleeding time
    b. platelet count
    c. prothrombin time
    d. activated partial thromboplastin time
    e. clot dissolution

11. Which of the following tests is used to detect deficiencies in the intrinsic clotting system?
    a. bleeding time
    b. platelet count

    c. prothrombin time

    d. activated partial thromboplastin time

    e. clot dissolution

12. Which of the following tests is used to check primary clot formation?

    a. bleeding time

    b. platelet count

    c. prothrombin time

    d. activated partial thromboplastin time

    e. clot dissolution

Lymphatics and the Immune Response

After completing this exercise, you should be able to

1. Perform the slide blood-typing procedure
2. Understand the basis of blood typing and cross-matching
3. Identify major blood group antigens in unknown blood samples
4. Define *electrophoresis*
5. Outline a general procedure for electrophoresis
6. Use electrophoresis to separate and to identify some specific serum proteins
7. Use electrophoresis to identify abnormal hemoglobin molecules

The human body defends itself against foreign and invading agents by means of its immune system, which consists of two closely allied forms of immune responses. One of these responses, known as *cellular (cell-mediated) immunity,* involves the formation of special lymphocytes, termed T cells, that are capable of attacking foreign factors and cells and destroying them. Cell-mediated immunity is particularly effective against various types of microorganisms, worms, virus-infected cells, cancer cells, and foreign (to the body) tissue transplants. The second type of immune response, known as *humoral (antibody-mediated) immunity* involves antibodies that are present in blood plasma and lymph (the body's fluids or *humors*). In humoral immunity, another type of specialized lymphocytes, termed B cells, develop into plasma cells that produce antibodies known as *immunoglobulins.* This form of immunity is particularly effective against bacterial and viral disease agents.

Both types of immunity are essential for an effective immune system, and both originate in lymphoid tissue located in the lymph nodes, spleen, gastrointestinal tract, and bone marrow (Color Plate 46). Lymphoid tissue is strategically located to intercept a foreign and/or invading agent before it can spread too far in the body and cause extensive injury. This exercise uses blood typing to show antibodies in action against foreign factors known as antigens. An *antigen* is any chemical substance recognized by the immune system as being foreign and which causes the body to produce specific immunoglobulins and/or T cells that react with the antigen.

The second portion of this exercise uses electrophoresis to show the presence of antibodies in a blood specimen. Antibodies belong to a group of plasma *globulins,* specifically termed gamma globulin.

It is for this reason that antibodies are also referred to as immunoglobulins *(Igs).* There are five different classes of immunoglobulins, designated as *IgG, IgA, IgM, IgD,* and *IgE.* Each has a distinct chemical structure and a specific biological role.

In 1900, Karl Landsteiner observed that the mixing of red blood cells and sera from two different human donors, followed by incubation at body temperature, resulted in the clumping or agglutination of the red cells (Figure 40-1). This phenomenon, known as *isohemagglutination,* enabled Landsteiner and others to demonstrate the presence of every individual's red blood cells' specific, genetically determined antigens, termed *agglutinogens.* At present, more than 300 blood factors are known to exist in the human species. Examples include A, B, O, AB, M, N, S, Rh_o, Hr_o, and rh'. The major classification of blood types is the ABO blood grouping system (Figure 40-2). Blood types are based on the presence of these factors. For example, an individual whose red cells contain only agglutinogen B (of the ABO group of factors) is said to have blood type B.

Figure 40-1 Blood typing results, showing agglutination (positive) or lack of agglutination (negative).

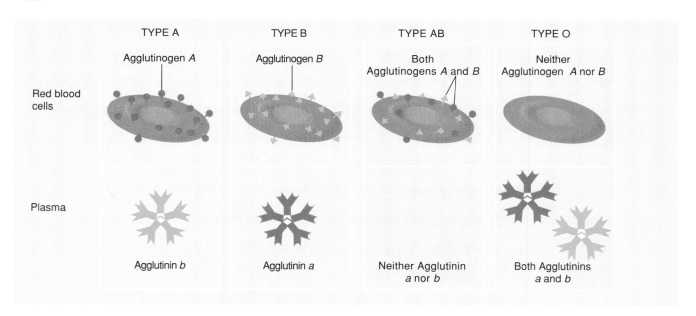

Figure 40-2 Agglutinogens (antigens) and agglutinins (antibodies) involved in the ABO blood-grouping system. (From G. J. Tortora and S. R. Grabowski, *Principles of Anatomy and Physiology,* 7th ed., New York: HarperCollins, 1993.)

In addition to having specific blood-group antigens on red cells, the blood of some humans contains antibodies that react with the erythrocytes from individuals of the other blood types, causing them either to agglutinate or to lyse. These antibodies, termed *agglutinins,* are found in the blood plasma and/or serum. Usually, the antibodies of an individual are not directed against the factors normally present in that individual's blood. Most individuals have antibodies against factors that are absent from their blood cells. For instance, a person of blood type O has antibodies against agglutinogen A (termed *anti-A, antibody a*) and against agglutinogen B (*anti-B, antibody b*). Figure 40-2 shows the characteristics of the major blood types. Antibody concentrations can be increased as a result of transfusions or in the case of women who have borne children possessing different blood characteristics.

Because AB individuals do not possess antibodies against any of the other major blood types, they are referred to as *universal recipients.* Because type O blood cells have no agglutinogens to serve as antigens in another person's blood, members of this group are referred to as *universal donors.*

In 1937, Landsteiner and A. S. Wiener uncovered antisera that clumped (agglutinated) rhesus monkey blood, as well as the cells of 85 percent of Caucasians. This phenomenon was independent of the reactions obtained with the four major blood types. The factor was designated *Rh* for "rhesus." Several other factors subsequently were found to be involved in a complex Rh–Hr system. These factors are now recognized as the principle cause of *erythroblastosis fetalis* (hemolytic disease of the newborn) and as the cause of some other transfusion reactions.

The determination of an individual's blood group can be made by testing his or her red cells with standardized anti-A or anti-B sera. The mixtures are observed for the presence or absence of agglutination. Table 40-1 shows the patterns of reactions that are used for blood-type determinations.

Table 40-1 Reaction Patterns for Blood-Type Determinations

Blood Type	Agglutination When Mixed with Antiserum	
	anti-A	anti-B
A	+	–
B	–	+
AB	+	+
O	–	–

Determining the Rh factor of an individual's red cells is obtained by agglutination tests using the appropriate antisera. The classic and clinically most important of the factors is Rh_o (also called "D," using a different method of categorizing blood groups).

Both false-positive and false-negative reactions may be encountered in testing. To prevent this, suitable control measures are used. These include (1) pretesting of sera for activity, (2) dilution of cells or sera in physiological saline if a reaction has not occurred after a sufficient length of time, and (3) incorporation of tests with known cells and sera.

The blood-group antibodies are of significant importance in cases of transfusions. The procedure known as *compatibility testing* or *cross-matching* is carried out to prevent a transfusion reaction as a consequence of blood incompatibilities between the recipient and the donor. Specifically, the cross-matching technique is designed to detect any incompatibility between the recipient's serum and the donor's cells (known as the *major cross-match*) or between the recipient's cells and the donor's serum (the *minor cross-match*). After the blood specimens have been typed as to the ABO, Rh (or D), and any other factors that appear to be indicated, different mixtures of donor's and recipient's cells and sera are made. Agglutination or lysis of the red cells from either the donor or the recipient is considered to be an incompatible situation.

In certain clinical conditions, which include hemolytic anemia, incompatible transfusions, and hemolytic disease of the newborn, unusual antibodies are present on the surface of blood cells. These globulins are not detected in major or minor cross-matches. The detection of such antibodies is accomplished by the Coombs' or antiglobulin test. The test employs the use of antihuman globulins (Coombs' reagent). This diagnostic material is prepared by successive injections of human serum into a suitable laboratory animal, such as a rabbit or goat. The serum of the immunized animal will form antibodies against the globulin of human serum. This material is collected and used in the detection of human antibodies on the surface of RBCs. The procedure known as the direct Coombs' test is performed by mixing RBCs with the Coombs' reagent. Agglutination of the cells is a positive test demonstrating the presence of human antibodies. The direct Coombs' test is demonstrated in this exercise.

Medicolegal aspects of blood grouping are important. Standard blood-grouping procedures provide information, such as the compatibility of blood donors and recipients, identification of blood stains, detection of rare blood types, potential dangers in pregnancy, and facts that help decide in cases of disputed parentage. In this exercise, the procedures for blood typing and cross-matching are considered.

Electrophoresis is an analytical technique in which charged molecules in solution, mainly proteins and nucleic acids, migrate in response to an electrical field (Figure 40-3). The rate of movement of the molecules depends on several factors, including the net electrical charge, the size and shape of the molecules, the strength of the electrical field, and the ionic, strength, thickness, and temperature of the *medium* (material substance) through which the molecules are moving. Electrophoresis is a simple, rapid, and highly sensitive technique used to identify and study the properties of a single charged molecule and to separate a mixture of charged molecules. Refer to Color Plate 47. Proteins or unusual molecules found in serum are the most frequently studied substances.

Electrophoresis may be carried out in a supporting medium or matrix, such as paper, cellulose acetate, or gels formed in tubes, slabs, or flat beds. In most electrophoretic units, the gel is mounted between two buffer chambers containing separate electrodes (see top left of Figure 40-4). With this arrangement, the only electrical connection between the chambers is through the gel.

After the electrophoresis of a protein or nucleic acid sample is completed, the gel or other supporting matrix can be analyzed by one of several methods. The most common analytical procedure is staining. Protein gels are frequently stained by Coomassie blue or Ponceau S. Nucleic acids are usually stained with ethidium bromide, a fluorescent dye that produces an orange glow when exposed to ultraviolet light. This exercise considers the use of electrophoresis with serum proteins.

CAUTION

Working with blood or other body fluids will not pose any danger if the following precautions are observed:

1. Place any and all blood or materials containing blood into the disinfectant containers provided.
2. Dispose of any blood-containing materials as indicated by the instructor.
3. In cases of blood spillage, wipe the area with a disinfectant-soaked paper towel, and dispose of all materials as indicated by the instructor.
4. Always wash your hands after handling blood and associated materials.
5. The use of disposable surgical gloves

Figure 40-3 Complete S&S EP Mini-Electrophoresis System, showing tank, electrodes, and precast gel. (Courtesy of Schleicher & Schuell Inc., Keene, New Hampshire.)

Figure 40-4 The Isophore® module and its components. This device is quite compact and straightforward in operation. Note the two chambers and their respective parts. The insert at the upper left-hand corner shows a side view of two gel cassettes held in place by clamping bars and screws. (After Isolab, Inc.; RESOLVE®, Isophore®, Isofocus®, and Isolytes® are registered trademarks of Isolab, Inc.)

BLOOD TYPING

Materials

1. The following materials should be provided per each four students:
   a. Commercially prepared sterile alcohol prep-pads
   b. Sterile gauze pads (2 × 2 inch)
   c. Sterile blood lancets
   d. Eight glass slides
   e. Wooden applicator sticks
   f. Container for the disposal of the lancets and applicator sticks

2. The following materials should be provided for class use:
   a. Blood typing antisera—anti-A, anti-B, and anti-D (Rh$_o$)
   b. Unknown blood specimens
   c. Disposable surgical gloves

Procedure 1: ABO and Rh$_o$ Determinations

This procedure is to be performed by students in groups of four.

1. Place 1 drop of each antiserum on separate areas of the blood-typing slide.

2. Obtain a blood sample from one of your laboratory partners (as described in Exercise 5).
3. Allow 1 full-sized drop of blood to fall directly into each drop of antisera.
4. Mix the components in each area, using different ends of the applicator sticks for each mixture.
5. If a reaction does not develop after 2 minutes, warm the slide over a desk lamp or similar device, and gently shake the slide from side to side.
6. Refer to Table 40-1 and Figure 40-1 to interpret the reactions obtained.
7. Record the reactions obtained with your blood and your laboratory partner's in Table 40-2 of the results and observations section.

Procedure 2: Unknown Blood Sample

1. Obtain an unknown blood sample from your instructor.
2. Determine the major blood type and Rh factor or factors, using Procedure 1 of this exercise and Figure 40-1 as a guide.
3. Record your findings in the results and observations section. Check your results with the instructor.

CROSS-MATCHING

Materials

The following materials should be provided for class use: (*Note:* All plasma or serum preparations should be heated to 56°C for 30 minutes prior to use.)

1. Donor blood cell preparation, type A, in physiological saline
2. Donor blood cell preparation, type O, in physiological saline
3. Recipient plasma or serum preparation, type A
4. Recipient plasma or serum preparation, type O
5. Two recipient plasma or serum preparations, blood type unknown
6. Recipient blood-cell preparation, type A, in physiological saline
7. Recipient blood-cell preparation, type O, in physiological saline
8. Donor plasma or serum preparation, type A
9. Donor plasma or serum preparation, type O
10. Two donor plasma or serum preparations, blood type unknown
11. Glass slides
12. Pasteur pipettes with rubber bulbs
13. Wooden applicator sticks

14. Container with disinfectant, for the disposal of blood-containing items.
15. Disposable surgical gloves

Procedure

This procedure is to be performed by students in groups of four.

1. Perform the four major cross-matches, using the combinations of donor-cell preparations and recipient serum preparations indicated in Table 40-3 in the results and observations section.
2. Observe for the presence of agglutination. Record your findings, and answer the questions in the results and observations section.
3. Perform the four minor cross-matches using the combinations of recipient cell preparations and donor serum preparations indicated in Table 40-4 in the results and observations section.
4. Observe for the presence of agglutination. Record your findings, and answer the questions in the results and observations section.

DEMONSTRATION OF THE DIRECT COOMBS' TEST

Materials

The following materials should be provided for class demonstration:

1. One saline-washed suspension of sensitized RBCs (coated with human gamma globulins)
2. One saline-washed suspension of nonsensitized RBCs
3. One saline-washed suspension of unknown RBCs
4. Coombs' reagent
5. Glass slides
6. Pasteur pipettes with bulbs
7. Wooden applicator sticks
8. Marking pen
9. Container for the disposal of applicator sticks and pipettes
10. Disposable surgical gloves

Procedure

The instructor will demonstrate this procedure.

1. Label three separate areas of a glass slide, " + ," " − ," and "X" (for an unknown). Make the markings near the edge of the slide.

2. Place 1 drop of each blood-cell suspension near its labeled area.
3. Add 1 drop of the Coombs's reagent to each blood drop.
4. Mix the components in each area, using a different applicator stick for each combination.
5. Examine the slide for agglutination during a 3-minute period.
6. Answer the questions in the results and observations section.

RESULTS AND OBSERVATIONS

Blood Typing

1. Enter your findings regarding the ABO and Rh_o determinations in Table 40-2. Indicate agglutination by a plus sign and no agglutination by a minus sign.
2. Enter your findings on the unknown blood sample.

   a. Sample number _____

   b. ABO type _____

   c. Rh_o type _____

Cross-Matching

1. Enter your findings for the major cross-matches in Table 40-3. Indicate an incompatibility by a plus sign and compatibility by a minus sign.
2. Which of the unknown combinations for the major cross-matches showed incompatibility? _____
3. Enter your findings for the minor cross-matches in Table 40-4.
4. Which of the combinations for the minor cross-matches showed incompatibility? _____
5. Give at least two possible causes for the incompatibilities observed. _____

Direct Coombs' Test

1. What reaction indicates a positive reaction? _____
2. Was the unknown specimen positive or negative? _____
3. What is the purpose of the Coombs' test? _____

Table 40-2

Student	Reactions with Antisera			Blood Type
	anti-A	anti-B	anti-D (Rh_o)	

Table 40-3

Donor Cells Blood Type	Recipient Blood Type			
	A	O	Unknown 1	Unknown 2
A				
O				

Table 40-4

Recipient Cells Blood Type	Donor Blood Type			
	A	O	Unknown 1	Unknown 2
A				
O				

PLASMA PROTEIN DETERMINATIONS

Materials

The following materials should be provided for general class use:

1. Isophore® Universal Gel Electrophoresis system or other electrophoresis unit
2. Silicon grease or similar lubricant
3. Electrophoresis basic buffer (Isophore®)
4. Electrophoresis acidic buffer (Isophore®)
5. Stacking buffer (Isophore®)
6. Application buffer (Isophore®)
7. Fixative Ponceau S dye solution
8. Acetic acid rinse solution
9. Cleaning solution
10. Phenol red pH indicator (0.1 percent)
11. Power source constant current
12. Separate trays for butter, dye, rinse, and cleaning solutions
13. Glass vials or tubes (small capacity)
14. Test tube racks
15. Disposal surgical gloves
16. 1-ml pipettes, graduated in 0.1 ml, with rubber bulbs
17. Pasteur pipettes with rubber bulbs
18. Single-edge razor blades or scalpels
19. Transparent tape (1-inch wide)
20. Spatulas
21. Flat, blunt-end forceps
22. Filter paper
23. Marking pencil

The following materials should be provided for students in groups of four, or as indicated by the instructor:

1. Normal plasma sample (0.5 ml)
2. Three unknown plasma samples (0.5 ml each)
3. Nine 5-lambda pipettes or microsyringes or micropipettes
4. One gel cassette (Isofocus® survey gel, pH 3–10)
5. One sample spacer comb (Isophore®)
6. One set of drying weights (to be used during drying of gel after electrophoresis run)

Procedure 1: Electrophoresis Unit and Accessories

This procedure is to be performed by students in groups of four.

1. Examine the Isophore® unit (or other electrophoresis system provided), power unit, and accessories.
2. Refer to Figure 40-4, and identify the following components of the Isophore® unit:
   a. Anode in the upper chamber
   b. Upper chamber support containing the cathode
   c. Safety-interlocking lid (prevents contact with the buffers and platinum electrodes)
   d. Gel cassettes that form the side walls of the upper chamber
3. Examine the power unit and accessories provided.

Procedure 2: Gel Preparation

This procedure is to be performed by students in groups of four.

1. Remove one Isofocus® gel from its package. Shake any excess liquid from the surface. (Do not separate the glass or plastic plate from the gel before performing the electrophoresis run. Note that the gel is held between a glass plate and a notched plastic plate. A notched line, which extends along the lower edge of the plastic plate, prevents the gel from sliding out of its position.
2. Examine the gel, to be sure that the plastic spacers on both sides of the gel are in direct contact with it. If they are not, make a small slit in the tape, and gently push the spacer with a spatula until it touches the gel (Figure 40-5a).
3. Use a Pasteur pipette to fill the space above the gel with a stacking buffer (see Figure 40-5b).
4. Place the sample spacer comb on top of the gel so that the tips of the comb enter the gel surface 1 to 2 mm (Figure 40-5c). When the sampler spacer is pressed into the gel, 9 separated sample wells are formed.

Figure 40-5 Steps in sample application and an electrophoresis run: (a) correcting the spacers; (b) applying the stacking buffer; (c) the Isofocus® gel and sample-containing comb; (d) applying a sample with a microsyringe—each end well and each narrow sample well receives application buffer only; (e) transfering gel to transporter and tray; (f) fixation and staining. (From G. A. Wistreich, *Laboratory Manual for Human Physiology,* New York: Harper & Row, 1986.)

5. Place the upper chamber of the electrophoresis system on a convenient working surface. Loosen the screws on each side of the machine so that there is space between the gaskets. (Refer to Figure 40-4.)
6. Coat the gaskets with silicon grease.
7. Place one gel cassette between each pair of gaskets. If only one gel is to be run, remove one clamping bar, and replace it with a blanking plug. The notched plastic plate should be placed so that it faces toward the center of the upper chamber.
8. Tighten the screws in the clamping bar until each cassette is held snugly in place. Use a light touch when clamping—too much pressure will crack the glass plate.
9. Fill the lower chamber with the cathode (basic) buffer. The buffer level should be about 8.0 cm from the top of the chamber. About 1500 ml of buffer will be needed.

10. Carefully slide the upper chamber into the lower chamber.
11. Check the top and bottom of the gel cassette for air bubbles. Remove any bubbles by stirring the buffer or flushing with buffer (a syringe works well for this). The buffer must be in direct contact with the entire bottom edge of the gel. *It is important that all air bubbles be eliminated.*
12. Pour the anode (acidic) buffer into the upper chamber.

Procedure 3: Sample Preparation

This procedure is to be performed by students in groups of four.

1. Obtain 4 small vials, and label them "N" (for normal plasma sample) and "1," "2," and "3" (for each of the unknown samples).

2. With a 1-ml pipette, introduce 0.2 ml of the application buffer to each vial.

3. With individual microsyringes or micropipettes, add 50 µl of each protein sample to its labeled tube.

4. With a new microsyringe, add 4 to 5 µl of application buffer to each narrow well and to both end wells of the sample comb (Figure 40-5d).

5. With individual micropipettes, introduce the protein sample into the sample wells, as follows: To introduce a sample, insert the micropipette or syringe between the teeth of the sample comb so that the tip of the micropipette or syringe is immediately above the gel. Gently express the sample onto the gel. In this electrophoretic gel, Samples 1, 2 and 3 will be tested in duplicate.

    a. Add 5 µl of Sample N to the first sample well on the left. (See Figure 40-5d. The end well and first narrow well have been filled with application buffer.)

    b. Add 5 µl of Sample 1 to the next sample well (second from left).

    c. Add 5 µl of Sample 2 to the third sample well from left.

    d. Add 5 µl of Sample 3 to the fourth sample well from left.

    e. Repeat 5-µl additions of Samples 1, 2, and 3 to the next three sample wells.

6. With a new syringe, apply 1 drop of the 0.1 percent phenol-red dye solution to the sample well containing sample N (the normal plasma sample). This chemical is a tracking dye, to show the movement in the system.

7. The samples are now ready for electrophoresis.

Procedure 4: Electrophoretic Run

This procedure is to be performed by students in groups of four.

1. Slide the safety interlock lid over the Isophore® unit, enclosing both the chambers (see Figure 40-4). Connect the power cables (generally, black = negative, cathode, and lower chamber; red = positive, anode, and upper chamber).

2. Turn on the power, and adjust the voltage to 200 volts on the power setting. The milliamperes (mA) should read 3.5 to 5.8 mA.

3. With no more than eight samples, constant voltage is applied for 60 minutes.

4. Continue the electrophoretic run until separation is complete. (Run times may generally be shortened by increasing the voltage.)

5. Upon completion of the run, turn off the power supply, remove the safety lid, and pull out the upper chamber.

6. Pour out the anode buffer, and remove the gel for fixation and staining.

Procedure 5: Fixation and Staining

This procedure is to be performed by students in groups of four.

1. After electrophoresis, open the cassette by slitting the tape on one side with a single-edge razor blade.

2. With the aid of a spatula, pry apart the two plates. Remove the plastic plate, leaving the gel on the glass plate.

3. Remove the sample spacer comb and the side spacers. To remove the comb, loosen the top screws on each clamp. Then pull up gently on the comb while rocking it back and forth. Air bubbles should be visible underneath the comb teeth.

4. Float the gel off the glass plate onto the transporter (Figure 40-5e). Place the holding screen over the gel to prevent it from floating away during the staining and destaining steps.

5. Place the gel in the tray containing the fixative Ponceau S dye solution for 7 to 10 minutes (Figure 40-5f).

6. Lift the transporter from the stain tray. Allow excess stain to drain back into the tray.

7. Transfer the transporter to the acetic acid rinse tray. Change the acetic acid 2 to 4 times, or until the dye no longer comes off the gel.

8. Transfer the gel to the clearing solution, and gently agitate the gel for about 1 minute.

9. Remove the gel and place it on a glass plate. Squeegee the gel gently to remove excess solution, and allow it to air dry. Place a drying weight on each edge of the gel, to prevent curling.

10. Place a clean glass plate over the dried gel, making a sandwich.

11. Place tape over the edges of the plates.

12. Label the tape to identify the different specimens used.

Procedure 6: Examination

This procedure is to be performed by students in groups of four.

1. Examine the stained gel, and identify in each sample the plasma proteins, albumin, and the alpha, beta, and gamma globulins. Refer to Color Plate 47.

2. Sketch and label a representative electrophoretic pattern for one plasma protein sample in the space provided in the results and observations section.

3. Answer the questions in the results and observations section.

RESULTS AND OBSERVATIONS

1. Sketch an electrophoretic pattern for one plasma protein sample in the space provided below. Label the respective fractions. Refer to Color Plate 47.

2. Were all plasma fractions similar in appearance in duplicate samples? If not, which ones were not?

_____

3. Which of the plasma protein fractions was the most

abundant? _____

LABORATORY REVIEW *40*

LYMPHATICS AND THE IMMUNE RESPONSE

COMPLETION, DEFINITION, AND INTERPRETATION

1. On each of the three slides shown in Figure Q40-1, the left side contains anti-A serum, the middle anti-B serum, and the right side anti-Rh_o(D) serum. One drop of an unknown blood sample was added to the antisera on each side. On the basis of the reactions shown, determine (i) the major blood type of each unknown blood sample and (ii) whether it contained the Rh_o factor.

(a)

a. Major blood factor _____

   Rh_o factor _____

(b)

b. Major blood factor _____

   Rh_o factor _____

(c)

c. Major blood factor _____

   Rh_o factor _____

Figure Q40-1

ANSWERS

2. Define or explain the following:

   a. Cross-matching _____

   _____

   b. Coombs' test _____

   c. Hemolytic disease of newborn (Rh baby) _____

   _____

   _____

   d. ABO incompatibility _____

   _____

   e. Rh factor _____

   f. Agglutination _____

   _____

   g. Hemolysis _____

3. List five major blood types other than the ABO and Rh groups.

   a. _____

   b. _____

   c. _____

   d. _____

   e. _____

4. What is the value of electrophoresis? _____

   _____

5. Define or explain the following.

   a. Anode _____

   b. Cathode _____

   c. Isoelectric point _____

   d. Amino group _____

   e. Carboxyl group _____

1. _____

2. _____

3. _____

4. _____

5. _____

6. _____

7. _____

8. _____

9. _____

10. _____

MULTIPLE CHOICE

Enter the correct answers in the spaces provided.

1. Which of the following are antibodies known to be present in the blood of individuals who are either blood type A, B, or O?
   a. agglutinogens
   b. agglutinins
   c. agglutinates

2. Against which of the following would an individual who is blood type A be expected to have antibodies?
   a. A
   b. B

   c. O
   d. Rh_O
   e. choices *a* and *b*
3. Against which of the following would an individual who is blood type B be expected to have antibodies?
   a. A
   b. B
   c. O
   d. Rh_O
   e. choices *a* and *b*
4. Against which of the following would an individual who is blood type O be expected to have antibodies?
   a. A
   b. B
   c. Rh_O
   d. choices *a* and *b*
   e. does not have antibodies against A, B, or O
5. Against which of the following would an individual who is blood type AB be expected to have antibodies?
   a. A
   b. B
   c. Rh_O
   d. choices *a* and *b*
   e. does not have antibodies against A, B, or O
6. Which of the following cells are known for their involvement with cell-mediated immune responses?
   a. B lymphocytes
   b. plasma cells
   c. erythrocytes
   d. lymphocytes
   e. T lymphocytes
7. How many different classes of immunoglobulins are recognized?
   a. 2
   b. 3
   c. 4
   d. 5
   e. 10
8. What is a substance that is recognized by the body as being foreign and that stimulates an immune response?
   a. immunoglobulin
   b. antibody
   c. antigen
   d. plasma cell
   e. foreign body
9. What is the term for cells from which plasma cells originate?
   a. T lymphocytes
   b. B lymphocytes
   c. antigens
   d. erythrocytes
   e. determinants
10. In which of the following do T lymphocytes originate?
   a. tonsils
   b. spleen
   c. liver
   d. thyroid gland
   e. thymus gland

The Respiratory System

W hen the term *respiration* is used alone, it generally refers to the overall metabolic activity that occurs within cells. When *respiration* is used in association with the *respiratory system,* it refers to the process of gaseous exchange between the atmosphere, the blood, and the cells.

During the exchange of gases, several processes are operating simultaneously. The movement of air between the atmosphere and the lungs is termed *ventilation.* It is achieved by *breathing,* a method of creating differences in pressure in the chest cavity. These pressure changes are a direct consequence of a coordinated contraction of muscles, especially the intercostals and those of the diaphragm and abdomen. *External respiration* is the gaseous exchange of carbon dioxide, oxygen, and some nitrogen that occurs between the lung cells and the lung capillary blood. *Internal respiration* is a similar exchange between the blood and cells throughout the body.

In addition to its major role in the process of gaseous exchange, the respiratory system plays a minor role in temperature regulation and water balance, based on the amount of water vapor lost during exhalation. It also participates—along with the kidney, liver, and gastrointestinal tract—in the overall regulation of blood pH.

The average normal individual has a lung capacity of about 6000 ml. This total is made up of several respiratory air components, including the expiratory reserve volume, inspiratory reserve volume, residual volume, and tidal volume (Figure XII-1). A variety of pulmonary-function tests are available to determine whether the lung tissues are functioning properly. In diseases such as carcinoma, emphysema, fibrosis, pneumonia, and pulmonary congestion, there is a substantial reduction in normally functioning lung tissue. Several pulmonary-function procedures are quite complex or require elaborate equipment. In Exercise 42, several of the more common tests are described and performed.

GENERAL REFERENCES

Lehrer, S. *Understanding Lung Sounds* (with audiocassette). Philadelphia: Saunders, 1984.
Levitzky, M. G. *Pulmonary Physiology.* New York: McGraw-Hill, 1982.
Tortora, G. J., and S. R. Grabowski. *Principles of Anatomy and Physiology,* 7th ed. New York: HarperCollins, 1993.

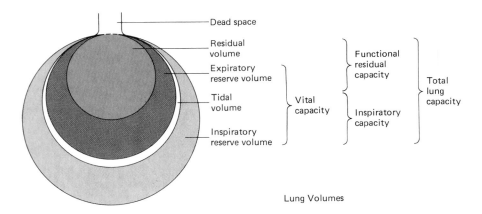

Lung Volumes

Figure XII-1 Diagrammatic representation of respiratory air components.

SPECIFIC REFERENCES

Laszlo, G., and Sudlow, M. F. (Eds.). *Measurement in Clinical Respiratory Physiology.* New York: Academic Press, 1983.

Naeye, R. L. "Sudden Infant Death." *Scientific American,* 242: 56–62, 1980.

Nero, A. V., Jr., "Controlling Indoor Air Pollution." *Scientific American,* 258: 42–48, 1988.

Schlesinger, R. B. "Defense Mechanisms of the Respiratory System." *BioScience,* 32: 45–50, 1982.

Weibel, E. R. "How Does Lung Structure Affect Gas Exchange?" *Chest,* 83: 657–665, 1983.

Whipp, B. J., and K. Wasserman (Eds.). *Lung Biology in Health and Disease: Vol. 52. Exercise—Pulmonary Physiology and Pathophysiology.* New York: Marcel Dekker, 1991.

Anatomy of the Respiratory System

After completing this exercise, you should be able to

1. Identify the organs of the respiratory tract of the fetal pig
2. Name the organs of the human respiratory tract from charts, models, and other demonstrations
3. Describe the functions of the organs in internal and external respiration

Materials

The following materials should be provided for class use:

1. Fetal-pig dissection specimens
2. Dissecting tools
3. Model of the human respiratory tract
4. Charts of the human respiratory system
5. Demonstration of a mammalian head cut in sagittal section

ORGANS OF THE FETAL-PIG RESPIRATORY SYSTEM

Locate the paired *external nares* on the large fetal-pig snout (Figure 41-1). Air is taken into the nares, passes through the *nasal chamber* and the *internal nares*, and then enters the *pharynx*. The pharynx is divided into three regions: (1) the *nasopharynx*, (2) the *oropharynx*, and (3) the *laryngopharynx*. Cut back at the angles of the jaw (Figure 41-2), and locate the *hard palate* and the *soft palate*. Place a probe into the opening of the *esophagus* and then into the *glottis*, the opening of the

larynx. Examine the tongue. Feel the rough papillae. The teeth may not have erupted as yet, but attempt to feel them along the gumline. From the laryngopharynx, the air passes through the glottis and enters the larynx.

Return to the thoracic cavity. Make a midsagittal cut into the ventral surface of the larynx and part of the trachea. Spread the two sides of the larynx, and observe the narrow *vocal cords* within it. Posterior to the larynx is the trachea, which runs the length of the neck into the thoracic cavity between the lungs. The trachea never flattens because there is a series of cartilaginous rings that keep it open. Run your finger over the trachea to feel these rings. Find the tracheal rings on your own body. At the lung, the trachea bifurcates to form paired *bronchi,* which then penetrate the lung tissue and subdivide many times (Figure 41-3).

Locate the right and left lungs. The right lung contains four lobes: anterior, medial, posterior, and mediastinal lobes. The left lung has three lobes: anterior, medial, and posterior lobes.

The *mediastinum* or *mediastinal cavity* is located between the lungs (review Exercise 8). This cavity contains the heart, esophagus, thymus gland, thyroid gland, trachea, and major blood vessels. Each lung is enclosed by membranes. The inner *pleural membrane* is the *visceral pleura,* which adheres closely to the lung. The outer pleural membrane is the *parietal pleura.* Between the two membranes is a potential space, the *pleural cavity.* The pleural membranes move with the lung as it expands and contracts during inhalation and exhalation.

Posterior and ventral to the larynx is a dark, roundish mass, the *thyroid gland,* which produces the hormones thyroxin and calcitonin. On either side of the trachea, there is a column-shaped whitish mass, the thymus gland. In the fetus and in young individuals, this gland is large. It produces certain T cells and prepares the body to resist disease organisms, thus setting up the immune system at a very early age. The thymus

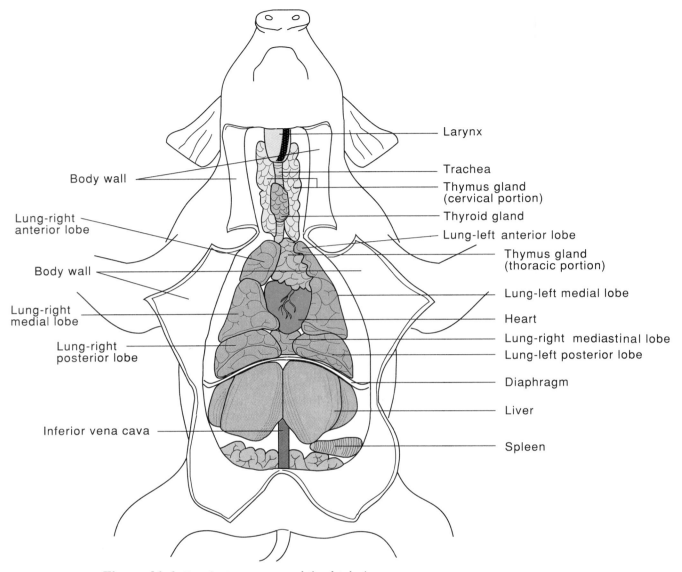

Figure 41-1 Respiratory organs of the fetal pig.

gland extends caudally and can be found as a more diffuse mass as far back as the heart. The anterior columnar masses, which are usually divided into three or four bundles (see Figures 41-1 and 41-3), are called the *cervical portion of the thymus gland.* The posterior diffuse region is the *thoracic portion of the thymus gland.*

THE HUMAN RESPIRATORY SYSTEM

Study models of the human respiratory system and all charts and photos available (see Figures 41-4 and 41-5).

Lung

The human lung has five lobes:

Right superior lobe
Right medial lobe
Right inferior lobe
Left superior lobe
Left inferior lobe

The relationship of the human pleura to the lungs, and the relationship between the human thoracic and mediastinal cavities is the same as it is in the pig. There are two pleural membranes (inner and outer), and the

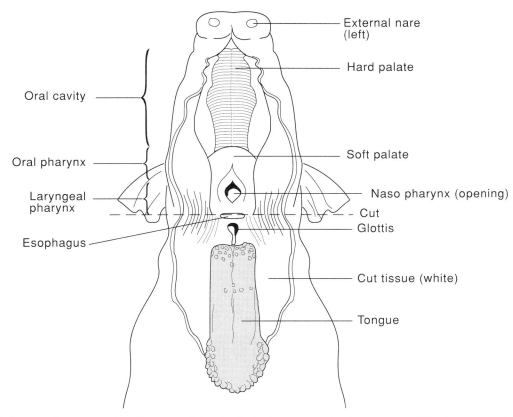

Oral cavity

Oral pharynx

Laryngeal
pharynx

Esophagus

External nare
(left)

Hard palate

Soft palate

Naso pharynx (opening)

Cut

Glottis

Cut tissue (white)

Tongue

Figure 41-2 The pharynx and oral cavity of the fetal pig (cut through at an angle of the jaw on each side).

potential space between these two layers is the pleural cavity. In this cavity is a small amount of fluid, which eases the friction during inspiratory and expiratory breathing movements (review Exercise 8).

Nares and Nasal Cavity

The *nostrils,* or *external nares,* open into a large bony cavity of the skull below the cranium and above the mouth (Figures 41-6 and 41-7). The bones involved are the *ethmoid, maxillae, palatines, vomer,* and the *inferior nasal conchae.* Review these bones in Exercises 10 and 11. The ethmoid bone (which contains the superior and middle nasal conchae) forms the lateral walls and roof of the nasal cavity, along with the inferior nasal conchae, which are separate bones. The floor of the nasal cavity is formed by the maxillae and the palatine bones; these four bones (paired palatines and maxillae) also form the *hard palate.* The cartilaginous part of the nasal septum that divides the nasal cavity into right and left sides is contiguous with the *perpendicular plate* of the ethmoid and vomer bones. In this cavity,

air is warmed, filtered, and moistened. Sensory olfactory nerve endings are found in parts of the epithelial lining, which is of the *pseudostratified ciliated* type (Figure 41-8).

Pharynx

Air from the nasal cavity passes through the *internal nares* (see Exercises 10 and 11) to enter a large cavity at the back of the mouth (the pharynx). It is a common passageway for both food and air and is highly muscular. The cavity is divided into three regions:

1. Nasopharynx
2. Oropharynx
3. Laryngopharynx

Lymphoid tissue—*tonsils*—is found in the pharynx. Two types—the *palatine tonsils* and the *lingual tonsils*—are located in the oropharynx, and the *pharyngeal tonsils* (adenoids) are located in the nasopharynx.

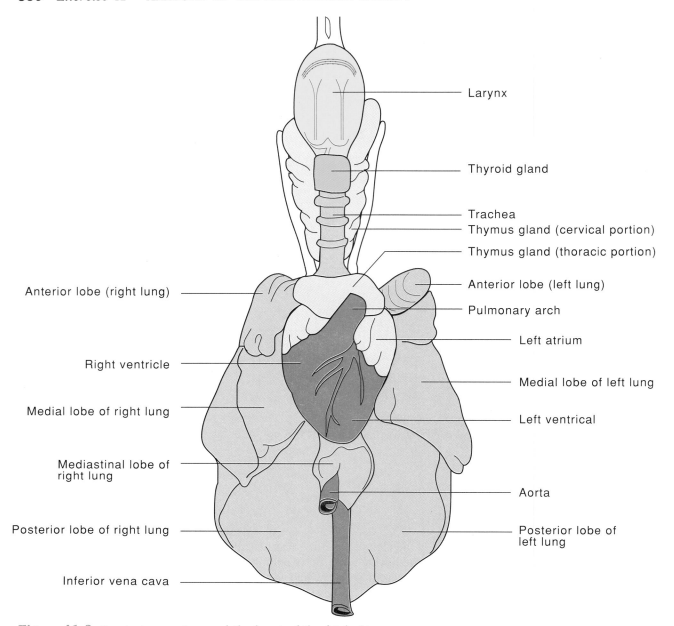

Larynx

Thyroid gland

Trachea
Thymus gland (cervical portion)
Thymus gland (thoracic portion)

Anterior lobe (right lung)

Anterior lobe (left lung)
Pulmonary arch
Left atrium

Right ventricle

Medial lobe of left lung

Medial lobe of right lung

Left ventrical

Mediastinal lobe of
right lung

Posterior lobe of right lung

Aorta

Posterior lobe of
left lung

Inferior vena cava

Figure 41-3 Respiratory system and the heart of the fetal pig.

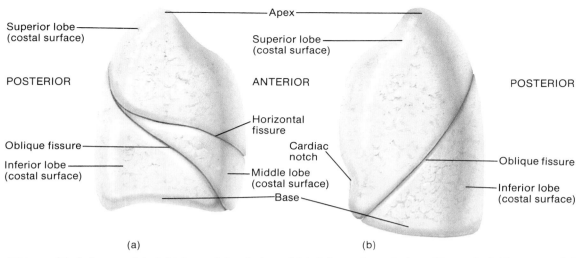

Figure 41-4 Lungs: (a) right lung, lateral view; (b) left lung, lateral view. (From G. J. Tortora and S. R. Grabowski, *Principles of Anatomy and Physiology,* 7th ed., New York: HarperCollins, 1993.)

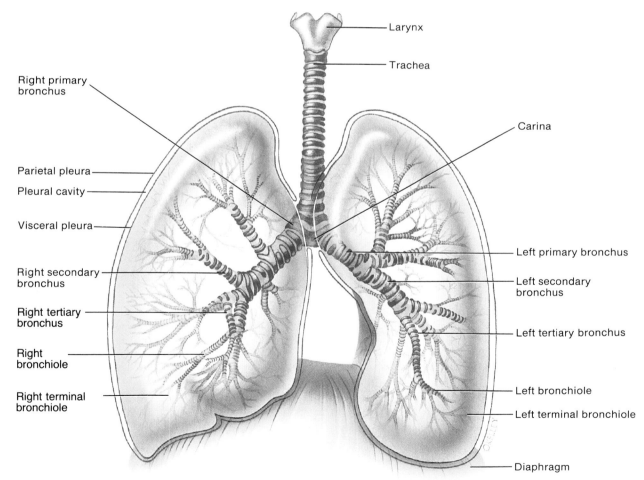

Figure 41-5 Air passageways to the lungs: the bronchial tree in relation to the lungs. (From G. J. Tortora and S. R. Grabowski, *Principles of Anatomy and Physiology,* 7th ed., New York: HarperCollins, 1993.)

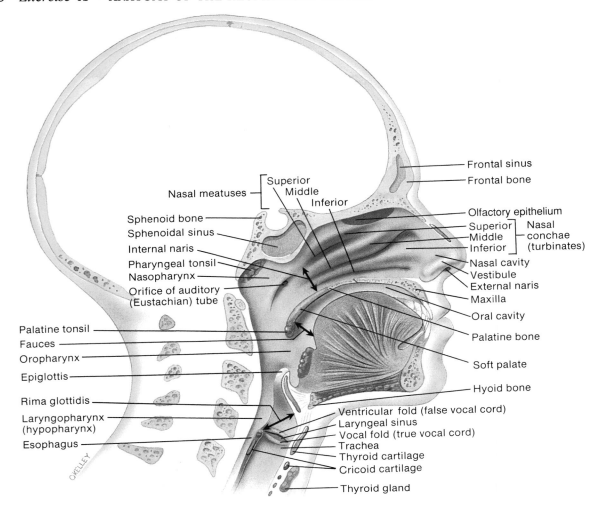

Figure 41-6 Human nose: left side of head and neck seen in sagittal section with the nasal septum. (From G. J. Tortora and S. R. Grabowski, *Principles of Anatomy and Physiology*, 7th ed., New York: HarperCollins, 1993.)

Larynx

The opening from the pharynx into the larynx is the *glottis*. Guarding this opening is a large laryngeal cartilage, the *epiglottis*. When one is swallowing, the epiglottis covers the opening to the glottis, and the food enters the esophagus. When inhaling or talking, the glottis is open, to permit air to flow through the larynx over the vocal cords to the lungs. Study Figure 41-9.

The vocal cords vibrate as air passes over them, producing sounds. The framework of the larynx consists of nine cartilages that are moved by muscles, many of which you have already observed in the fetal pig.

Human Laryngeal Cartilages

> Thyroid cartilage (forms the Adam's apple in males)
> Epiglottis
> Cricoid cartilage
> Paired arytenoid cartilages
> Paired corniculate cartilages
> Paired cuneiform cartilages

Trachea, Bronchi, and Bronchioles

The trachea is an air-conducting tube located in the mediastinal cavity. It is supported by cartilaginous rings

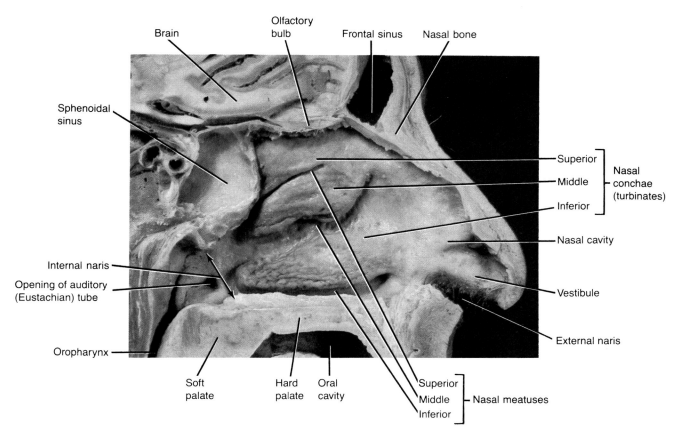

Figure 41-7 Midsagittal section of the human head, showing upper respiratory tract. (From G. J. Tortora and S. R. Grabowski, *Principles of Anatomy and Physiology,* 7th ed., New York: HarperCollins, 1993.)

Figure 41-8 Photomicrograph of enlarged aspect of the tracheal pseudostratified columnar epithelium at a magnification of 600×. (Courtesy of Andrew J. Kuntzman.)

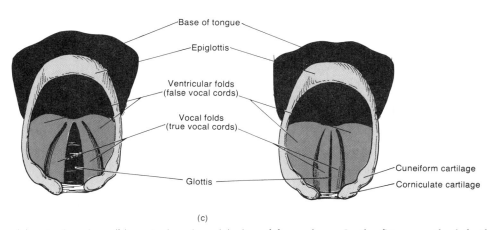

Figure 41-9 Larynx: (a) anterior view; (b) posterior view; (c) viewed from above. In the figure on the left, the true vocal cords, are relaxed. In the figure on the right, the vocal cords are pulled taut. (From G. J. Tortora and N. P. Anagnostakos, *Principles of Anatomy and Physiology,* 3rd, 4th, 5th, 6th eds., New York: HarperCollins, 1981, 1984, 1987, 1990.)

to keep it from collapsing. The trachea begins at the larynx and terminates at its bifurcation, the *carina,* into the bronchial tubes. There are many mucus-producing *goblet cells* in the trachea's pseudostratified ciliated epithelial lining. The bronchi and bronchioles branch about 24 times within the lung tissue and end in tiny air sacs, *alveoli,* through which gases are exchanged. This exchange of gases between alveoli and blood is *external respiration.* Study Figure 41-10 and define all parts indicated by your instructor.

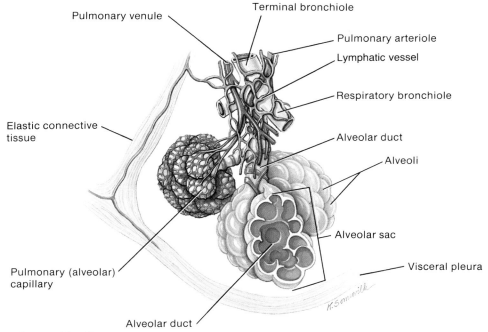

Figure 41-10 Portion of a human lung lobule showing alveoli. (From G. J. Tortora, *Introduction to the Human Body,* 2nd ed., New York: HarperCollins, 1991.)

LABORATORY REVIEW *41*

THE ANATOMY OF THE RESPIRATORY SYSTEM

MATCHING AND FIGURE IDENTIFICATION

1. Match the phrases on the left with the correct term on the right. Use numbers.

   a. _____ Cavity between membranes surrounding the lung

   b. _____ Name for two branches of the trachea

   c. _____ Respiratory passage that contains the vocal cords

   d. _____ Terminal part of respiratory-duct system in contact with blood

   e. _____ Opening into the larynx from the pharynx

   f. _____ Number of lobes in human lung

   g. _____ Name for the membranes surrounding the lungs

   h. _____ A bone that makes up the hard palate

   i. _____ Another bone that makes up the hard palate

   j. _____ A prominent bone within the nasal chamber

2. Trace a molecule of oxygen from the air until it reaches the blood.

   a. Air _____

   b. _____

   c. Nasal chamber _____

   d. _____

   e. _____

   f. Glottis _____

   g. _____

   h. _____

   i. _____

   j. Bronchioles _____

   k. _____

   l. _____

   m. Blood capillaries _____

3. Answer the following questions:

   a. _____ How is pseudostratified ciliated epithelium able to filter particles entering the nasal cavity?

   b. _____ Name two different body openings for air (b, c).

   c. _____

ANSWERS

1. Bronchioles
2. Pharynx
3. Maxillary
4. 5
5. Mediastinum
6. Palatine
7. Ethmoid
8. Larynx
9. Epiglottis
10. Pleural
11. Glottis
12. 4
13. Bronchi
14. Alveoli
15. Pericardial
16. 6
17. Mandible

d. _____ Can items (b) and (c) both be open, and thus used at the same time?

e. _____ In which specific body cavity is the lung located in both pigs and humans? (Review Exercise 8.)

f. _____ Is any organ other than the lung located in the cavity in question (e)?

g. _____ In which body cavity is the thymus gland located?

h. _____ Is any organ other than the thymus gland located in the cavity in question (g)? If so, which organ(s)?

i. _____ Which body cavity contains both lung and thymus?

j. _____ In which specific part of the lung does external respiration occur?

4. Label the items shown in Figure Q41-1.

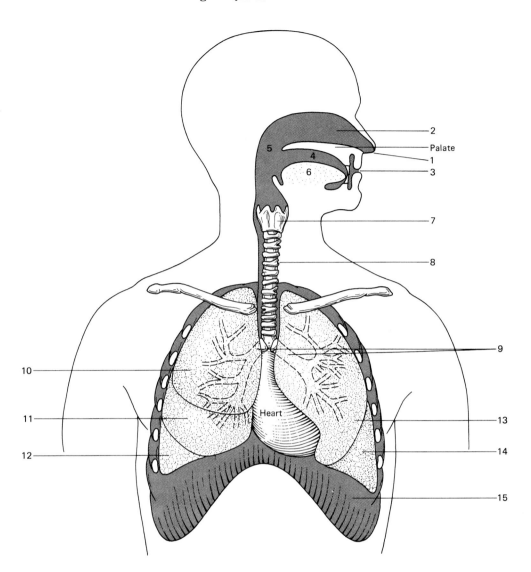

Figure Q41-1 Human respiratory tract.

1. _____   9. _____
2. _____   10. _____
3. _____   11. _____
4. _____   12. _____
5. _____   13. _____
6. _____   14. _____
7. _____   15. _____
8. _____

Pulmonary Function Tests

1. Define *atmospheric pressure, intrapulmonary pressure,* and *intrapleural pressure*
2. Perform standard measurements and calculations used in evaluating the functioning of the respiratory system
3. Define *tidal volume, minute volume, vital capacity, residual air, total lung capacity, inspiratory reserve, expiratory reserve,* and *dead space*
4. Interpret respiratory volume measurements as being indicative of normal or abnormal functioning

The respiratory system consists of a number of anatomic passageways through which the air we breathe travels to the delicately structured lungs. During *inspiration* (the movement of air from the atmosphere into the lungs), air is moved through the nose, pharynx, larynx, trachea, and successively smaller divisions of the bronchial tree until it reaches the numerous tiny air sacs, the *pulmonary alveoli.* Conditions at the alveoli are ideal for the physiological exchanges of gases. During *expiration* (the movement of air from the lungs into the atmosphere), the path is reversed. Air flows through these passageways during respiration because of a pressure gradient involving three important pressures:

1. *Atmospheric pressure*—the pressure exerted by the air surrounding the body, which at sea level is equal to 760 mm of mercury (mm Hg)
2. *Intrapulmonary pressure*—the pressure within the respiratory space of the lungs (bronchi and alveoli)

3. *Intrapleural pressure*—the pressure within the pleural space between the lungs and the walls of the thoracic cage

Various factors are known to regulate respiration. For example, the temperature of blood flowing through the respiratory center may influence breathing. Above-normal body temperature, as during fever or severe muscular exercise, results in an increase in the respiratory rate (number of breaths per minute). Subnormal temperature, as during hypothermia, results in a decreased respiratory rate. Other respiration-regulating factors include pain and emotional state.

Before birth, the lungs contain no air, but after birth, once they have been filled with air, they are never completely empty. No expiration ever completely empties the alveoli; neither are they completely filled. The ability of an individual's respiratory system to exchange air with the outside environment is important in evaluating normal function or the presence of respiratory disease. A number of different measurements can be made in studying the various properties of the respiratory system (Figure 42-1). These are described in Table 42-1 (a graphic display of the relationships among these is shown in Figure XII-1). The instrument used to measure lung capacities, volumes of air exchanged, and rates of respiration is the *spirometer.* The spirometer can play an important role in the detection of restrictive or obstructive lung defects. This exercise describes procedures for making some of the respiratory measurements.

Materials

The following materials should be provided for class use:

1. Spirometers, such as the Collins Survey Spirometer, the Breon Spirometer, or other appropriate device (with their operator's manuals, if possible)

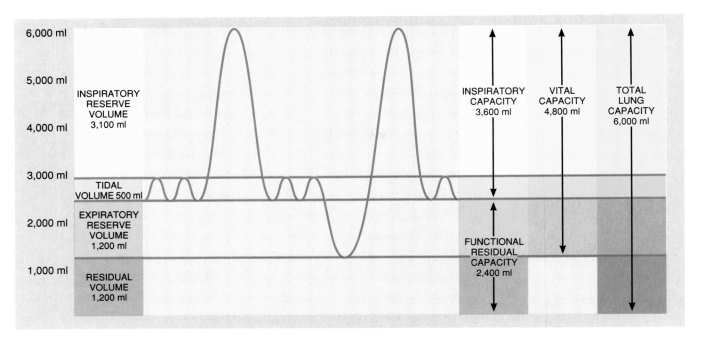

Figure 42-1 Spirogram of pulmonary volumes and capacities. (From G. J. Tortora and S. R. Grabowski, *Principles of Anatomy and Physiology,* 7th ed., New York: HarperCollins, 1993.)

Table 42-1   Respiratory Volume Measurements

Measurement	Brief Description	Normal Values for Adults
Tidal volume (TV)	Quantity of air moved into and out of the lungs during a normal breath	500 ml
Respiratory rate (RR)	Number of breaths per minute	12–15
Minute volume (MV)	Quantity of air moved into and out of the lungs in 1 minute (MV = TV × RR)	6000 ml/min
Inspiratory reserve volume (IR)	Quantity of air above tidal volume that can be forcefully inspired after a normal inhalation	3000 ml
Expiratory reserve volume (ERV)	Quantity of air above tidal volume that can be forcefully expired after a normal inhalation	1200 ml
Vital capacity (VC)	Maximum quantity of air that can be moved into and out of the lungs (VC = IR + TV + ER)	4500–5500 ml
Residual air (RA)	Quantity of air left in the lungs after maximum expiration	1200 ml
Total lung capacity (TLC)	Quantity of air the lungs can hold (TLC = VC + RA)	5900 ml
Dead space	Air that is not available for gas exchange	100 ml (female) 150 ml (male)

2. Breathing tube attachment
3. Disposable mouthpieces (adult)
4. Nose clips
5. Disinfectant and sterile gauze pads
6. Recording charts or paper
7. Recording pens and ink

Procedure 1: Identification of Spirometer Parts and Accessories

This procedure is to be performed by students individually.

1. Examine the spirometer provided.
2. With the aid of Figure 42-2 and the operator's manual, identify the typical parts and accessories of the instrument.
3. Your instructor will demonstrate the correct usage of the spirometer.

Procedure 2: Respiratory Rate

This procedure is to be performed by students in pairs.

1. Have your partner sit quietly for 5 minutes.

2. After the rest period, count the number of complete respiratory cycles (inspiration and expiration) in 60 seconds. Make two additional counts while your partner is still seated.
3. Make three counts with your partner standing and three counts while your partner is in the supine position.
4. Determine the average respiratory rate in each position by dividing the number of times counts were taken into the number of complete cycles. Enter your findings in the results and observations section.
5. Change places with your partner, and repeat Steps 1–4.

Procedure 3: Tidal Volume (TV)

This procedure is to be performed by students in pairs.

1. Have your partner sit quietly for 5 minutes.
2. Disinfect a nose clip. Place it on your partner's nose so that the nasal passages are closed.
3. Place the disposable mouthpiece into your partner's mouth.

(a)

Mouthpiece

(b)

Figure 42-2 Two instruments used for the measurement of lung volumes and capacities (spirometry)—note the various components of each: (a) Collins Respirometer (courtesy of Warren E. Collins, Inc., Braintree, MA); (b) Breon Spirometer (courtesy of Breon Laboratories, Inc., New York, NY).

4. Start the recording system, and have your partner breathe normally into the spirometer system for five complete respiratory cycles.
5. Record the time (in seconds) for each inspiration and the volume of air inspired. The average value for the inspiratory volume is the *tidal volume* at rest.
6. Repeat this procedure with your partner in the standing position and in the supine position.
7. Change places with your partner, and repeat Steps 1–6.
8. Record the measurements in the results and observations section.

Procedure 4: Expiratory Reserve Volume (ERV)

This procedure is to be performed by students in pairs.

1. Have your partner sit quietly and breathe normally.
2. Attach a nose clip, as described in Procedure 3.
3. Turn on the recording system.
4. Instruct your partner to breathe normally for four respiratory cycles, and then on the fifth normal expiration to exhale as much as possible into the spirometer mouthpiece. (Help your partner by counting the inspirations and expirations of the complete cycles.)
5. Record the volume.

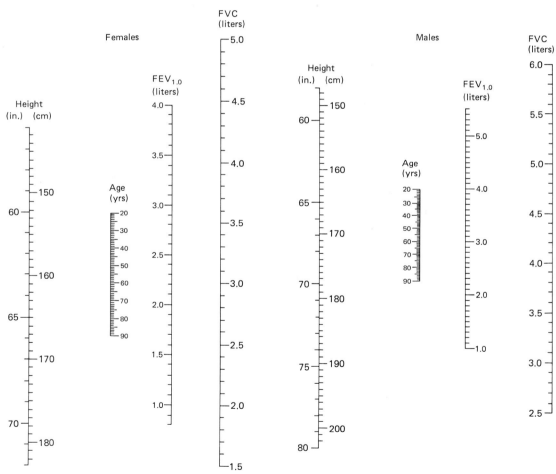

Figure 42-3 Scales for the determination of normal spirometric values for females and males. These values are expressed in terms of a normal range for age, height, and sex. To use these scales, place the edge of a ruler at the correct point for the subject on the height scale and at the correct point on the age scale. The points at which the ruler intersects the $FEV_{1.0}$ (forced expiratory volume in 1 second) and FVC (forced vital capacity) scales give the normal values for those measurements. (Modified from T. L. Petty, *Office Spirometry*. New York: Breon Laboratories, 1980.)

6. Repeat Steps 4 and 5 two more times, and calculate an average value.
7. Repeat the procedure with your partner in the standing position and in the prone position.
8. Change places with your partner, and repeat Steps 1–7.
9. Record the measurements in the results and observation section.

Procedure 5: Forced Expiratory Vital Capacity (FVC)

This procedure is to be performed by students in pairs.

1. Have your partner sit quietly and breathe normally.
2. Attach a nose clip, and turn on the recording system.
3. Instruct your partner to breathe normally for four complete respiratory cycles and then on the fifth cycle to inhale as deeply as possible and to exhale as forcefully and steadily as possible into the spirometer mouthpiece.

4. Record the volume, and time the expiratory portion of the last cycle. Determine the average FVC value.
5. Repeat Steps 3 and 4 two more times.
6. Change places with your partner, and repeat Steps 1–5.
7. Find the normal values for FVC on Figure 42-3.
8. Record the measurements in the results and observations section.

RESULTS AND OBSERVATIONS

1. Enter your RR determinations in Table 42-2 (A = average).
2. Enter your TV determinations in Table 42-3 (T = time, V = volume, and A = average).
3. Enter your ERV determinations in Table 42-4 (A = average).
4. Enter your FVC determinations in Table 42-5 (V = volume and T = time for forced expiration).

Table 42-2 RR

Student	Sitting				Standing				Prone			
	1	2	3	A	1	2	3	A	1	2	3	A

Table 42-3 TV

Student	Sitting								Standing								Prone							
	1		2		3		A		1		2		3		A		1		2		3		A	
	T	V	T	V	T	V	T	V	T	V	T	V	T	V	T	V	T	V	T	V	T	V	T	V

Table 42-4 ERV

Student	Sitting				Standing				Prone			
	1	2	3	A	1	2	3	A	1	2	3	A

Table 42-5 FVC

Student	1		2		3		Average	
	V	T	V	T	V	T	V	T

LABORATORY REVIEW *42*

PULMONARY FUNCTION TESTS

COMPLETION, DEFINITION, AND INTERPRETATION

1. Distinguish between inspiration and expiration. _____

   _____

   _____

2. Complete Table Q42-1 by briefly defining or explaining the terms listed and giving the normal values for each.

Table Q42-1

Measurement	Definition	Normal Value
Tidal volume	a.	b.
Minute volume	c.	d.
Inspiratory reserve volume	e.	f.
Expiratory reserve volume	g.	h.
Vital capacity	i.	j.
Residual air	k.	l.
Total lung capacity	m.	n.
Dead space	o.	p.
Respiratory rate	q.	r.

3. Define or explain the following:

   a. Atmospheric pressure _____

   _____

   b. Intrapulmonary pressure _____

   _____

   c. Intrapleural pressure _____

   _____

4. Indicate whether the measurement values in Table Q42-2 are normal or abnormal for adults.

Table Q42-2

Measurement	Value(s)	Normal/Abnormal
Tidal volume	200 ml	a.
	500 ml	b.
Inspiratory reserve	1000 ml	c.
	2000 ml	d.
	3000 ml	e.
Vital capacity	2000 ml	f.
	4500 ml	g.
	5000 ml	h.
	5500 ml	i.
Total lung capacity	3500 ml	j.
	6000 ml	k.

5. List three major physiological functions of the respiratory system.

   a. _____

   b. _____

   c. _____

6. How would the respiratory rate be affected by the following situations?

   a. High fever _____

   b. Running _____

   c. Hypothermia _____

7. What is *hypoxia?* _____

   _____

8. Define or explain *hyperventilation.* _____

   _____

9. List four factors that regulate respiration.

   a. _____

   b. _____

   c. _____

   d. _____

10. With the aid of your textbook or other reference, complete the descriptions of respiratory disorders in Table Q42-3.

Table Q42-3

Disorder	Description
a.	Collapse of lung tissue caused by extrinsic pressure or intrinsic pressure in pulmonary airways
Bronchitis	b.
Tuberculosis	c.
d.	Allergic reaction to proteins in foreign substances localized in respiratory membranes
Emphysema	e.
Pneumonia	f.
Pneumocystis infection	g.

MULTIPLE CHOICE

Enter the correct answers in the spaces provided.

1. Which of the following describes the movement of air between the atmosphere and the lungs?
   a. external respiration
   b. internal respiration
   c. breathing
   d. ventilation
   e. choices *a* and *c* only

2. What is the term for the gaseous exchange of carbon dioxide, oxygen, and some nitrogen that occurs between lung cells and lung capillary blood?
   a. external respiration
   b. internal respiration
   c. breathing
   d. ventilation
   e. choices *a* and *c* only

3. What is the term for the gaseous exchange between the blood and cells throughout the body?
   a. external respiration
   b. internal respiration
   c. breathing
   d. ventilation
   e. choices *a* and *c* only

1. _____
2. _____
3. _____
4. _____
5. _____
6. _____
7. _____
8. _____
9. _____
10. _____

4. Which of the following represents the pressure within the respiratory space of the lungs (bronchi and alveoli)?
   a. atmospheric pressure
   b. intrapulmonary pressure
   c. intrapleural pressure
   d. expiration
   e. inspiration

5. Which of the following represents the pressure exerted by the air surrounding the body, which at sea level is equal to 760 mm Hg?
   a. atmospheric pressure
   b. intrapulmonary pressure
   c. intrapleural pressure
   d. expiration
   e. inspiration

6. What is the term for the quantity of air moved into and out of the lungs during a normal breath?
   a. residual air
   b. total lung capacity
   c. vital capacity
   d. tidal volume
   e. minute volume

7. What would a tidal volume of 300 ml be considered?
   a. normal
   b. abnormal

8. What is the quantity of air moved into and out of the lungs in 1 minute?
   a. residual air
   b. total lung capacity
   c. vital capacity
   d. tidal volume
   e. minute volume

9. What is the term describing the quantity of air that the lungs can hold?
   a. residual air
   b. total lung capacity
   c. vital capacity
   d. tidal volume
   e. minute volume

10. What is the term for the quantity of air left in the lungs after maximum expiration?
    a. total lung capacity
    b. residual air
    c. vital capacity
    d. minute volume
    e. tidal volume

SECTION XIII

The Digestive System

As a group, nutrients such as carbohydrates, lipids, proteins, vitamins, and minerals are obtained from foods. Three processes take place in the body to convert the food consumed into usable forms. These processes are digestion, absorption, and metabolism. Exercises in this section consider selected components of the digestive system and emphasize the major types of normal activities, as well as some abnormal activities, associated with each one. Attention also is given to the role of microorganisms in tooth decay (caries).

Also examined are enzymes and the effects of factors such as pH and temperature on enzymatic activity. Tests used in the identification of proteins, carbohydrates, lipids, cholesterol, and other chemicals of importance in foods are demonstrated as well.

GENERAL REFERENCES

Davenport, H. W. *Physiology of the Digestive Tract,* 5th ed. Chicago: Year Book Medical Publishers, 1982.

Netler, F. H. *The CIBA Collection of Medical Illustrations, Vol. 3, Part I, Upper Digestive System.* Summit, NJ: CIBA Pharmaceutical Products, 1959.

Sleisenger, M. H., and J. S. Fordtran (Eds.). *Gastrointestinal Disease: Pathophysiology, Diagnosis, Management,* 3rd ed., Vols. 1 and 2. Philadelphia: Saunders, 1984.

Tortora, G. J., and S. R. Grabowski. *Principles of Anatomy and Physiology,* 7th ed. New York: HarperCollins, 1993.

SPECIFIC REFERENCES

Davenport, H. W. "Why the Stomach Does Not Digest Itself." *Scientific American,* 226: 86–93, 1972.

Martin, R. J., B. D. White, and M. G. Hulsey. "The Regulation of Body Weight." *American Scientist,* 79: 528–541, 1991.

Merlo, A., and S. Cohen. "Swallowing Disorders." *Annual Reviews Medicine,* 39: 17–28, 1988.

Moog, F. "The Lining of the Small Intestine." *Scientific American,* 245: 154–176, 1981.

Peters, M. N., and C. T. Richardson. "Stressful Life Events, Acid Hypersecretion and Ulcer Disease." *Gastroenterology,* 84: 114–119, 1983.

Sacks, F. M., and W. W. Willett. "More on Chewing the Fat—The Good Fat and the Good Cholesterol." *The New England Journal of Medicine,* 325: 1740–1742, 1991.

Track, N. S. "The Gastrointestinal Endocrine System." *Canadian Medical Association Journal,* 122: 287–292, 1980.

Wilkins, R. A., and H. B. Nunnerley. *Imaging of the Liver, Pancreas, and Spleen.* Boston: Blackwell Scientific, 1990.

The Organs of the Digestive System

After completing this exercise, you should be able to

1. List the components of the mouth
2. List the areas of specific taste reception on the tongue
3. Identify and give the functions of the four groups of teeth
4. List the organs in the digestive tract
5. Point out the associated digestive organs and the ducts connecting them to the digestive tract

Materials

The following materials should be provided for class use:

1. Fetal-pig dissection specimens
2. Dissecting tools
3. Model of the human torso
4. Chart of the human digestive system

INTRODUCTION: DIGESTION

Most foods we consume cannot be utilized directly and must be digested or chemically broken down into smaller components that can be absorbed into the blood and cells. This process is the major function of the *digestive system*. The chemical processes involved in the splitting of complex chemicals into simpler ones are *hydrolytic activity* (splitting in the presence of water molecules) and *enzymatic activity*. Hydrolysis is the basic mechanism for carbohydrate, protein, and fat digestion (the three major components of human diets). Specific enzymes are involved for each type of food substance. Digestion includes all the processes by which food is converted into end products that may be absorbed from the lumen of the intestine into the blood. Digestion begins in the mouth and continues in the stomach and small intestine. Elimination is the main function of the large intestine.

Many cell types contribute to digestion. Some produce mucus, which lubricates the tract and provides the medium for action of hydrolytic enzymes, while others secrete various chemicals including enzymes.

THE MAMMALIAN DIGESTIVE TRACT

Digestion and the subsequent absorption and elimination of nonabsorbed foodstuffs involve the organs in the digestive tract and the associated digestive glands. Each organ in it is specifically constructed and adapted for different stages in the digestive process.

The digestive system includes the digestive tract and associated organs such as the salivary glands, liver, pancreas, and gallbladder. The salivary glands, liver, and pancreas are glandular and produce secretions. The gallbladder is a storage sac for bile.

ORGANS OF THE DIGESTIVE TRACT—FETAL-PIG DISSECTION

Oral Cavity

Return to the dissection of the pig's oral cavity and pharynx (Figure 43-1), and review the organs studied in the previous exercise. Find both the *hard palate* beneath which the *oral cavity* is located and the *soft palate*, which separates the nasal cavity from the phar-

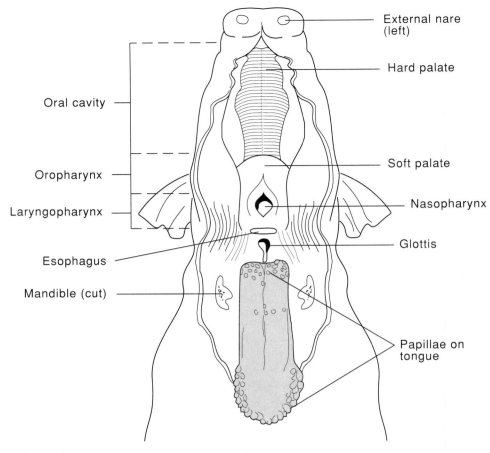

Figure 43-1 The oral cavity of the fetal pig.

ynx. The mouth (oral) cavity contains a muscular tongue and teeth. Make the following observations:

1. *Tongue*—Run your finger over the surface of the tongue. Notice the papillae of various shapes located on its surface. Move your finger in both directions (front to back and vice versa). In which direction do these papillae point? What do you think is their function?
2. *Teeth*—Examine the gums in the upper and lower jaws. Determine whether any teeth have erupted. There will be differences among the pigs in the class specimens. If you cannot see any teeth, determine by rubbing your finger gently over the gum line whether there are bulges indicating that teeth are just below the surface.

Pharynx

The pharynx (throat) is divided into the following regions:

1. The *oral pharynx* or *oropharynx* is the area you have observed beneath the soft palate.

2. The *nasal pharynx* or *nasopharynx* is the portion lying above the soft palate. If you slit the soft palate, you can look into the nasopharynx and see that there is a septum (the *nasal septum*), which divides the nasal chamber into right and left sides. The openings from these chambers are the *internal nares.* Also look for two small slits, which are the openings to the *eustachian tubes,* which connect the pharynx and the middle ear cavity.

Digestive Organs in the Thoracic Cavity and Neck of the Fetal Pig

At this time, return to the body cavity of the pig. Examine the area beneath the larynx and the trachea. Locate the *esophagus,* a flat muscular tube that begins at the pharynx and continues back to the diaphragm. Place a probe or similar tool into the opening to the esophagus in the oropharynx. Push the probe as far back as possible, and observe its presence in the esophagus. Separate the esophagus from the connective tissue surrounding it until you see it penetrating the diaphragm. It merges into the stomach just below the

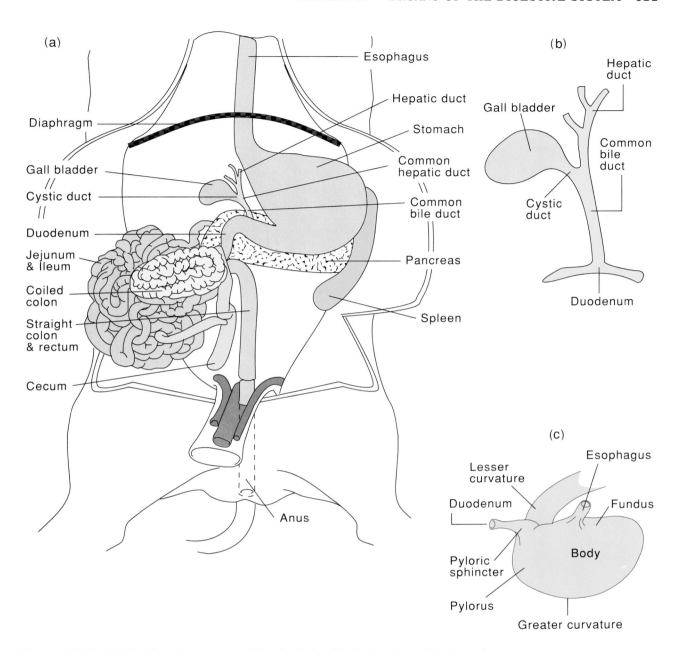

Figure 43-2 (a) The digestive system of the fetal pig; (b) duct system; (c) stomach.

diaphragm. Before leaving the thoracic cavity, note once more the thyroid gland, thymus tissue, heart, and major blood vessels.

Digestive Organs in the Abdominal Cavity (Figures 43-2 and 43-3)

The *digestive tract* (alimentary tract) is the series of organs through which the food passes between the mouth and the anus. Continue tracing the series of organs of the digestive tract into the abdominal cavity.

Stomach

Below the diaphragm, locate the point at which the esophagus penetrates the diaphragm. It immediately expands into a large saclike food reservoir, the *stomach* (see Figure 43-2c). Lift the left lobes of the liver, and observe this thin-walled organ (Fig. 43-3). The stomach is actually lying on its right side, so that you are looking at the anatomical left side.

A part of the stomach near the opening of the esophagus at the *esophageal (cardiac) valve* rises above

Figure 43-3 The fetal-pig's digestive system, with the liver lobes raised to expose the underside.

the esophageal opening. This small region is the *fundus* of the stomach. In this part of the stomach, note a small triangular diverticulum (outpocketing), which probably represents a vestigial additional stomach.

Next to the fundus, and making up most of the stomach, is the *body* of the stomach. The far end of the organ near the small intestine, is the *pyloric* region of the stomach, or the *pylorus*. This terminates in the *pyloric sphincter*, which regulates the passage of food into the small intestine. The somewhat concave anterior border of the stomach (anatomical ventral side) is the *lesser curvature*, and its posterior convex border (anatomical dorsal side) is the *greater curvature*.

The *greater omentum*, a large mesentery attached to the greater curvature of the stomach, suspends the stomach from the dorsal body wall. Due to embryonic torsion of the stomach, the greater omentum becomes a double-walled sac. Make a longitudinal slit in the stomach from the cardiac sphincter to the pyloric sphincter. Remove the greenish mucus *(meconium)* from the interior. Wash it out, and observe the corrugated ridges—termed *rugae*—that make up the inner lining. These rugae are characteristic of mammalian stomachs.

Small Intestine

The *small intestine,* the longest part of the digestive tract, begins at the pyloric sphincter and consists of three regions:

1. *Duodenum*—This relatively short portion of the small intestine extends from the pyloric sphincter to the first backward bend of the small intestine. Trace the duodenum until you locate this backward bend. Demonstrate the entire length of the duodenum to your instructor.

2. *Jejunum*—Begin at the first backward bend of the small intestine. This marks the beginning of the jejunum, which extends for approximately half the remaining distance of the small intestine. It is the same diameter as the duodenum and the next part of the small intestine. One cannot distinguish them externally, but there are definite differences histologically.

3. *Ileum*—The small intestine continues, twisting many times as the ileum. Trace it along its length. When you notice a distinct increase in the diameter, you have reached the end of the ileum. At this time, cut out about 1 inch in length from the small intestine. Slit it open and wash it out. Gently rub your finger over the surface, which forms a pile much like that of velvet cloth. These tiny elevations are *villi,* and they increase the surface area of the small intestine. These villi are a feature of all mammalian small intestines, including those of humans.

Large Intestine (Colon)

At the juncture with the small intestine, there is a sphincter valve, the *ileocecal sphincter* (Figure 43-3). It joins the ileum to a diverticulum of the colon, the *cecum.* The cecum of the fetal pig is proportionately longer than the human cecum. There is no *appendix* in the fetal pig. The colon begins as a tightly packed, coiled mass. Follow it until it straightens out and becomes a tube that turns freely. It terminates as a relatively straight *rectum.* The parts of the large intestine are often referred to as the *coiled colon,* the *straight colon,* and the *descending colon (rectum).* The *anus* opens to the outside from the rectum or descending colon.

Organs Associated with the Digestive Tract

The following associated digestive organs contribute to the digestive process but are not a part of the alimentary tract: *three glandular organs* (salivary glands, liver, and pancreas), and *one storage sac* (gallbladder).

Salivary Glands

The following paired salivary glands have already been noted and dissected. Review them once more:

Parotid glands
Submandibular glands
Sublingual glands
Buccal glands

Liver

The liver (Figure 43-4) is the largest mass in the abdominal cavity and has five lobes. In addition to its many other functions, the liver produces *bile,* which aids in digestion. Bile has no enzymes, but its salts neutralize chyme, facilitate the digestion of fats, and stimulate peristalsis. The bile also contains bile pigments, a waste product. With the aid of Figure 43-3, find each of the five liver lobes on your pig specimen:

1. Left medial lobe
2. Left lateral lobe
3. Right medial lobe
4. Right lateral lobe
5. Right caudate lobe

Gallbladder

Gently lift the liver with your fingers, and examine the underside of the organ. The *gallbladder* (Figure 43-2b) is a small colorless sac, which is attached to the underside of the right medial lobe of the liver by connective tissue. With small pointed scissors, free it and its duct, the *cystic duct.* Pull the gallbladder down far enough to see its attachment to the duct from the liver, the *common hepatic duct* (Figure 43-2b). The liver tissue is frangible, and the gallbladder and cystic duct are small and without the usual bile-green color. Be very careful. Dissect and demonstrate this duct system to your instructor.

Pancreas

The pancreas (Figure 43-2) is light-colored, lobulated tissue lying between the stomach and the duodenum. It produces *pancreatic juice,* which contains powerful digestive enzymes. The two parts to the organ are the *head of the pancreas* and the *tail of the pancreas.* The head lies along the medial side of the duodenum, and the tail passes laterally to the animal's left under the stomach to the *spleen.* The spleen, which belongs to the lymphatic system (see Exercise 35) is located on the left side of the abdominal cavity below the stomach. Notice that it lies in a part of the greater omentum. The *common bile duct* lies in another small mesentery, the *lesser omentum.* Along with the common bile duct, in the lesser omentum, are the *portal vein* and several blood vessels.

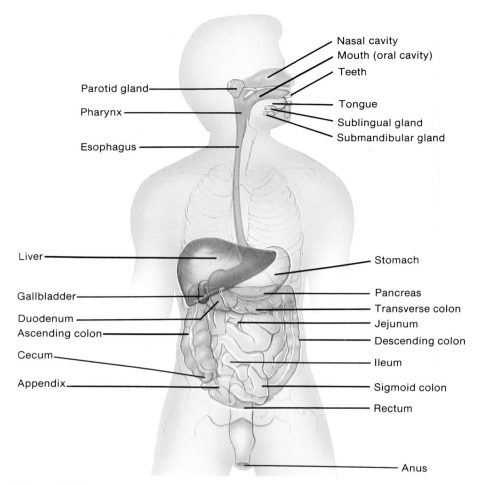

Figure 43-4 Organs of the human digestive system and related structures. (From G. J. Tortora and S. R. Grabowski, *Principles of Anatomy and Physiology,* 7th ed., New York: HarperCollins, 1993.)

HUMAN DIGESTIVE SYSTEM

The human digestive system (Figure 43-4) is remarkably similar to the one you have just dissected in the fetal pig. Compare the organs in Figure 43-4 with those in the fetal pig. The digestive system includes the *digestive (alimentary) tract* and the *associated digestive organs* (salivary glands, liver, pancreas, and gallbladder).

THE HUMAN MOUTH

The floor of the human mouth (see Figure 43-5) lies in a horseshoe around the tongue and is continuous laterally with the gum, or *gingiva,* and *centrally* with the tongue (Color Plate 48). Near the front end are the openings of the submandibular and sublingual salivary glands (Figure 43-5 and Color Plate 49). These glands empty their secretions into the mouth. Here, the fluids aid in swallowing, chewing, and digestion.

The tongue is made up of skeletal muscle. In addition to its role in chewing and swallowing, the tongue is important to humans because it assists in speech and contains receptors for taste. The rough surface of the tongue is due to several types of *papillae* (elevations) that are heavily supplied with taste buds. Taste buds are accumulations of taste cells that carry stimuli to the central nervous system (see Color Plate 33). The sense of taste provided by the taste buds appears to be limited to four basic types—bitter, salty, sour, and sweet. These tastes are localized in various areas of the tongue. (Refer to Exercise 32.)

The human is supplied with two sets of teeth: 20 deciduous teeth (baby or milk teeth) and 32 permanent teeth. Teeth are categorized into four groups: *Incisors* are used for cutting food, *canines* for tearing food, and

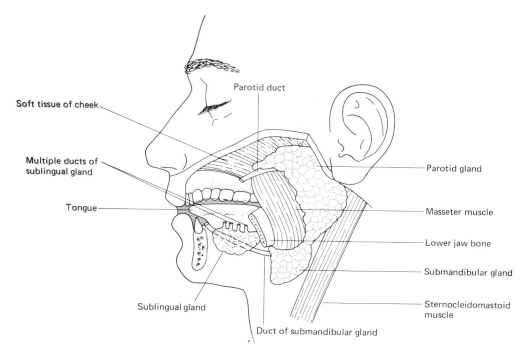

Figure 43-5 General organization of the human mouth.

premolars (bicuspids) and *molars* for grinding. The parts of each tooth are the *crown,* the portion above the gum; the *root,* the structure embedded in the jaw; and the *neck,* the constricted region between the crown and the root.

Materials

Identify the human mouth parts using the following materials:

1. Models of human adult skull with teeth
2. Hand mirrors
3. Wooden tongue depressors

Procedure

This procedure is to be performed by students in pairs.

1. Examine a human skull model, and identify the major categories of adult teeth.
2. Use a hand mirror to examine your own mouth. Identify the major categories of teeth present. Do you have a full set of 32 teeth?
3. Examine your laboratory partner's mouth, using a tongue depressor, and identify the base of the tongue, gum tissue, visible parts of teeth, and major categories of teeth. Use Figure 43-5 for reference. Does your laboratory partner have a full set of teeth?

Pharynx

Like the pharynx of the pig, the human pharynx (Figure 43-6) has three divisions:

1. *Oropharynx*—Find this region on Figure 43-6. It lies posterior to the oral cavity from the soft palate to the hyoid bone. It is lined by stratified squamous epithelium and forms a common passageway for food and air. Both the *palatine tonsils* and *lingual tonsils* are located in this division.
2. *Nasopharynx*—This portion extends from the nasal cavity to its juncture with the *oropharynx*. It contains two pairs of openings (like the fetal pig): the *internal nares* and the openings to the *eustachian tubes*.
3. *Laryngopharynx*—This small cavity behind the *epiglottis* extends to the esophagus posteriorly and the larynx anteriorly.

Esophagus

This collapsed muscular tube lies dorsal to the trachea and transports food to the stomach.

Stomach

The human stomach (Figures 43-4 and 43-7) is similar to that of the pig, in that it has basically the same parts.

Pharyngeal tonsil

Nasopharynx

Opening of auditory (Eustachian) tube

Soft palate

Oropharynx

Epiglottis

Laryngopharynx (hypopharynx)

Cricoid cartilage

Esophagus

Inferior nasal concha

Hard palate

Oral cavity

Tongue

Mandible

Hyoid bone

Thyroid cartilage (Adam's apple)

Trachea

Sagittal section

Figure 43-6 Photograph of a portion of the left side of the human head and neck, with the nasal septum removed. (Courtesy of J. A. Gosling, P. F. Harris, et al., *Atlas of Human Anatomy,* Gower Medical Publishing Ltd., 1985.)

It lies on its anatomical right side. There is an *esophageal (cardiac) sphincter* between it and the esophagus. The stomach itself is composed of a *fundus, cardia, body,* and *pylorus.* Locate these on the diagram of the stomach. There is a *pyloric sphincter* between the stomach and the duodenum. It is lined with columnar epithelium (Figure 43-7) and has three layers of smooth muscles:

1. Longitudinal layer
2. Circular layer
3. Oblique layer

A gross examination of the stomach lining reveals the same muscular corrugations, the *rugae,* seen in the pig stomach.

Small Intestine

The length (approximately 21 feet) of the small intestine (Figure 43-4) is consistent with the human's omnivorous diet. The small intestine is divided into three parts, distinctive from one another by histological examination:

1. *Duodenum*—Approximately the first 10 to 12 inches

to the first backward bend. The major ducts open into this part of the small intestine.
2. *Jejunum*—This next region (about 8 feet long) is beyond the duodenum.
3. *Ileum*—The ileum (about 12 feet long) is the last part of the small intestine. The end of the jejunum and the beginning of the ileum cannot be distinguished by any external marking. They are different histologically, however.

Large Intestine or Colon

Two vestigial parts of the colon (Figure 43-8)—the *cecum* and the *vermiform appendix*—are located inferior to the point of entry of the ileum, where the *ileocecal Sphincter* is located. Recall that the pig had a longer cecum but had no appendix. Other parts of the colon:

Ascending colon—The digestive tube has enlarged in diameter from approximately 1 inch in the small intestine to about 2½ inches in the colon. The ascending colon occupies the right side of the abdominal cavity between the cecum and a sharp bend to the left under the liver. This sharp bend is the

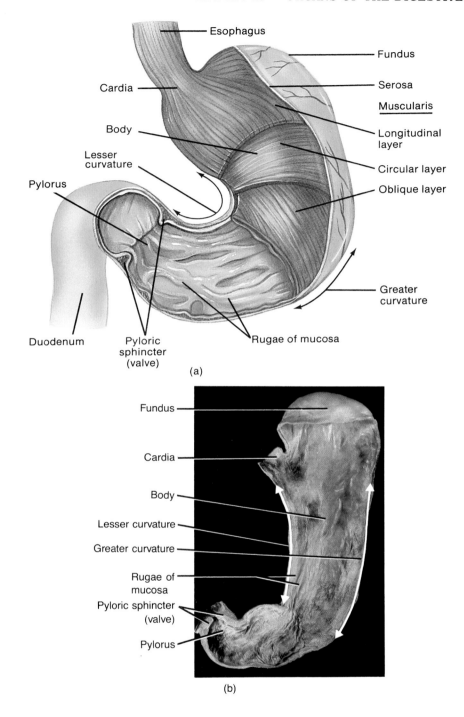

Figure 43-7 External and internal antomy of the human stomach: (a) diagram (from G. J. Tortora, *Principles of Human Anatomy,* 6th ed., New York: HarperCollins, 1992); (b) photograph of the internal surface showing rugae (courtesy of J. A. Gosling, P. F. Harris, et al., *Atlas of Human Anatomy,* Gower Medical Publishing Ltd., 1985).

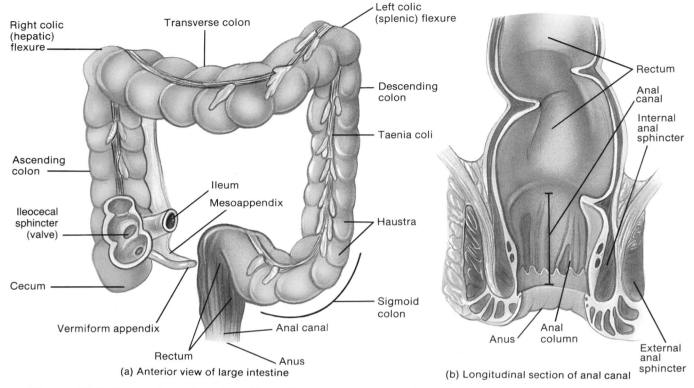

Figure 43-8 Human large intestine. (From G. J. Tortora, *Principles of Human Anatomy*, 6th ed., New York: HarperCollins, 1992.)

right colic flexure or *right hepatic flexure*. The entire colon is gathered into pouches *(haustra)* by *tenia coli*, which are muscular bands. The waste is passed along slowly from one haustrum to the next by peristaltic contractions. Water is removed and returned to the blood, and the waste is compacted as it passes along.

Transverse colon—This is the part of the colon from the *right colic flexure* to a point on the left side of the body where it flexes downward beneath the spleen. (This is the *left colic flexure (left splenic flexure)*.

Descending colon—This part of the colon passes down the left side of the abdominal cavity between the *left colic flexure* and the *sigmoid colon.*

Sigmoid colon—The sigmoid (S-shaped) colon swings back to the midline by taking an S-shaped turn to the right.

Rectum—This highly muscular straight tube (without haustra) conveys waste to the *anus*.

The Associated Digestive Glands

These glands produce secretions that reach the digestive tract through ducts and that perform vital digestive functions.

Salivary Glands

Three pairs of salivary glands (Figure 43-9) produce *saliva*, which lubricates the upper digestive tract and facilitates chewing, swallowing, and speech. These paired glands, together with their ducts, also play a small part in enzymatic digestion:

Parotids—parotid or Duct of Stenson

Submandibulars—submandibular duct or Duct of Wharton

Sublinguals—sublingual or Duct of Rivinus

These glands are similarly arranged in the pig. The fourth pair of salivary glands in the pig, the *buccal glands*, are usually not discussed in human anatomy. In fact, however, there are numerous tiny buccal glands in the human mucous membrane that lines the oral cavity. They add small quantities of saliva to the total.

Liver

The very large human liver (Figure 43-10) is composed of four lobes—three on the right side of the body and one on the left side:

Right lobe proper

Right quadrate lobe

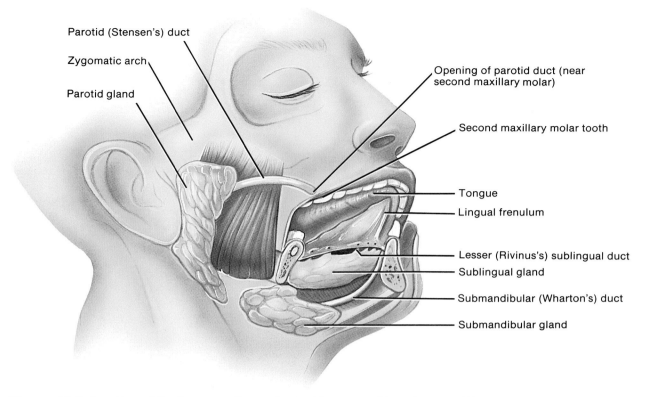

Parotid (Stensen's) duct

Zygomatic arch

Parotid gland

Opening of parotid duct (near second maxillary molar)

Second maxillary molar tooth

Tongue

Lingual frenulum

Lesser (Rivinus's) sublingual duct

Sublingual gland

Submandibular (Wharton's) duct

Submandibular gland

Figure 43-9 Location of the human salivary glands. (Courtesy of Douglas Merrill.)

Right caudate lobe

Left lobe

The three lobes on the right are usually considered as one, so anatomists usually refer to the *right liver* and the *left liver.* Separating right and left sides in the midline is the *falciform ligament,* a remnant of the embryonic ventral mesentery. As mentioned in regard to the pig, the human liver also secretes bile, composed of bile salts and bile pigments—the former are important in digestion and the latter are wastes. Bile is emptied into various *hepatic ducts,* which unite to form *right* and *left hepatic ducts* (Figure 43-10). The hepatic ducts from right and left sides unite to form a *common hepatic duct,* which joins the *cystic duct* from the *gallbladder.* Near the duodenum, the *main pancreatic duct* empties into the common bile duct (Figure 43-10).

Gallbladder

Located beneath the right liver is a pear-shaped storage sac, the *gallbladder.* It stores *bile* produced by the liver. The duct that drains the gallbladder is the *cystic duct.*

Pancreas

This small gland (Figure 43-10) is composed of two histological regions:

1. *Pancreatic duct cells,* which produce powerful digestive enzymes that break down all food types
2. *Islet cells,* located between the ducts, which produce *hormones* that are secreted directly into the capillaries.

Two large pancreatic ducts transport pancreatic juice away from the pancreas to the duodenum:

1. *Main pancreatic duct (duct of Wirsung)*—This duct (Figure 43-10) joins the common bile duct close to the duodenum. Before entering the duodenum, there is a swelling in the common bile duct, the *ampulla of Vater.* Beyond this, the common bile duct opens into the duodenum at the *sphincter of Oddi.*
2. *Accessory pancreatic duct (duct of Santorini)* enters the duodenum about one inch above the ampulla of Vater (Figure 43-10).

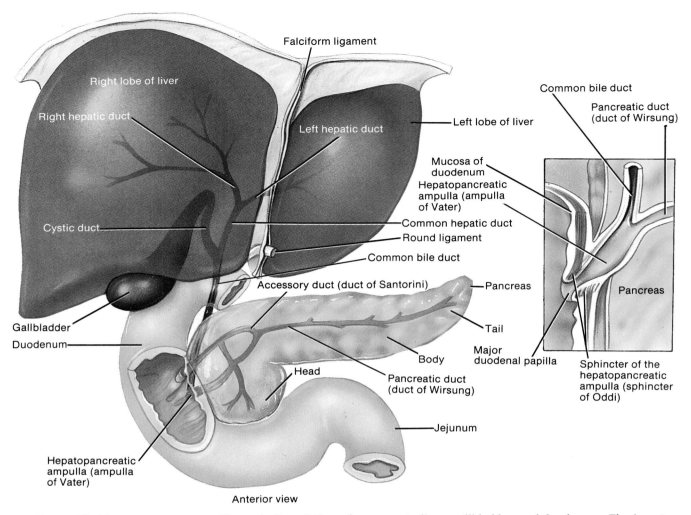

Anterior view

Figure 43-10 Human pancreas: Shown is the relation of pancreas to liver, gallbladder, and duodenum. The insert shows details of the common bile duct and pancreatic duct forming the hepatopancreatic ampulla (of Vater) emptying into the duodenum. (From G. J. Tortora, *Principles of Human Anatomy,* 6th ed., New York: HarperCollins, 1992.)

LABORATORY REVIEW *43*

THE ORGANS OF THE DIGESTIVE SYSTEM

COMPLETION, LABELING, AND MATCHING

1. Name the differences between the fetal-pig and the human digestive tracts regarding the following.

Digestive Tract Part	Fetal Pig	Human
a. Size of Cecum	_____	_____
b. Overall length of the gastrointestinal tract	_____	_____
c. Number of liver lobes	_____	_____
d. Number of pancreatic lobes	_____	_____
e. Number of salivary glands	_____	_____

2. Label the parts shown in Figure Q43-1 in the spaces provided.

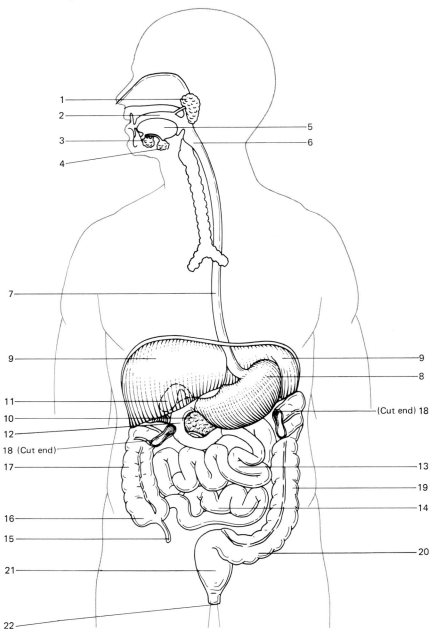

Figure Q43-1 Digestive system of the human.

1. _____   9. _____   17. _____
2. _____   10. _____   18. _____
3. _____   11. _____   19. _____
4. _____   12. _____   20. _____
5. _____   13. _____   21. _____
6. _____   14. _____   22. _____
7. _____   15. _____
8. _____   16. _____

3. Match the definitions on the left with the parts of the human digestive system listed on the right.

a. _____ A salivary gland

b. _____ Organ that stores bile

c. _____ Duct that opens into the duodenum from the liver

d. _____ Duct leading from the gallbladder

e. _____ Ventral mesentery supporting the liver

f. _____ Mesentery supporting the large intestine

g. _____ Tube that transports food to the stomach

h. _____ Part of the digestive tract that opens into the large intestine

i. _____ Pouches within the large intestine

j. _____ Part of digestive tract through which food passes from the mouth

k. _____ Lymph gland located in the mesentery of the small intestine

l. _____ Middle part of the small intestine

m. _____ Vestigial structure attached to the cecum

n. _____ Four-lobed organ ventral to stomach

o. _____ Striated lining of the stomach

ANSWERS

1. Pancreatic duct
2. Falciform ligament
3. Esophagus
4. Haustra
5. Parotid
6. Rugae
7. Ileum
8. Gallbladder
9. Pancreas
10. Cystic duct
11. Appendix
12. Hepatic duct
13. Gastrosplenic ligament
14. Liver
15. Pancreas of Aselli
16. Jejunum
17. Pharynx
18. Teniae coli
19. Mesocolon
20. Common bile duct

Metabolism: Enzyme Action

After completing this exercise, you should be able to

1. Define or explain *optimum, enzyme, substrate, hydrolysis,* and *end product,* as they relate to metabolism
2. List three specific enzymes and the substrates and end products associated with each one
3. Perform experiments to determine the effect of pH, temperature, and substrate concentration on enzymatic activity
4. Summarize the effects and importance of pH and temperature on enzymatic activity

The carbohydrates, fats, and proteins ingested as food must undergo specific changes before they can be absorbed from the digestive tract. Other substances—such as water, vitamins, and minerals—are ready for immediate absorption. Complex foods are prepared for absorption by *mechanical* and *chemical* activities. Mechanical activities include chewing, swallowing, and several different types of peristaltic contractions that break down large particles, mix the food with mucus and enzymes, and propel it through the digestive tract. Chemical activities break down the large molecules (substrates) that we eat into molecules that are small enough so that they can be absorbed through the walls of the digestive tract, transported through the body, and eventually used by body cells.

This exercise is concerned with selected tissues and chemical actions associated with metabolic enzymes (Figure 44-1). Several different enzymes and the factors that affect their activities are also considered. The procedures include specific tests used in the detection and identification of carbohydrates, lipids, and proteins.

HISTOLOGY OF THE ORGANS AND STRUCTURES ASSOCIATED WITH DIGESTION

Materials

The following materials should be provided for class use:
1. Prepared slides of
   a. Salivary glands
   b. Stomach, showing mucosa and chief parietal and mucous cells
   c. Small intestine, showing epithelium and goblet cells
   d. Large intestine, showing mucosa
   e. Pancreas
2. Anatomical charts of digestive system components

Procedure

This procedure should be performed by students individually.

1. Examine the prepared histological slides.
2. Sketch and label a representative field showing the characteristic cells of each organ or related structure in the results and observations section. Use Color Plates 31 and 49–52 as guides.
3. Answer the questions in the results and observations section.

DEMONSTRATION OF EXTRACELLULAR DIGESTION

Microorganisms are used here to show the extracellular nature of digestive processes.

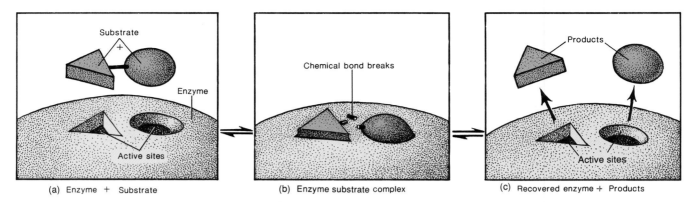

(a) Enzyme + Substrate   (b) Enzyme substrate complex   (c) Recovered enzyme + Products

Figure 44-1 Enzyme action in a decomposition reaction: The enzyme and substrate molecules combine to form an enzyme–substrate (ES) complex. During combination, the substrate is changed into products. Once the products are formed, the enzyme is recovered and may be used again to catalyze a similar reaction. (From G. J. Tortora, *Introduction to the Human Body,* 2nd ed., New York: HarperCollins, 1991.)

Materials

The following materials should be provided for class demonstration:

1. One uninoculated starch agar plate
2. One starch agar plate centrally inoculated with the bacterium *Bacillus subtilis,* but not incubated
3. One starch agar plate inoculated with *B. subtilis* and incubated for 24 hours
4. Lugol's iodine solution in a container with a dropper
5. Container with disinfectant for disposal of cultures

Procedure

This procedure is to be demonstrated by the instructor.

1. Examine the starch agar plates. Sketch the appearance, and indicate the color of each on one side of the appropriate circle in the results and observations section.
2. Your instructor will flood the plates with an iodine solution.
3. After 15 seconds, examine the plates. Sketch their appearance, and indicate the colors formed on the other side of the appropriate circles in the results and observations section.
4. Answer the questions in the results and observations section.

SALIVARY AMYLASE ACTIVITY

Starch is one of the major carbohydrates in the human diet. To use starch, the body must first break down the large starch molecule into smaller glucose molecules.

The process by which starch is hydrolyzed into smaller units begins in the mouth, by the action of an enzyme *amylase*. In the next set of experiments, the progress of this enzymatically catalyzed reaction and the various factors that can effect it are followed by observing the disappearance of the reacting substrate (starch), using the iodine test.

Materials

The following materials should be provided for class use:

1. Three stock starch solutions—5 percent, 1 percent, and 0.5 percent
2. Three buffer solutions, with pH of 4, 7, and 9
3. Lugol's iodine or other similar iodine reagent in a dropper bottle
4. Pieces of paraffin or sugarless gum
5. Pyrex test tubes
6. Test-tube racks
7. 12-ml Erlenmeyer flasks
8. Porcelain spot plates
9. Gauze
10. Glass or plastic funnels
11. 50-ml graduated cylinders
12. Water baths—4°C, 37°C, and 70°C
13. Distilled water
14. Container for the disposal of saliva and related materials

Procedure 1: Salivary Amylase Stock Preparation

This procedure is to be performed by students in groups of four.

1. Two students should produce and collect 20 ml of saliva each. The sample should be measured and pooled in a clean Erlenmeyer flask. (Note that saliva excretion can be stimulated by drinking a glass of water and then chewing on a small piece of paraffin or sugarless gum.)
2. Add an equal amount of room temperature distilled water, and mix the contents of the flask. Pass the mixture through a gauze filter system, to remove most of the mucus. The resulting preparation is your amylase stock solution for the remaining procedures in this exercise. (If additional enzyme material is needed, repeat this procedure.)

Procedure 2: Effects of Temperature on Amylase Activity

This procedure is to be performed by students in pairs.

1. Label four test tubes with numerals 1–4, respectively, and place them in a test-tube rack.
2. Add 5 ml of amylase stock solution to Tubes 1, 2, and 3.
3. Place the tubes in the temperature environment specified next. Keep the tubes at these temperatures for at least 10 minutes (to allow all the solution to reach the desired temperature).

   Tube 1: 4°C

   Tube 2: 37°C

   Tube 3: 70°C

   Tube 4 (control): room temperature (25°C). This is the control tube for the starch–iodine reaction.

4. Add 5 ml of the 1 percent starch solution to each tube, shake gently, and return the mixtures to their water baths to incubate for 5 minutes.
5. Remove a small sample from each tube, place each into a separate depression on a spot plate, and test immediately for starch hydrolysis by adding 2 drops of the iodine reagent.
6. Enter your findings in the results and observations section.
7. Wash and rinse the tubes used.

Procedure 3: Effects of pH on Amylase Activity

This procedure is to be performed by students in pairs.

1. Number the four test tubes, and place them in a rack.
2. Add 3 ml of the amylase stock solution to each tube.
3. Add the following substances to the tubes, as indicated:

   Tube 1: 3 ml buffer solution, pH 4

Tube 2: 3 ml buffer solution, pH 7

Tube 3: 3 ml buffer solution, pH 9

Tube 4: 3 ml distilled water, pH 7

4. Shake and mix the contents, and place them in a 37°C water bath for 10 minutes.
5. Add 3 ml of 1 percent starch solution to each tube, mix, and continue to incubate at 37°C.
6. Almost immediately, test the contents of each tube for starch hydrolysis, as described in Procedure 2, Step 5.
7. Continue to incubate the preparations at 37°C, and take additional samples at 15 and 30 minutes. Test for starch hydrolysis, as in Step 6.
8. Enter your findings in the results and observations section.
9. Wash and rinse the tubes used.

Procedure 4: Effect of Substrate Concentration on Amylase Activity

This procedure is to be performed by students in pairs.

1. Number 12 test tubes, and place them in a rack in two rows of 6 tubes each.
2. Add 1 ml of the amylase stock solution to each tube.
3. Add the following substances to the tubes, as specified:

Row I

Tube 1: 1 ml 1 percent starch solution
Tube 2: 2 ml 1 percent starch solution
Tube 3: 3 ml 1 percent starch solution
Tube 4: 5 ml 1 percent starch solution
Tube 5: 10 ml 1 percent starch solution
Tube 6: 5 ml distilled water

Row II

Tube 1: 1 ml 5 percent starch solution
Tube 2: 2 ml 5 percent starch solution
Tube 3: 3 ml 5 percent starch solution
Tube 4: 5 ml 5 percent starch solution
Tube 5: 10 ml 5 percent starch solution
Tube 6: 5 ml distilled water

4. Shake the tubes gently to mix the contents, and place them in a 37°C water bath for 10 minutes.
5. Take samples and test for starch hydrolysis at 1 minute, 15 minutes, 30 minutes, and 60 minutes, as described in Procedure 2, Step 5.
6. Enter your findings in the results and observations section.
7. Wash and rinse the tubes used.

DIGESTION OF FATS

Materials

The following materials should be provided for class use:

1. The following substances in appropriate containers
   a. Safflower oil
   b. Distilled water
   c. Ethyl alcohol
   d. Xylene
   e. Bile salts (5 percent solution)
2. 5-ml pipettes with bulbs
3. Test tubes
4. Test-tube racks

Procedure

This procedure is to be performed by students in pairs.

1. Number five test tubes and place them in a rack.
2. Add 2 ml of safflower oil to each tube.
3. Carefully add the following substances to the tubes, as specified:

   Tube 1: 5 ml distilled water

   Tube 2: 5 ml ethyl alcohol

   Tube 3: 3 ml xylene

   Tube 4: 3 ml bile salts

   Tube 5: no additions

4. Shake the tubes gently to mix the contents, and examine all tubes immediately. Look for a complete mixing of the contents of the tubes.
5. Repeat your examinations at 15 and 30 minutes.
6. Enter your findings in the results and observations section.

DIGESTION OF PROTEIN

Materials

The following materials should be provided for class use:

1. Small pieces of coagulated egg albumin
2. The following solutions in appropriate containers
   a. 1 percent trypsin
   b. 1 percent pepsin
   c. 0.5 percent sodium carbonate
   d. 0.8 percent hydrochloric acid
3. Distilled water
4. Biuret test reagents
5. Test tubes
6. Test-tube racks
7. 5-ml pipettes with bulbs
8. 37°C water bath

Procedure: The Actions of Pepsin and Trypsin on Egg Albumin

This procedure is to be performed by students in pairs.

1. Prepare a test-tube rack containing two rows of six tubes each. Number the tubes accordingly (1–6 in each row).
2. Add the following substances to the tubes, as specified:

Row I

Tube 1: 5 ml pepsin + 5 ml distilled water
Tube 2: 5 ml pepsin + 5 ml sodium carbonate
Tube 3: 5 ml pepsin + 5 ml hydrochloric acid
Tube 4: 5 ml distilled water + 5 ml sodium carbonate
Tube 5: 5 ml distilled water + 5 ml hydrochloric acid
Tube 6: 5 ml distilled water

Row II

Tube 1: 5 ml trypsin + 5 ml distilled water
Tube 2: 5 ml trypsin + 5 ml sodium carbonate
Tube 3: 5 ml trypsin + 5 ml hydrochloric acid
Tube 4: 5 ml distilled water + 5 ml sodium carbonate
Tube 5: 5 ml distilled water + 5 ml hydrochloric acid
Tube 6: 5 ml distilled water

3. Add several pieces of coagulated egg albumin to each tube, shake gently to mix, and place all tubes in a 37°C water bath.
4. Examine each tube at 15-minute intervals during a 90-minute period, for evidence of the digestion of the egg albumin. This type of reaction is indicated by a disintegration of the protein and a general clearing of the protein material.
5. At the end of 90 minutes, add 4 drops of the Biuret solution to each tube, to test for the presence of the end products of protein digestion. The formation of a pink to violet color indicates the presence of polypeptides. (Refer to Color Plate 53.)
6. Record your findings in the results and observations section.
7. Wash and rinse the tubes used.

RESULTS AND OBSERVATIONS

Histology of the Organs and Tissues Associated with Digestion

1. Sketch and label representative portions of the tissues indicated.

Salivary glands

Large intestine

Stomach

Pancreas

Small intestine

2. Indicate the specific metabolically associated product(s) of the structures or cells listed.

   a. Salivary glands _____

   b. Chief cells _____

   c. Parietal cells _____

   d. Mucous cells _____

   e. Brunner's glands _____

   f. Goblet cells (small intestine) _____

   g. Pancreas _____

   _____

Demonstration of Extracellular Digestion

1. Sketch the appearance, and indicate the color of the starch agar plates before and after the application of the iodine reagent.

Before   After
Uninoculated
plate

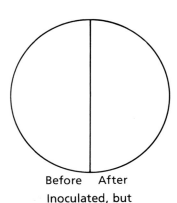

Before   After
Inoculated, but
not incubated plate

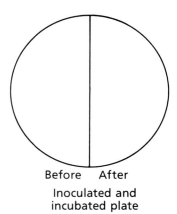

Before   After
Inoculated and
incubated plate

2. In which of the plates was starch hydrolysis observed? _____

3. Why must an inoculated plate be incubated to produce a positive reaction? _____
_____

4. What color develops after the addition of iodine in the cases of

   a. Starch hydrolysis _____

   b. Presence of starch _____

Salivary Amylase Activity

1. Enter in Table 44-1 the reaction to iodine at different temperatures, and indicate whether or not starch hydrolysis occurred. Indicate a positive reaction with a " + " and a negative reaction with a " − ."

2. Enter in Table 44-2 the results of the tests for starch hydrolysis on substances at different pHs. Indicate a positive reaction with a " + " and a negative reaction with a " − ."

3. Enter in Table 44-3 the results of the tests for starch hydrolysis on substances of different substrate concentration. Indicate a positive reaction with a " + " and a negative reaction with a " − ."

4. Summarize your conclusions regarding the influence of temperature, pH, and substrate concentration on enzyme activity. Use an additional page, if needed. _____
_____
_____

5. What would have happened in the tests of amylase activity if alcohol had been used as the diluting fluid, instead of distilled water? _____
_____
_____

Digestion of Fats

1. Enter in Table 44-4 your observations of the degree to which the contents of the tubes mixed.

2. If there were any observable differences in any of the tubes at 15 or 30 minutes, explain why. _____
_____
_____

3. Where is bile produced in the body? _____

4. What is the function of bile? _____
_____

Digestion of Protein

1. Enter your observations and test results for protein digestion in Table 44-5. Indicate evidence of protein and a positive test reaction with a " + " and no evidence and a negative test reaction with a " − ."

Table 44-1

Tube	Temperature	Color After Addition of Iodine	Starch Hydrolysis
1	4°C		
2	37°C		
3	70°C		
4	25°C		

Table 44-2

Tube	Substance	Immediate	15 Minutes	30 Minutes
1	Buffer, pH 4			
2	Buffer, pH 7			
3	Buffer, pH 9			
4	Distilled water, pH 7			

Table 44-3

Row	Tube	Concentration Amount (ml)	Solution	1 Minute	15 Minutes	30 Minutes	60 Minutes
I	1	1	1% starch				
I	2	2	1% starch				
I	3	3	1% starch				
I	4	5	1% starch				
I	5	10	1% starch				
I	6	5	distilled water				
II	1	1	5% starch				
II	2	2	5% starch				
II	3	3	5% starch				
II	4	5	5% starch				
II	5	10	5% starch				
II	6	5	distilled water				

2. Which of the preparations produced a positive Biuret reaction? _____

_____

3. What is digestion? _____

_____

Table 44-4

Tube	Substance Mixed with Oil	Immediate	15 Minutes	30 Minutes
1	Distilled water			
2	Ethyl alcohol			
3	Xylene			
4	Bile salts			
5	None			

Table 44-5

Row	Tube	Solution	15 Minutes P	15 Minutes T	30 Minutes P	30 Minutes T	45 Minutes P	45 Minutes T	60 Minutes P	60 Minutes T	75 Minutes P	75 Minutes T	90 Minutes P	90 Minutes T
I	1	Distilled water												
I	2	Sodium carbonate												
I	3	HCl												
I	4	Sodium carbonate												
I	5	HCl												
I	6	—												
II	1	Distilled water												
II	2	Sodium carbonate												
II	3	HCl												
II	4	Sodium carbonate												
II	5	HCl												
II	6	—												

LABORATORY REVIEW *44*

METABOLISM: ENZYME ACTION

NAME _____

LAB SECTION _____ DATE _____

COMPLETION, DEFINITION, AND INTERPRETATION

1. Are the pH optimums the same for most enzymes? Explain. _____
   _____
   _____

2. What is an enzyme–substrate complex? Use labeled diagrams to explain your answer.

3. List three major factors that affect enzyme action. Briefly explain how each affects enzymatic activity.

   a. _____
      _____

   b. _____
      _____

   c. _____
      _____

4. How are enzymes named? _____
   _____
   _____

5. Define or explain the following:

   a. Enzyme _____

   b. Substrate _____

   c. Optimum _____

   d. Active site _____

ANSWERS

1. _____

2. _____

3. _____

4. _____

5. _____

MULTIPLE CHOICE

Enter the correct answers in the spaces provided.

1. What color formation indicates a positive starch hydrolysis reaction after the addition of an iodine solution?
   a. purple
   b. yellow
   c. green
   d. black
   e. blue

2. Which of the following can affect enzymatic activity?
   a. temperature
   b. pH
   c. substrate concentration
   d. enzyme concentration
   e. choices *a* through *d*

3. In which of the following would fats be dissolvable?
   a. distilled water
   b. bile salts
   c. xylene
   d. ethyl alcohol
   e. choices *b, c,* and *d* only

4. If a pink color formed when Biuret solution was added to an unknown specimen, which of the following would this color be indicating to be present?
   a. distilled water
   b. starch
   c. fat
   d. polypeptides
   e. sodium carbonate

5. If iodine has been added to an unknown solution, which rapidly turns purple, which of the following does the solution probably contain?
   a. amylase
   b. albumin
   c. starch
   d. glucose
   e. bile

Digestion and Chemistry

After completing this exercise, you should be able to

1. Perform general procedures and chemical tests to demonstrate the presence of carbohydrates, lipids, and proteins
2. Apply these procedures to identify the contents of an unknown specimen

Several tests and procedures, which are used in the identification of carbohydrates, lipids, proteins, and other substances of dietary importance, are demonstrated in this exercise. Be sure to handle all chemicals carefully. Observe safety rules.

CARBOHYDRATE IDENTIFICATION

Materials

The following materials should be provided for class use:

1. 1 percent known solutions in dropper bottles
   a. Glucose
   b. Lactose
   c. Maltose
   d. Albumin
   e. Starch
   f. Gelatin
2. Test materials in dropper bottles
   a. Honey
   b. Sugar-free drink
   c. Nondiet cola drink
   d. Distilled water
   e. Syrup
   f. Milk
   g. Saccharin
   h. Orange juice
3. Test reagents
   a. Benedict's reagent
   b. Lugol's iodine
   c. Molisch test reagents—5 percent alpha-naphthol alcoholic solution and concentrated sulfuric acid
4. Pyrex test tubes, test-tube racks, and test-tube holders
5. 1-ml and 10-ml pipettes with bulbs
6. Boiling water bath
7. Wax pencils or other marking devices

Procedure 1: Molisch Reaction

The Molisch reaction is a general test for carbohydrates. The formation of a purple or pink color, upon the addition of the Molisch reagents indicates the presence of a carbohydrate.

This procedure is to be performed by students in pairs.

1. Label a test tube for each of the 1 percent known solutions. Place the tubes in a rack.
2. Introduce 10 drops of each known solution into its labeled tube. Use a separate pipette for each. Examine each preparation for any distinctive color, smell, or other noticeable property. Enter your findings in the results and observations section.
3. Add 2 drops of the 5 percent alpha-naphthol solution to each tube. Mix the contents by gently tapping the tube.
4. Tilt each tube, and carefully add 5 ml of the concentrated sulfuric acid (H_2SO_4) so that it runs slowly down the inside surface. The acid is heavier and should form a layer beneath the test solutions.

5. Examine each tube for the formation of a pink or purple color at the boundary between the test solution and the sulfuric acid. Record your findings in the results and observations section.
6. Dispose of the chemical solutions as directed by your instructor.
7. Repeat Steps 1–6 with the test materials.

Procedure 2: Reducing Sugars Test

Benedict's reagent is used to detect the presence of a reducing sugar. When the cupric ion (Cu^{2+}) in the reagent reacts with a reducing sugar, a red precipitate of cuprous oxide forms.

This procedure is to be performed by students in pairs.

1. Label a test tube for each of the 1 percent known solutions. Place the tubes in a rack.
2. Introduce 10 drops of each known solution into its labeled tube. Use a separate pipette for each.
3. Using a 10-ml pipette, add 5 ml of Benedict's reagent to each test tube. Mix the contents by gently tapping the tubes.
4. Place all the tubes in a boiling water bath for 5 minutes.
5. Examine each tube for evidence of the presence of a reducing sugar (red precipitate) (see Color Plate 55). Enter your findings in the results and observations section.
6. Dispose of the chemical solutions as directed by your instructor.
7. Repeat Steps 1–6 with the test materials.

Procedure 3: Starch Test

The presence of starch generally is indicated by the formation of a blue-black color, upon the addition of an iodine reagent.

This procedure is to be performed by students in pairs.

1. Label a test tube for each of the 1 percent known solutions. Place the tubes in a rack.
2. Introduce 10 drops of each known solution into its labeled tube. Use a separate pipette for each.
3. Add 3 drops of the iodine reagent to each tube, and mix.
4. Examine each tube for the formation of a blue-black color. Enter your findings in the results and observations section.
5. Dispose of the chemical solutions as directed by your instructor.
6. Repeat Steps 1–5 with the test materials.

PROTEIN IDENTIFICATION

Materials

The following materials should be provided for class use:

1. 1 percent known solutions
   a. Albumin
   b. Cysteine
   c. Glucose
   d. Glycine
   e. Phenylalanine
   f. Starch
   g. Tryptophan
2. Test materials
   a. Casein
   b. Hair
   c. Gelatin
3. Test reagents
   a. Biuret reagents—6 molar (M) sodium hydroxide (NaOH) and 0.5 percent copper sulfate ($CuSO_4$)
   b. 1 molar (M) lead acetate ($PbAc_2$) solution
   c. Concentrated nitric acid (HNO_3)
   d. Glacial acetic acid
   e. Litmus paper
4. 1-ml and 5-ml pipettes with bulbs
5. Pyrex test tubes, test-tube racks, and test-tube holders
6. Boiling water bath
7. Wax pencils or other marking devices

Procedure 1: Biuret Test

The Biuret test is named for the purple compound that forms when a dilute alkaline solution of copper sulfate reacts with substances containing two or more peptide bonds.

This procedure is to be performed by students in pairs.

1. Label a test tube for each of the 1 percent known solutions and test materials. Place the tubes in a rack.
2. Introduce 1 ml of each solution into its labeled tube. Use a separate pipette for each.
3. Add 2 ml of the sodium hydroxide solution and 10 drops of the copper sulfate to each tube. Shake the tube gently to mix the contents.
4. Look for the presence of a pink to purple color (Color Plate 53). Enter your findings in the results and observations section.
5. Dispose of the chemical solutions as directed by your instructor.

Procedure 2: Sulfur Test

The sulfur test is used to detect the presence of amino acids in proteins that contain sulfur. When lead acetate is added to compounds containing sulfur, lead sulfide is formed. Lead sulfide is black.

This procedure is to be performed by students in pairs.

1. Label a test tube for each of the 1 percent known solutions. Put the tubes in a rack.
2. Introduce 1 ml of each known solution into its labeled tube. Use a separate pipette for each.
3. Add 2 ml of sodium hydroxide to each tube, and place the tubes in a boiling water bath for 60 minutes.
4. Remove the tubes.
5. Neutralize the contents of each tube by carefully adding glacial acetic acid until blue litmus paper (pH indicator) turns red.
6. Add 10 drops of lead acetate.
7. Look for a black color. Enter your findings in the results and observations section.
8. Dispose of the chemical solutions as directed by your instructor.

Procedure 3: Xanthoproteic Test

This test detects the presence of benzene rings in amino acids. Such compounds produce yellow to red colors when exposed to nitric acid.

This procedure is to be performed by students in pairs.

1. Label a test tube for each of the 1 percent known solutions and the test materials. Place the tubes in a rack.
2. Introduce 1 ml of each known solution into its labeled tube. Use a separate pipette for each.
3. Carefully add 1 ml of concentrated nitric acid by letting the acid run down the side of each tube.
4. Place the tubes in a boiling water bath for 10 seconds.
5. Remove the tubes, and observe the reactions. Enter your findings in the results and observations section.
6. Dispose of the chemical solutions as directed by your instructor.

Procedure 4: Denaturation by Heat

Some proteins become *denatured* (permanently altered) and precipitate out of solution when exposed to boiling temperatures.

This procedure is to be performed by students in pairs.

1. Label a test tube for each of the 1 percent known solutions. Place the tubes in a rack.
2. Introduce 5 ml of each known solution into its labeled tube. Use a separate pipette for each.
3. Place the tubes in a boiling water bath for 3 minutes.
4. Examine each tube for the presence of a solid mass or precipitate. Enter your findings in the results and observations section.
5. Dispose of the chemical solutions as directed by your instructor.

PROPERTIES OF FATS

Materials

The following materials should be provided for class use:

1. Solvents
   a. Distilled water
   b. Ethyl alcohol
   c. 0.55 percent sodium chloride
   d. Xylene
2. Test solutions in dropper bottles
   a. 1 percent albumin
   b. Cottonseed or immersion oil
   c. Gelatin
   d. 2 percent glucose
   e. 2 percent starch
3. Test tubes and test-tube racks
4. 5-ml pipettes with bulbs
5. Brown paper strips (7.5 × 5 cm)
6. Wooden applicator sticks
7. Wax pencils or other marking devices

Procedure 1: Solubility Testing

This procedure is to be performed by students in pairs.

> **CAUTION**
> (*Note:* Make certain that there are no open flames in the area where the following tests are being performed.)

1. Label a test tube for each of the solvents. Place the tubes in a rack.
2. Introduce 5 ml of each solvent into its labeled tube. Use a separate pipette for each.

3. Add 6 drops of the test-material albumin to each tube and shake.
4. Examine each tube immediately and after 15 minutes, to see whether the test material dissolved. Enter your findings in the results and observations section.
5. Dispose of the chemical solutions as directed by your instructor.
6. Repeat Steps 1–5 with each of the other test materials.

Procedure 2: Grease Spot Test

This procedure is to be performed by students in pairs.

1. On a brown paper strip, label a small area for each of the test materials.
2. Using a separate applicator stick for each, apply a small amount of each test material to its labeled area.
3. Examine each immediately and after 30 minutes for the presence of a grease spot. Enter your findings in the results and observations section.

IDENTIFICATION OF CONTENTS OF UNKNOWN SAMPLE

Materials

The following materials should be provided for class use:

1. Test reagents provided earlier for carbohydrate, lipid, and protein tests
2. Pyrex test tubes, test-tube racks, and test-tube holders
3. 1-ml and 10-ml pipettes with bulbs
4. Boiling water bath
5. One unknown sample (10 ml) per student

Procedure

This procedure is to be performed by students individually.

1. Obtain an unknown sample from your instructor.
2. Examine the unknown specimen for color, smell, and other distinctive properties.
3. Divide your specimen into several samples, and identify its contents by performing whichever of the preceding procedures you think are necessary.
4. Enter your findings in the results and observations section.
5. Dispose of all chemical solutions as directed by your instructor.

RESULTS AND OBSERVATIONS

Carbohydrate Identification

1. In Table 45-1, enter your observations of physical properties of the known solutions and test materials.
2. In Table 45-2, enter the chemical test results.
3. Which of the materials tested showed the presence of a reducing sugar? _____

_____

Protein Identification

1. Enter your observations of the reactions to the chemical tests in Table 45-3.
2. Which of the materials tested contain peptide bonds?

_____

3. Which of the amino acids tested contain a benzene ring? _____

_____

4. How many of the materials tested showed the presence of sulfur? Which ones? _____

_____

5. Which of the materials tested solidifed when heated? _____

Properties of Fats

1. Indicate in Table 45-4 the results of the fat-solubility test.
2. Which of the substances tested produced an obvious grease spot on the brown paper strip?_____

_____

3. What other types of reactions besides those of solubility did you find? _____

_____

Identification of Contents of Unknown Sample

1. List the tests used, and enter your findings in Table 45-5.
2. On the basis of the tests performed and the results obtained, the unknown sample _____

_____ contains _____

_____

Table 45-1

Materials	Color	Smell	Other Properties
Glucose			
Lactose			
Maltose			
Albumin			
Starch			
Gelatin			
Honey			
Sugar-free drink			
Cola drink			
Distilled water			
Syrup			
Milk			
Saccharin			
Orange juice			

Table 45-2

Materials	Carbohydrate Test	Reducing Sugars Test		Starch Test	
	Molisch Reaction	Reaction with Benedict's Reagent	Reducing Sugar Present	Reaction with Iodine	Starch Present
Glucose					
Lactose					
Maltose					
Albumin					
Starch					
Gelatin					
Honey					
Sugar-free drink					
Cola drink					
Distilled water					
Syrup					
Milk					
Saccharin					
Orange juice					

Table 45-3

Material	Biuret Reaction	Sulfur Reaction	Xanthoproteic Reaction
Albumin			
Cysteine			
Glucose			
Glycine			
Phenylalanine			
Starch			
Tryptophan			
Casein			
Hair			
Gelatin			

Table 45-4

Solvent	Albumin		Cottonseed Oil		Gelatin		Glucose		Starch	
	0 Time*	After 15 Minutes	0 Time*	After 15 Minutes	0 Time*	After 15 Minutes	0 Time*	After 15 Minutes	0 Time*	After 15 Minutes
Distilled water										
Ethyl alcohol										
Sodium chloride										
Xylene										

*Immediate reaction.

Table 45-5

Test or Procedure	Description of Results
General appearance	

LABORATORY REVIEW *45*

DIGESTION AND CHEMISTRY

COMPLETION, DEFINITION, AND INTERPRETATION

1. Complete Table Q45-1.

Table Q45-1

Chemical Test Reagent	Test Substance	Reaction Produced
Iodine	Glucose	a.
Iodine	Starch	b.
Benedict's solution	c.	Red precipitate
Benedict's solution	Galactose	d.
Molisch reagents (alphanaphthol and concentrated sulfuric acid)	Albumin	e.
Molisch reagents	Sucrose	f.
Ninhydrin	g.	Purple color
Ninhydrin	Starch	h.
Biuret test reagent	Starch	i.
Biuret test reagent	Albumin	j.
Iodine	Orange juice and starch solution	k.

2. Define or explain the following:

   a. Reducing sugar _____

   _____

   b. Carbohydrate _____

   _____

   c. Vitamin _____

   _____

   d. Hydrolysis _____

   _____

   e. Lipase _____

3. Differentiate between the following:

    a. Monosaccharide and polysaccharide _____ _____

    _____

    b. Amino acid and peptide _____ _____

    _____

    c. Exocrine and endocrine _____ _____

    _____

4. What chemical elements are present in the compounds listed?

    a. All carbohydrates _____;

    b. All proteins _____;

    c. All lipids _____;

    d. All nucleic acids _____.

Cholesterol Testing

**After completing this exercise,
you should be able to**

1. Differentiate between *atherosclerosis* and *arteriosclerosis*
2. Describe several properties of cholesterol and its role in atherosclerosis
3. Distinguish among *triglycerides, low-density lipoproteins,* and *high-density lipoproteins*
4. List conditions that can cause elevated total serum cholesterol levels
5. List conditions that can cause decreased levels of total serum cholesterol
6. Perform a laboratory test for the determination of total serum cholesterol levels

Cholesterol is a fatlike substance derived from either the breakdown of ingested fats or formed by the liver. It is found throughout the body, especially in blood, bile, and brain tissue. Most of the body's cholesterol is formed in the liver and flows with bile into the small intestine, where it mixes with its dietary sources. Cholesterol and triglycerides cannot dissolve in water and thus cannot travel in the blood in their unaltered forms. They are made water-soluble by combining with proteins produced by the liver and intestine *(apoproteins),* and the complexes thus formed are termed *lipoproteins,* which vary in size, weight, and density.

Two of the major classes are *low-density lipopro-*

Atherosclerosis is a vascular disease associated with lesions, known as *atherosclerotic plaques* (Figure 46-1), on and in the walls of high-pressure arteries. These include the vessels in the heart, brain, kidneys, arms, and legs. The disease is a form of *arteriosclerosis,* which includes several conditions characterized by abnormal thickening or hardening of the arteries. The two conditions differ, in that atherosclerosis involves the depositing of fatty substances, especially *cholesterol* and *triglycerides* (ingested fats), while arteriosclerosis is associated with individual or combination deposits of calcium, lipid, and connective tissue.

Atherosclerotic vascular disease and its complications, which include cerebral infarction, coronary artery disease, coronary or myocardial infarction (heart attack), and pulmonary obstruction, are the leading causes of death in Western society. Most epidemiological analyses show a strong correlation of serum cholesterol levels with coronary disease.

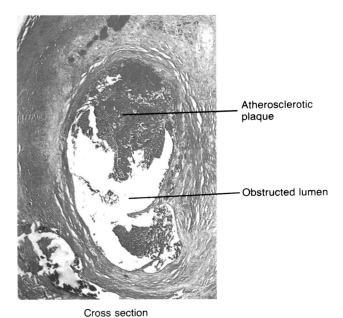

Atherosclerotic plaque

Obstructed lumen

Cross section

Figure 46-1 Photomicrograph of the cross-section of an artery partially obstructed by an atherosclerotic plaque. (Courtesy of Sklar/Photo Researchers.)

teins *(LDLs)* and *high-density lipoproteins (HDLs).* There is an important difference between the two: *LDL* contains 60 to 70 percent of the *total serum cholesterol.* It also appears to pick up cholesterol and deposit it in body cells, including—under abnormal conditions—smooth-muscle fibers in arteries. Thus, as blood levels of LDL increase, the risk of coronary artery disease and related conditions *increases.* In contrast, as blood levels of HDL increase, the risk of coronary artery disease and other related conditions *decreases.* A third type of lipoprotein, known as a *very low-density lipoprotein (VLDL),* contains about 10 to 15 percent of total serum cholesterol and a large amount of triglyceride (fat). Its role in the development of atherosclerotic plaques is not clear. The amount of *total cholesterol (HDL, LDL, and VLDL,* combined) in blood is one commonly used indicator of risk for coronary artery disease.

Total cholesterol concentrations are reliable indicators of an individual's tendency toward atherosclerosis. Elevated cholesterol levels *(hypercholesterolemia)* are considered to be one of the factors accompanying vascular disorders. Although cholesterol itself does not cause atherosclerosis, excess amounts of lipoproteins (hyperlipoproteinemia) containing cholesterol do contribute to the formation of plaque deposits that cause artery blockage (Figure 46-1). Thus, cholesterol levels can be used to evaluate the risk of future heart attack (myocardial infarction). Determinations of total cholesterol, such as the one considered in this exercise, are used mainly in the study of individuals subject to cardiovascular disease and may occasionally be included with tests dealing with liver function and disease diagnosis.

The range of normal cholesterol values varies in adults and appears to be influenced by several factors, including age, diet, sex, race, physical activity, stress, and seasonal changes. Table 46-1 shows the total serum cholesterol values according to age. Elevated total cholesterol levels may occur in conditions such as ather-

osclerosis; infections involving the kidney, liver, and pancreas; diabetes mellitus; hypothyroidism; leukemia; pregnancy; early starvation; possible hereditary disorders associated with large amounts of cholesterol in the blood *(hypercholesterolemia);* and increased levels of lipoproteins in the blood *(hyperlipoproteinemia).* In general, as the total cholesterol level increases above 150 mg/dl (a *deciliter* [dl] is $\frac{1}{10}$ of a liter), the risk of coronary and related diseases slowly begins to rise. Above 200 mg/dl, the risk increases even more rapidly. The chance of a heart attack has been found to double with every 50 mg/dl increase in total cholesterol once the level goes over 200 mg/dl.

Decreased total cholesterol levels may develop with conditions such as extensive liver disease, malnutrition, some forms of anemia, severe infections, terminal cancer, tuberculosis, intestinal abstruction, and hepatitis.

This exercise uses the AccuMeter® Cholesterol Test. The procedure incorporates a handheld cassette that consists of three plastic parts (Figure 46-2): *a cover, a movable tab, and a base.* The cover holds the *measurement scale,* the *blood application well,* and *two windows* through which the functioning of the test reagents and the completion of the assay can be observed. The movable tab and base operate together to facilitate a sequence of steps in the procedure and the associated chemical reactions. The assay is performed in four simple steps (see Figure 46-3):

1. The blood sample is obtained from the finger.
2. The blood is applied to the well and allowed to filter.
3. The tab on the cassette is pulled, to start the reactions.
4. The results are then read directly from the cassette.

The reactions involved can be summarized as follows. The blood sample first comes in contact with a buffered protein solution containing horseradish peroxide (HRP). Cholesterol within the sample is converted to hydrogen peroxide by two enzymes (cholesterol esterase, and cholesterol oxidase) on the chromatography paper. The hydrogen peroxide makes contact with a uniformly coated dye in a distal region of the cassette (the measurement scale), and the hydrogen peroxide and the dye react with the HRP to form a purple peak.

Once the chemical reactions begin, a color change is visible at the proximal end of the paper strip (the first window), to show that the device is functioning properly. When the entire solution has reached the top of the measurement scale, an indicator produces a green color (seen through the second window), to signal that the assay is complete. The height of the color peak is proportional to the amount of hydrogen peroxide and thus the concentration of cholesterol within the sample. The height (in millimeters) is converted to cholesterol (in mg/dl) by means of a conversion table (Table 46-2).

Table 46-1 Total Normal Cholesterol Values, According to Age

Age	Normal Cholesterol Values (in mg/dl)
Newborn	50–100
1 year	70–175
Adolescence years	135–240
20–29	144–275
30–39	165–295
40–49	170–315
50–69	175–340
Over 70	130–245

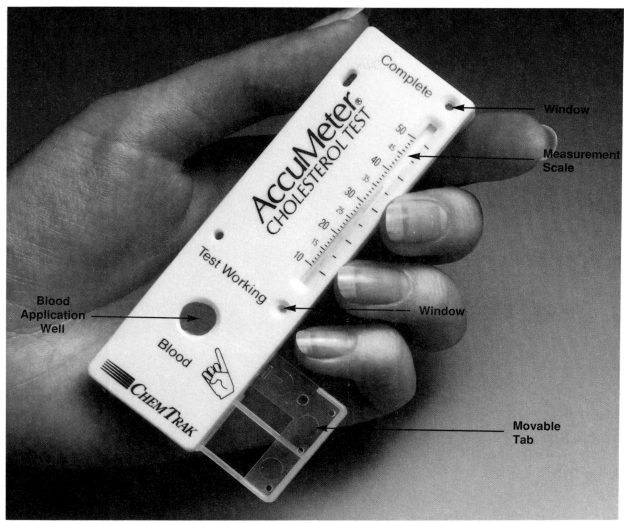

Figure 46-2 The AccuMeter® is a handheld cassette consisting of three plastic parts: a *cover,* a *movable tab,* and a *base*. (Courtesy of CHEM TRAK, Sunnyvale, California.)

Materials

The following materials should be provided for class use:

1. Sterile, disposable blood lancets
2. Commercial alcohol prep-pads
3. Sterile cotton squares
4. AccuMeter® Cholesterol Test cassettes (1 per student)
5. Disinfectant solution for disposal of blood-containing materials
6. Disposable surgical gloves

Procedure

This procedure should be performed by students in pairs.

CAUTION

Working with blood or other body fluids will not pose any danger if the following precautions are observed:

1. Place any and all blood or materials containing blood into the disinfectant containers provided.
2. Dispose of any blood-containing materials as indicated by the instructor.
3. In cases of blood spillage, wipe the area with a disinfectant-soaked paper towel, and dispose of all materials as indicated by the instructor.
4. Always wash your hands after handling blood and associated materials.
5. The use of disposable surgical gloves.

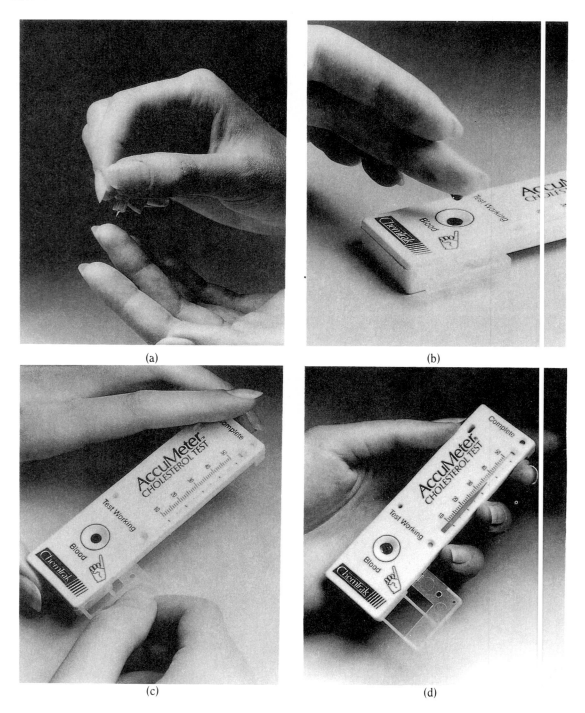

(a)

(b)

(c)

(d)

Figure 46-3 Steps in the use of the AccuMeter®: (a) obtaining the specimen; (b) applying the blood specimen to the reaction well; (c) starting the reaction by pulling the "tab"; (d) reading the result. (Courtesy of CHEM TRAK, Sunnyvale, California.)

Table 46-2 AccuMeter® Cholesterol Result Chart[a]

Cassette Reading	Cholesterol mg/dl[b]	Cassette Reading	Cholesterol mg/dl
16.0	123	31.0	248
19.0	128	31.5	254
19.5	132	32.0	259
19.5	136	32.5	265
20.0	141	33.0	270
20.5	145	33.5	276
21.0	150	34.0	281
21.5	155	34.5	287
22.0	159	35.0	293
22.5	164	35.5	298
23.0	169	36.0	304
23.5	173	36.5	310
24.0	178	37.0	316
24.5	183	37.5	322
25.0	188	38.0	328
25.5	192	38.5	334
26.0	197	39.0	341
26.5	202	39.5	347
27.0	207	40.0	353
27.5	212	40.5	360
28.0	217	41.0	367
28.5	222	41.5	373
29.0	227	42.0	380
29.5	233	42.5	387
30.0	238	43.0	394
30.5	243	43.5	401

[a]National Cholesterol Education Program Guidelines:

Total Cholesterol	Classification
Less than 200 mg/dl	Desirable
200 to 239 mg/dl	Borderline high
240 mg/dl or greater	High

[b]mg/dl = milligrams per ceciliter

1. Examine the AccuMeter®, and locate its various parts (Figure 46-2).
2. Select the finger that will be used, and wipe it with a cotton square that has been moistened, but not saturated, with alcohol. Do not use a finger that will be involved later in holding a pencil or other such activity.
3. Unwrap the blood lancet. Note that the lancet is to be used only once and then discarded as indicated by the instructor.
4. Hold the finger in a downward position, and gently squeeze it.
5. Quickly puncture the ball of the finger with the lancet (Figure 46-3a).
6. Keeping the finger in a downward position, drip 3 drops of blood into the cassette well (Figure 46-3b).
7. Pull the tab on the right side of the AccuMeter® to start the reaction (Figure 46-3c).
8. Wait 12 minutes until the "completion-indicator window" turns green.
9. Read the result on the scale, and refer to Table 46-2 for your cholesterol reading.
10. Enter your scale reading and cholesterol value in the results and observations section.
11. Enter the readings and values of three other students in the class, for comparison purposes.
12. Dispose of all blood-containing materials as indicated by your instructor.
13. Wash your hands.
14. Answer the questions in the results and observations section.

RESULTS AND OBSERVATIONS

1. Enter the AccuMeter® readings and cholesterol values in Table 46-3.
2. Compare the cholesterol values to those in Table 46-2

a. Were all values within the normal range? _____

b. If not, what factors could account for the values observed? _____ _____

Table 46-3

Student	AccuMeter® Reading	Cholesterol Value (mg/dl)	Approximate Age (in years)

LABORATORY REVIEW 46

CHOLESTEROL TESTING

LISTING, COMPLETION, AND DEFINITION

1. List seven factors that may influence adult serum-cholesterol levels.

   a. _____

   b. _____

   c. _____

   d. _____

   d. _____

   f. _____

   g. _____

2. List eight conditions in which elevated serum-cholesterol levels may occur.

   a. _____

   b. _____

   c. _____

   d. _____

   e. _____

   f. _____

   g. _____

   h. _____

3. List eight conditions in which decreased serum-cholesterol levels may occur.

   a. _____

   b. _____

   c. _____

   d. _____

   e. _____

   f. _____

   g. _____

   h. _____

4. Define or explain the following:

   a. Cholesterol _____

   _____

   b. Hypercholesterolemia _____

   _____

c. Hyperlipoproteinemia _____

_____

d. Atherosclerotic plaque _____

_____

MULTIPLE CHOICE

Insert the correct answers in the spaces provided.

1. _____

2. _____

3. _____

4. _____

5. _____

6. _____

7. _____

1. In atherosclerosis, which of the following substances are deposited in arterial walls?
   a. protein
   b. polysaccharides
   c. nucleic acids
   d. triglycerides
   e. amino acids

2. The lesion in atherosclerosis is called an "atherosclerotic [*what?*]"
   a. LDL
   b. plaque
   c. HDL
   d. VLDL
   e. choices *a* and *c*

3. What are two major classes of lipoproteins that form the total cholesterol of an individual?
   a. VLDL
   b. HDL
   c. LDL
   d. choices *a* and *b*
   e. choices *b* and *c*

4. As blood levels of LDL increase, what happens to the risk of coronary artery disease?
   a. remains the same
   b. increases
   c. decreases

5. When the risk of coronary artery disease decreases, what are likely to be the associated blood levels of HDL?
   a. remains the same
   b. increases
   c. decreases

6. What is the factor considered in ratio to total cholesterol that is used for predicting the risk of developing coronary artery disease and related conditions?
   a. HDL
   b. LDL
   c. plaque
   d. VLDL
   e. choices *a* and *d* only

7. Which of the following terms describe a condition of elevated cholesterol levels?
   a. hypocholesterolemia
   b. hyperlipoproteinemia
   c. atherosclerosis
   d. arteriosclerosis
   e. hypercholesterolemia

The Urinary and Reproductive Systems

THE URINARY SYSTEM

The major function of the urinary system is to regulate the volume and composition of the extracellular fluid. The accumulation of unused or toxic by-products of metabolism and changes in the individual's environment, food ingestion, or metabolic activities may upset the balance or constancy of the body's internal environment. The kidneys—major components of the urinary system—are essential homeostatic organs because of their ability to remove wastes and to alter the volume and composition of blood through the production and excretion of urine.

The composition of urine can vary greatly. Some variation in the amount of substances normally found in the urine may reflect altered body activity or food and fluid intake. Elevated levels of these common components or the presence of substances normally not found in urine may indicate disturbed metabolic functions or abnormal renal activity.

Disorders and diseases of the kidneys may result in severe impairment of function. If the kidneys cannot adequately regulate the composition of blood, an artificial device must be used in order to maintain life. The so-called artificial kidney uses a dialysis mechanism involving a membrane through which substances from the blood may diffuse.

Because of the importance of the kidneys and related structures, the functioning of the urinary system is routinely tested. Evaluation of kidney function usually involves determination of the composition and volume of urine. Such tests also may reveal the presence of microbial infection.

This section considers the anatomy of the two systems, some specific physiological principles of renal function, the normal and abnormal components of urine, and the principle of hemodialysis. It examines the procedures involved in routine urinalysis tests and detection of bacterial infection. It also demonstrates the pH value and buffering capacity of several substances involved in the acid–base balance of body fluids.

THE REPRODUCTIVE SYSTEM

Exercises in this section describe the structures and functions of organs associated with sex-cell production and the stages of pregnancy and birth.

All eukaryotic cells exhibit the complex mechanism of *mitosis,* which ensures the even distribution of hereditary material. This process is quite efficient, and errors rarely occur. Mitosis results in the formation of two new cells, each containing the same number and types of chromosomes as in the original cell. The chromosome number normally remains constant from one generation of body, or somatic, cells to the next. In *meiosis,* or reduction division, the chromosome number is reduced by one-half. The process is associated with the sexual process of fertilization. In *fertilization,* the nuclei of two different cells, each containing one-half the normal number of chromosomes (haploid), fuse and establish the normal number of chromosomes (diploid). Errors during the stages of chromosome separation can result in genetic abnormalities. *Karyotyping,* a laboratory procedure, can be used to detect a number of chromosomal abnormalities. Exercises in this section also deal with meiosis, stages of early development, ovulation detection, pregnancy testing, and karoytyping.

GENERAL REFERENCES

Greep, R. O. (Ed.). *Reproductive Physiology IV.* Baltimore: University Park Press, 1983.

Lenin, A. *The Cycling Female.* San Francisco: Freeman, 1979.

Stamey, T. A., and R. W. Kindrachuk. *Urinary Sediment and Urinalysis.* Philadelphia: Saunders, 1984.

Sullivan, L. D., and J. J. Grantham. *Physiology of the Kidney,* 2nd ed. Philadelphia: Lea & Febiger, 1982.

Tortora, G. J., and S. R. Grabowski. *Principles of Anatomy and Physiology,* 7th ed. New York: HarperCollins Publishers, 1993.

SPECIFIC REFERENCES

Asman, J. (Ed.). *Infertility: Diagnosis and Management.* New York: Springer-Verlag, 1984.

Beaconsfield, P., G. Birdwood, and R. Beaconsfield. "The Placenta." *Scientific American,* 243: 94–102, 1980.

Fuchs, F., and A. Klopper (Eds.). *Endocrinology of Pregnancy,* 3rd ed. Philadelphia: Harper & Row, 1983.

Hirschhorn, N., and W. B. Greenough III. "Progress in Oral Rehydration Therapy." *Scientific American,* 264: 50–56, 1991.

Iannucci, L. "The Perplexities of Pregnancy." *FDA Consumer,* 24: 16–19, 1990.

Lewis, R. "Genetic Screening—Fetal Signposts on a Journey of Discovery." *FDA Consumer* 24: 17–23, 1990.

Nightingale, E. L., and M. Goodman. *Before Birth: Prenatal Testing for Genetic Disease.* Cambridge, MA: Harvard University Press, 1990.

Wassarman, P. M. "Fertilization in Mammals." *Scientific American,* 259: 78–84, 1988.

Anatomy of the Urogenital System

After completing this exercise, you should be able to

1. Name the parts of the kidney
2. Note the retroperitoneal position of the kidney
3. Point out the major regions seen in a coronal section of the kidney
4. Trace the pathway of wastes from the blood through the microscopic nephrons and gross structures of the urinary system
5. Name the parts of the reproductive systems of both sexes
6. Recognize the similarity of the location in the body of the reproductive organs of each sex

Materials

The following materials should be provided for class use:

1. Fetal-pig dissection specimens
2. Dissecting tools
3. Model of human kidney
4. Charts of human kidney
5. Bone forceps
6. Models of human male and female reproductive organs
7. Charts of reproductive organs

The organs of the *urogenital system* (the urinary system plus the genital system) are closely related during embryonic development. In the lower chordates, the urinary, reproductive, and digestive systems have a common outlet from the body, the *cloaca*. During the course of its development, the human embryo passes through the cloacal stage until a subdivision of this region takes place. The result is a separate opening for fecal wastes (the anus) and for genital and urinary products (the urogenital sinus). In the human female, there has been a further division of the urogenital region, resulting in separate openings of the urinary tract and the genital tract. There are three separate openings in the human female. In the male, reproductive products and urine share the same duct and open together.

DISSECTION OF THE PIG URINARY SYSTEM

If you have not already done so, remove the *parietal peritoneum,* which covers the kidney (Figure 47-1). Because the kidneys are not actually inside the abdominal cavity, they are said to be located *retroperitoneally* (behind the peritoneum).

Once again, identify the *renal artery* and *renal vein* associated with the concave medial surface of the kidney, termed the *hilum* or *hilus* (Figure 47-1). Emerging from the hilum is the broad yellow *ureter.* Trace the ureter posteriorly to the *urinary bladder* (Figure 47-2). The fetal pig is in the late stage of embryonic development, and the urinary bladder has not yet lost its connection with the *extraembryonic membranes.* The urinary bladder is still a part of the embryonic *allantois,* from which it will not separate until birth, when the *umbilical cord* is severed. The urinary bladder is flanked by the large and prominent *umbilical arteries.* Note the *umbilical vein,* which was transected and examined in another exercise. The urinary bladder stores urine, which then leaves the body through the urethra.

The urinary organs in the body cavity are the same in both sexes. For the study of the reproductive system, follow the instructions for the sex of the pig that you are dissecting. When you have completed this dissec-

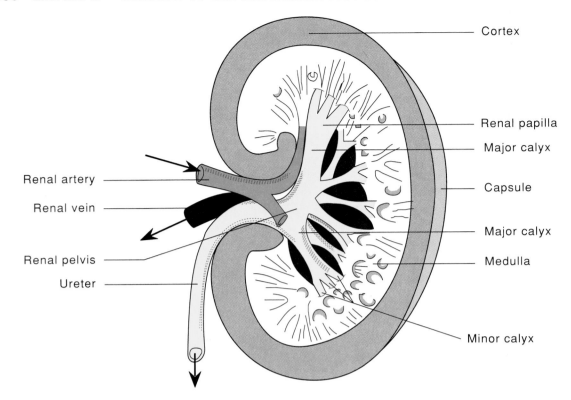

Cortex

Renal papilla

Major calyx

Renal artery

Capsule

Renal vein

Major calyx

Renal pelvis

Medulla

Ureter

Minor calyx

Figure 47-1 Pig kidney—coronal section.

Inferior vena cava

Aorta

Adrenal gland

Renal artery

Renal vein

Kidney

Ureter

Urinary bladder

Umbilical vein

Umbilical artery

Urethra

Urogenital sinus

Genital papilla

Figure 47-2 Urinary system of the fetal pig.

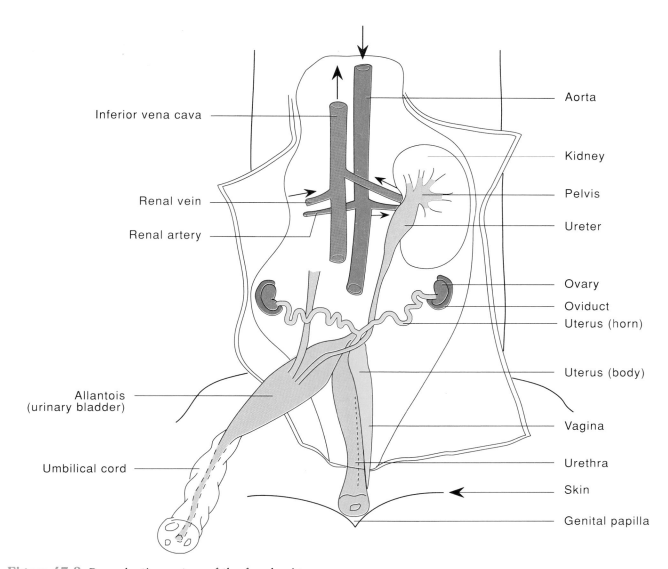

Inferior vena cava

Renal vein

Renal artery

Allantois
(urinary bladder)

Umbilical cord

Aorta

Kidney

Pelvis

Ureter

Ovary

Oviduct

Uterus (horn)

Uterus (body)

Vagina

Urethra

Skin

Genital papilla

Figure 47-3 Reproductive system of the female pig.

tion, compare it with a pig of the opposite sex. You are responsible for recognizing the anatomy of both male and female systems.

DISSECTION OF THE REPRODUCTIVE SYSTEM OF THE FEMALE PIG

On the surface of the pig, locate the *genital papilla* and the opening of the *urogenital sinus*. Return to the posterior end of the abdominal cavity, and look for the small, oval, paired *ovaries*. They are attached to a convoluted tube, the *oviduct* (also called the *uterine tube* or *fallopian tube*). The oviduct flares out at its proximal

end and partially encloses the ovary. Both the oviducts and the ovaries are so small that they should be examined with a 3× lens. The oviduct merges with the coiled *uterine horn,* which terminates in the centrally positioned *uterine body* (Figure 47-3). The distal end of the body of the uterus ends in a constricted neck (Figure 47-4), the *cervix* of the *uterus*. Covering the cervix is the most posterior organ of the reproductive tract, the *vagina*. Do not attempt to locate the cervix on your fetal pig. It is small and undeveloped. If you have a demonstration preparation of a larger, mature animal, such as the pregnant cat, make the observation of the cervix. The vagina opens into the *urogenital sinus*, which terminates at the *genital papilla* on the surface of the body (Figure 47-3).

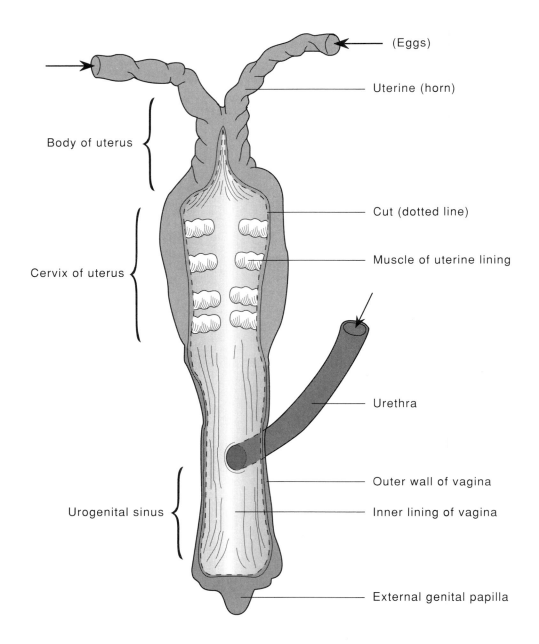

(Eggs)

Uterine (horn)

Body of uterus

Cut (dotted line)

Muscle of uterine lining

Cervix of uterus

Urethra

Outer wall of vagina

Urogenital sinus

Inner lining of vagina

External genital papilla

Figure 47-4 Posterior end of the female pig's reproductive system.

DISSECTION OF THE REPRODUCTIVE SYSTEM OF THE MALE PIG

Locate the scrotal sacs once again. They can be seen in Figure 47-5. They should now be dissected. Carefully remove the testes from the *scrotal sacs* in which they lie. Work on one side of the body or both sides, as your instructor wishes. Lying along the medial surface of each testis is an *epididymis,* a coiled tubule. Sperm are temporarily stored in this tubule. Emerging from the tubule is the *vas deferens (ductus deferens),* which passes into the body cavity from the scrotum through the *inguinal canal,* an opening in the body wall. Locate this canal, and observe how it has the potential to weaken the wall of the abdomen in that region. The human male has the same anatomy, and *inguinal hernias* are not uncommon. Running parallel with the vas deferens are the *spermatic artery* (red), the *spermatic vein* (blue), and the *spermatic nerve* (yellow). These three ducts and a nerve are bound together as the *spermatic cord,* which enters the body cavity. The artery, vein, and nerve con-

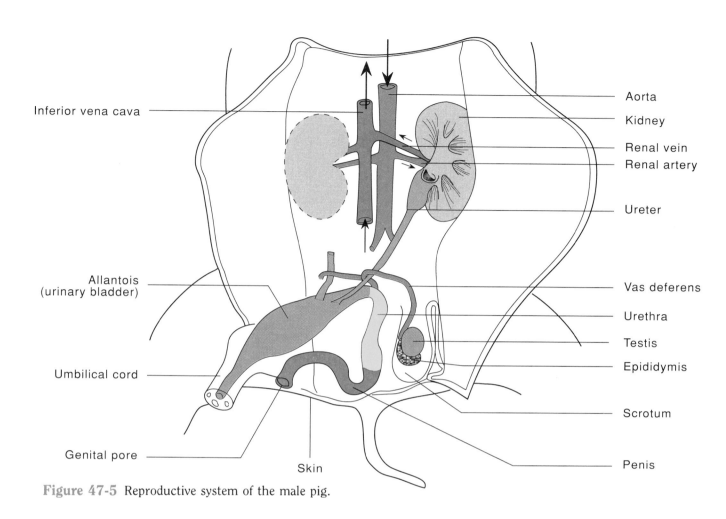

Inferior vena cava

Aorta

Kidney

Renal vein

Renal artery

Ureter

Allantois
(urinary bladder)

Vas deferens

Urethra

Testis

Umbilical cord

Epididymis

Scrotum

Genital pore

Skin

Penis

Figure 47-5 Reproductive system of the male pig.

tinue anteriorly to just below the kidney, but the vas deferens separates from the cord, loops over the ureter, and enters the *urethra* posterior to the urinary bladder. This means that sperm from the testes and urine from the urinary bladder share the same duct as they exit the body. This is the reproductive anatomy in all male mammals, including the human male.

The urethra passes to the outside of the body within the *penis,* a long, tough, whitish tube. In the fetal pig, the penis is internal. Locate the urethra by carefully splitting the *pubic symphysis* of the pelvic girdle. Follow the urethra by removing all muscle and connective tissue around it. The penis empties to the outside of the body at the *urogenital orifice* (review Exercise 15) on the surface of the body just posterior to the umbilical cord. Carefully dissect this duct system.

There are three pairs of male *associated reproductive glands,* which produce semen, a sticky, nutritious, semifluid compound, which empties into the urethra. To find these glands, return to the place where the vas deferens empties into the urethra just posterior to the urinary bladder. The paired *seminal vesicles* are on either

side of the urethra at this point, and they empty into the urethra. In this same region in the midline, find a single dark gland lying over the urethra. This is the *prostate gland* (Figure 47-6). Now move posteriorly down the urethra to the region of the pubic symphysis. There you will find large paired *bulbourethral glands* (also called "Cowper's glands") lateral to the urethra. These reproductive glands all contribute to the semen, which mixes with the sperm and provides a fluid medium for their transfer to the female reproductive system.

HUMAN URINARY SYSTEM

The relationship of parts of the urinary system is the same in all mammals. A study of the human kidney reveals that the kidneys are located retroperitoneally and that the left kidney is slightly superior to the right. An examination of a coronal section of a human kidney (Figure 47-7) reveals two distinct tissue types, an outer *cortex* and an inner *medulla*. Within the kidney is the

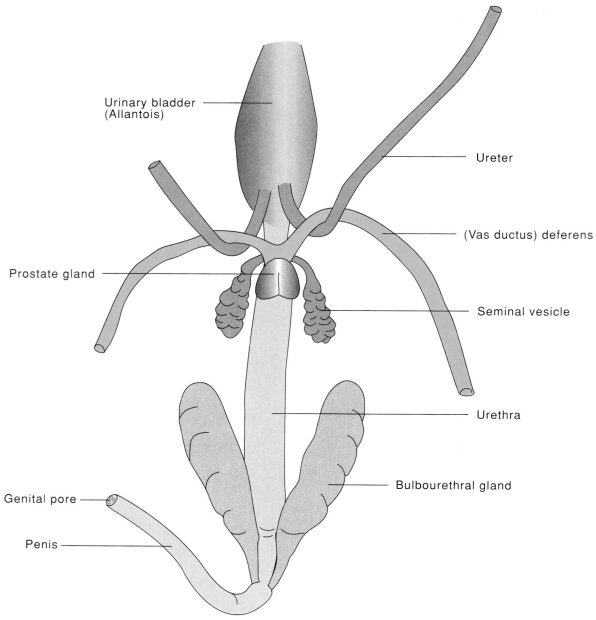

Urinary bladder
(Allantois)

Ureter

(Vas ductus) deferens

Prostate gland

Seminal vesicle

Urethra

Bulbourethral gland

Genital pore

Penis

Figure 47-6 Male pig reproductive glands.

pelvis, a large cavity where urine is collected. At the concave medial surface (hilum) there are a *ureter,* a *renal artery,* and a *renal vein* (Figure 47-7). Wastes are brought to the kidney in the renal artery, then urine is formed in the nephrons. From there, urine flows from the collecting duct system into the pelvis, then through the ureter to the urinary bladder (Figure 47-6), where it is temporarily stored. Urine leaves the body through the urethra. Examine a model of the human kidney, and compare it with Figure 47-8.

Urine is produced by millions of *nephrons,* tiny

microscopic units (Figure 47-9). Most of each nephron is located within the *cortex* area of the kidney, and each has an intimate association with the vascular system. Wastes are taken from the blood at the nephron's *glomerulus* into *Bowman's capsule.* They pass along the tubules (refer to your textbook for the physiology of urine formation) from the *proximal tubule,* the *loop of Henle,* and the *distal tubule* to the collecting duct system, which is composed of *collecting tubules, papillary ducts, minor calyces,* and *major calyces,* and on to the pelvis (Figure 47-9). These parts of the kidney can be

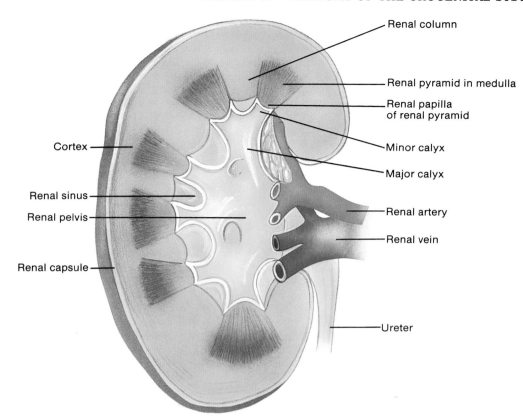

Figure 47-7 Coronal section through human kidney. (From G. J. Tortora and S. R. Grabowski, *Principles of Anatomy and Physiology,* 7th ed., New York: HarperCollins, 1993.)

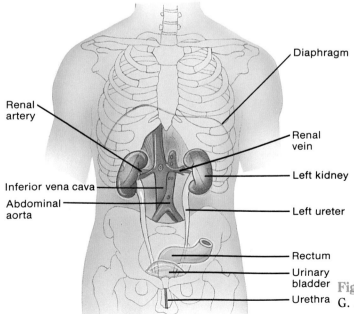

Anterior view

Figure 47-8 Location of human urinary organs. (From G. J. Tortora and S. R. Grabowski, *Principles of Anatomy and Physiology,* 7th ed., New York: HarperCollins, 1993.)

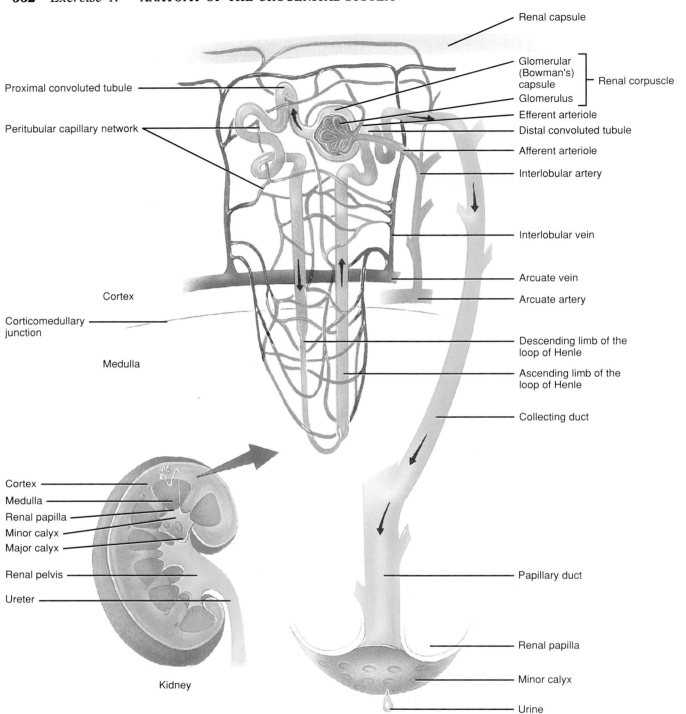

Figure 47-9 The human nephron. (From G. J. Tortora and S. R. Grabowski, *Principles of Anatomy and Physiology,* 7th ed., New York: HarperCollins, 1993.)

found on most models and on the diagram. The human urinary bladder has a triangular area internally between the emptying ureters and the urethra called the *trigone* (Figure 47-10).

THE HUMAN REPRODUCTIVE SYSTEM

Female

The human female reproductive tract is similar to that of the pig. The main difference is the shape of the uterus. The human female has a *simplex* uterus, which usually accommodates one fetus at a time. Note its shape in Figure 47-11. The pig has a *bicornuate* (two-horned) uterus with the horns on each side capable of accommodating up to a dozen embryos. The human may occasionally have a multiple birth, but usually the uterus accommodates one fetus at a time.

In the human, the oviduct is usually called the *Fallopian tube* or the *uterine tube.* It is more prominent in the human than in the pig. Study Figures 47-11 and 47-12 and compare the size and shape of the various organs in the two species.

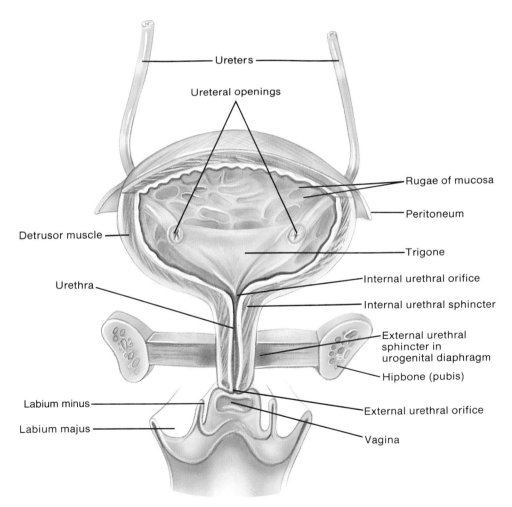

Figure 47-10 Female urinary bladder showing trigone. (From G. J. Tortora and S. R. Grabowski, *Principles of Anatomy and Physiology,* 7th ed., New York: HarperCollins, 1993.)

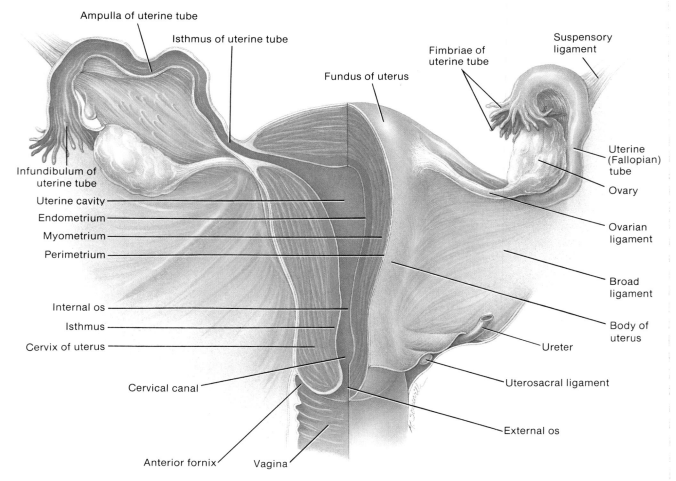

Figure 47-11 Human female reproductive system—posterior view. (From G. J. Tortora and S. R. Grabowski, *Principles of Anatomy and Physiology,* 7th ed., New York: HarperCollins, 1993.)

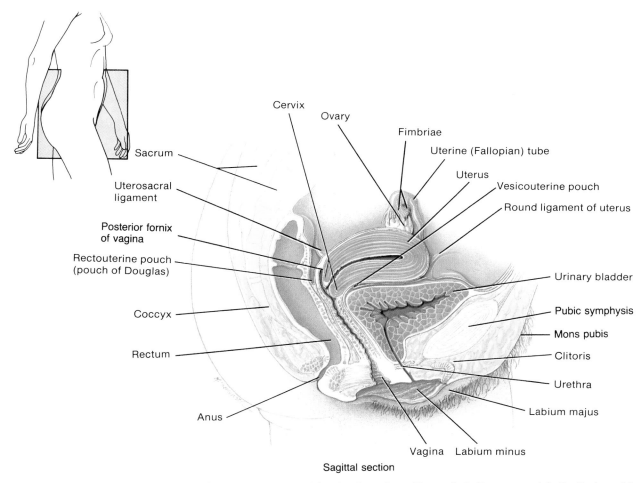

Sagittal section

Figure 47-12 Human female reproductive system—midsagittal section. (From G. J. Tortora and S. R. Grabowski, *Principles of Anatomy and Physiology,* 7th ed., New York: HarperCollins, 1993.)

Male

The anatomy of the male reproductive system of the human is almost the same as the fetal pig (Figures 47-13 and 47-14). The location of the testes and the anatomy of the duct system are almost identical. Even the associated reproductive glands are the same. In the diagrams locate the human *prostate gland,* the paired *seminal vesicles* and the paired *bulbourethral glands.*

The only major difference between your dissection of the pig and the anatomy of the human is the location of the *penis* which is external in the human male and lies internal to the genital pore in the pig.

Compare as many organs as you can between fetal pig and human using Figures 47-13 and 47-14 and human models.

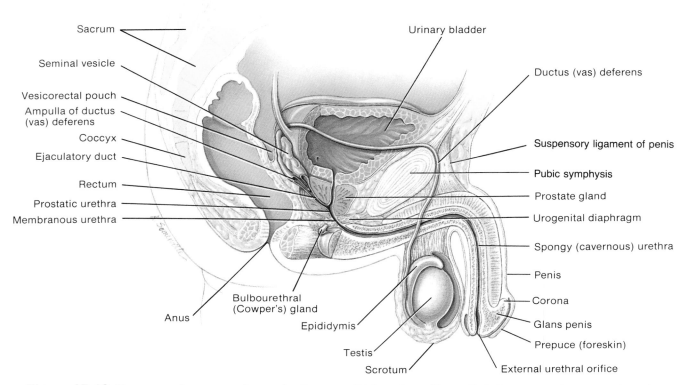

Figure 47-13 Human male organs of reproduction—sagittal section. (From G. J. Tortora and S. R. Grabowski, *Principles of Anatomy and Physiology,* 7th ed., New York: HarperCollins, 1993.)

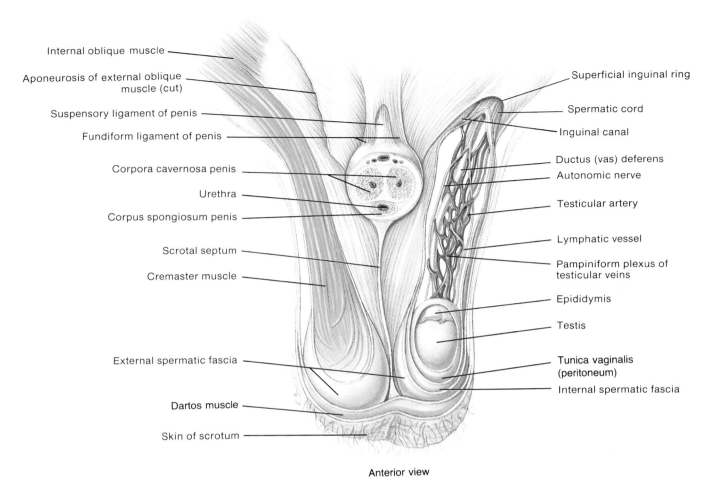

Internal oblique muscle

Aponeurosis of external oblique muscle (cut)

Suspensory ligament of penis

Fundiform ligament of penis

Corpora cavernosa penis

Urethra

Corpus spongiosum penis

Scrotal septum

Cremaster muscle

External spermatic fascia

Dartos muscle

Skin of scrotum

Superficial inguinal ring

Spermatic cord

Inguinal canal

Ductus (vas) deferens

Autonomic nerve

Testicular artery

Lymphatic vessel

Pampiniform plexus of testicular veins

Epididymis

Testis

Tunica vaginalis (peritoneum)

Internal spermatic fascia

Anterior view

Figure 47-14 Spermatic cord and inguinal canal. (From G. J. Tortora and S. R. Grabowski, *Principles of Anatomy and Physiology,* 7th ed., New York: HarperCollins, 1993.)

LABORATORY REVIEW *47*

ANATOMY OF THE UROGENITAL SYSTEM

TRACING, COMPLETION, AND
FIGURE IDENTIFICATION

1. Trace the pathway of a waste product from the bloodstream through the kidney to its elimination from the body. Some of the steps are provided. You fill in the rest.

   a. <u>Afferent arteriole</u>

   b. _____

   c. _____

   d. <u>Proximal convoluted tubule</u>

   e. _____

   f. _____

   g. <u>Collecting duct</u>

   h. _____

   i. _____

   j. <u>Renal pelvis</u>

   k. _____

   l. _____

   m. <u>Urethra</u>

2. Identify the parts of the human kidney in Figure Q47-1.

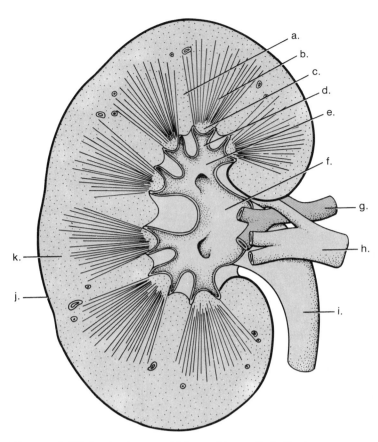

Figure Q47-1 The human kidney. (From G. J. Tortora, *Principles of Human Anatomy,* 3rd, 4th, 5th eds., New York: HarperCollins, 1983, 1986, 1989.)

a. _____

b. _____

c. _____

d. _____

e. _____

f. _____

g. _____

h. _____

i. _____

j. _____

k. _____

3. What anatomical feature of the human female reproductive tract makes it difficult for her to have multiple births? Explain.

_____

_____

_____

4. The anatomy of the male reproductive tract in both the pig and the human is still tied to the urinary system, whereas in the female, the two systems have separated. Explain the relationship of the reproductive system and the urinary system in both male and female.

_____

_____

_____

_____

_____

_____

_____

5. Trace the pathway of a human spermatozoan from its place of origin to the site of fertilization of an ovum. Some of the answers are provided. You fill in the rest.

a. <u>Seminiferous tubules</u>

b. _____

c. _____

d. <u>Urethra</u>

e. <u>Vagina</u>

f. _____

g. _____

h. <u>Fallopian tube or oviduct</u>

Fluid and Electrolyte Balance and a Demonstration of Dialysis

After completing this exercise, you should be able to

1. Describe the process of urine formation
2. List the general parts of a nephron
3. Measure the rate of urine formation after ingesting a known quantity of a test solution
4. Determine selected physical and chemical characteristics of urine samples after ingesting a known quantity of a test solution
5. Describe the dialysis process
6. Perform a procedure to demonstrate dialysis
7. List three conditions in which dialysis can be used

URINE FORMATION, FILTRATION, AND PHYSIOLOGICAL REGULATION OF FLUID AND ELECTROLYTE BALANCE

Urine is an extremely complex water solution of various inorganic and organic substances that are the waste products of metabolism or are derived from ingested foods and fluids. The most important *inorganic* constituents of urine include ammonia, chlorides, phosphates, and sulfates. The chief *organic* components are creatinine, urea, and uric acid. Urine is a valuable indicator of various normal and abnormal situations because most pathological conditions originating in the kidneys and various diseases involving other body organs alter its composition. Certain chemical substances and large numbers of microorganisms appear in urine only when pathological conditions are present. (Exercise 49 describes representative disease conditions and the substances that are associated with them.)

Urine formation is a process essential to the body's capability to rid itself of metabolic waste products. The kidneys are among the organs that assume a major responsibility for the regulation of the volume, pH, and chemistry of body fluids. Functional kidney performance depends on its basic units, the *nephrons* (Figure 48-1). These delicate, microscopic structures vary the concentration, volume, and composition of urine, according to the needs of the body. Each nephron consists of a *glomerulus* (a tuft of capillaries enclosed by a double-walled cup termed the *renal* or *malpighian corpuscle*), the *proximal convoluted tubule,* the *loop of Henle,* the *distal convoluted tubule,* and the *collecting duct,* which leads to the *renal pelvis.* Each kidney contains approximately 1 million nephrons.

Urine formation begins in the glomerulus, where water and small molecules are filtered from the blood. The blood supply to this filtering unit is provided by the afferent arteriole, which in turn is supplied by a renal artery. The resulting filtrate passes through long tubules. Physiologically important substances (such as glucose, water, electrolytes, including sodium, chloride, and phosphates, and some ions) are selectively reabsorbed, while other substances (such as ammonia, potassium, hydrogen ions, and foreign materials such as penicillin) are secreted into the urine. The final tubule fluid, urine, that enters the bladder is quite different in composition from the glomerular filtrate. The amount of filtrate that forms in both kidneys every minute is known as the *glomerular filtration rate (GFR).* An average, normal adult secretes about 180 liters (48 gallons) each day. This rate amounts to approximately 125 ml/minute.

The GFR is regulated by two major mechanisms: *renal autoregulation* and *neural regulation. Renal autoregulation* refers to the ability of the kidneys to maintain a constant blood pressure and GFR, in spite of

Proximal convoluted tubule

Efferent arteriole

Peritubular capillary

Glomerular (Bowman's) capsule

Distal convoluted tubule

Glomerulus

Afferent arteriole

Blood to renal vein

Blood from renal artery

Collecting duct

Urine

1. Glomerular filtration
2. Tubular reabsorption
3. Tubular secretion

Figure 48-1 A summary of nephron function. Note the numbered parts of the diagram, as well as the arrows. ① As blood flows from the glomerulus into the glomerular (Bowman's) capsule, it is filtered by the endothelial–capsular membrane. As the filtrate passes through the nephron ②, some components are selectively reabsorbed into blood, and ③ other substances are secreted from the blood into the filtrate for elimination in urine. (From G. J. Tortora and S. R. Grabowski, *Principles of Anatomy and Physiology,* 7th ed., New York: HarperCollins, 1993.)

changes in the body's arterial pressure. *Neural regulation* involves the sympathetic division of the autonomic nervous system. Under strong sympathetic stimulation, the adrenal medulla secretes epinephrine, which causes a decrease in the GFR.

The continuous adjustment of urine volume and electrolyte content within the tubules is controlled by several substances and conditions, including (a) *antidiuretic hormone (ADH)* from the neurohypophysis; (b) the sodium-retaining hormone, *aldosterone,* from the adrenal cortex blood flow; and (c) the acid–base balance.

ADH, also known as *vasopressin,* is regulated by osmoreceptors in the hypothalamus. When the body is dehydrated, the concentration of water in the blood falls below normal levels and consequently stimulates these receptors to bring about the release of ADH into the bloodstream and its transportation to the kidneys. The kidneys respond by decreasing the urine output, thereby conserving water. If, on the other hand, the blood contains higher-than-normal water concentrations, the os-

moreceptors stop ADH secretion. The kidneys then release large quantities of urine, and the volume of water in body fluids is brought back to normal. Angiotensin II is an active hormone associated with the release of ADH and water retention by the kidneys. Secretion of ADH also can be influenced by various chemical factors, such as alcohol, nicotine, and other drugs, and by physical and mental factors, such as pain, trauma, blood loss, and stress.

Aldosterone acts on some specific cells in the kidneys, to increase their reabsorption of sodium (Na^+) ions from the urine, return them to the blood, and thus prevent rapid depletion of sodium from the body. At the same time, aldosterone stimulates excretion of potassium (K^+) ions, so that large amounts of potassium are lost in the urine.

The control of aldosterone secretion involves several mechanisms. One of these is the *renin–angiotensin pathway.* A decrease in blood pressure—caused by a drop in blood volume due to dehydration, sodium (Na^+)

ion deficiency, or severe blood loss—stimulates some specific kidney cells to secrete the enzyme *renin* into the bloodstream. This enzyme converts *angiotensinogen,* a protein produced by the liver, into *angiotensin I,* which is then converted into *angiotensin II* by an enzyme in the lungs. Angiotensin II stimulates the adrenal cortex to produce more aldosterone. Aldosterone brings about increased Na^+ reabsorption, and water is also reabsorbed, thus increasing fluid volume and restoring blood pressure to normal.

Another mechanism for the control of aldosterone is potassium (K^+) ion concentration. An increased K^+ concentration in extracellular fluid directly stimulates aldosterone secretion and causes the kidneys to eliminate excess K^+. A decreased K^+ concentration in extracellular fluid reverses the process.

In the first part of this exercise, the effects of salt, sugar, grapefruit juice (a source of potassium), and coffee or tea (sources of caffeine) on the volume, pH, specific gravity, and chloride content of urine are considered. Students who are diabetic or are under medical treatment for a kidney or related problem should not participate in the first portion of this exercise. The principle of dialysis and factors that can affect the process are considered in the second portion of this exercise.

HEMODIALYSIS

Dialysis is a process by which certain substances in solution can be separated from one another by virtue of the difference in their rates of diffusion through a semipermeable membrane. *Hemodialysis* represents one type of dialysis that is used to correct a variety of conditions in the body that affect renal excretion. In situations such as drug overdose, temporary stoppage of urine formation, and extensive kidney destruction, hemodialysis may mean the difference between life and death.

Two types of commercial hemodialysis systems, frequently referred to as *artificial kidneys,* are in general use. Both operate on the same principle. Blood from the patient is circulated by a pump past one side of a semipermeable membrane, termed a *dialyzing membrane* and usually made of cellophane. Circulating on the other side of the membrane is a *dialyzing solution,* containing electrolytes in the same concentration as in normal plasma. As the blood and solution pass on either side of the membrane, excess electrolytes in the blood diffuse through the membrane into the solution because of the concentration gradient.

In one type of hemodialysis system—the *coil type*—the dialyzing membranes are single or parallel flat cellophane tubes wrapped around a central core, supported by plastic screening. The dialyzing or bathing solution

is pumped through the concentric layers and over the flattened tubing of the system. The coil assembly is presterilized and is generally discarded after use.

The second type of system, the *plate* or *layer dialyzer,* contains cellophane sheets instead of tubes (Figure 48-2). The sheets are held apart within plastic frames to allow passage of the fluid to be dialyzed between them. The dialyzing solution flows on the outside of the cellophane sheets. All parts of this system are sterilized before use.

The dialyzing solution is temperature controlled and has a composition that chemically approximates normal extracellular fluid. Specific needs of individuals on hemodialysis may require alterations in the extracellular fluid. Dialysis begins when blood from one of the patient's arteries is pumped through the system. After removal of the unwanted substances, the blood passing through the dialysis system is returned to the patient's circulation by a vein.

Figure 48-2 The operation of an artificial kidney. The blood route is indicated in red and blue. The route of the dialysis is indicated in gold. (From G. J. Tortora and S. R. Grabowski, *Principles of Anatomy and Physiology,* 7th ed., New York: HarperCollins, 1993.)

EFFECTS OF SODIUM, POTASSIUM, CAFFEINE, AND WATER ON URINE COMPOSITION

Materials

1. The following materials should be provided for class use:
   a. Urinometers (cylinder and hydrometer float device)

b. pH indicator paper strips (pHydrion or nitrazine test papers)
c. Thermometers
d. 250-ml beakers
e. Filter paper
f. Test tubes
g. Eyedroppers
h. 20 percent potassium chromate solution in dropper bottle
i. 2.9 percent silver nitrate solution
j. Disposable drinking cups
k. 1-ml pipettes, graduated in 0.1 ml, with bulbs
l. Test-tube racks
m. Liquid soap and brushes for washing containers
n. Containers of disinfectant

2. The following materials should be provided for students in groups of five:
a. Five urine-specimen containers
b. Five 500-ml graduated cylinders
c. Five 250-ml beakers
d. Test solutions in the volumes indicated:
   Cola drink (nondietetic), 750 ml
   Distilled water, 1250 ml (500 ml is to be used with a salt tablet)
   Unsweetened, strong coffee or tea, 750 ml
   Unsweetened grapefruit juice, 750 ml
e. One salt tablet (4.5g)

3. Disposable surgical gloves

CAUTION

Working with urine or other body fluids will not pose any danger if the following precautions are observed:

1. Place any and all urine or materials containing urine into the disinfectant containers provided.
2. Dispose of any urine-containing materials as indicated by the instructor.
3. In cases of urine spillage, wipe the area with a disinfectant-soaked paper towel, and dispose of all materials as indicated by the instructor.
4. Always wash your hands after handling urine and associated materials.
5. The use of disposable surgical gloves.

Procedure 1: Volume, Color, Turbidity, and Odor

This procedure is to be performed by students individually, but they are to work in groups of five. The instructor will indicate the facilities to use in order to obtain urine samples.

1. Obtain a urine specimen container, and empty your bladder completely. Note the time, and enter it in the appropriate space of the results and observations section. This sample is your normal urine specimen.

2. Use one of the 500-ml graduated cylinders to determine the volume of the normal specimen. Record the information in the results and observations section. The normal volume per 24 hours is 1 to 2 liters.

3. Note the color, the presence of turbidity (cloudiness), and any odor. Record your findings in Table 48-3 of the results and observations section. Normal, freshly voided urine is generally yellow to amber, is clear, and does not have a disagreeable odor.

Procedure 2: Specific Gravity and pH

This procedure is to be performed by students individually, but they are to work in groups of five.

1. Pour about 30 ml of the urine sample into a small beaker. Carefully pour the urine from the beaker into a clean urinometer cylinder until it is three-fourths full. Obtain more of the specimen if necessary. If foam forms on the surface, remove it with a piece of filter paper. Discard the paper as indicated by your instructor.

2. Obtain a thermometer, and check the reading on it. If the reading is not 37°C, (carefully) shake it down to that level. Insert the thermometer in the urinometer, and read the temperature after 3 minutes. The temperature of the sample is used in the determination of specific gravity in the next step. Clean the thermometer as indicated by your instructor.

3. Float the hydrometer in the cylinder. Make certain that it is not touching the bottom or sides of the cylinder. Adjust the value of the hydrometer reading according to the temperature, to obtain the specific gravity. For each 3° above 25°C, add 0.001 to the hydrometer reading; for each 3° below 25°C, subtract 0.001. Record your findings in the results and observations section. The specific gravity of normal urine is 1.002 to 1.030. Remove the hydrometer, and clean it as indicated by your instructor.

4. Dip a pH indicator test strip paper in the urinometer cylinder three consecutive times. Remove excess urine by touching the paper to the inside surface of the cylinder. After 1 minute, compare the color of the strip with the color guide on the strip container or other available guide. Normal urine has an average pH of 6 but ranges from 4.8 to 7.5. Record your findings in the results and observations section.

5. Clean the urinometer as indicated by your instructor.

Procedure 3: Chloride Concentration Determination

This procedure is to be performed by students individually, but they are to work in groups of five.

1. Place two test tubes in a rack.
2. Using a 1-ml pipette, introduce 0.5 ml of the urine sample into each tube.
3. Add 1 drop of 20 percent potassium chromate to each tube.
4. Gently shake the tubes. The solution in the tubes should be yellow.
5. With an eyedropper, add 1 drop at a time of the silver nitrate solution provided to the first tube. Gently shake the tube after each drop. Count the number of drops required to change the color of the urine solution from yellow to brown. *Note:* One drop of the silver nitrate solution added is equivalent to 61 mg of chloride ion (Cl^-) per 100 ml of urine.
6. Calculate the chloride concentration of the urine sample by multiplying the number of silver nitrate drops added by 61. The value should be reported in milligrams/100 ml of urine.
7. Repeat Steps 5 and 6 with the second tube. Average your findings, and enter them in Table 48-3 of the results and observations section.
8. Empty the tubes as indicated by your instructor, and clean all glassware and other materials.

Procedure 4: Effects of Various Test Solutions on Urine Composition

This procedure is to be performed by students in groups of five.

1. Determine within your group the individuals who will drink a particular test solution or material. The test solutions include distilled water, cola drink, unsweetened grapefruit juice, and unsweetened strong coffee or tea. A salt tablet with water also will be used.

2. All members of the group should obtain the necessary volume of their test solution; 750 ml should be consumed. The individual selecting the salt tablet should swallow it with 500 ml of distilled water.
3. Drink the test solution as quickly as possible. If an individual is unable to consume the entire volume designated, the quantity consumed should be recorded in Table 48-2. Your instructor will indicate the pH, specific gravity, and chloride concentration of each test solution. Enter this information in Table 48-2.
4. Begin timing as soon as you have consumed the test solution entirely. Be prepared to obtain a urine sample at four 30-minute intervals. A total of four urine specimens will be collected.
5. Perform Procedures 1, 2, and 3 with each urine specimen collected. Enter the findings obtained with your test solution in Table 48-3 of the results and observations section.
6. Compare your findings with those obtained by other members of your group, and answer the questions in the results and observations section.

RESULTS AND OBSERVATIONS

1. In Table 48-1, list the food and drink items you consumed approximately 2 hours before starting the experiment. Indicate the approximate quantity of each item.
2. a. Approximate time of urinary bladder emptying prior to the laboratory exercise: _____

   b. Time at which normal urine specimen was obtained at the beginning of the laboratory exercise: _____

3. a. Enter the pH, odor, specific gravity, and relative chloride content of each test solution provided for your group in Table 48-2.

   b. Which test solution did you use? _____

Table 48-1

Food or Drink Item	Quantity	Food or Drink Item	Quantity

Table 48-2

Test Solution	pH	Odor	Specific Gravity	Relative Chloride Content	Volume Consumed
Distilled water					
Unsweetened cola drink					
Unsweetened coffee or tea					
Unsweetened grapefruit juice					
Salt tablet and water					

4. Enter the findings obtained with your urine samples in Table 48-3.
   a. To calculate your rate of urine production for the normal sample, find the difference in time between emptying the bladder before the exercise and taking the normal urine sample. Then calculate the number of milliliters produced per minute. Enter the result in Table 48-3.
   b. Calculate the rate of urine formed for each interval after drinking the test solution, in ml/minute. Enter the results in Table 48-3.

5. a. Were any of the properties observed with the normal urine sample changed by the test solution consumed? If so, which ones? _____
   _____

   b. Offer an explanation for any changes observed.
   _____
   _____

6. a. Were any of the properties observed with the

Table 48-3

Property	Experimental Samples				
	Normal Sample	30 Minutes	60 Minutes	90 Minutes	120 Minutes
Volume (in ml)					
Color					
Odor					
Turbidity (clouding)					
pH					
Specific gravity					
Relative chloride content (in mg/ml)					
Rate of urine formation (in ml/minute)					

normal urine samples of your group changed by the test solutions they consumed? If so, which ones? _____

_____

b. Offer an explanation for any changes observed.

_____

_____

DEMONSTRATION OF DIALYSIS

Materials

1. The following materials should be provided per four students:
   a. Two pieces of dialysis tubing, each 20 cm (about 8 inches) long
   b. One 50-ml beaker and two 250-ml beakers
   c. One small funnel
   d. Four Pyrex test tubes (10–15 ml capacity)
   e. One 100-ml graduated cylinder
2. The following materials should be provided for class use:
   a. 2 percent glucose solution
   b. 5 percent starch solution
   c. 10 percent sodium chloride (NaCl) solution
   d. 10 percent egg albumin solution
   e. Benedict's reagent (for glucose determinations)
   f. Silver nitrate reagent (for Cl^- determinations)
   g. Iodine reagent (for starch determinations)
   h. Sodium hydroxide (NaOH) and copper sulfate ($CuSO_4$) reagents (for protein determinations)
   i. Distilled water, kept at 4°C, room temperature (25°C), and 37°C
   j. String for dialysis bags
   k. Boiling water bath

Procedure 1: Preparation of Dialysis Bags

This procedure is to be performed by students in groups of four.

1. Soak the dialysis tubing in distilled water (25°C) until soft.
2. While the dialysis tubing is softening, mix in a 50-ml beaker with 10 ml each of the albumin, glucose, sodium chloride, and starch solutions.
3. When the tubing material is soft, prepare a dialysis bag by tying one end of the tube with a small piece of string. Make certain that the knot is tight enough so that the system will not leak when filled.
4. Have a laboratory partner hold a small funnel in the open end of the dialysis bag. Fill the bag with the prepared chemical mixture. Leave only enough

space at the top of the bag so that the tube can be tied again to seal the bag.
5. Close the open end of the dialysis bag by tying a knot with another piece of string.
6. Carefully rinse the filled dialysis bag with distilled water (25°C).
7. Place the bag in a 250-ml beaker containing distilled water at 4°C. The bag should be completely submerged.
8. Prepare another dialysis bag by repeating Steps 1–6. Submerge the second bag in a beaker containing distilled water at 37°C.

Procedure 2: Testing the Dialysate

This procedure is to be performed by students in groups of four. The instructor will demonstrate the testing procedures for the *dialysate* (the chemicals passing out of the dialysis bag) before the students perform the procedure.

1. Immediately after submerging the dialysis bags in the beakers, carefully pour about 20 ml of the water from each beaker into separate graduated cylinders.
2. Divide the sample of the 4°C water into four separate test tubes (about 5 ml each). To determine which of the substances added to the dialysis mixture are present in the dialysate, perform the following tests, using one test tube for each test.
   a. The *Biuret* or *protein (egg albumin) test*—Add approximately 5 ml of the sodium hydroxide and 2 ml of the copper sulfate reagents. The formation of a pink-to-violet color is a positive reaction. Refer to Color Plate 53.
   b. *Chloride ion (Cl^-) test*—Add 8 drops of the silver nitrate solution. The formation of a white precipitate is a positive result.
   c. *Glucose test*—Add 10 ml of the Benedict's reagent, and place tube in a boiling water bath for about 5 minutes. The formation of an orange color is a positive result (see Color Plate 55).
   d. *Starch test*—Add 3 drops of the iodine reagent. The formation of a purple-to-black color is a positive result.
3. Remove 20-ml samples at 15 minutes, 30 minutes, 1 hour, 2 hours, and 48 hours (or during the next laboratory period).
4. With each sample, perform the tests listed in Step 2.
5. Enter your findings in the results and observations section.
6. Repeat Steps 2–5 with the second dialysis system (37°C water).
7. Answer the questions in the results and observations section.

RESULTS AND OBSERVATIONS

1. Describe the test results that would indicate the presence or absence of the substances listed in Table 48-4.
2. Enter in Table 48-5 the results of the tests performed. Indicate the presence of a substance with a " + " or its absence with a " − ."
3. What other methods could be used to test for the following substances?

   a. Glucose _____

   b. Albumin _____

4. Briefly explain the relationship between dialysis and kidney function. _____

   _____

   _____

5. Briefly describe the appearance of the dialysis bags at the beginning and at the end of the experiment.

   _____

   _____

Table 48-4

Substance	Positive Result	Negative Result
Chloride ion		
Egg albumin		
Glucose		
Starch		

6. What general conclusions can you draw from the results of this experiment? _____

   _____

   _____

Table 48-5

Time of Sampling	Chloride Ion		Egg Albumin		Glucose		Starch	
	4°C	37°C	4°C	37°C	4°C	37°C	4°C	37°C
Immediate								
15 minutes								
30 minutes								
1 hour								
2 hours								
48 hours								

LABORATORY REVIEW *48*

FLUID AND ELECTROLYTE BALANCE AND A DEMONSTRATION OF DIALYSIS

COMPLETION AND DEFINITION

1. Briefly describe the general events involved with urine formation. ———————

   ————————————————————————————————

   ————————————————————————————————

   ————————————————————————————————

2. List the three distinct processes involved with urine formation.

   a. ————————————————————

   b. ————————————————————

   c. ————————————————————

3. Define or explain the following:

   a. Glomerulus ————————————————————————

   ————————————————————————————————

   b. Nephron ————————————————————————

   ————————————————————————————————

   c. Antidiuretic hormone ————————————————————

   ————————————————————————————————

   d. Aldosterone ————————————————————————

   ————————————————————————————————

4. Give the function of the following structures in urine formation:

   a. Glomerulus ————————————————————————

   ————————————————————————————————

   b. Proximal tubule ————————————————————————

   ————————————————————————————————

DEMONSTRATION OF DIALYSIS

5. Define *dialysis*. ————————————————————————

   ————————————————————————————————

   ————————————————————————————————

6. What is hemodialysis? ————————————————————————

   ————————————————————————————————

ANSWERS

7. List and briefly describe two types of systems used for hemodialysis.

a. _____

_____

b. _____

_____

8. List three conditions in which hemodialysis is necessary.

a. _____

b. _____

c. _____

9. Define or explain the following:

a. Electrolyte _____

_____

b. Osmotic pressure _____

_____

c. Artificial kidney _____

_____

d. Reagent _____

_____

MULTIPLE CHOICE

Insert the correct answers in the spaces provided.

1. Which of the following are not among the chief organic components of urine?
   a. creatinine
   b. chloride
   c. uric acid
   d. urea
2. Which of the following are not among the most important inorganic constituents of urine?
   a. chlorides
   b. ammonia
   c. uric acid
   d. phosphates
   e. sulfates
3. Which of the following is not a component of a nephron?
   a. glomerulus
   b. loop of Henle
   c. glomerular capsule
   d. collecting duct
   e. proximal convoluted tubule
4. Which of the following structures leads to the renal pelvis?
   a. glomerulus
   b. loop of Henle
   c. glomerular capsule
   d. collecting duct
   e. proximal convoluted tubule

1. _____
2. _____
3. _____
4. _____
5. _____
6. _____
7. _____
8. _____
9. _____
10. _____

5. In which of the following structures does urine formation begin?
    a. glomerulus
    b. loop of Henle
    c. capsule
    d. collecting duct
    e. proximal convoluted tubule
6. Which of the following substances is or are not selectively reabsorbed during urine formation?
    a. glucose
    b. water
    c. any of the electrolytes
    d. ammonia
    e. choices *c* and *d* only
7. Which of the following is or are secreted in urine?
    a. glucose
    b. water
    c. any of the electrolytes
    d. ammonia
    e. choices *c* and *d* only
8. Which of the following is or are known to stimulate the release of ADH into the bloodstream?
    a. body dehydration
    b. higher-than-normal water concentration in the blood
    c. normal water concentration in the blood
9. Which of the following substances increases aldosterone release?
    a. increased K^+
    b. decrease in Na^+
    c. increased ammonia levels
    c. choices *a* and *c* only
    e. choices *a* and *b* only
10. Which of the following specifically stimulates the adrenal cortex to produce aldosterone?
    a. renin
    b. angiotensin I
    c. angiotensin II
    d. angiotensinogen
    e. Na^+

Routine Urinalysis

1. List eight substances normally found in urine
2. List five abnormal constituents of urine and their associated causes
3. Describe, perform, and interpret the basic tests included in a routine urinalysis
4. Perform and interpret the results of a microbiological examination of a urine specimen
5. Describe the appearance of urine in cases of diabetes insipidus and diabetes mellitus
6. Define or explain *albuminuria, glycosuria,* and *glomerulonephritis*

Urine is a watery solution containing nitrogenous waste and inorganic salts that are removed from blood plasma and eliminated by the kidneys. Table 49-1 lists the general properties of normal urine. Table 49-2 lists the average amounts of solid constituents of normal urine.

Urine and blood are analyzed more frequently than any other body fluid. As with blood, the composition of urine is a functional indicator of the state of body health at any given period of time. Some abnormal constituents and the types of conditions responsible for their appearance in urine are listed in Table 49-3. The detection of several such abnormal constituents can be achieved by tests that are based on color changes caused by enzymes or chemical reagents. These tests may be qualitative (showing the presence of an abnormal constituent) or quantitative (showing the concentration of the constituent). For most of these tests (Table 49-4), the color-reacting substance has been impregnated on plastic strips so that a urine sample can be easily applied.

An example of a diagnostic urinary test strip demonstrating the features described here is Phenistix. It is used for the detection of the hereditary metabolic defect, phenylketonuria (PKU). A baby born with this defect is unable to synthesize adequate amounts of the

Table 49-1 Physical Properties of Urine

Property	Normal Feature	Abnormal Features
Color	Generally light yellow to amber	Dark red or brown
Odor	Aromatic	Fishy, ammonialike
Turbidity (cloudiness)	Clear	Hazy, may contain obvious shreds of mucus
Specific gravity (at 25°C)	1.002–1.030	Any major change from normal range
pH (hydrogen ion concentration)	6.0 average, but ranges from 4.8 to 7.5	Any major variation from normal range
Volume	1–2 liters per 24 hours	Any major variation from normal range

Table 49-2   Solid Constituents of Normal Urine

Constituent	Daily Normal Values
Bilirubin	0.02 mg
Calcium	30–75 mg
Creatinine	1 g
Glucose	130 mg
Sodium	125–260 mEq (milliequivalents)
Urea nitrogen	6–17 g
Uric acid	250–750 mg
Average amount of solids	60 g

Table 49-3   Abnormal Constituents of Urine and Associated Causes

Abnormal Constituent and Term Used to Indicate Its Presence	Associated Causes
Acetone bodies or ketone bodies *(acetonuria* or *ketosis)*	Diabetes mellitus Starvation Diarrhea in children Pernicious vomiting of pregnancy
Albumin *(albuminuria)*	Damaged kidney tubules Chemical poisoning Prolonged exercise Cold baths Altered glomeruli
Bile	*Obstructive jaundice* (blockage of bile duct resulting in the yellowing of tissues)
Bilirubin	Liver disorders Obstructive jaundice
Glucose *(glycosuria)*	Diabetes mellitus
Hemoglobin	Conditions involving marked blood destruction, such as Chemical poisoning Poisonous snake bites Malaria Transfusion with incompatible blood
Leukocytes and other components of pus *(pyuria)*	*Cystitis* (infection of the urinary bladder) Infection of the kidney *Urethritis* (infection of the urethra) Tuberculosis of the kidney
Red blood cells *(hematuria)*	Infection of glomeruli Tumors of the urinary tract Kidney and bladder stones Tuberculosis Circulatory system disorders
Sugars, galactose, and fructose	Severe liver disease Inherited metabolic defect

Table 49-4 Examples of Commercially Available Urinary Test Strips

Product Name and Manufacturer	Tests for the Presence of	Color Reaction for Definite Positive Reaction	Significance of Results	Limitations
Chemstrip (BioDynamics)	Leukocytes Nitrates	Purple Red-violet	Both tests used in combination detect possible urinary tract infection	Certain antibiotic drugs interfere Large amounts of ascorbic acid (orange juice, vitamin C tablets) decrease sensitivity
Clinistix (Ames)	Glucose	Purple-blue	Results can be used to adjust diet and/or medication for diabetics and others	Vitamin C, aspirin, and some drugs may produce a false-negative result
Combistix[a] (Ames)	pH Glucose Protein	Orange-yellow and green-blue Green-brown Green-blue	Test results may provide information concerning kidney function, acid–base balance, and carbohydrate metabolism	False-positive reactions may occur with alkaline, highly buffered specimens
Diastix (Ames)	Glucose	Green-brown	Highly specific for glucose	Generally none
Ketostix (Ames)	Urinary ketones (acetoacetates and acetone)	Lavender-purple	Detects ketoacidosis	Phthalein medications may produce false-positive reactions
Multistix (Ames)	pH	Broad range of colors, depending on pH	Provides information on acid–base balance	Contamination (runoff) from other reaction patches causes color changes
	Glucose	Green-brown	Provides information on carbohydrate metabolism	Vitamin C and aspirin may cause false reactions
	Protein	Green-blue	Provides information on kidney function	Highly alkaline urine and some drugs interfere with test results
	Bilirubin	Various shades of tan	Provides information on liver disorders and diseases	Various drugs and vitamin C may cause false reactions
	Ketone	Maroon	Detects ketoacidosis, indication of carbohydrate metabolism disorders	Certain drugs interfere with test results
	Nitrite	Pink	Detection of urinary tract infection	Drugs and vitamin C may cause false positive results
	Urobilinogen	Brown-orange	Sensitive to elevated urobilinogen secretion	Certain drugs may produce false positive reactions
	Blood	Green spots	Highly sensitive to the presence of hemoglobin and indicates possible infection	Vitamin C, microorganisms, and elevated protein may produce false reactions
Phenistix (Ames)	Urinary phenylketones	Blue-gray	Test used for PKU screening	Aspirin, excess bilirubin, and phenothiazines interfere with the test

[a]A variation of this product, Hema-Combistix, incorporates a test for the detection of occult blood in urine.

enzyme phenylalanine hydroxylase. As a consequence, the amino acid phenylalanine is not converted to the compound tyrosine but instead is converted to phenylpyruvate. Excessive amounts of phenylpyruvate in the baby's body can damage its central nervous system. If PKU is detected early, foods that contain phenylalanine can be eliminated from the diet and brain damage can be avoided. The Phenistix is based on a reaction between a ferric ion reagent and phenylpyruvic acid, to produce a gray-green color. The diagnostic reagent strip with a test patch containing the reagent is immersed in the urine specimen, tapped to remove excess urine, and in 30 seconds compared to a color chart on the strip container. As little as 8 mg of phenylpyruvate per deciliter can be detected. A blood test generally is recommended in association with positive Phenistix results. Table 49-4 describes several commercially available urine diagnostic test strips. Tests using reagent strips are performed in this exercise.

Microscopic examination of a drop of urine that has been centrifuged may show the abnormal presence of blood and other cells, casts, and renal calculi (Figure 49-1). *Casts* are small masses of material that have hardened in the lumens of tubules and have assumed the shape of the lumen. Renal calculi are tiny stones formed within the kidney.

Normal urine is generally sterile or contains 1000 or fewer bacteria per milliliter, these few being the microorganisms making up the normal urethral flora. The presence of a microbial infection in any portion of the urinary system usually produces significant changes in urine specimens. The presence of large numbers and kinds of bacteria in urine, or *bacteriuria,* can be determined by specific bacteriological techniques and the use of commercially produced dipsticks or related devices (see Figure 49-2).

To be of value, a urine analysis must be done on a representative specimen, which should be examined within 30 minutes after it is passed or be stored under refrigeration. Because an early morning specimen is concentrated, it is more likely to reveal abnormalities than urine collected at some other time. In some laboratories, standard determinations are semiautomated.

Routine analysis of a urine specimen involves a description of physical properties (color, turbidity, odor, specific gravity, and pH; see Table 49-1) and determination of the presence or absence of abnormal constituents. A microscopic examination of urinary sediment and tests for the detection and identification of microorganisms such as bacteria also may be required. Several of these aspects of urinalysis are considered in this exercise.

DESCRIPTION OF PHYSICAL PROPERTIES

Materials

The following materials should be provided for class use:

1. Urinometer (cylinder and hydrometer float device)
2. pH indicator paper strips (pHydrion or nitrazine test papers)
3. Two freshly voided urine samples
4. Thermometers
5. 250-ml beakers
6. Filter paper
7. Liquid soap and brushes for washing of containers
8. Containers with disinfectant
9. Disposable surgical gloves

CAUTION

Working with urine or other body fluids will not pose any danger if the following precautions are observed:

1. Place any and all urine or materials containing urine into the disinfectant containers provided.
2. Dispose of any urine-containing materials as indicated by the instructor.
3. In cases of urine spillage, wipe the area with a disinfectant-soaked paper towel, and dispose of all materials as indicated by the instructor.
4. Always wash your hands after handling urine and associated materials.
5. The use of disposable surgical gloves.

Procedure

This procedure is to be performed by students individually.

1. Pour about 30 ml of one of the samples into a beaker. Carefully pour the urine from the beaker into a clean urinometer cylinder until it is three-fourths full. Obtain more of the specimen if necessary. If foam forms on the surface, remove it with a piece of filter paper. Discard the paper as directed by your instructor.
2. Note the color, odor, and turbidity of the sample. (Refer to Table 49-1.) Record your findings in the results and observations section.

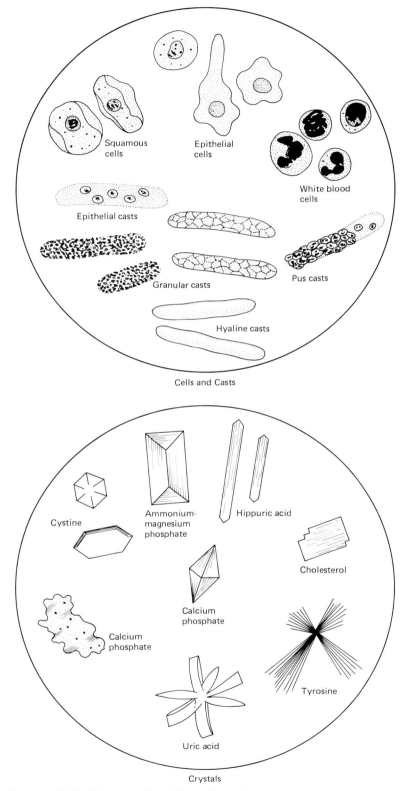

Figure 49-1 Elements found in urine sediments.

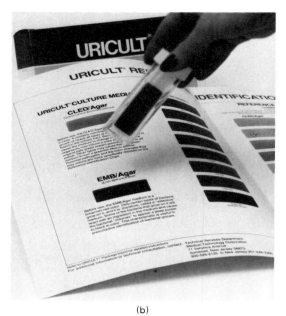

(a)   (b)

Figure 49-2 One approach to testing urine for the presence of an infectious disease caused by bacteria: (a) This dip paddle device contains nutrients on which bacteria can grow; it is dipped into a urine specimen to collect the microorganisms. (b) After an incubation period, evidence of an infection is clear; each of the small clumps on the dipstick is a bacterial colony containing 1000 organisms. (Courtesy of Medical Technology Corporation.)

3. Check the reading on a thermometer. If it is not at 37°C, shake it down to that level. Insert the thermometer in the urinometer, and read the temperature after 3 minutes. The temperature of the sample is used in the determination of specific gravity in the next step. Clean the thermometer as directed by your instructor.

4. Float the hydrometer in the cylinder. Make certain that it is not touching the bottom or the sides of the cylinder. Adjust the value of the hydrometer reading according to temperature to obtain the specific gravity. For each 3° above 25°C, add 0.001 to the hydrometer reading; for each 3° below 25°C, subtract 0.001. Record your findings in the results and observations section. Remove the hydrometer, and clean it as directed by your instructor.

5. Dip a pH indicator test strip paper in the urinometer cylinder three consecutive times. Remove excess urine by touching the paper to the inside surface of the cylinder. After 1 minute, compare the color of the strip with the color guide on the strip container or other available guide. Compare your findings with the values in Table 49-1. Record your findings in the results and observations section.

6. Clean the urinometer as directed by your instructor.

7. Repeat Steps 1–6 with the second urine sample.

DETERMINATION OF PRESENCE OF ABNORMAL CONSTITUENTS

Materials

The following materials should be provided for class use:

1. N-Multistix (Ames) and Phenistix (Ames) reagent strips with their respective containers showing test results or individual or other combined test reagent strips for urinalysis, such as Albustix, Clinistix, Ketostix, Hemastix, and Bili-Labstix.
2. Two freshly voided urine samples
3. 250-ml Pyrex beakers
4. 10-ml pipettes with bulbs
5. Benedict's reagent in dropper bottles
6. Beta-phenylpyruvic acid test solution in a dropper bottle
7. Pyrex test tubes
8. Electric hot plate or other heating system
9. Liquid soap and brushes
10. Pasteur pipettes with bulbs

Procedure 1: N-Multistix Reagent Strips

This procedure is to be performed by students individually.

1. Pour about 125 ml of one of the urine samples into a beaker.
2. Remove one N-Multistix reagent strip from its container. Be careful not to handle the specific color squares. Examine the strip and the identification or test-interpretation portion of the label on the container. Note that this strip contains material for determinations of pH, protein, glucose, ketones, bilirubin, blood, nitrite, and urobilinogen.
3. Dip the test strip into the urine sample, and tap the excess liquid off on the inside top surface of the beaker. Wait 15–30 seconds, and compare the color reactions with the test interpretation label. (Refer to Color Plate 56.) Record your findings in the results and observations section. Discard the test strip as directed by your instructor.
4. Repeat Steps 1–3 with the second urine sample.

Procedure 2: Phenistix Reagent Strips

This procedure is to be performed by students individually.

1. Remove one Phenistix reagent strip from its container. Be careful not to handle the specific color squares. Examine the strip and the identification or test-interpretation portion of the label on the container. Note that this strip contains material for the determination of phenylpyruvate.
2. Place 2 drops of the beta-phenylpyruvic acid test solution on the strip reaction side. Wait 30 seconds, and compare the results with the color chart on the container. Dispose of the strip as indicated by your instructor.
3. Pour about 25 ml of one of the urine samples into a clean beaker.
4. Obtain a fresh Phenistix strip, and dip it into the urine sample. Tap the excess liquid off on the inside

top surface of the beaker. Wait 30 seconds, and compare the color reactions with the test interpretation label. Record your findings in the results and observations section. Discard the test strip as directed by your instructor.
5. Repeat Steps 3 and 4 with the second urine sample.

Procedure 3: Quantitative Test for Glucose

This procedure is to be performed by students individually.

1. Pipette 5 ml of Benedict's reagent into two separate Pyrex test tubes.
2. Add 10 drops of the urine sample to each tube and mix by tapping the tubes. Place the tubes in a beaker of water on a hot plate or other heating system and boil for 5 minutes. Allow the tubes to cool, and determine the quantity of glucose according to Table 49-5 (see Color Plate 55). Record your findings in the results and observations section.
3. Clean the equipment used as directed by your instructor.
4. Repeat Steps 1–3 with the second urine sample.

URINARY SEDIMENT EXAMINATION

Materials

The following materials should be provided for class use:

1. Sedi-stain (concentrated stain preparation for urinary sediment examination) with dropper
2. Centrifuge and centrifuge tubes

Table 49-5   Benedict's Reagent Test Results

Color	Glucose	Result Reported
Clear, white turbidity, or green turbidity	Zero	Negative
Yellow-green	0.5–1.5	Weak positive
Yellow-orange	1.5–4.0	Moderately positive
Red	4.10 and over	Strongly positive

3. Urine disposal container
4. Two freshly voided urine samples
5. Glass slides and cover slips
6. Microscopes
7. 100-ml capacity graduated cylinders
8. Disposable Pasteur pipettes with bulbs
9. Containers with disinfectant

Procedure

This procedure is to be performed by students individually.

1. Use a graduated cylinder to transfer a urine sample to a centrifuge tube.
2. Centrifuge your sample as directed by your instructor. Usually, a speed of 1500 revolutions per minute (rpm) for 5 minutes is sufficient.
3. Without disturbing the sediment in the tube, pour off the fluid into the urine-disposal container or as directed by your instructor.
4. Add 2 drops of the Sedi-stain preparation to the urine sediment, and mix by tapping the tube with your index finger.
5. Transfer 1 drop of the stained material to a glass slide. Cover the specimen with a cover slip.
6. Examine the preparation under low and high power. With the aid of Figure 49-1, identify the various items in the sediment. Sketch a representative field in the results and observations section.
7. Repeat Steps 1–6 with the second urine sample.

DETECTION OF BACTERIURIA

Materials

The following materials should be provided for class use:

1. Bacturcult tubes with counting strip (one per student)
2. Urine specimen with low bacterial count
3. Urine specimen with high bacterial count
4. Sterile small beakers
5. Urine disposal container with disinfectant

Procedure

This procedure is to be performed by students in pairs.

1. Pour a small amount of one of the urine specimens into a beaker.
2. Open one Bacturcult tube, and pour the urine specimen into it.

3. *Immediately* pour the urine back into the beaker, and allow all the fluid to drain completely from the Bacturcult tube for a few seconds.
4. Replace the screw cap snugly on the tube, loosen the cap about a half turn, and incubate the tube in your laboratory drawer for at least 48 hours.
5. Dispose of the urine specimen as directed by your instructor.
6. Have your laboratory partner repeat Steps 1–5 with the other urine specimen.
7. After incubation, place the counting strip around an area of the tube where there appears to be an even distribution of bacterial colonies. Such colonies appear as variable-sized specks on the sides of the tube. (Refer to Color Plate 57.)
8. Count the number of colonies within the circle, and enter your findings in the results and observations section. If the number of colonies is fewer than 25, no urinary infection exists. A count of greater than 50 is a definite indication of a urinary tract infection.

UNKNOWN SPECIMENS

Materials

The following materials should be provided for class use:

1. All materials listed for the other procedures
2. Containers for urine specimen collection (one per student)
3. Unknown urine specimen (one per student)

Procedure

This procedure is to be performed by students individually.

1. Collect a sample of your own urine in the container provided by your instructor.
2. Obtain an unknown specimen from your instructor.
3. Perform the following procedures and tests on both specimens, and time yourself while carrying out the entire analysis. Enter your findings, and indicate your time in the results and observations section.
   a. Description of physical properties (color, odor, turbidity, pH, and specific gravity)
   b. Chemical determinations for bilirubin, blood, glucose (qualitative and quantitative), ketones, nitrite, and urobilinogen
   c. Bacteriuria determination
   d. Sediment examination
4. Discard the specimens and materials used as directed by your instructor. Clean all glassware.

RESULTS AND OBSERVATIONS

Description of Physical Properties

1. Enter your findings in Table 49-6.
2. Were there any unusual properties associated with the urine samples? If so list them. _____
   _____
   _____

Determination of Presence of Abnormal Constituents

1. Enter your findings in Table 49-7. Indicate the following items: the color (C) of the reagent test strip; the units (U) represented by the color reaction if appropriate; the color (NC) of a normal result; and the units (NU) that correspond to the normal result if appropriate.
2. Were any abnormal results observed with the urine specimens? Explain. _____
   _____
   _____

Table 49-6

Property	Urine Sample 1	Urine Sample 2
Color		
Odor		
Turbidity		
Specific gravity		
pH		

Table 49-7

Test	Urine Sample 1				Urine Sample 2			
	C	U	NC	NU	C	U	NC	NU
Protein								
Glucose								
Ketone								
Bilirubin								
Blood								
Nitrite								
Phenylpyruvic acid								
Urobilinogen								
Benedict's								

3. Why should urine specimens be refrigerated if tests cannot be performed soon after collection?

_____

_____

Urinary Sediment Examination

1. Sketch the results of your urinary sediment examination.

Specimen 1

Specimen 2

2. What did your urine specimens contain? _____

_____

_____

Detection of Bacteriuria

1. Indicate the bacterial counts for each of the urine specimens tested.

   a. Specimen 1 _____

   b. Specimen 2 _____

2. Which of the specimens showed signs of an infection? _____

3. Was this procedure simple to perform? _____

_____

Unknown Specimens

1. Enter your findings in Table 49-8.
2. Were any of the results obtained with the unknown specimen abnormally high? If so, indicate which ones. _____

_____

_____

3. How long did it take you to perform the urinalysis?

_____

4. How did your time compare with others in your laboratory? _____

_____

Table 49-8

Procedure or Test	Results for Student Specimen	Results for Unknown Specimen #_____
Color		
Turbidity		
pH		
Bilirubin[a]		
Blood[a]		
Glucose (qualitative)		
Glucose (quantitative)		
Ketone[a]		
Nitrite[a]		
Phenylpyruvic acid[a]		
Protein[a]		
Urobilinogen[a]		
Bacterial count[b]		
Urinary sediment examination		

[a]These determinations are to be made with reagent strips.
[b]This result can be determined only after 24 hours.

LABORATORY REVIEW *49*

ROUTINE URINALYSIS

COMPLETION, INTERPRETATION, AND DEFINITION

1. In Table Q49-1, distinguish between diabetes mellitus and diabetes insipidus as to the properties of urine.

Table Q49-1

Property of Urine	Diabetes Mellitus	Diabetes Insipidus
Volume	a.	b.
Specific gravity (low, normal, high)	c.	d.
Presence of sugar	e.	f.
Color	g.	h.

2. What would a count of 200,000 bacterial colonies on a dipstick device indicate?

_____

3. List five abnormal constituents of urine, and indicate the possible cause for their appearance.

ABNORMAL CONSTITUENT   POSSIBLE CAUSE

a. _____   _____

b. _____   _____

c. _____   _____

d. _____   _____

e. _____   _____

4. Define or explain the following:

a. Albuminuria _____

b. Casts _____

   _____

c. Urea _____

d. Uric acid _____

e. Creatinine _____

f. Glycosuria _____

g. Jaundice _____

h. Phenylketonuria _____

ANSWERS

1. _____

2. _____

3. _____

4. _____

5. _____

6. _____

7. _____

8. _____

9. _____

10. _____

11. _____

5. List three microbial diseases of the urinary system.

   a. _____

   b. _____

   c. _____

MULTIPLE CHOICE

Insert the correct answers in the spaces provided.

1. In the hereditary metabolic disorder known as PKU, foods containing which of the following amino acids cannot be metabolized?
   a. phenylalanine hydroxylase
   b. tyrosine
   c. phenylpyruvate
   d. glycine
   e. phenylalanine

2. Which of the following is an individual with PKU unable to synthesize?
   a. tyrosine
   b. phenylalanine
   c. phenylpyruvate
   d. phenylalanine hydroxylase
   e. Phenistix

3. What would be the normal volume of urine per 24 hours in liters for an average adult?
   a. 1–2
   b. 2–3
   c. under 1
   d. 3–4
   e. 5

4. What would a brown-colored urine specimen be considered?
   a. normal
   b. abnormal
   c. a variation of a normal value

5. What should be the specific gravity value of normal urine?
   a. 6
   b. 4.8–7.5
   c. 1.002–1.030
   d. 24
   e. 25

6. The pH of urine normally can range from a low of 4.8 to a high of what?
   a. 6.0
   b. 7.5
   c. 8.0
   d. 6.5
   e. 10

7. Which of the following is not a normal constituent of urine?
   a. sodium
   b. calcium
   c. acetone bodies
   d. uric acid
   e. glucose

8. Hematuria indicates the presence of which of the following in urine?
   a. fructose
   b. red blood cells
   c. bile
   d. leukocytes
   e. urea
9. Which of the following conditions may result in the presence of large amounts of albumin in urine?
   a. damaged kidney tubules
   b. diabetes mellitus
   c. chemical poisoning
   d. malaria
   e. choices *a* and *c* only
10. What is the term for the small masses of body secretions and other materials that harden in tubules and that assume various shapes and appear in urine?
   a. bacteria
   b. casts
   c. renal calculi
   d. urethral flora
   e. choices *a* and *b* only
11. In each milliliter of normal urine, up to what number of bacteria may be found?
   a. 1000
   b. 5000
   c. 10,000
   d. 100,000
   e. 1,000,000

Gamete Formation, Stages of Early Development, and Karyotyping

After completing this exercise, you should be able to

1. Distinguish between *mitosis* and *meiosis*
2. Describe and identify the stages of meiosis
3. Recognize characteristic structures of ovarian tissue and the general appearance of spermatozoa
4. Identify early stages of division of the zygote
5. Prepare a karyotype from a metaphase spread of chromosomes
6. List and describe four structural changes of chromosomes
7. Recognize normal and selected abnormal karyotypes

As indicated in Exercise 3, chromosomes carry the hereditary chemical material, deoxyribonucleic acid (DNA). All cells of the body, except *gametes* (the sex cells), contain two complete chromosome sets. For humans, the total cellular number is 46, or 23 pairs. The 46 is referred to as the *diploid number* or *2n*. Gametes contain one set of 23 chromosomes; this *haploid number* is *n*.

Reproduction includes the process by which genetic material is passed from generation to generation. One chromosome from each pair comes from the mother, and the other comes from the father. *Homologous chromosomes,* the two chromosomes of a pair, contain genes that control the same genetic traits. If a chromosome holds a gene for eye color, its homologous chromosome will have a gene for eye color.

A gene that dominates or controls is termed the *dominant gene,* and the associated trait expressed is said to be a *dominant trait.* The gene that is inhibited is referred to as the *recessive gene,* and the trait it controls is the *recessive trait.* Genes that control the same inherited trait, such as height, eye color, or hair color, and that occupy the same position on homologous (paired) chromosomes are known as *alleles.*

Twenty-two (44) of the 23 pairs are referred to as the *autosomes* and the remaining pair as the sex chromosomes. In the female, one chromosome—the *X chromosome*—has a homologous X chromosome, while in the male, one *X chromosome* is replaced by a smaller one, the *Y chromosome.* Thus, under normal conditions, human body cells contain 46 chromosomes, including either two X chromosomes for a female, or one X and one Y for a male.

In preparation for the sexual reproduction (*fertilization*) process, diploid cells divide into haploid cells through the special nuclear division process, termed *meiosis. Meiosis,* or reduction division, is a process by which the chromosome number is reduced by half. Whereas mitosis involves one division of the nucleus following a single duplication of the normal or diploid number of chromosomes (described in Exercise 3), meiosis consists of two distinct nuclear divisions, resulting in four new nuclei. Meiosis is associated with the production of sex cells.

In the testes of the male, meiosis results in the production of gametes termed *spermatozoa* through the process of *spermatogenesis* (Color Plate 59). In the ovaries of the female, meiosis results in the production of gametes termed ova (singular is *ovum*), through the process of *oogenesis* (Color Plate 58). Figure 50-1 diagrammatically shows meiosis in the formation of female sex cells (oogenesis) and male sex cells (spermatogenesis). Only three chromosome pairs are used in this figure, to show how they are distributed during the process.

During spermatogenesis, *spermatogonia* (diploid cells) undergo two stages of nuclear division. During the first division, each pair of homologous chromo-

Figure 50-1 Meiosis in female sex-cell production *(oogenesis)* and male sex-cell production *(spermatogenesis)*. Each diploid spermatogonium produces four haploid sperm cells. Meiosis is essentially the same in the female, except that the division of the cytoplasm is unequal, and only one functional haploid ootid is produced from each diploid oogonium.

somes (each consisting of two halves, termed *chromatids*) form a *tetrad* (4). During this time, the cells are known as *primary spermatocytes*. The pairs of chromosomes next migrate to opposite ends of the cell, thereby completing the first nuclear (maturation) division and resulting in the formation of two new cells

termed *secondary spermatocytes*. These cells are haploid. It is important to note that at this stage, while the chromosomes are haploid, each chromosome consists of two halves (chromatids). In the second nuclear (maturation) division, the chromatids separate, forming *spermatids*. These cells also are haploid but contain half the number of chromosomes as found in the spermatogonia. When the spermatids mature, they form *spermatozoa*. Figure 50-1 shows that at the completion of spermatogenesis, four spermatozoa have developed from one primary spermatocyte.

During the first stage of oogenesis, the *oogonium* forms tetrads and develops into a *primary oocyte*. Once the primary oocyte is released by the ovary (ovulation), and is fertilized by a spermatozoan, it will undergo division resulting in the formation of two cells of unequal size. The larger of the two received most of the cytoplasm. This cell is termed the *secondary oocyte*. The smaller one, which is known as the *first polar body* eventually disintegrates. During ovulation, the secondary oocyte is released into the uterine tube. If fertilization occurs, the secondary oocyte undergoes the second nuclear (maturation) division, forming two haploid cells, again each of unequal size. The larger of the two cells is known as an *ootid,* and the smaller one is the *second polar body*. The ootid develops into one ovum. The process of oogenesis is similar to spermatogenesis, however, it results in only one gamete, the *ovum*.

In organisms that reproduce sexually, the union of the haploid sperm produced by a male parent and the haploid ovum produced by the female parent forms the diploid fertilized egg or *zygote*. The zygote undergoes a type of mitotic division known as *cleavage*. The cell division is so rapid that there is insufficient time for a cell to enlarge between divisions. The number of cells increases, but the cells become smaller and smaller, so that the mass of cells that develops is about the same size as the original zygote and known as the *blastula* (Figure 50-2). The blastula generally consists of cells organized inside a hollow sphere that surrounds a fluid-filled cavity, the *blastocoel* (Figure 50-2b). Processes collectively known as *gastrulation* convert the single-layered blastula into a multilayered structure with functionally specialized areas. These changes in the developing cellularly specialized *embryo* depend on a series of cellular movements and the segregation of specific groups of cells. Demonstration slides are used in this exercise to show these early stages of development.

Since the discovery that human somatic (body) cells contain 46 chromosomes, a number of diseases have been found to be related to abnormalities in chromosome structure and number. Detailed cytogenetic procedures have been established for studying chromosomes. *Karyotyping* is one method commonly used to demonstrate many chromosomal *aberrations* (deviations from the normal). The method (Figure 50-3) involves the growing of cells, such as leukocytes or skin cells, in tissue culture for a defined period, stopping cell division during metaphase by the use of a drug such as colchicine, staining the chromosomes, and photographing the metaphase spread of chromosomes (Figure 50-4a). Once the photograph is made, the individual chromosomes are cut out and placed in order, according to size, shape, and staining pattern. The longest chromosome is 7 to 8 μm, while the shortest one is about 1 to 2 μm. In a routine analysis, a chromosomal count is done. There may be more than the normal number

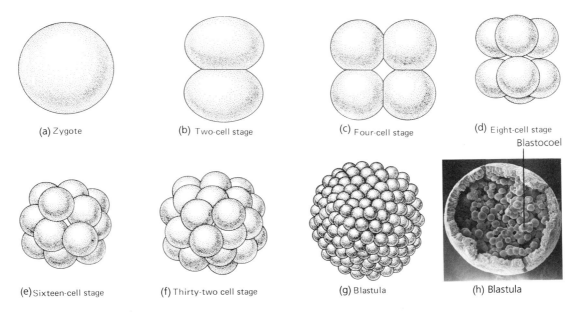

(a) Zygote   (b) Two-cell stage   (c) Four-cell stage   (d) Eight-cell stage Blastocoel

(e) Sixteen-cell stage   (f) Thirty-two cell stage   (g) Blastula   (h) Blastula

Figure 50-2 Early stages of development (a) through (g): the repeated mitotic divisions (known as *cleavage*) until the formation of the blastula. (h) A scanning photomicrograph of the internal features of a blastula (from S. Komazaki, *Cell Differentiation*, 24: 25–32, 1988).

White blood cells for culture

Tissue culture medium with stimulator for mitosis

Incubate; later add colchicine to stop mitosis at metaphase

White blood cell

Transfer to centrifuge tube and spin down

Remove culture medium

Add hypotonic solution to cause cell swelling and to spread chromosomes

Add fixative, to kill and to preserve cells

Gently resuspend cells

Prepare slide

Stain preparation

Microscopic examination of slide

Photograph chromosomes in metaphase

Cut out each chromosome

Prepare karyotype with chromosomes arranged in order

Figure 50-3 The steps in a karyotyping procedure.

or a rearrangement of those present. Differences from the normal number or chromosome arrangement are usually indicative of clinically important genetic defects.

When paired, numbered, and grouped in a standard way, the pattern formed by the chromosomes is termed a *karyotype* (Figure 50-4b and Table 50-1). The grouping system makes use of the location of the centromere associated with the chromosome. A chromosome with a centromere in the middle, which divides it into two equal arms, is termed *metacentric*. If the centromere is somewhat nearer one end, creating a distinct long and short arm, the chromosome is *submetacentric*. When the centromere is very near to one end, so that the chromosome has a very short arm and longer long arm, the chromosome is termed *acrocentric*.

Staining methods termed *banding techniques* are used in cytogenetics to identify chromosomes and abnormalities. *Banding* refers to the horizontal, alternating light and dark crossbands that result from the staining. Different staining techniques are used for different purposes. Some identify and differentiate the chromosomal pairs. Such techniques routinely are used

for detecting and defining chromosomal aberrations. Other procedures stain specific chromosome areas to identify special problem situations. The most popular banding technique is G (Giemsa stain) banding. Other techniques are Q (quinacrine dye) bands and C (centromere staining) bands. Figure 50-5 is a diagrammatic representation of banding patterns revealed by staining and of the system used to specify location on the chromosome. The chromosome pairs are identified by number. The short and long arms are labeled *p* and *q*, respectively. The bands are specified by a two-digit numbering system. Thus, if the chromosome shown in Figure 50-5 were chromosome 1, the light band at the tip of the short arm would be designated 1 *p*36, and the light band at the bottom end of the long arm would be 1 *q*44.

Chromosomal aberrations or abnormalities are of two types: numerical changes and structural changes. Numerical changes are deviations from the total complement number of 46. Such changes may result from a misdivision during metaphase called *nondisjunction*. In this situation the sister chromatids fail to separate, resulting in one daughter cell with an extra chromo-

Figure 50-4 Karyotyping: (a) a metaphase spread from a normal male cell; (b) a karyotype of a normal male. In determining the karyotype, the chromosomes are grouped and arranged according to their shapes and sizes. The X chromosome is located at the right of group C. The Y chromosome is located at the right of group G. (From L. Larson, *American Journal of Medical Technology,* 49: 687–697, 1983).

Table 50-1   The Grouping System for Chromosomes in a Karyotyping Procedure

Group	Description
Group 1–3 or A	The three largest pairs of chromosomes, all of which are metacentric
Group 4–5 or B	Two slightly smaller submetacentric pairs
Group 6–12 and X or C	Different-sized autosomes, some metacentric and some submetacentric, and X chromosome (sex chromosomes are sometimes placed separately at the lower right of a karyotype as in Figure 50-6)
Group 13–15 or D	Three pairs, nearly as large as those in Group 6–12, which are submetacentric and have satellites (tiny chromatic masses attached to the short arm)
Group 16–18 or E	Three smaller pairs, which are metacentric or submetacentric and lack satellites
Group 19–20 or F	The two smallest metacentric pairs
Group 21–22 and Y or G	The two smallest acrocentric autosomes, which have satellites, and the Y chromosome, which is also acrocentric and typically larger than the others, but the size varies

some and one with a missing chromosome. One extra chromosome at a pair location is called *trisomy;* one missing is referred to as *monosomy.* The consequences depend on the chromosomes involved. An example of a trisomy involving an autosome is Down's syndrome or trisomy 21, that is, an extra chromosome at pair 21 (Figure 50-6). Abnormalities involving sex chromosomes include Klinefelter's syndrome, an XXY combination in a male, and Turner's syndrome, a single X chromosome in a female (Table 50-2).

In a karyotype description numerical changes are indicated by a "+" or a "−" placed before a chromosome number to indicate an extra or missing whole chromosome (see the caption for Figure 50-5).

There are several types of structural changes in chromosomes. A *deletion* (del) occurs when a chromosome is broken and a particular fragment is lost. This form of aberration usually results in severe malformations in the individual.

A *translocation* (t) involves a break in a chromosome and the transfer of part or all of one chromosome to another. In a karyotype description, translocations are indicated by a "t" placed before the numbers of chromosomes involved, given in parenthesis. The break points are also indicated. For example, t(2:10) (*p*24;*q*26) specifies a translocation involving chromosomes 2 and 10, breaks occurring on the short arm (*p*) at band 24 and on the long arm (*q*) at band 26.

An *inversion* (inv) results when a section breaks off a chromosome and reattaches in an inverted position. The section may be one end of a chromosome or a fragment created by two breaks within the chromosome.

An *isochromosome* (i) is caused by an unequal division of the centromere.

A *ring chromosome* is formed when the ends of both the long and short arms of a chromosome break off and then reattach to each other to form a ringlike structure.

This exercise includes the preparation of karyotypes and identification of chromosomal parts and abnormalities. Table 50-2 lists several human chromosomal aberrations.

Figure 50-5 A diagrammatic representation of the human chromosome's banding patterns revealed by the different kinds of staining techniques. The letter *p* represents the short arm, *q* the long arm. Two numbers specify the bands.

Figure 50-6 Karyotypes: (a) a karyotype of a patient with trisomy 21 *(arrow)*— G *(Giemsa)* bands can be seen on the individual chromosomes; (b) a G-stained karyotype of a patient, showing a translocation (t) involving pairs 8 and 21. Note also that the Y chromosome is missing (−Y) and that therefore the individual has only 45 chromosomes. This condition is found in individuals with a form of acute leukemia. The karyotype is expressed as 45,XY, t(8:21). (From A. Cork, *American Journal of Medical Technology,* 49: 703–714, 1983.)

Table 50-2   Human Chromosomal Aberrations

Abnormality	Type of Chromosome	Total Number of Chromosomes and Abnormal Pair	Resulting Conditions
Down's syndrome	Autosome	47, trisomy 21 (shown in Figure 50-6)	Growth stunted: characteristic facial changes; mental retardation; increased susceptibility to infection; affects both sexes
Klinefelter's syndrome	Sex	47,XXY	Affects males; *atrophic* (abnormally small) testes; enlarged breasts; secondary sex characteristics underdeveloped or lacking; usually sterile
Turner's syndrome	Sex	45,X–	Affects females; small, webbed neck; secondary sex characteristics lacking; ovaries underdeveloped; absence of menstruation
XXX female (superfemale)	Sex	47,XXX	Generally appears normal; slightly depressed mental ability; can produce normal children

DEMONSTRATIONS OF MEIOSIS— SEX CELLS AND CLEAVAGE STAGES

Materials

The following materials should be provided for class use:

1. Prepared slides of roundworm (*Ascaris equorum*) eggs showing meiosis
2. Models or charts showing stages of meiosis
3. Prepared slides of human spermatozoa
4. Prepared slides of ovarian tissue showing graafian follicles
5. Prepared slides of starfish cleavage stages
6. Microscopes for demonstration slides
7. Charts and diagrams showing cleavage stages and general appearance of sex cells

Procedure 1: Meiosis

This procedure is to be performed by students individually.

1. Examine a prepared slide of portions of the oviducts of an *Ascaris*. The preparation should show eggs undergoing meiosis.

2. With the aid of Figure 50-1 and other material showing the stages of meiosis, identify metaphase I, anaphase I, and telophase I of the first meiotic division, and anaphase II and telophase II of the second meiotic division.
3. Sketch and label these stages in the results and observations section.
4. Count the number of chromosomes (as best you can) in the anaphase of both meiotic divisions.
5. Answer the questions in the results and observations section.

Procedure 2: Sex-Cell Demonstration

This procedure is to be performed by students individually.

1. Examine the demonstration slide of ovarian tissue.
2. With the aid of Color Plate 58 and other references, locate the following structures: graafian or vesicular follicle, the fluid-filled cavity of the follicle, and a primary oocyte within a follicle.
3. Sketch and label a representative view in the results and observations section.
4. Examine the slide containing the spermatozoa.
5. With the aid of Color Plate 59 and other references,

identify the head and thinner tail portions. Look for abnormalities such as two heads or no tail. Sketch and label a normal and an abnormal spermatozoon in the results and observations section.

6. Answer the questions in the results and observations section.

Procedure 3: Cleavage Stages Demonstration

This procedure is to be performed by students individually.

1. Examine the demonstration slide showing cleavage stages in the starfish.
2. Identify the following stages: zygote, 2-cell, 4-cell, 8-cell, 16-cell, and blastula. (See Figure 50-2.)
3. Sketch and label the stages as indicated in the results and observations section.
4. Answer the questions in the results and observations section.

KARYOTYPING

Materials

The following materials should be provided per each two students:

1. Two copies of Figure 50-7 (both removed from students' manuals) or similar reproductions of photographs of metaphase chromosome spreads

2. One pair of scissors
3. A millimeter ruler
4. Rubber cement or similar material with which to paste

Procedure

This procedure is to be performed by students individually.

1. Examine the chromosome spread in Figure 50-7a.
2. Carefully cut out the chromosomes, and arrange them in pairs according to approximate length and general appearance. Refer to Figure 50-4b.
3. Paste the chromosomes on one of the karyotype analysis sheets provided in the results and observations section. Use Figure 50-4b as a guide.
4. Compare your karyotype with Figure 50-4b, and determine the number of chromosomes; whether the karyotype is for a female or male, and whether the karyotype is an abnormal one for a human.
5. Record your findings in the results and observations section.
6. Repeat this entire procedure with the chromosome spread in Figure 50-7b.

(a)

(b)

Figure 50-7 Metaphase chromosome spreads.

RESULTS AND OBSERVATIONS

Meiosis

1. Sketch and label stages of meiotic division.

Meiosis I

Metaphase

Meiosis I

Anaphase, Telophase

Meiosis II

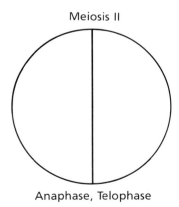

Anaphase, Telophase

2. Were the same numbers of chromosomes in anaphase also in meiosis I and meiosis II? If not, explain. _____

_____

_____

Sex Cells

1. Sketch representative views of a graafian follicle and representative spermatozoa.

Graafian follicle

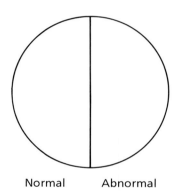

Normal   Abnormal

2. Were abnormal spermatozoa present? If so, what percentage of the cells observed were abnormal?

_____

3. What factors are responsible for abnormal spermatozoa formation? _____

_____

4. Complete the following diagram of possible meiotic divisions.

Figure 50-8 Possible meiotic divisions in spermatogenesis. Only two pairs of chromosomes are shown.

Cleavage Stages

1. Sketch stages of cleavage as indicated.

Zygote   Two-cell stage

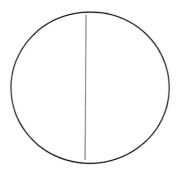

Eight-cell stage   Blastula

2. Compare the zygote and blastula stages. Which one is larger? _____

3. Are each of the cells in the eight-cell and blastula stages the same size? _____

Karyotyping

1. Examine the karyotypes you prepared from Figures 50-7a and 50-7b.

2. Which is the karyotype of a human female? _____ Which is the karyotype of a human male? _____

3. Are the karyotypes normal? If not, list the aberrations. _____

_____

_____

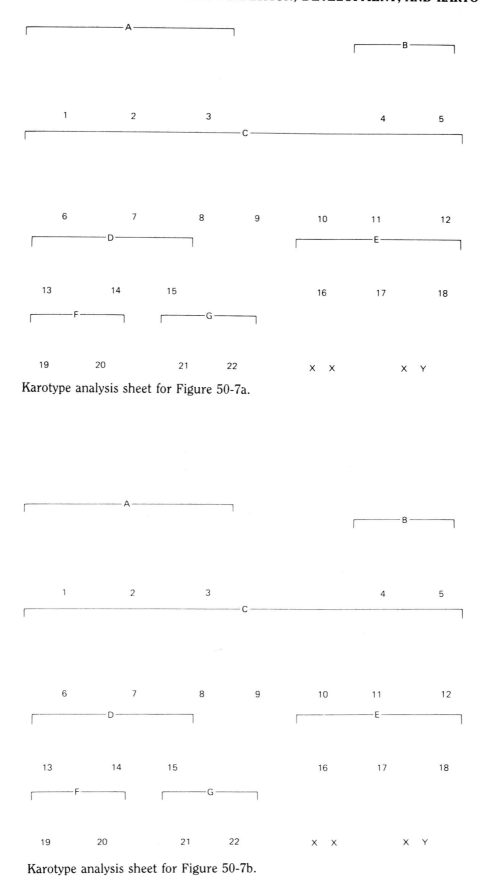

Karotype analysis sheet for Figure 50-7a.

Karotype analysis sheet for Figure 50-7b.

LABORATORY REVIEW *50*

NAME _____

LAB SECTION _____ DATE _____

GAMETE FORMATION, STAGES OF EARLY DEVELOPMENT, AND KARYOTYPING

COMPLETION, DEFINITION, AND INTERPRETATION

1. List three features of meiosis that distinguish it from mitosis.

   a. _____

   b. _____

   c. _____

   d. _____

2. Define or explain the following:

   a. Chromosome _____

   b. Chromatid _____

   c. Haploid _____

   d. Diploid _____

   e. Meiosis _____

   f. Down's syndrome _____

   g. Nondisjunction _____

3. List and describe four specific structural chromosomal aberrations.

   a. _____

   b. _____

   c. _____

   d. _____

4. Examine the karyotype in Figure Q50-1.

Figure Q50-1 (From L. Larson, *American Journal of Medical Technology,* 49: 687–697, 1983.)

a. Is the arrangement of chromosomes correct for a normal human? _____

b. Is the karyotype typical of a normal human? _____

MULTIPLE CHOICE

Insert the correct answers in the spaces provided.

1. What is the total number of chromosomes for all cells of the body except gametes?
   a. 23
   b. 46
   c. 92
   d. 184
   e. 48

2. What is the total number of chromosomes in each human gamete?
   a. 23
   b. 46
   c. 92
   d. 184
   e. 48

ANSWERS

3. Which of the following is the term for X and Y chromosomes?
   a. dominant chromosomes
   b. sex chromosomes
   c. autosomes
   d. gametes
   e. body chromosomes

4. Which of the following chromosome arrangements do the body cells of a normal female contain?
   a. 1X
   b. 2Xs
   c. XY
   d. Y
   e. either choice *a* or *b*

5. Which of the following chromosome arrangements do the body cells of a normal male contain?
   a. 1X
   b. 2Xs
   c. XY
   d. Y
   e. either choice *a* or *b*

6. What are the genes that control the same inherited trait and occupy the same site on homologous chromosomes?
   a. recessive
   b. dominant
   c. homologous
   d. chromatids
   e. alleles

7. What are the halves of chromosomes that develop during spermatogenesis?
   a. recessive
   b. dominant
   c. homologous
   d. chromatids
   e. alleles

8. In the meiotic process, how many spermatozoa does one primary spermatocyte normally give rise to?
   a. 2
   b. 3
   c. 4
   d. 5
   e. 6

9. In the meiotic process, how many sex cells does one primary oocyte normally give rise to?
   a. 1
   b. 2
   c. 3
   d. 4
   e. 6

10. What is the term for when one extra chromosome is added to a chromosome pair?
    a. monosomy
    b. trisomy
    c. translocation
    d. karyotype
    e. deletion

1. _____
2. _____
3. _____
4. _____
5. _____
6. _____
7. _____
8. _____
9. _____
10. _____

Ovulation Prediction, Human Chorionic Gonadotropin, and Pregnancy Testing

After completing this exercise, you should be able to

1. Perform and interpret the results of a luteinizing hormone detection test
2. Perform and interpret the results of a slide pregnancy test
3. List three conditions that could produce false-positive pregnancy test results
4. Explain the immunological basis of a slide test for pregnancy

Ovulation is the process during which a secondary oocyte (refer to Exercise 50) is released from an ovary. The few days around ovulation (two days before, the day of ovulation, and one day after) mark the most likely time for fertilization to take place. A woman's body normally produces a small quantity of *luteinizing hormone (LH)*. However, the LH level greatly increases for a short time in the middle of the menstrual and ovarian cycles. This LH increase or *surge* causes ovulation. If ovulation is to take place, it will occur within 40 hours after the LH surge begins.

In this exercise, a test kit that can be used as an aid in predicting the time of ovulation is examined. The *OvuGen test* uses an enzyme-linked immunoassay (ELISA) and is designed to produce two visually detectable purple dots: one for a urine sample and one for a reference standard control (Figure 51-1). The control provides a standard for a visual comparison each time the test for the LH surge is performed. If the urine sample develops a purple dot that is darker than or as dark as that of the reference standard, this is interpreted as an LH surge associated with ovulation.

Another hormone, human chorionic gonadotropin (HCG), appears in the urine and serum of pregnant women in large concentrations. The hormone is a glycoprotein produced by placental trophoblastic cells and plays a major role in maintaining the developing corpus luteum. The corpus luteum maintains the state of pregnancy by the production of the hormones progesterone and estradiol until the placenta and fetus begin the secretion of steroid and protein hormones. Following implantation of a fertilized ovum, HCG levels increase rapidly in urine and may reach detectable levels shortly after conception. Laboratory test demonstration of the hormone's presence is generally indicative of pregnancy. However, false-positive results may be obtained if the urine and serum contain high levels of LH, if the individual is menopausal, or if some specific cancerous states exist, such as choriocarcinoma and chorioepithelioma, or in situations of a spontaneous abortion. Following spontaneous abortion, the results remain positive for approximately one week.

The immunoenzymatic test for pregnancy considered in this exercise is an accurate, rapid, and simple procedure. It clearly replaces the various tests that require specialized equipment, radioactive reagents, or the use of animals such as frogs, rats, mice, or rabbits.

The Pregnospia test is an immunoenzymatic assay. It uses a revolutionary type of diagnostic tool, the monoclonal immunoglobulin (antibody). The production of pure or monoclonal (single-type) antibodies is achieved by an in vitro hybridization process in which an antibody-producing cell is fused with a myeloma (particular cancer) cell under special experimental conditions. In this process, the genetic materials of both cells are combined to create a hybrid cell that contains desirable characteristics from both of the parent cells.

The Prognosia test is an agglutination immunoassay. The monoclonal antibodies mentioned earlier are coated onto special gold particles, which have an intense

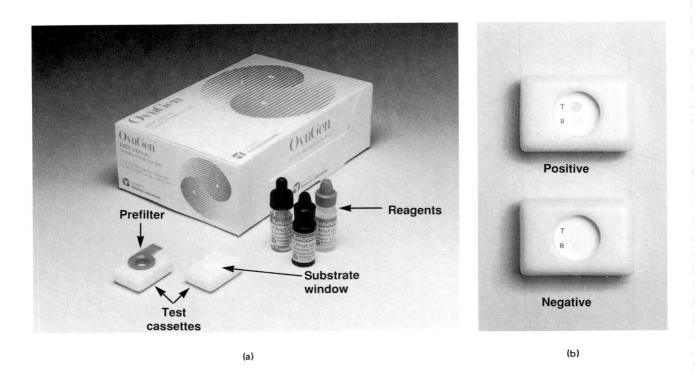

(a)

(b)

Figure 51-1 Luteinizing hormone (LH) detection materials, including the test cassette. (Courtesy of BioGenex Laboratories, San Ramon, California).

reddish-purple color in solution. These antibodies are noted for their specificity for special sites on the intact HCG molecule and one of its polypeptide subunits. When the HCG antibody preparation reacts with HCG present in a urine specimen, connecting molecular bridges are formed, binding the gold particles together and thereby causing them to clump or agglutinate. The resulting agglutination is evidenced by a distinct color change from the original reddish-purple color of the HCG antibody preparation. If there is no HCG present in the urine sample, the agglutination reaction will not change color. This result is interpreted as a negative reaction.

CAUTION

Working with urine or other body fluids will not pose any danger if the following precautions are observed:

1. Place any and all urine or materials containing urine into the disinfectant containers provided.
2. Dispose of any urine-containing materials as indicated by the instructor.
3. In cases of urine spillage, wipe the area with a disinfectant-soaked paper towel, and dispose of all materials as indicated by the instructor.
4. Always wash your hands after handling urine and associated materials.
5. The use of disposable surgical gloves.

LUTEINIZING HORMONE DETECTION (OVUGEN OVULATION PREDICTION TEST)

Materials

The following materials should be provided per two students. All items should be available in the commercial kit.

1. Four test cassettes in individual pouches
2. Four disposable droppers (with bulbs if appropriate)
3. OvuGen solution A (Vial A)
4. Conjugate solution (Vial B)
5. Substrate solution (Vial C)
6. Stop solution (Vial D)
7. One known positive urine control (labeled " + ")
8. One known negative urine control (labeled " − ")
9. Two unknown specimens (one per student) (labeled "1" and "2")
10. One marking pen or similar device
11. Container for disposing of droppers and prefilters
12. Container of disinfectant
13. Disposable surgical gloves

Procedure

The procedure is to be performed by students in pairs.

General

1. Obtain four test cassettes and other materials needed to perform the procedure (Figure 51-1).
2. Examine the cassettes, and note the (removable) purple prefilter.
3. Label each cassette for each specimen (positive control, negative control, and each unknown).

Solution Preparation

1. Add OvuGen solution (Vial A) to the conjugate solution.
2. Close the vial tightly, and mix the contents by turning it upside down at least five times.
3. Allow the mixed solution to stand for at least 10 minutes before use.

Test

1. Using a disposable dropper, place 7 drops of one of the unknown urine samples onto the prefilter area of the labeled test cassettes (Figure 51-2a). Allow the system to drain.

2. Add 7 drops of conjugate solution by means of the vial's dropper into the prefilter (Figure 51-2b). Allow the system to drain.
3. Remove the prefilter, and place it in the container for disposal.
4. Add 7 drops of the substrate solution into the cassette (test window) (Figure 51-2c). Wait 3 minutes.
5. Add 7 drops of the stop solution (Figure 51-2d).
6. Examine the cassettes for a color change. A test dot that is darker purple than or that matches the reference dot indicates an LH surge. A test dot that is lighter than the reference dot or that does not appear is a negative result (Figure 51-2e).
7. Repeat Steps 1 through 6 with each of the controls and with the second unknown urine specimen.
8. Interpret your findings, and enter them in the results and observations section.
9. Check your results with the instructor.
10. Answer any questions pertaining to this exercise in the laboratory review section.

RESULTS AND OBSERVATIONS

Enter your findings in Table 51-1.

HUMAN CHORIONIC GONADOTROPIN DETECTION

Materials

The following materials should be provided per each two students: (All reagents are available in the commercial kit.)

1. Four tubes of Organon Accusphere reagent (gold solid particles bonded with monoclonal antibodies against HCG)
2. Buffer (2 ml)
3. One 5-ml pipette, calibrated in 0.1-ml units, and a rubber bulb
4. Three 1-ml pipettes, calibrated in 0.1-ml units, and rubber bulbs
5. One known positive urine control
6. One known negative urine control
7. Two unknown specimens (one per student)
8. One marking pen or similar device
9. Container for disposing of pipettes

Figure 51-2 The luteinizing hormone detection procedure.

Table 51-1

Specimen	Color Formed		LH Present	Indicate of LH Surge
	Reference	Test		
Known positive				
Known negative				
Unknown 1				
Unknown 2				

Procedure

This procedure is to be performed by students in pairs.

1. Obtain four tubes containing the Accusphere reagent.
2. Number the tubes accordingly: 1 (known positive), 2 (known negative), 3 (unknown), and 4 (unknown).
3. Remove the rubber stoppers from the four tubes.
4. With the 5-ml pipette, introduce 0.4 ml of the buffer provided to each tube.
5. Replace the stoppers.
6. Gently shake each tube for approximately 15–30 seconds until the Accusphere reagent is completely resuspended.
7. Using separate 1-ml pipettes, add 0.1 ml of the positive, negative, and each of the unknown samples to their respectively marked tubes.
8. Gently shake the tubes for about 30 seconds to achieve thorough mixing.
9. Place the tubes in a rack or on a flat surface, and wait 5 minutes.
10. Examine the tubes for a color change. If a change does not occur within this time period, continue to observe the tubes for another 25 minutes.
11. Compare the unknown tubes with the results shown in the positive and negative samples (Tubes 1 and 2).
12. Interpret your findings, and enter them in the results and observations section.
13. Check your results with the instructor.
14. Answer any questions pertaining to this exercise in the laboratory review section.

RESULTS AND OBSERVATIONS

Enter your findings in Table 51-2.

Table 51-2

Specimen	Agglutination	HCG Present	Indicative of Pregnancy (positive/negative)
Known positive			
Known negative			
Unknown 1			
Unknown 2			

LABORATORY REVIEW *51*

OVULATION PREDICTION, HUMAN CHORIONIC GONADOTROPIN, AND PREGNANCY TESTING

COMPLETION, DEFINITION, AND INTERPRETATION

1. What does an LH surge indicate? _____

_____

2. a. Is the ovulation prediction test simple to perform? _____

   b. Should LH-surge detection testing be performed only once or more often during a several-day period? Explain your answer. _____

_____

3. What is HCG? _____

_____

4. Briefly explain the principle involved in the urinary chorionic gonadotropin (UCG) in vitro test. _____

_____

_____

_____

5. List three advantages of the in vitro immunoenzymatic test for the detection of pregnancy, when compared to the use of laboratory animals for the same purpose.

   a. _____

   b. _____

   c. _____

6. What is the appearance of the results of the UCG test or a comparable test?

   a. Negative result _____

   b. Positive result _____

7. What conditions produce a false-positive result? _____

_____

_____

8. List two limitations of the in vitro test.

   a. _____

   b. _____

ANSWERS

1. ⎯⎯⎯⎯⎯⎯⎯

2. ⎯⎯⎯⎯⎯⎯⎯

3. ⎯⎯⎯⎯⎯⎯⎯

4. ⎯⎯⎯⎯⎯⎯⎯

MULTIPLE CHOICE

Insert the correct answers in the spaces provided.

1. To which of the following does ovulation refer?
   a. fertilization
   b. increased production of LH
   c. secondary oocyte release
   d. HCG production
   e. choices *b* and *c* only
2. From what does ovulation result?
   a. fertilization
   b. increased production of LH
   c. secondary oocyte release
   d. HCG production
   e. choices *b* and *c* only
3. Which of the following appears in the urine and serum of a pregnant woman?
   a. LH
   b. secondary oocyte
   c. primary oocyte
   d. HCG
   e. corpus luteum
4. Which of the following increases rapidly in urine, following the implantation of a fertilized ovum?
   a. LH
   b. secondary oocyte
   c. primary oocyte
   d. HCG
   e. corpus luteum

Instructions for Physiograph Applications

A variety of devices are currently in use to demonstrate and record physiological activities and data accurately. Among these instruments are multichannel recording devices, known as physiographs, and the classic old-style kymograph.

The *physiograph* is a modern device designed to be an expandable recording system. The basic components of a physiograph system generally include (1) the input transducer (converts the energy of the physiological event into a proportional electrical signal), (2) the amplifier (provides the required increase in signal strength to permit visualization), (3) the coupler (renders the input energy from the transducer or the signal from the subject compatible with the channel amplifier), and (4) the output transducer (provides visual or audio representation of information). In studying physiological events such as heart action, muscle contraction, blood pressure, or cerebral activity, a detection device first is used to pick up the electrical signal from the tissue preparation or the experimental subject. Examples of such devices include an electrode for electrical signals, as in cases of neural activity, and an input transducer for nonelectrical signals, as in situations of temperature measurement.

A transducer converts the physiological event into an electrical signal. Because physiological events often produce weak signals, an amplifier is used to enlarge the magnitude of the event. The amplified signals are converted by an output transducer into some form of visual or audio display that can be observed and measured. The display can be in the form of an electrical meter, an oscilloscope, or a chart recorder.

Various arrangements of physiograph components are possible. Figure A-1 shows an arrangement example. The components of a physiograph are contained in a main frame assembly, which provides power, devices for the control and monitoring of circuitry and instrument calibration, and locations for mounting accessory plug-in units (Figure A-2). Table A-1 lists and describes both the basic components and the accessories (Figure A-3) used with physiographs.

The *kymograph* is a moving-drum recording instrument. It is the classical device used for the measurement of movements produced by muscle contractions. The kymograph is not considered here.

In this appendix, the features of physiological recording devices are demonstrated. This material may serve as a general frame of reference for procedures

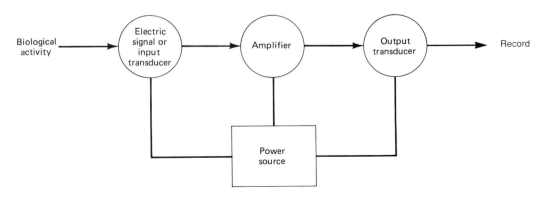

Figure A-1 The general arrangement of physiograph instrumentation components.

(a)

(b)

Figure A-2 Examples of physiographs: (a) a Gilson Biophysical and Industrial Recorder and its components (courtesy of Gilson Medical Electronics, Inc.); (b) physiograph (courtesy of Nacro Bio-Systems).

Table A-1   Physiograph Components and Their Functions

Physiograph Component	Functions
Amplifier	Increases the strength of the electrical signal while keeping the proportional relationship of the electrical signal obtained from the transduced physiological activity.
	Amplifiers contain power supply units to convert alternating current (AC) to direct current (DC) and to either increase or decrease voltage requirements.
Amplifier controls	The types of control are determined by the function of an amplifier. Controls include 1. *Gain control:* adjusts sensitivity of amplifier to the strength of the incoming signal 2. *Mode control:* enables instrument to accept either AC or DC input signals 3. *Centering (zero) control:* moves the zero line on a recording system in either a positive or a negative position 4. *Calibration control:* used for checking or establishing the sensitivity of the amplifier
Bellows pneumograph	Photoelectric transducer used for recording respiration
Blood pressure transducer	Photoelectric transducer for the measurement of direct blood pressure
Coupler	Renders the input energy from the transducer (or the signal) from the subject compatible with the channel amplifier
Electrode	Device used to carry an electrical current into or away from a biological specimen
Myograph	Transducer used primarily for the quantitative measurement of muscular contractions
Oscilloscope (cathode ray tube)	Output transducer that provides a visual display of voltage signals on a fluorescent screen
Power supply (for amplifier)	Converts alternating current (AC) to direct current (DC); either increases or decreases the voltage needed by the amplifier
Recording device	Converts electrical energy into rotary motion; receives electrical signals from the amplifier and drives the pen across a moving chart, thereby providing a record of the event being measured
Stimulator	Device used to produce electrical shocks of desired voltages in experimental situations
Stylus	Writing instrument used to indicate time, beginning and ending of an event, etc.
Thermistor	Transducer that monitors temperature changes
Transducer	Converts a physiological event into a proportional electrical signal; the most sensitive components in a transducer are a light bulb and photoelectric cell

dealing with muscle measurements of laboratory animals.

Identification of Physiograph Components

Although physiographs from different manufacturers vary in appearance or design, they generally consist of the same basic components. The parts of a physiograph and their functions are described in Table A-1. Use this information as you examine and locate the parts of the physiograph.

Materials

The following materials should be provided for class examination and demonstration:

1. Physiographs
2. Accessory devices and attachments, including those used for ECG, pulse detection, respiratory measurements, and electromyograms
3. Reference diagrams or instruction manuals

Figure A-3 Various accessories used in physiological measurement. The display includes a coupler component (C), bellows pneumograph (P), general-purpose transducer (T), ECG accessory pulse-rate detector (D), ear/nose transducer for pulse pickup (PP), respiration belt using a strain gauge bridge (RB), and photoelectric pulse detector (PD). (Courtesy of Sensor Medics Corporation.)

Procedure

This procedure is to be performed by students individually.

1. Examine the physiograph and accessories provided. With the aid of Figures A-1 and A-2, locate and learn the functions of each component listed in Table A-1, and others indicated by your instructor.
2. Your instructor will indicate the following control and other appropriate switches associated with the physiological functions:

a. Turning on the instrument
b. Changing the paper speed
c. Increasing the intensity of electrical signal or recording
d. Recording the time
e. Recording the event
f. Balancing the transducer
g. Increasing the amplifier sensitivity
h. Stopping or starting the chart movement

Laboratory Animal Preparation

Procedure: Pithing

This procedure is to be performed by students in pairs.

1. Hold the frog in your left hand, with its hind limbs extended and its dorsal surface facing toward you (Figure B-1a).
2. Place your forefinger over the frog's snout (nose), and bend the frog's head downward until it forms a sharp angle with the trunk of its body. Use your thumb and remaining fingers to hold the frog's limbs securely.

3. Take the dissecting needle in your right hand and run the tip down the midline of the frog's skull, to locate a depression or groove between the skull and the first vertebra. This depression is about 2–3 mm behind the posterior border of the animal's eardrums (Figure B-1a). It marks the location of the *foramen magnum,* which is a large opening of the skull, through which the spinal cord emerges.
4. Push the needle in a forward and downward direction with a sharp movement through the skin, the depression, the underlying tissues, and the foramen

(a)

(b)

Figure B-1 Pithing: (a) the single-pithing procedure—shows the holding position for the process, and the location of the foramen magnum of the skull (note the bent appearance of the legs); (b) the double-pithing procedure—shows the insertion of the dissection needle into the spinal cord, which causes the frog's legs to extend and stiffen initially.

magnum into the brain. Twist and turn the needle, to destroy the brain. Remove the needle, and stop any bleeding by pressing a gauze pad over the depression. This completes the *single-pithing process*.

5. To double-pith the frog, insert the needle again between the front vertebra and the skull, but in a downward or caudal direction into the spinal cord.

Twist and turn the needle to destroy the spinal cord (Figure B-1b).

The legs of the frog should extend during Step 5, indicating entry of the needle into the spinal cord. Successful double pithing results in the animal's limbs becoming limp soon thereafter. The frog is now ready for procedures requiring pithed animals.

Laboratory Animal Blood-Specimen Preparation Procedures

When living animals are to be used in obtaining blood specimens, care always must be taken to avoid causing unnecessary irritation or injury.

Materials

The following items should be provided for the bleeding procedures:

1. One live healthy laboratory mouse or rat
2. One pair of animal handling gloves (optional)
3. One mouse or rat restrainer (The size of the restrainer is determined by the weight of the laboratory animal to be used; the restrainer should be made of a transparent plastic material, such as acrylic, to allow for maximum visibility.) (Acrylic restrainers are obtainable from Harvard Bioscience, An Ealing Division, Pleasant Street, South Natick, MA 01760.)
4. One pair of sterile scissors (for mice)
5. Sterile, individually wrapped blood stylets (for larger animals, such as rats)
6. Sterile cotton or gauze pads
7. Alcohol prep-pads or similar sterilizing gauze pads (obtainable from MED PAK Corporation, Poco, WY 25159)
8. Slides or test tubes with sterile saline, in which to collect blood specimens
9. Container with disinfectant for the disposal of blood-containing items.

Procedure

This procedure is to be performed by the instructor or by students under the direct supervision of an instructor.

1. Carefully remove the laboratory animal from its cage or container by grasping it firmly behind the neck. If the animal is a laboratory rat, a pair of animal gloves should be worn.
2. Insert the animal into the restrainer (see Figure C-1).
3. Position the adjustable tail plate in the restrainer so that the animal's tail is exposed.
4. Unwrap an alcohol-containing prep-pad package, remove the alcohol pad, and wipe the end of the laboratory animal's tail. Allow the tail to dry.
5. If the laboratory animal to be used is a rat, proceed to Step 8; if a mouse, take a sterile pair of scissors, and—while squeezing the tail—snip off approximately ⅛ inch. Continue to squeeze the tail until drops of blood appear.
6. The blood may be used to make blood smears directly or may be allowed to drip into a test tube containing sterile saline. In the event that blood drops do not appear with the initial cut of the mouse's tail, an additional ⅛ inch of tail material can be removed.
7. After enough blood has been obtained, pressure should be applied with a sterile gauze pad, to stop any unnecessary bleeding. Once the bleeding has

(a)

(b)

Figure C-1 Obtaining blood specimens from laboratory animals (mice or rats): (a) an example of an animal restrainer; (b) a mouse positioned in a restrainer, in order to obtain a blood specimen from its tail.

been stopped, the animal should be returned to its cage. This procedure should provide enough blood for 4–6 blood smears. Additional mice should be used if more smears are needed. This is the end of the procedure when using mice.

8. If the laboratory animal to be used is a rat, use a sterile stylet to puncture the animal's tail near the tip (about 1 inch from the end). Rubbing the area with a fresh gauze pad will help to bring blood to the surface of the tail. Squeeze the tail until drops

of blood appear. The blood may be used for the same types of specimens as mentioned for the mouse (see Step 6).

9. After enough blood has been obtained, pressure should be applied with a sterile gauze pad, to stop any unnecessary bleeding.

10. With the aid of laboratory gloves, carefully remove the rat from its restrainer, and return the animal to its cage.

Color Atlas

Plate 1 The letter "e" under low power.

Plate 2 Cheek cells: Note the presence of a nucleus and a cytoplasmic membrane in these cells.

Plate 3 A human blood smear showing erythrocytes (red blood cells) and a neutrophil.

Plate 4 Frog blood: Here both red blood cells (RBCs) and white blood cells are nucleated. Compare this photo with that of human blood cells.

Plate 5 A preparation of smooth muscle cells. (Courtesy Carolina Biological Supply Company.)

Plate 6 A view of tracheal epithelium: Pseudostratified epithelium bearing cilia and underlying connective tissue are shown in this micrograph. (Courtesy Carolina Biological Supply Company.)

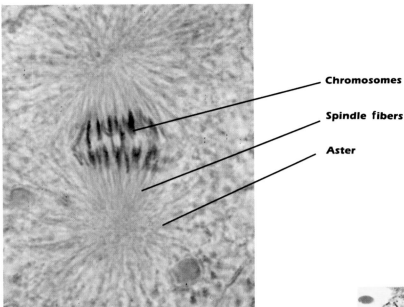

Chromosomes

Spindle fibers

Aster

Plate 7 The appearance of an aster: Note the strand composition of the cellular component.

Plate 8 The liver.

Nucleus

Plate 9 DNA and RNA in animal cells, stained with Azure B: The DNA in the nucleus appears bluish-green and the RNA in the smaller, nucleolus appears purple. (Courtesy of Ward's Natural Science Establishment, Inc.)

Nucleus

Plate 10 Feulgen reaction showing the purple appearance of DNA in cell nuclei.

Plate 11 A living cell undergoing mitosis as seen under phase contrast microscopy.

Plate 13 The appearance of a facial rash.

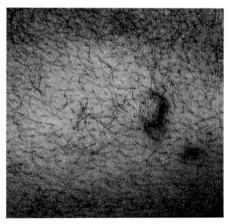

Plate 14 Kaposi's sarcoma, a tumor associated at times with the disease acquired immune deficiency syndrome (AIDS).

Plate 12 Dye diffusion zones: crystal violet (CR), methylene blue (MB), and safranin (S).

Plate 15 The features of a local infection.

Connective
tissue
sheath

External root
sheath

Hair root

a

b

Plate 16 Human scalp, showing the features of hair follicles: (a) longitudinal section—particular structures included here are the hair follicle, sebaceous glands, and arrector pili; (b) this transverse section shows the central cortex, the surrounding cuticle layer, internal root sheath, external root sheath, and the connective tissue sheath. (Courtesy of Fisher Scientific Company.)

Plate 17 The blackening of a nail due to a physical injury. The blackening is the result of hemorrhaging.

Plate 18 The discoloration of the nails shown is a side effect of certain medications.

Plate 19 The white lines are associated with a genetic factor.

Plate 20 Skeletal muscle. A longitudinal section. (Courtesy Carolina Biological Supply Company.)

Plate 21 A cross-section of skeletal muscle. (Courtesy Carolina Biological Supply Company.)

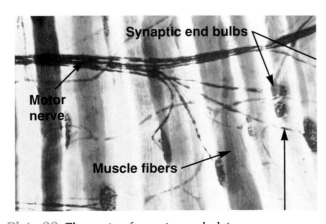

Plate 22 The parts of a motor end plate.

Synaptic end bulbs

Motor nerve

Muscle fibers

Plate 23 The thymus gland: Both the lighter inner medulla region and the outer cortex region of a lobule are shown. A centrally located thymic (T) corpuscle also is present.

Plate 24 Appearance of cells in a normal thyroid gland.

Plate 25 Appearance of cells in a thyroid gland specimen from a goiter.

Gland Tissue

Fat Cell

Gland Tissue

Plate 26 Cellular structure of a portion of the parathyroid.

Plate 27 A higher magnification of the cells in the parathyroid gland.

Capsule

Plate 28 The adrenal gland: Capsule and the zone glomerulosa. This zone contains secretory cells, arranged in irregular ovoid clumps, separated by connective tissue strands.

Plate 29 The zona fasciculata, the intermediate and broadest of the three zones of the adrenal cortex. This zone consists of narrow cords of large secretory cells, with lightly staining cytoplasm.

Plate 30 The adrenal medulla consists of closely packed, irregular clumps of secretory cells. These cells have large granular nuclei and dark-blue- to purple-staining cytoplasm.

Plate 32 Actual view of the retina.

Plate 31 The pancreas: The centrally located islets of Langerhans (L) and the surrounding acini are shown.

Plate 33 Microscopic view of taste buds (T).

Plate 34 Meissner's corpuscles are small, oval, encapsulated sensory organs located in the dermal papillae immediately beneath the skin epithelium. (Courtesy of Ward's Natural Science Establishment, Inc.)

Plate 35 The pacinian corpuscle, an encapsulated nerve ending associated with the sensations of touch and pressure. Dendritic endings are enclosed within many onionlike layers of connective tissue. Such layers absorb some of the pressure when the stimulus is applied and maintained.

Plate 36 A blood smear, showing eosinophils.

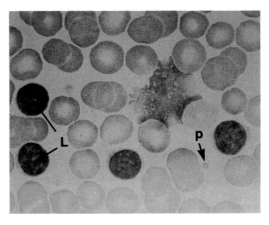

Plate 37 A number of lymphocytes (L) and platelets (P) in a blood smear.

Plate 38 A bone-marrow preparation, showing the features of an acute monocytic leukemia. A cell in mitosis (arrow) can be seen in this smear. Bone marrow is usually obtained by aspiration, which involves inserting a needle into the marrow cavity and withdrawing a sample. The sternum, ilium, iliac crest, and vertebral spinous process are the most common sites used for aspiration. (Courtesy of Dr. I. Mirchandani, Wayne State University.)

Plate 39 Chronic granulocytic leukemia. (Courtesy of Dr. I. Mirchandani, Wayne State University.)

Plate 40 Chronic lymphocytic leukemia. (Courtesy of Dr. I. Mirchandani, Wayne State University.)

Plate 41 Acute myelomonocytic leukemia. (Courtesy of Dr. I. Mirchandani, Wayne State University.)

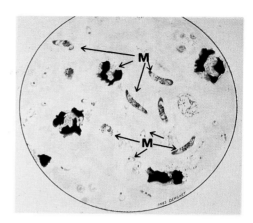

Plate 42 Stages of the malarial (M) parasite *Plasmodium falciparum*.

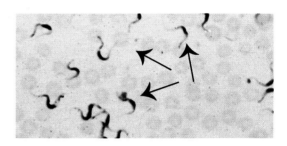

Plate 43 The causative agents (arrows) of African sleeping sickness.

Plate 44 The intracellular location of *Neisseria gonorrhoeae* (arrow), the causative agent of the sexually transmitted disease gonorrhea.

Plate 45 A Tallquist reaction showing a normal hemoglobulin content.

Plate 46 Bone-marrow smear.

SERUM PROTEINS

8 Liver Disease

7 Amyloidosis

6 Idiopathic paraproteinemia

5 Gamma G multiple myeloma

4 Gamma M macroglobulinemia

3 Gamma A multiple myeloma

2 Hyperglobulinemic purpura

1 Control serum

Gamma | Beta | Alpha 2 | Alpha 1 | Albumin

Plate 47 An electrophoretic pattern, showing the appearance of serum proteins of a normal specimen, cases of liver disease, and other abnormal conditions.

Plate 48 A severe form of the gum disease, gingivitis.

Plate 49 The submandibular salivary gland.

Plate 50 The submaxillary salivary gland.

Plate 51 The parotid gland.

Plate 52 A portion of intestine showing secretory cells (goblet cells) and columnar cells.

Plate 53 The appearance of a positive (purple) Biuret test.

Plate 54 A microscopic view of the kidney.

Plate 55 The results of a Benedict's test. Left tube shows no glucose, while the one in the center and the one on the right clearly contain glucose.

Plate 56 Multistix™ reactions: The strip on the left is normal.

Plate 58 Human ovary showing a follicle (arrow) in the first stages of growth. (Courtesy of Carolina Biological Supply Company.)

Plate 57 Bacturcult results showing a bacterial count under 100,000/ml. Note the appearance of the counting strip.

Plate 59 Microscopic view of human testes.

Index